PROGRESS IN COLLOID & POLYMER SCIENCE

Editors: H.-G. Kilian (Ulm) and G. Lagaly (Kiel)

Volume 81 (1990)

# Trends in Colloid and Interface Science IV

Guest Editors: M. Zulauf (Basel), P. Lindner and P. Terech (Grenoble)

# Springer-Verlag Berlin Heidelberg GmbH

ISBN 978-3-662-16024-4    ISBN 978-3-7985-1687-8 (eBook)
DOI 10.1007/978-3-7985-1687-8
ISSN 0340-255-X

© 1990 by Springer-Verlag Berlin Heidelberg

Originally published by Dr. Dietrich Steinkopff Verlag GmbH & Co. KG, Darmstadt in 1990

Softcover reprint of the hardcover 1st edition 1990

Chemistry editor: Dr. Maria Magdalena Nabbe; English editor: James Willis; Production: Holger Frey.

Type-Setting: Graphische Texterfassung, Hans Vilhard, D-6126 Brombachtal

# Preface

When the executives of the European Colloid and Interface Society (ECIS) decided to have the Third ECIS Meeting in Switzerland, they found that their collegues at the Institut Laue-Langevin (ILL) had announced a Workshop on "Structure and Dynamics of Colloid Systems" in Grenoble. It seemed natural for both parties to join forces, and thus a combined ECIS Conference & ILL Workshop was held in Basel. It was attended by nearly 250 scientists representing the major groups working in the field in Europe, but also some groups from other countries. Roughly 90% of the oral contributions given at the meeting are published here in their written version. They appear in the order that they were presented in the meeting, within the loosely defined framework of topics. We hope that both the attendants in Basel and those who could not be there read and reread them with pleasure and profit!

About 80 posters are covered in this volume by very short abstracts. They appear in alphabetical order of the submitting author's name. We regret that those works had to be abbreviated due to space restrictions.

The conference was hold at the Congress Center of F. Hoffmann-La Roche. We are much indebted to those responsible and to the executives of the company for the generous support extended to us in preparing and running the conference. We gratefully acknowledge financial help from: Institut Laue-Langevin, Hoffmann-La Roche, Ciba-Geigy, and Sandoz. The scientific program was established under the wise guidance of an efficient program committee comprising M. Corti, H.-F. Eicke, D. Langevin, D. Richter, and A. Vrij.

On behalf of ECIS and ILL, we would like to thank all the participants of the meeting for their brilliant contributions and stimulating discussions. Certainly they inspired a lot of cooperation among colloid scientists, which is, of course, one of the aims of ECIS.


Martin Zulauf
Peter Lindner
Pierre Terech

# Contents

## Aggregates, ordering and structural transitions

## Films, membranes, surfaces and wetting

## Abstracts

# The effect of electric fields on nonaqueous dispersions

R. H. Ottewill[1]), A. R. Rennie[2]) and A. Schofield[1])

[1]) School of Chemistry, University of Bristol, UK
[2]) Institut Laue Langevin, Grenoble, France

*Abstract:* Small angle neutron scattering has been used to investigate the structure of poly(methylmethacrylate) dispersions in dodecane in the presence of an electric field. In dodecane alone field effects were not observed. In the presence of calcium octanoate, however, the particles apparently become charged and increased ordering was observed. The effect of an electric field on the latter dispersions produced an anisotropy of structure which suggested that the particles were forming loose "strings of beads" in the direction of the applied field.

*Key words:* Neutron scattering; polymer latices; nonaqueous dispersions; electric fields; microstructure

## Introduction

The effects of electric fields on nonaqueous dispersions have been rather sparsely studied in comparison to studies carried out on aqueous systems. However, the availability of monodisperse spherical polymer particles at high volume fractions in media of low relative permittivity [1, 2] has provided excellent systems for fundamental studies [3, 4]. Moreover, the development of small angle neutron scattering has provided a means of studying the structure of concentrated colloidal systems in applied fields and has already been used to examine systems of this type under shear fields [5, 9].

In this communication we provide a brief report on the use of small angle neutron scattering to investigate the effect of applied electric fields on the microstructure of dispersions of poly(methylmethacrylate) particles stabilised by poly (12-hydroxystearic acid) in a hydrocarbon medium. In the first instance the particles were examined in dodecane alone and then in dodecane with the addition of small amounts of calcium octanoate to provide some conductance to the dispersion medium.

## Experimental

### Materials

Dodecane was B.D.H. material.

NUOSYN CALCIUM 10% was obtained from Durham Chemicals Limited, Birtley, Chester-le-Street, Co. Durham, U.K. According to the manufacturers this is a solution of calcium iso-octanoates in a hydrocarbon medium. It is reported in this communication as calcium octanoate.

The polymer latex, SPSO90, was prepared by the dispersion polymerisation technique previously described [2]. The number average particle diameter determined by electron microscopy was 0.20 μm and the weight average diameter determined by small angle neutron scattering was 0.22 μm [5].

### Small angle neutron scattering

The small angle scattering measurements were carried out using the neutron diffractometer D11, at the Institut Laue Langevin, Grenoble [10], with a sample-detector distance of 35.7 m and a neutron beam wavelength of 0.80 nm. For elastic scattering measurements, the magnitude of the scattering vector, $Q$, is defined by,

$$Q = 4\pi \sin(\theta/2)/\lambda$$

with $\theta$ = the angle between the incident and the scattered beams.

The measurements were carried out in an optical-standard quartz cell with a pathlength of 1 mm. Bright platinum rods were inserted into the cell at a separation distance of 6 mm. An A.C. field of frequency 50 Hz was applied across the electrodes. The applied voltage was varied from zero to 1800 V. The experimental arrangement is illustrated schematically in Fig. 1. The neutron beam, of diameter 4 mm, passed between the electrodes in a direction perpendicular to the applied electric field.

The scattered neutrons were measured using a two-dimensional detector, a square matrix of 64 × 64 elements.

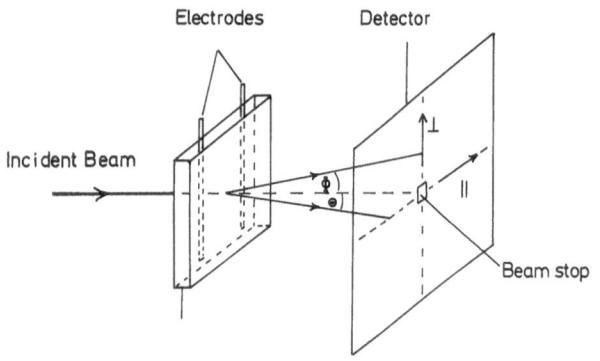

Fig. 1. Experimental arrangement used for examining the small angle neutron scattering from a polymer colloid dispersion during the application of an electric field

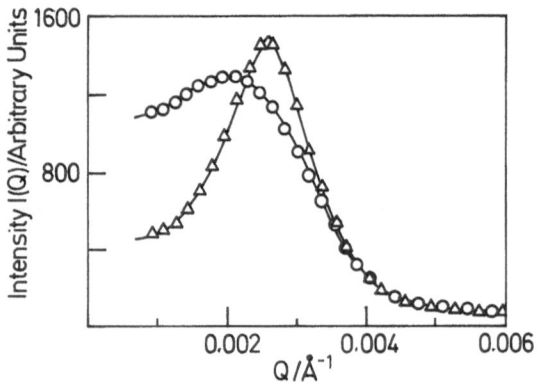

Fig. 2. Intensity, $I(Q)$, against scattering vector, $Q$, for a 40% w/w poly(methylmethacrylate) latex dispersion: $-\bigcirc-$, in dodecane; $-\triangle-$, in dodecane containing 0.1% calcium octanoate

This allowed either two-dimensional contour plots to be obtained showing lines of equal intensity or three-dimensional plots showing the intensity across the detector. For isotropic two-dimensional patterns the intensities can readily be radially averaged to give one-dimensional plots of intensity, $I(Q)$, against $Q$. Anisotropy of the sample can be clearly recognised in a contour plot. In this case in order to obtain plots of $I(Q)$ against $Q$ the averaging procedure can be restricted to small sectors, for example, parallel $(0 \pm 15°)$ and perpendicular $(90 \pm 15°)$ to the direction of the applied field [11].

## Results

*Intensity measurements without an electric field*

The intensity of scattering at a particular value of $Q$ is, for spherical particles, given by [3, 4],

$$I(Q) = A \phi V_p P(Q) S(Q)$$

where for spheres of the core-shell type $A$ contains instrumental constants, and terms in coherent scattering length density described elsewhere, [3], $V_p =$ the particle volume, $\phi =$ volume fraction of the particles, $P(Q) =$ the particle shape factor and $S(Q) =$ the structure factor.

Figure 2 shows a curve for a 40% w/w dispersion of the latex SPSO90 in dodecane. This shows a broad peak consistent with short-range interactions of the hard sphere type examined in earlier work on polymethylmethacrylate latices [3–5]. The addition of calcium octanoate to the sample caused a substantial change in the form of the curve. As can be seen from Fig. 2, the broad peak of the original latex became very much narrower and sharper. Since the basic latex was the same in both cases, i.e. $P(Q)$ was the same,

and the samples were essentially in the same medium, $A$ also remained constant. Thus the changes in the scattering curve were attributed to changes in $S(Q)$. The changes in shape indicated an increase in interaction between the particles with the implication that this arose because the addition of calcium octanoate caused the particles to acquire a weak electrical charge. This led to the form of the interaction changing from that between sterically stabilised particles, a short-range repulsion, to that between electrically charged particles, a long-range repulsion. In the latter case the system becomes more ordered at the same volume fraction, and hence the peak in $I(Q)$ against $Q$ becomes more clearly defined. Electrophoresis measurements on dilute dispersions confirmed that in the presence of calcium octanoate the particles had a positive mobility.

## The effect of an electric field

*Calcium octanoate absent*

Figure 3 shows the 2-dimensional contour plots obtained using an 18% w/w dispersion of poly(methylmethacrylate) particles in dodecane alone. The pattern obtained from latex in the absence of the field is shown in Fig. 3a, and that with a field strength of 3800 V cm$^{-1}$ applied is shown in Fig. 3b. No discernable change in the pattern was observed on the application of the field.

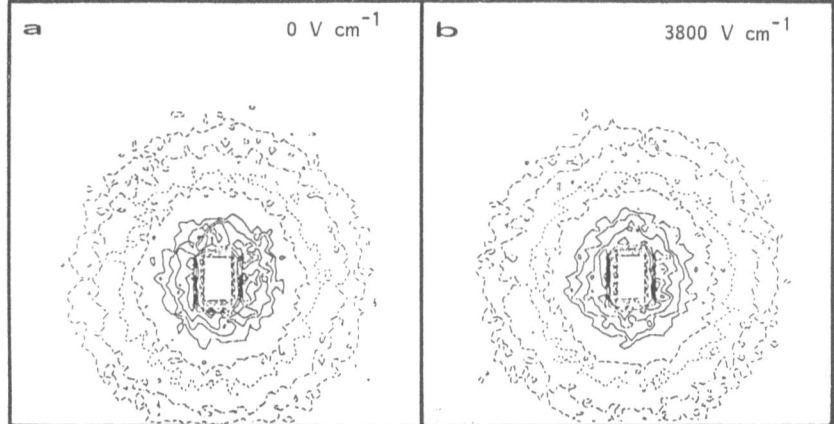

Fig. 3. Contour plots of constant intensity, $I(Q)$, lines against $Q$ for an 18% w/w poly(methylmethacrylate) dispersion in dodecane in the absence and presence of an electric field

*Calcium octanoate present*

Figure 4 shows the 2-dimensional pattern obtaining using a 10% w/w dispersion of poly(methylmethacrylate) containing 0.025% calcium octanoate. Figure 4a shows the absence of an applied field. The contours are symmetrical showing that the sample is behaving as an isotropic dispersion with the peak in the structure factor showing up as a well-defined ring. Figures 4b, 4c and 4d illustrate the contour plots obtained at field strengths of 900, 1800 and 3800 V cm$^{-1}$. The anisotropy of the sample increases as the field strength increases and is most marked in Fig. 4d the system at the highest field strength.

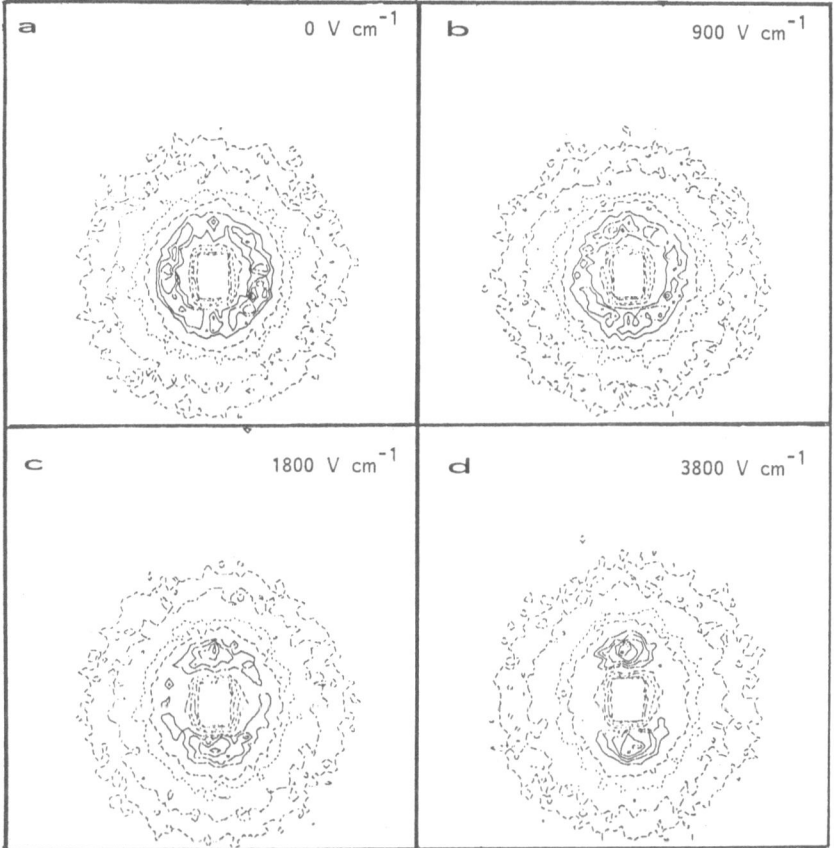

Fig. 4. Contour plots of constant intensity, $I(Q)$, lines against $Q$ for a 10% w/w poly(methylmethacrylate) dispersion in dodecane containing 0.025% calcium octanoate at various applied field strengths

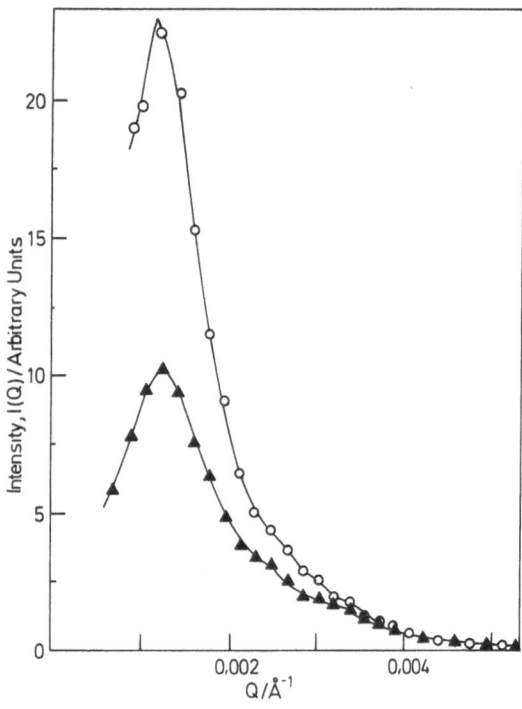

Fig. 5. Intensity, $I(Q)$, against scattering vector, $Q$, for a 10% w/w poly(methylmethacrylate) dispersion in dodecane containing 0.025% calcium octanoate at a field strength of 3800 V cm$^{-1}$. $-\bigcirc-$, perpendicular to the applied field; $-\blacktriangle-$, parallel to the applied field

Using the information obtained on the 2-dimensional detector, radial averages were taken in sectors at $0° \pm 15°$ and $90° \pm 15°$. This gave $I(Q)$ against $Q$ plots for the sample in directions parallel and perpendicular to the applied electric field. These are illustrated in Fig. 5. In the direction of the field the intensity has been reduced in comparison to that in the absence of the field. The peak in intensity is very well defined suggesting a significant peak in the curve of $S(Q)$ against $Q$ thus indicating strong interaction between the particles. In a direction perpendicular to the applied electric field the intensity has markedly increased compared to that in the absence of the field. The peak is still present, although less well defined, and has moved marginally to a lower $Q$ value.

## Discussion

Three separate pieces of evidence suggest that particles of poly(methylmethacrylate) stabilised by poly(12-hydroxystearic acid) become electrically charged in the presence of low concentrations of cal-

cium octanoate in a low relative permittivity medium, dodecane. These are:

i) the formation of a well-defined peak in the curves of $I(Q)$ against $Q$ indicating a well-defined peak in $S(Q)$ against $Q$
ii) the development of a positive electrophoretic mobility in the presence of calcium octanoate
iii) the development of anisotropy, as detected by small angle neutron scattering in the presence of an applied electric field.

The calcium octanoate used in these experiments has a similarity to the calcium disopropyl salicylate used in earlier work by van der Minne and Hermanie [12–14]. Since the conductivity of the dodecane increases on the addition of calcium octanoate [15] weak dissociation appears to occur to give,

$$Ca(Oct)_2 \gtrless CaOct^+ + Oct^-$$

and possibly

$$CaOct^+ \gtrless Ca^{2+} + Oct^-.$$

Since the Oct$^-$ ion is essentially oleophilic, it appears likely that this will remain in the dodecane whilst the Ca$^{2+}$ or CaOct$^+$ associates with the dipole on the ester groups of the poly(12-hydroxy stearic acid), leading to a positive change on the particles by ion-dipole association, viz:

$$-C=O\text{---}Ca^{2+}$$
$$\longmapsto$$

Only a small number of charges would be needed to produce a positive charge on the particles.

The results shown in Fig. 2 for a 40% w/w dispersion suggest that in the absence of added calcium octanoate the interaction between the particles is relatively short range consistent with the hard sphere behaviour observed in earlier work [3]. Addition of calcium octanoate in small quantities causes the peak in $I(Q)$ against $Q$ to sharpen considerably suggesting much stronger interaction and short range ordering of the particles, due to electrostatic interactions. The fact that the latter are of long range is substantiated by experiments on the 10% w/w dispersion where a well-defined peak is found with calcium octanoate present (Fig. 4a), but is essentially absent for the dispersion without calcium octanoate. On the application of the electric field an interesting effect occurs in that the intensity increases in the direction perpendicular to the field and decreases in the direction parallel to the field. This shows clearly with a field of 3800 V cm$^{-1}$ as illustrated in Fig. 5. In the direction of the field the peak is very sharp. It is somewhat less so in the perpendicular direction and has moved slightly to a lower $Q$ value. The complete interpretation of these results requires further analysis and separation of the curves into interaction ($S(Q)$) and form factors ($P(Q)$). A possible interpretation of the results is that the dipoles formed by the charges on the particles line up in the electric field and cause a preferential alignment in the direction of the field which causes increased interaction in that direction without essentially changing the number concentration, hence sharpening $S(Q)$. In view of the sensitivity of the $I(Q)$ against $Q$ curves to the rapid falling off of $P(Q)$ with $Q$ this could lead to a major drop in intensity. Similarly, a weaker interaction in the perpendicular direction at the same number concentration would diminish the structure and hence give an increase in intensity.

A measurement of electrophoretic mobility on a very dilute dispersion in the presence of calcium octanoate gave a value of 0.1 μm cm V$^{-1}$ sec$^{-1}$ and hence the time for the particle to travel one particle radius (0.1 μm) in a field of 3800 V cm$^{-1}$ can be calculated to be 0.3 msec. A similar calculation for the particle to diffuse translationally one radius gives the time as 0.7 msec. Hence, the ratio:

$$\frac{\text{time to travel one radius in the electric field}}{\text{time to diffuse one radius}}$$

gives a value of 0.43. This rather simplistic calculation would suggest that appreciable Brownian motion of the particles is still occurring even in the presence of the electric field of 3800 V cm$^{-1}$. This would suggest that the average number concentrations in the directions parallel and perpendicular to the field are probably not appreciably different, although the particle-particle interactions, as suggested above, are significantly different.

*Acknowledgements*

We wish to acknowledge with thanks the continued support of the Science and Engineering Research Council and the Institute Laue Langevin for neutron facilities.

## References

1. Barrett KEJ (1975) Dispersion Polymerisation in Organic Media, John Wiley and Sons, Inc., London
2. Antl L, Goodwin JW, Hill RD, Ottewill RH, Owen SM, Papworth S, Waters J (1986) Colloids Surfaces 17:67
3. Marković I, Ottewill RH, Underwood SM, Tadros TF (1986) Langmuir 2:625
4. Ottewill RH (1989) Langmuir 5:4
5. Lindner P, Marković I, Oberthür RC, Ottewill RH, Rennie AR (1988) Prog Colloid Polym Sci 76:47
6. Ashdown S, Marković I, Ottewill RH, Lindner P, Oberthür RC, Rennie AR (1990) Langmuir 6:303
7. Rennie AR, Ashdown S, Lindner P, Marković I, Oberthür RC, Ottewill RH (1990) Polymer Colloids NATO Advanced Study Institute, Strasbourg, Kluwer
8. Ottewill RH, Rennie AR (1989) J Applied Hydrodynamics, in press
9. Ackerson BJ, Hayter JB, Clark NA, Cotter L (1986) J Chem Phys 84:2344
10. (1988) Neutron Beam Facilities at the High Flux Reactor, Institut Laue Langevin, Grenoble
11. Ghosh RE (1989) A Computing Guide for Small Angle Scattering Experiments, Institut Laue Langevin, 89 GH 02 T
12. van der Minne JL, Hermanie PHJ (1952) J Colloid Sci 7:600
13. van der Minne JL, Hermanie PHJ (1953) J Colloid Sci 8:38
14. Klinkenberg AK, van der Minne JL (1958) Electrostatics in the Petroleum Industry, Elsevier, Amsterdam
15. Marković I, Ottewill RH (1989) to be published

Authors' address

Prof. Dr. R. H. Ottewill
School of Chemistry
University of Bristol
Cantock's Close
Bristol, BS8 1TS, England

# Crystals of fluid films in systems of amphiphiles*

J. Charvolin

Laboratoire de Physique des Solides, University Paris Sud, Orsay, France and
Institut Max von Laue − Paul Langevin, Grenoble, France

*Abstract:* Crystalline structures built by amphiphilic molecules in presence of water are different from classical molecular crystals. We present here an approach to these systems, structures formed by films of amphiphiles, which leads to consider them as crystals of defects, disclinations.

*Key words:* Crystals of fluid films, frustration, interfacial constraints, disclination, defects

Crystals are classically described as periodic organizations of point objects, atoms or molecules, regularly disposed at the knots of lattices of various symmetries. Their long range crystalline order is generally considered as resulting from the propagation of the short range order of the individual point objects, which are therefore the building blocks of the structures. However, recent studies show that a long range crystalline order can exist without any "affine" molecular order at short range. Such a situation was encountered investigating liquid crystalline materials, particularly those formed by amphiphilic molecules such as soaps, detergents and lipids in presence of water. They build crystalline structures although they have a short range disorder of liquid-like type [1]. Typical structures are shown in Fig. 1. X-ray and neutron scattering data,

NMR data exemplify this apparently paradoxical behaviour and clearly show that the building blocks of these structures can not be individual molecules and that their understanding requires the definition of more operative elements of structures.

If it is remembered that amphiphiles stabilize liquid/liquid interfaces, an obvious candidate is the infinite fluid film built by two facing amphiphile/water interfaces, as defined in Fig. 2. This direction leads to consider a new class of crystals, where the element of structure is no longer a point object but a bi-dimensional one. In order to bring out a general basis for the understanding of this polymorphism, we investigated the possible periodic configurations of model symmetric films considering that the interactions between the interfaces maintain constant distances them and that the constraints parallel to the interfaces control their curvatures. The configurations must therefore conciliate the constant interfacial distances and

---

*) Extended abstract, a guide to the published works at the basis of the oral presentation given at the Conference.

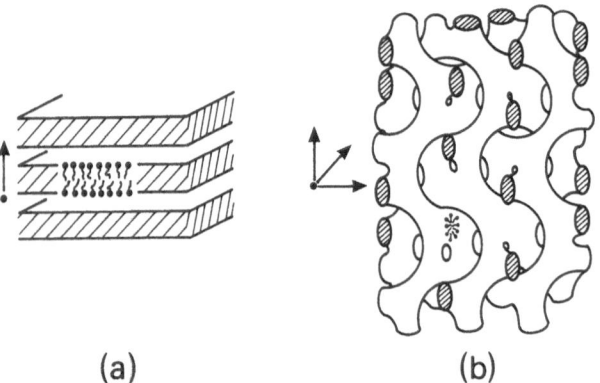

(a)                    (b)

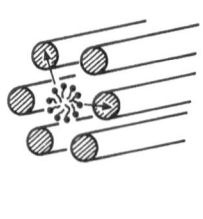

(c)

Fig. 1. Drawings of the lamellar (a), bicontinuous cubic Ia3d viewed along a ⟨100⟩ direction (b) and hexagonal (c) structures

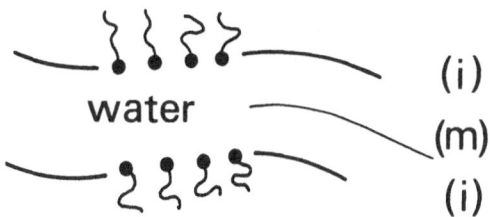

Fig. 2. The film built by two facing interfaces (i) is supported by the ideal middle surface (m)

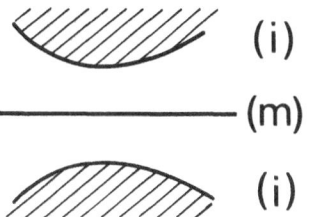

Fig. 3. A frustrated film, constant interfacial distances and non zero interfacial curvatures are not compatible

curvatures imposed by the thermodynamical parameters. When the interfacial curvature is zero, the obvious solution is a periodic stacking of parallel layers. When the curvature are not zero, adjacent interfaces must have the same concavity relative to the two media and the constant distances between them can no longer be maintained if the lamellar geometry is kept, as shown in Fig. 3. This is a typical case of frustration which implies a change of structure [2]. We determined the possible solutions [3] as well as their sequence in the phase diagrams [4], following a geometrical approach similar to those developed for other cases of frustration in condensed matter physics. Some

of these solutions, belonging to three main topological classes, are shown in Fig. 4. The agreement with the observations, as far as topologies, symmetries and sequences are concerned, is quite satisfactory and emphasizes the role of spatial constraints in this problem. It also leads to consider non-lamellar structures as structures of disclinations, defects of rotation affecting curvatures, which partly relax the frustration.

Recent X-ray scattering data, obtained with monocrystalline samples, suggest however that the classical description of the structures and their modelization in terms of structures of disinclinations are indeed limited representations of real structures. The broadlines of the preceding paragraph are still valid, but additional details must be included, which are necessary for understanding the stabilities of phases and the phase transitions. This is apparent on Fig. 5, where a X-ray scattering pattern observed in an experiment on a monocrystal of lamellar phase is shown [5]. The Bragg's reflections, the intense spots, are at the expected places in the pattern but rather well-characterized diffuse scatterings are visible away from the Bragg's spots. These diffuse scatterings have been observed in other system [6, 7]. They reveal the existence of departures from the structural description given above. They are to be associated with the appearance of a fragmentation of the lamella when the boundaries of lamellar domains in the phase diagrams are approached. Such fragmentations, as those shown in Fig. 6, can be described in terms of associations of defects of translation, or dislocations. Their presence can be understood considering the fact that the structures of disclinations considered above are global solutions optimizing the frustration for discrete values of it only [4] so that local solutions, the above fragmentations or defects, are needed when the frustration value moves away from these discrecte values [8].

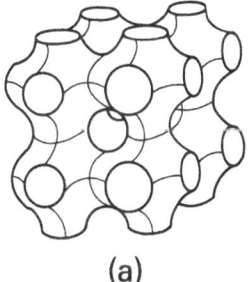

(a)

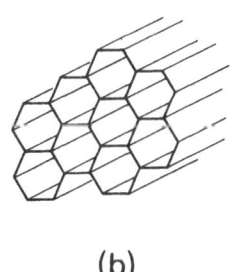

(b)

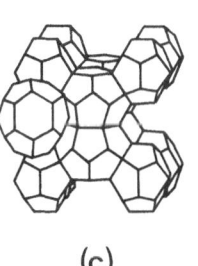

(c)

Fig. 4. Three ordered configurations of the middle surface of the film obtained following the frustration approach: the Schwarz's surface separating the two labyrinths of the bicontinuous Im3m cubic structure (a), the honeycomb configuration containing the cylinders of the hexagonal structure (b), the polyedral configuration of the Pm3n micellar cubic structure (c)

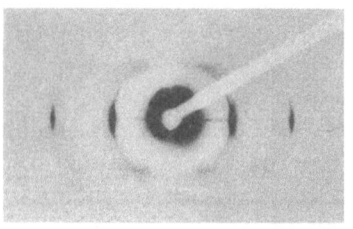

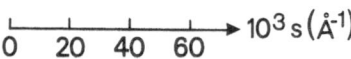

Fig. 5. A X-ray pattern of a monocrystal of lamellar phase of $C_{12}EO_6/H_2O$

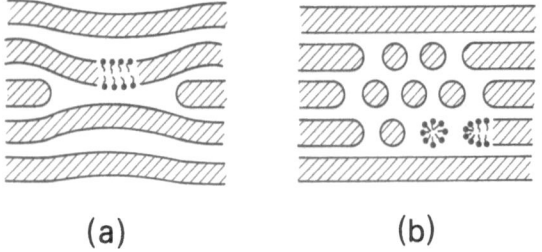

(a)                              (b)

Fig. 6. Schematic representation of a fragmented lamellar structure (a) with a short range hexagonal fluctuation (b)

Their density increases when phase transitions are approached.

Thus, the perfectly ordered structures of defects described previously and presented in Fig. 4 exist in lim-

ited area of the domains of existence in the phase diagram. The structures are most often imperfect with variable densities of structural defects such as those described just above. This is a situation reminiscent of that known in the physics of alloys where imperfectly ordered structures exist around the ordered structures formed by stoechiometric compounds [9]. The energies of these defects are certainly important parts of the thermodynamics of the structures. They also intervene in the phase transition process.

### References

1. Luzzati V (1968) Biological Membranes 1:71
   Mariani P, Luzzati V, Delacroix H (1988) J Mol Biol 204:165
2. Sadoc JF, Charvolin J (1986) J de Physique 48:1067
3. Charvolin J, Sadoc JF (1987) J de Physique 48:1559
   Charvolin J, Sadoc JF (1988) J de Physique 49:521
4. Charvolin J, Sadoc JF (1988) J Phys Chem 92:5787
5. Rançon Y, Charvolin J (1988) J Phys Chem 92:6339
6. Kekicheff P, Cabane B, Rawiso M (1984) J de Physique Lettres 45:L-813
7. Hendrikx Y, Charvolin J, Kekicheff P, Roth M (1987) Liquid Crystals 2:677
8. Charvolin J, Contemporary Physics, to be published
9. Cottrel AM (1948) Theoretical structural metallurgy, Edward Arnold and Co, London

Author's address:

Dr. J. Charvolin
Institut Max von Laue — Paul Langevin, 156X
38042 Grenoble Cedex, France

**Progress in Colloid & Polymer Science**                    Progr Colloid Polym Sci 81:9–12 (1990)

# Fluorescence quenching in the $C_{12}E_6$-water system: Diffusion-control in three to zero dimensions *

M. Almgren and J. Alsins

Department of Physical Chemistry, Uppsala University, Uppsala

*Abstract:* The quenching of fluorescence from pyrene by hexylbenzophenone in the nonionic surfactant $C_{12}E_6$ has been examined, both with the surfactant in neat form and with water added to produce lamellar, cubic, and hexagonal liquid crystalline phases, and the isotropic micellar solution. The efficiency of the quenching decreases in this order, due to that less quenchers are available within a certain distance from the excited state when the dimensionality is decreased. The decay curves were analysed using the appropriate models for diffusion-controlled quenching in three to one dimensions, and for quenching without exchange in small micelles. The values obtained for the mutual diffusion-coefficient of the probe-quencher pair vary between 3.5 and 6.5 $\times$ $10^{-11}$ m² s⁻¹ in a nonsystematic way.

*Key words:* Fluorescence quenching, diffusion-control, dimensionality, nonionic micelles, liquid-crystal phase

## Introduction

In time-resolved fluorescence quenching experiments a brief pulse of light is used to create a small number of excited molecules in the sample, and the fluorescence is used to follow their deactivation due to the interaction with the quenchers in addition to the intramolecular decay processes. If the excited state is inherently long-lived and the quenchers mobile and efficient, so that every encounter between an excited state and a quencher leads to deactivation, then the quenching rate is determined by the diffusion of the quenchers and probes. This situation has been well studied in homogeneous solutions, and is a textbook example of a diffusion-controlled process. Neglecting an initial transient term the decay is exponential and given by

$$F(t) = F(0)\exp(-k_0 t - k_2[Q]t) \qquad (1)$$

where $F(t)$ is the fluorescence intensity at time $t$ after the excitation, $k_0$ the first order decay constant with-

out quenching, $[Q]$ the quencher concentration, and $k_2$ the diffusion-controlled rate constant given by

$$k_2 = 4\pi DRN_A \qquad (2)$$

where $D = D_A + D_Q$ is the sum of the diffusion coefficients for the quencher and the excited probe, $R$ their encounter distance, and $N_A$ Avogadro's number.

Consider now the following thought experiment: A homogeneous solution, with probes and quenchers randomly dispersed, is cut into small cells with equal size and impenetrable walls, and so many that there is on average about one quencher per cell. The distribution of quenchers over the cells is then poissonian. The distribution of quenchers around the excited molecules within the cells should remain similar to the distribution in the homogeneous solution over the length-scale $l_c$ which characterizes the small cells. The deactivation would proceed in about the same way, therefore, during the time, $t_c = 2D/l_c^2$, which corresponds to a diffusive displacement fully within the cell. At much longer times, however, all surviving excited states will be found in cells without quenchers, and the decay rate will be given by only $k_0$. This can be considered as the zero-dimensional case.

*) This is a preliminary report. A full account is planned for publication in Israel J Chem (1990)

The one- and two-dimensional cases are found correspondingly by slicing the homogeneous solution into infinite rods and disks, respectively, with thickness $l_c$. Also in these systems the initial decay over a time in the order of $t_c$ should follow that in the original solution, and slow down thereafter — and more so in one than in two dimensions — due to that fewer quenchers will be available within the distance of an average diffusive displacement.

### An experimental realization

It is virtually impossible to realize an exact experimental comparison between quenching in systems that only differ with respect to their dimensionality; the experiment which will be desribed briefly is probably as close as one may come. Pyrene and hexylbenzophenone (1-(4-benzoyl)phenylhexane) were chosen as probe and quencher, respectively. These molecules are very little soluble in water, but may readily be dissolved in the neat, fluid surfactant $C_{12}E_6$, hexaethylene-glycol-mono-n-dodecyl ether (used as supplied by Nikko Chemicals, Tokyo, Japan). Addition of water to the neat surfactant (with appropriate concentrations of probe and quencher) brings the system successively into the lamellar phase, a cubic phase, the hexagonal phase, and the micellar solution, as shown in the phase diagram [1] in Fig. 1. The reactants are expected to stay in the hydrophobic environment of the surfactant tails, and are thus to be found in the 2-dimensional lamellae of the $L_\alpha$-phase, in the hydrophobic minimal surface [2] of the $V_1$-phase, in the long rods of the hexagonal phase, and in the globular small micelles in the dilute $L_1$-phase. The liquid crystalline samples could easily be prepared from the isotropic $L_1$-phase, which is always present at moderate temperatures above the $l_c$-phases.

### Results and discussion

For each composition samples without quencher were prepared for the determination of $k_0$. This was necessary since some variation in fluorescence lifetime occurred, most noticeably between the small micelles at 2% by weight of surfactant, and the large micelles just outside the hexagonal phase, at 33% surfactant. The decays were always nicely exponential when quencher were not present, but the lifetime increased from 200 nsec in the small micelles to about 300 nsec in the large ones, and remained with smaller variations

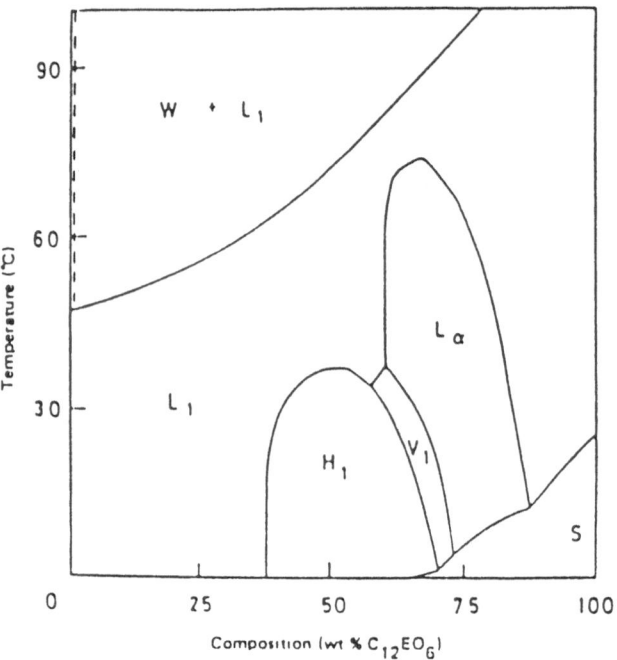

Fig. 1. The phase diagram of the water — $C_{12}E_6$-system. $L_1$ is the isotropic micellar solution, continuously connected to solutions of water in the liquid surfactant. $H_1$, $V_1$, and $L_\alpha$ are liquid crystalline phases, hexagonal, cubic, and lamellar, respectively. $S$ is sold surfactant, $W$ water. (Reproduced with permission of the Royal Society of Chemistry)

at this high value in the lc-phases. This seems to imply that the excited pyrene molecule is more exposed to the aqueous surroundings in the small micelles than in the other structures, which is reasonable since these expose less of the hydrocarbon core to the water. No change of the III to I fluorescence ratio [3] was detected, however.

The variation of the lifetime complicates the comparison of the quenching behaviour somewhat. The decay curves presented in Fig. 2 have been multiplied by $\exp(k_0 t)$, therefore, to remove the natural decay. They have also been normalized at a point several channels after the maximum. This was necessary because of an initial fast decay which was particularily dominant in the samples from the lamellar region. The origin of this effect is not fully understood; it is probably an impurity effect.

The results are obviously in good qualitative agreement with the introductory expectations. The decays start in about the same way, but make company only during a very short time; the curve for the neat solution is the first to deviate, whereas the decays for the rods and lamellae follow each other over a rather long period of time. The same applies to the decay curve

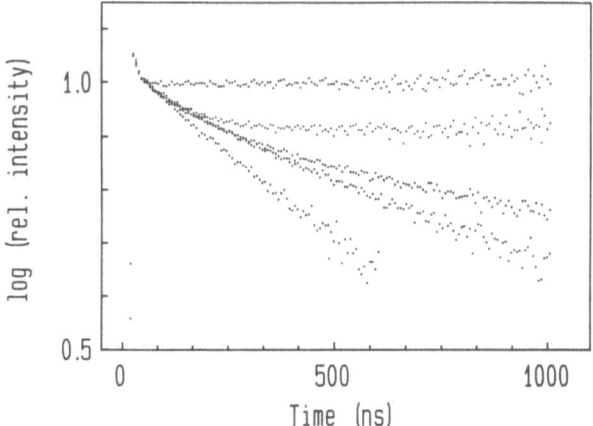

Fig. 2. Fluoresence decay curves, displayed as $\log(F(t) + k_0t$ vs $t$, for pyrene, $8 \times 10^{-6}$ M, quenched by hexylbenzophenone, both dissolved in $C_{12}E_6$, with various additions of water. The topmost curve is for 2% surfactant and no quencher. The rest of the curves are for a fixed concentration of quencher in the surfactant medium, $1.98 \times 10^{-2}$ M, and with the composition, from above, 2%, 33%, 75%, and 100% surfactant. — The curves were normalized at a point about 30 ns from the initial maximum

from the cubic phase, which fall in between these two; closest to the one for the lamellar phase.

In the quantitative evaluation the following equations were used: For the *zero-dimensional case* the model was quenching in small monodisperse micelles, without migration of probe and quencher [4, 5]:

$$F(t) = F(0)\exp\{-k_0t - n[\exp(-k_Qt) - 1]\} \quad (3)$$

where $k_Q$ is the first-order quenching constant and $n$ the average number of quenchers per micelle. The fitting in this case gives primarily $n$, from which the average aggregation number can be calculated, and the quenching constant, which can be related to the diffusion coefficient only through a theoretical model, or empirical correlations.

In the *one-dimensional case* of the rods Eq. (4) was used [6, 7].

$$F(t) = F(0)\exp\left\{-k_0t - \frac{2c_0}{h}\right.$$
$$\left.\left[e^{h^2Dt}\,\mathrm{crfc}(h\sqrt{Dt}) - 1 + \frac{2}{\sqrt{\pi}}h\sqrt{Dt}\right]\right\} \quad (4)$$

where $c_0$ is the number of quencher per unit length of the rod, $h = 2k_qR/3D$ is a parameter that weights reaction against diffusion, $R$ is the cylinder radius, and

$k_q$ the first order rate constant for reaction when a quencher is in the "reaction zone" of length $4R/3$ surrounding the excited state [6].

The reaction zone construction is a way to account for the finite reaction rate when probe and quencher meet. For very rapid reactions, or at long times ($h^2Dt \gg 1$) the Eq. (4) simplifies to Eq. (5)

$$F(t) = F(0)\exp\left\{-k_0t - 4c_0\sqrt{\frac{Dt}{\pi}}\right\} \quad (5)$$

which is on the same level of approximation as Eq. (6) for the 2-dimensional case. The fittings with the one-dimensional model were actually performed with Eq. (4), but in such a way (utilizing only the long-time part of the decay because of the initial perturbation of the decay cuves) that Eq. (5) would have given almost the same results.

In the *two-dimensional case* the equations due to Owen [8] was adopted, and applied to the long-time behaviour.

$$\ln\frac{F(t)}{F(0)} = -k_0t - zc_Qa^2Q_2\left(\frac{t}{\tau_Q}\right). \quad (6)$$

Here, $z$ is the half-width of the bilayer, $a$ the encounter radius $\tau_Q = D/a^2$, and $Q_2$ given by

$$Q_2\left(\frac{t}{\tau_Q}\right) = \frac{16}{\pi}\int_0^\infty \frac{dx}{x^3}\frac{1 - \exp\left(-\dfrac{Dt}{a^2}x^2\right)}{J_0^2(x) + Y_0^2(x)} \quad (7)$$

where $J_0$ and $Y_0$ are zero order Bessel functions of the first and second kind. Owen showed that a useful approximation to $Q_2$ is given by

$$Q_2' = 14.180(t/\tau_Q)^{1/2} + 3.17t/\tau_Q \quad (8)$$

which was used in the fittings.

*The three-dimensional case* is the decays in the neat surfactant, evaluated with Eqs. (1) and (2).

In order to obtain estimates for $D$ from the fitting parameters, values for the cylinder radius, the bilayer thickness, and the encounter distance are needed. According to Clunie et al. [9] the bilayer thickness is $2z = 13.6$ Å, and the cylinder radius in the hexagonal phase $R = 14.6$ Å. Although these values may vary somewhat with temperature and composition we have

Table 1. Estimates of the mutual diffusion coefficient for excited pyrene and hexylbenzophenone at 13 °C in the $C_{12}E_6$-water system from fluorescence decay curves

| wt % | $D \times 10^{11}$ m$^2$ s$^{-1}$ | | | |
|------|----------------|----------------|----------------|-------|
|      | 1-dim model | 2-dim model | 3-dim model | phase |
| 33   | 6.5    | (0.02) |     | $L_1$ |
| 50   | 6.5    | (0.1)  |     | $H_1$ |
| 63   | 6.4    | (0.5)  |     | $H_1$ |
| 68   | (22.4) | 3.5    |     | $V_1$ |
| 75   | (47.3) | 4.1    |     | $L_\alpha$ |
| 100  |        |        | 4.5 | neat |

used them throughout. For the encounter radius we have adopted a value of 8 Å.

The results from the fits are collected in Table 1. It is evident that the values of the diffusion coefficients, when estimated with the appropriate model, are very similar for all phases. The rod-model fits to the data for the solution just outside the hexagonal phase; rod like micelles are expected to be found here. The values are very sensitive to the values chosen for $a$, $z$, and $R$, and a discussion of the small differences that can be noted are not warranted at this stage.

The value for the cubic phase that conforms best with the rest was obtained with the two-dimensional model. The structure of this phase is not known with certainty; a periodic minimal surface (Schwarz type $P$) has been suggested [1]. The observation time in the measurements was about a microsecond, which means a root mean square displacement of about 85 Å for a diffusion coefficient as that obtained. Thus, the local geometry of the structure in the cubic phase should be close to two-dimensional on this length scale. This proposition seems reasonable.

The decay curve recorded for the most dilute solutions was typical for small micelles with stationary probes and quenchers, and the fitting to this model gave an aggregation number of 144 and a quenching constant $k_Q = 8.5 \times 10^6$ s$^{-1}$. The micelles cannot be quite spherical at this aggregation number — it corresponds to a hydrophobic radius of 23 Å. Using this radius and a correlation suggested previously, $k_q = 1.1\ D/R^2$ [6], one obtains $D = 5 \times 10^{-11}$ m$^2$ s$^{-1}$, which falls in well with the values given in Table 1.

## Conclusions

The results presented in Fig. 2 show that the reduction of dimensionality strongly reduces the quenching of fluorescence, when the mobility and local concentration of quenchers are kept the same. They also show, however, that only the exponential decay in the $3D$-case, and the typically biphasic one in the $0D$-case, have distinct features that allow a qualitative conclusion to be drawn. The 1 and $2D$-decay curves are similar and only the quantitative comparison of estimated parameter values can tell the difference. A hypothetical system with fractal geometry between 0- and $3D$ would probably give a decay curve with features similar to those of the 1 and $2D$ cases, that is without a signature. It would be very difficult to distinguish such a system from an 1 or $2D$ structure with slightly changed thickness. It would be even worse if polydisperse micelles and exchange possibilities had to be considered as well. As a model system, however, for the realization of various reaction geometries the $C_{12}E_6$-water system seems excellent.

## References

1. Mitchell DJ, Tiddy GJT, Waring L, Bostock T, MacDonald MP (1983) J Chem Soc Faraday Trans I 79:975
2. Sjöblom J, Stenius P, Danielsson I (1987) In: Schick MJ (ed) Nonionic Surfactants. Physical Chemistry. Surfactant Science Series vol. 22. Marcel Dekker, New York and Basel, p 369
3. Kalyanasundaram K, Thomas JK (1977) J Am Chem Soc 99:2039
4. Infelta PP (1979) Chem Phys Lett 61:88
5. Almgren M, Löfroth J-E (1981) J Coll Interface Sci 81:486
6. Almgren M, Alsins J, Mukhtar E, van Stam J (1988) J Phys Chem 92:4479
7. Alsins J, Almgren M (1989) J Phys Chem in the press
8. Owen CS (1975) J Chem Phys 62:3204
9. Clunie JS, Goodman JF, Symons PC (1969) Trans Faraday Soc 65:287

Authors' address:

Prof. M. Almgren
Department of Physical Chemistry
Uppsala University
Box 532
75121 Uppsala, Sweden

**Progress in Colloid & Polymer Science**                    Progr Colloid Polym Sci 81: 13 – 18 (1990)

# Polymorphism in dilute surfactant solutions: A neutron scattering study

J. Appell, P. Bassereau, J. Marignan and G. Porte

Groupe de Dynamique des Phases Condenses (unité associée au C.N.R.S. n⁰ 233)
Université des Sciences et Techniques du Languedoc Montpellier, France

*Abstract:* We present and discuss the oscillations observed in the small angle neutron scattering pattern of brine rich phases of surfactant for scattering vectors ranging from 0.1 to 0.5 Å$^{-1}$. These oscillations are expected to reflect the form factor of the local structure of the surfactant aggregates. Adjustement of the data to the form factors for the previously assumed local structure indeed confirms the assumption and indicates a certain polydispersity in the characteristic dimension (radius for the spherical or cylindrical local structure, or thickness for the planar local structure). The scattering pattern from solutions of large elongated micelles for scattering vectors in the range 0 to 0.05 Å$^{-1}$ displays features characteristic of semi-flexible coils: the persistence length is found to be 180 ± 50 Å.

*Key words:* Surfactant aggregates, structure, small angle neutron scattering

## Introduction

Surfactant molecules aggregate in dilute water or brine solutions and form different phases. Among these, the solutions of large elongated micelles and the so-called anomalous isotropic phase ($L_3$) arise currently much interest owing to their peculiar properties:

— the large elongated micelles are pictured [1, 2, 3] as long semi-flexible cylinders with a large distribution of lengths in dynamical equilibrium. They are thus expected to display a polymer-like behavior. This has been indeed observed in semi dilute solutions by Candau et al. [4]. Furthermore they ought to display rheological properties characteristic of "living polymers" [5]. This polymer-like representation of the elongated micelles implies a large flexibility of the elongated micelles; it can best be tested by a measure of the radius of the locally cylindrical micelle and of its persistence length: their relative magnitude is an indication of the validity of the assumed representation.

— the anomalous isotropic phase $L_3$ has been encountered in many systems; it is optically isotropic at rest but birefringent upon shaking. A bicontinuous structure, where an infinite bilayer separates two interwoven selfconnected brine domains, has been proposed [6, 7]. In this description, the bilayer in the $L_3$

phase has the same topology as the monolayer in bicontinuous microemulsions. The spontaneous curvature of the two films can however be different: in the $L_3$ phase it is everywhere zero (the two interwoven domains are identical); in a bicontinuous microemulsion the spontaneous curvature can have a finite value (when an asymmetry of the interwoven oil and water domains exists). Thus the $L_3$ phase could be used as a simplified model for a bicontinuous microemulsion if the proposed structure is correct. This implies first that the local structure of the aggregates is a thin bilayer (of very small thickness compared to its extension).

The detailed study [1, 8 – 10] of the phases formed in some water (or brine) rich binary (surfactant/water) or ternary (surfactant/alcohol/water) systems revealed a rich polymorphism in these systems in agreement with other groups work [3, 11, 12]. The phase behavior was interpreted in terms of the morphological transformations of the elementary objects induced by addition of salt to the binary systems [2] or by the variation of the alcohol to surfactant ratio in the ternary systems [10]. The local structure of the elementary objects was first deduced from the nature of the phases and from more or less indirect experimental evidences [1, 8, 9]. We found the, now classical, sequence for the local structure of the surfactant aggregates: spherical or cylindrical in micellar solutions and locally planar in the lamellar lyotropic liquid crystal and in the anomalous isotropic phase.

In order to confirm the validity of these assumptions and measure the characteristic dimensions of these local structures, we undertook on some representative samples, a small angle neutron scattering study on the largest possible range of scattering vectors. Most of the results obtained have been interpreted and discussed in a preceding paper [13]. They confirm the existence of the three local structures and support the idea of the potential use of the dilute surfactant phases as model systems (see above). The purpose of the present paper is to reinterpret more precisely a part of these results namely the oscillations which appear at the highest scattering vectors. These oscillations reflect the form factor of the aggregates on the scale of their local structure and are the best signature of these local structures.

## Experimental

The small angle neutron scattering experiments were performed at the cameras D11 and D16 of the Institut Laue Langevin in Grenoble. The experimental procedures are described at length in [13]. The results are interpreted using the well established procedures for the structural studies of colloical or micellar aggregates [14–16].

## Global structure

This structure is expected in micellar solutions of cetylpyridinium chloride (CPCl) in NaCl brine. The form factor oscillations are shown, in a plot $q^4 I(q)$ versus $q$, in Fig. 1. The best agreement in the position of the extrema is first obtained for spheres with $r = 25$ Å; the positions of the extrema is very sensitive to the choice of the radius. As can be seen in Fig. 1 the experimental damping of the oscillations can be only partly accounted for, either by assuming a rather large gaussian distribution of the radius of spheres or by assuming a slight shape anisometry of the micelles. These results are indeed in favor of globular micelles (trials with the form factors for cylinders or planes led to no agreement); the globules must be anisometric and polydisperse; a refinement of the adjustement to the experimental data taking both factors into account is however pointless (as is well known [16], the choice of parameters will then be non unique). The results for

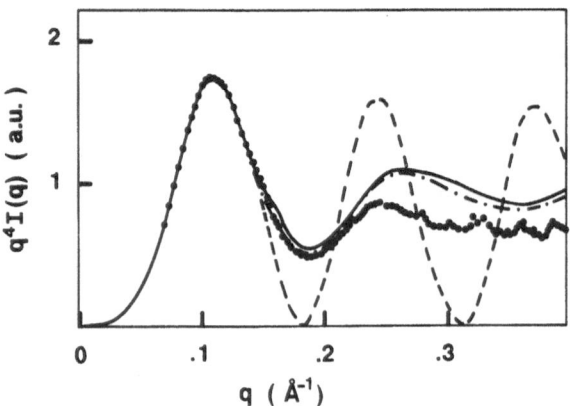

Fig. 1. Form factor oscillations in the Porod region: – – – = calculation for spheres with $r = 25$ Å; ——— = calculation for a distribution of spheres with $\bar{r} = 23$ Å and $\sqrt{(r - \bar{r})^2} = 4$ Å; –·–·– = calculation for ellipsoids with $a = 28$ Å and $b/a = 0.67$; dots = experimental result for the sample: Micellar solution 5% CPCl in 0.2 M NaCl brine

Table 1. The radius of the spherical local structure: the micelles of CPCl in 0.2 M NaCl brine

| Method of determination | Radius (Å) |
|---|---|
| $I(q)$ Guinier plot in the low $q$-range: (a) | |
| from the radius of gyration | $26 \pm 2$ Å |
| from the mass | $25.5 \pm 4$ Å |
| S/V ratio from the Porod limit (a) | $26 \pm 4$ Å |
| from the oscillations in the Porod region spheres (a) | $25 \pm 1$ Å |
| polydisperse spheres (b) | $\bar{r} = 23$ Å; $\sqrt{(r - \bar{r})^2} = 4$ Å |
| ellipsoid (b) | $a = 28$ Å; $b/a = 0.67$ |

(a) from reference [13] (b) this work

the dimension of the globular micelles obtained at all scattering vectors are summarized in Table 1 where they can be seen to be in good agreement.

## Planar structure

This structure has been probed in the $L_3$ anomalous isotropic phase of the ternary system cetylpyridinium chloride/Hexanol/0.2 M NaCl brine (hexanol/CPCl = 1.1 in weight). The form factor oscillations are shown, in a plot $q^4 I(q)$ versus $q$, in Fig. 2. An excellent agreement is obtained if the scattering objects are assumed to be locally planar bilayers with a gaussian

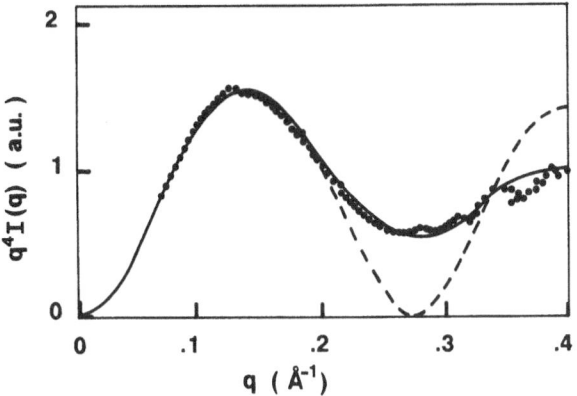

Fig. 2. Form factor oscillations in the Porod region: – – – = calculation for flat bilayers with $d = 22$ Å; —— = calculation for flat bilayers with a distribution of thicknesses $\bar{d} = 22$ Å and $\sqrt{(d - \bar{d})^2} = 5$ Å dots = experimental result for the sample: CPCl + hexanol in 0.2 M NaCl brine (90% brine)

Table 2. The thickness of the planar local structure: $L_3$ in the CPCl-hexanol-0.2 M NaCl brine system

| Method of determination | Thickness $d$ (Å) |
| --- | --- |
| $q^2 I(q)$ Guinier-like plot in the low $q$-range (a) | |
| from the radius of gyration | $24 \pm 2$ |
| from the mas/unit area | $23 \pm 4$ Å |
| S/V ratio from the Porod limit (a) | $22 \pm 4$ Å |
| From the oscillations in the Porod region | |
| bilayers of unique thickness (a) | $22 \pm 1$ Å |
| bilayers with a distribution of thicknesses (b) | $\bar{d} = 22$ Å and $\sqrt{(d - \bar{d})^2} = 5$ Å |

(a) from reference [13] (b) this work

distribution of thicknesses: $\bar{d} = 22$ Å and $\sqrt{(d - \bar{d})^2} = 5$ Å. The adjustment to the experimental data is very sensitive to the chosen parameters and no agreement could be found with form factors for other structures. The agreement with the thickness determined from other parts of the scattering pattern can be seen to be very good on Table 2.

Rather large fluctuations of the thickness are found; can one wonder on the possible origin of these? They could be due to the existence of a peristaltic mode in the bilayer or to the variation of the thickness of the bilayer with the variation of the film curvature from one point to the other. In the first case these fluctuations should also appear in the form factor of the lamellae of the adjacent lamellar phase $L_\alpha$ and they

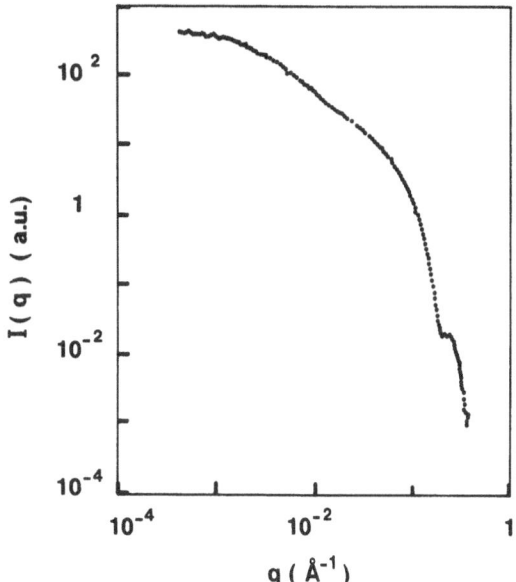

Fig. 3. Log/Log display of the spectral pattern. For samples representative of large elongated micelles: 1% CPBr in 0.8 M NaBr brine for the lowest $q$-range (up to $3 \times 10^{-3}$ Å$^{-1}$ light scattering results and up to $1.5 \times 10^{-1}$ SANS results) and 5% CPBr in 0.8 M NaBr brine (from $7 \times 10^{-2}$ Å$^{-1}$ to $5 \times 10^{-1}$ Å$^{-1}$ SANS results). The different sets of data have been shifted vertically to coincide

should disappear in the second case. Answers to these questions should be obtained from high spatial resolution SANS we plan to make on $L_\alpha$ samples. This study should also solve the following puzzle: the mean thickness found for $L_3$ is significantly smaller than the thickness of the lamellae in $L_\alpha$, derived indirectly from the evolution of the Bragg peak with the dilution of the phase [17]. This difference is possibly real however another explanation is more plausible namely that the thickness of the lamellae in the $L_\alpha$ phase appears artificially larger as a result of the large ondulations of the lamellae [13].

**Cylindrical structure**

This structure is expected in micellar solutions of large elongated micelles [1, 2]. It has been studied here on solutions of cetylpyridinium bromide in 0.8 M NaBr brine. The overall scattering pattern is shown in a log-log representation in Fig. 3. Going from lower to higher scattering vectors two linear regions are particularly distinct; one of zero slope corresponds to the overall aggregates; one of slope 1 is characteristic of a unidimensional structure (long thin rods). At the highest scattering vectors, oscillations are superim-

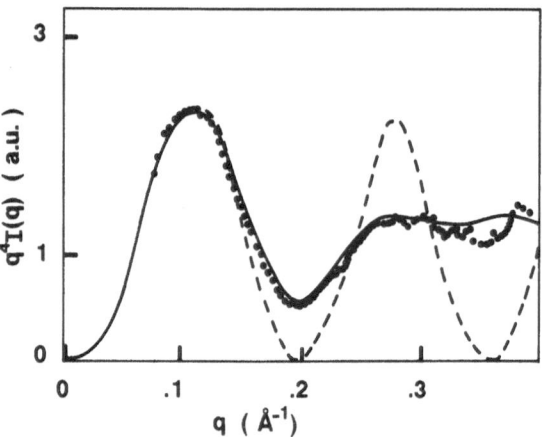

Fig. 4. Form factor oscillations in the Porod region: $- - -$ = calculation for cylinders with a circular cross section $r =$ 19.5 Å; ——— = calculation for cylinders with a distribution of circular cross section $\bar{r} =$ 19.5 Å and $\sqrt{(r - \bar{r})^2} =$ 1.5 Å; dots = experimental result for the sample: 5% CPBr in 0.8 M NaBr brine

Table 3. The radius of the cylindrical local structure: the micelles of CPBr in 0.8 M NaBr brine

| Method of determination | Radius $r$ (Å) |
|---|---|
| $qI(q)$ Guinier-like plot in the intermediate $q$-range: (a) | |
| from the radius of gyration | 22 ± 2 Å |
| from the mass/unit length | 22 ± 4 Å |
| S/V ratio from the Porod limit (a) | 20.5 ± 4 Å |
| From the oscillations in the Porod region | |
| circular cross section with a unique radius (a) | 19.5 ± 1 Å |
| circular with a distribution of radii (b) | $\bar{r} = 19.5$ Å and $\sqrt{(r - \bar{r})^2} = 1.5$ Å |

(a) from reference [13] (b) this work

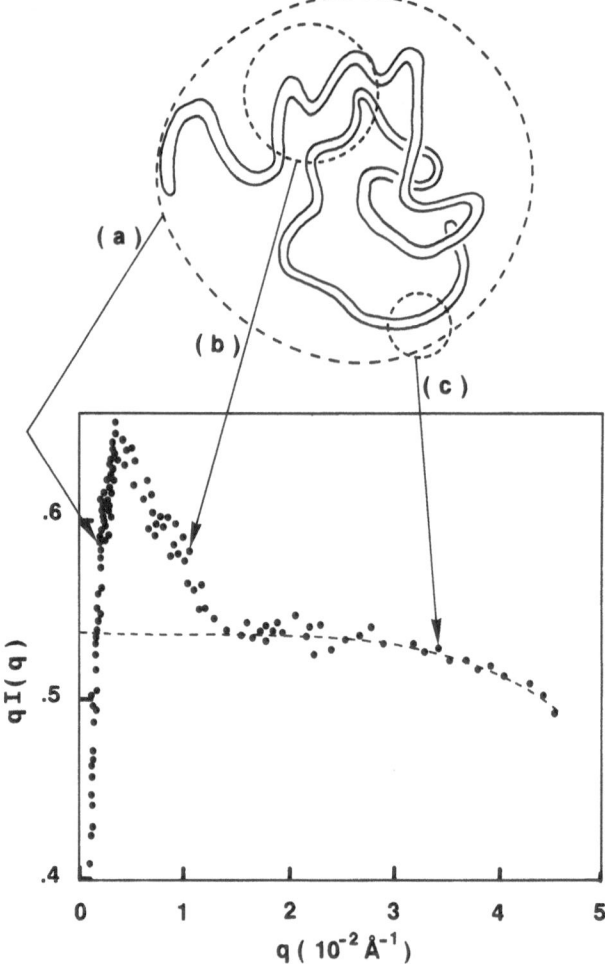

Fig. 5. Manifestation of the flexibility of the cylinders in the $qI(q)$ plot for the sample 1% CPBr in 0.8 M NaBr brine. (see text)

posed on the $q^{-4}$ decrease characteristic of a sharp-interface.

These oscillations are best seen in the $q^4 I(q)$ versus $q$ plot in Fig. 4. They are well mimicked assuming locally a circular cross section with a very narrow gaussian distribution of radii: $\bar{r} =$ 19.5 Å and $\sqrt{(r - \bar{r})^2} =$ 1.5 Å. A similar agreement has been obtained assuming a slight ellipticity of the cross section but we have no way to discriminate between these two possibilities. This mean radius compares well with the values determined from the other parts of the scattering pattern as shown in Table 3 and discussed in [13].

A somewhat larger radius is deduced from the intermediate-$q$ scattering pattern but, as discussed below, the elongated micelles are flexible on a scale which is not much larger than their diameter so that the influence of the flexibility can already be perceptible in this intermediate $q$ range.

## Some evidence of the flexibility of the elongated micelles

The intermediate and low-$q$ scattering pattern is shown in Fig. 5 in a $qI(q)$ versus $q$ plot which has a peculiar appearance. On this plot if the cylindrical micelles were infinite and rigid (= long thin rods) at all scale the curve should follow the dotted line while if they were rigid but with a finite length $L$ the curve

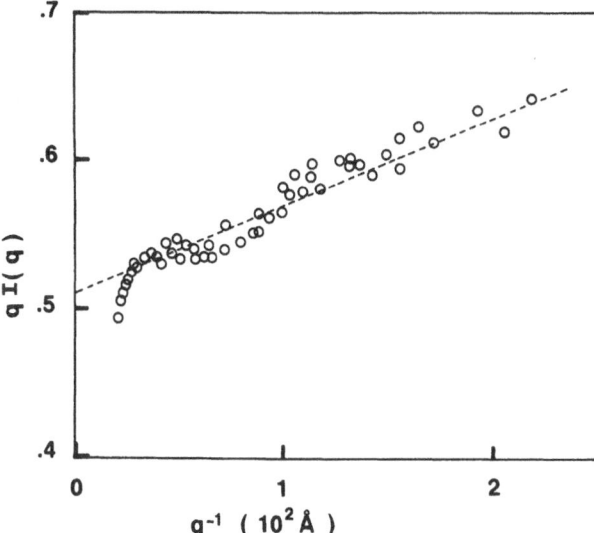

Fig. 6. Approximately linear part of $qI(q)$ as a function of $q^{-1}$ which allows for the determination of the persistence length characteristic of the flexibility of the cylinders for the sample 1% CPBr in 0.8 M NaBr brine. (see text)

should fall directly to zero from $q \approx L^{-1}$ to $q = 0$. Clearly we have none of these behaviors and as suggested in Fig. 5 we observe successively, going from lower to higher $q$: a behavior characteristic of finite size objects (a); then a decrease corresponding to semi flexible coils (b) and then a very slow decrease corresponding to thin rod-like parts (c). In this interpretation we can derive two estimates of the persistence length ($l_p$) characteristic of the flexibility of the elongated micelles. First the cross over from the semi-flexible coil pattern to the rigid rod pattern occurs at $q_c \approx 1.9/l_p$ [18] here $q_c \approx 1 \times 10^{-2}$ Å$^{-1}$ giving $l_p = 190 \pm 50$ Å. Then the scattering pattern for a semi-flexible coil can be represented by the following relation [19]:

$$qI(q) \alpha \frac{\pi}{L_{cyl}} + \frac{2}{3 l_p L_{cyl} q}.$$

We find a reasonably linear plot of part b of the scattering pattern in this representation as shown on Fig. 6 and deduce from it another estimate of $l_p = 180 \pm 50$ Å. These estimates are in reasonable agreement with those obtained previously ($l_p \approx 200 \pm 50$ Å) from entirely different methods [20].

Thus the present results confirm the representation of elongated micelles as long flexible cylinders; in our systems the flexibility is found to be large ($l_p/2r \approx 4$ to 5). However the flexibility is system dependent [21,

22] because it is a local property, which must depend on the detailed balance of the interactions between neighbouring surfactant molecules.

*Acknowledgements*

We gratefully acknowledge the assistance of E. Pebay-Peroula (ILL Grenoble) and R. May (ILL Grenoble) for the neutron experiments.

**References**

1. Porte G, Appell J (1981) J Phys Chem. 85:2511
2. Porte G (1983) J Phys Chem 87:3541
3. Young CY, Missel PJ, Mazer NA, Benedek GB, Carey MC (1978) J Phys Chem 82:1375
   Missel P, Mazer NA, Benedek GB, Young CY, Carey M (1980) J Phys Chem. 84:1044
4. Candau SJ, Hirsch E, Zana R (1984) J Phys France 45:1263
   (1985) J Colloid Interface Sci 105:521
   (1987) In: Safran S, Clark N (eds) Physics of complex and supramolecular fluids (Wiley New York), 569
5. Cates ME (1987) Macromolecules 20:2289
6. Cates ME, Roux D, Andelman D, Milner ST, Safran SA (1988) Europhys Lett 5:733
7. Porte G, Appell J, Bassereau P, Marignan J (1989) J Phys France 50:1335
8. Gomati R, Appell J, Bassereau P, Marignan J, Porte G (1987) J Phys Chem 91:6203
9. Porte G, Marignan J, Bassereau P, May R (1988) J Phys France 49:511
10. Porte G, Gomati R, El Haitamy O, Appell J. Marignan J (1986) J Phys Chem 90:5746
11. Benton WJ, Miller CA (1983) J Phys Chem 87:4981
    Natoli J, Benton WJ, Miller CA, Ford Jr TJ (1986) J Disp Sci Technol 7:215
12. Hoffmann H, Schwandner B, Ulbricht W, Zana R (1985) In: Degiorgio V, Corti M (eds) Physics of amphiphiles: Micelles Vesicles and microemulsions (North Holland Amsterdam) 261
13. Marignan J, Appell J, Bassereau P, Porte G, May RP (1989) J Phys France 51:3553
14. Guinier A, Fournet G (1955) Small angle Scattering of X-rays, Wiley, New York
15. Glatter O, Kratky O (1982) Small angle X-rays Scattering, Academic Press, New York
16. Cabane B, Duplessix R, Zemb T (1985) J Phys 46:2161
    Cabane B, In: Zana R (ed) Surfactant solutions, new methods of investigation, M. Dekker, New York 57
17. Bassereau P, Marignan J, Porte G (1987) J Phys France 48:673
18. Kirste RG, Oberthur RC in [15] 387
19. des Cloizeaux (1973) Macromolecules 6:403
20. Porte G, Appell J (1984) In. Mittal KL, Lindman B (eds) Surfactants in solution, Plenum press, New York 2:1
    Appell J, Porte G, Poggi Y (1982) J Colloid Interface Sci 51:18
21. Missel PJ, Mazer NA, Benedek GB, Carey MC (1982) In: Fendler EJ, Mittal KL (eds) Solution behavior of

surfactants; Plenum Press, New York 1:263 and (1983) J Phys Chem 87:1264
22. Van de Sande W, Persoons A (1985) J Phys Chem. 89:404

Authors' address:

Dr. J. Appell G.D.P.C. case 26
Université Sciences et Techniques du Languedoc
34060 Montpellier, France

**Progress in Colloid & Polymer Science**                    Progr Colloid Polym Sci 81:19 (1990)

# Structure and dynamics of polymerlike reverse micelles

P. Schurtenberger, R. Scartazzini, L. J. Magid[1), M. Leser[1), and P. L. Luisi

Institut für Polymere, ETH Zentrum, Zürich, Switzerland, and
1) Department of Chemistry, University of Tennessee, Knoxville, TN, USA

*Key words:* Lecithin, gel, viscoelasticity, network

We have recently described the formation of highly viscous, gel-like and optically transparent solutions in the system lecithin/organic solvent/water [1]. We found a dramatic increase of the viscosity of reverse micellar solutions of lecithin to values as high as $10^4$ poise upon the addition of a very small amount of water. Here we present evidence that the formation of viscoelastic solutions of lecithin in isooctane is due to the water-induced aggregation of lecithin molecules into flexible cylindrical reverse micelles and the subsequent formation of a transient network of entangled micelles.

The existence of cylindrical aggregates in these solutions is in agreement with the results from systematic small angle neutron scattering (SANS) measurements at different water to lecithin molar ratios ($w_0$) and different lecithin volume fractions ($\Phi$). A crossectional radius of gyration $R_{g,c}$ for both the reverse micellar crossection (headgroup + hydrophobic tail region) as well as for the headgroup region only has been determined, respectively, using contrast variation. The values of $R_{g,c} = 21.3$ Å for $d_{18}$-isooctane and $H_2O$, and $R_{g,c} = 14$ Å for $h_{18}$-isooctane and $D_2O$, respectively, are independent of $w_0$ and $\Phi$ and quantitatively consistent with the predictions of a model of one-dimensional cylindrical growth for lecithin reverse micelles.

The viscoelastic behavior of these solutions and the static and dynamic properties of the transient network formed at lecithin concentrations above the overlap concentration $\Phi^*$ were then characterized using a combination of dynamic light scattering (QLS), SANS and rheological measurements. Measurements of the hydrodynamic correlation length $\xi_h$ and the static correlation length $\xi_s$ at different values of $\Phi$ and $w_0$, re-spectively, were performed by means of QLS and SANS. The dependence of $\xi_h$ and $\xi_s$ upon $\Phi$ is well characterized by a power law of the form

$$\xi \sim \Phi^{-x} \qquad (1)$$

with $x = 0.65 \pm 0.07$ for $\xi_h$ and $x = 0.7 \pm 0.05$ for $\xi_s$, respectively. These values are in quantitative agreement with recent theoretical and experimental results obtained for transient networks present in semidilute polymer solutions [2, 3].

Our results suggest that viscoelastic solutions of lecithin in organic solvents may serve as good model systems for studies of dynamic and static properties of transient networks formed by entangled micelles. Analogies to polymers and "living polymers" can be made without the additional difficulties of ionic intermicellar interactions and salt effects usually present in aqueous viscoelastic micellar solutions.

A full account of this work has been published in J Phys Chem (1990, in press) and Rheol Acta (1989) 28:372.

## References

1. Scartazzini R, Luisi PL (1988) J Phys Chem 92:829
2. De Gennes PG (1979) Scaling Concepts in Polymer Physics, Cornell Univ Press, Ithaca, N.Y.
3. Brown W, Mortensen K (1988) Macromolecules 21:420

Authors' address:

Dr. P. Schurtenberger
Institut für Polymere
ETH-Zentrum
CH-8092 Zürich, Switzerland

**Progress in Colloid & Polymer Science**                    Progr Colloid Polym Sci 81:20 – 29 (1990)

# Scattering as a critical test of microemulsion structural models

T. N. Zemb, I. S. Barnes, P.-J. Derian, and B. W. Ninham

*Abstract:* Current models of microemulsion microstructure are reviewed and their predictive power tested by comparison between their predictions and scattering from some typical concentrated ternary microemulsions.

*Key words:* Microemulsion; microstructure; curvature; DDAB; Voronoï

## Introduction

The understanding of microemulsion microstructure is a prerequisite for gaining control of such sometimes counter-intuitive properties as conductivity, solubilising power or reactivity between molecules dissolved in microemulsions. The problem of microemulsion structure would be solved when, for any given composition, the stability, translational diffusion constants and small angle scattering curve could be predicted by some kind of parameter-free model. Unfortunately, this is not yet the case, and the keywords "microemulsion structure" correspond to more than a thousand publications over the last ten years. Some models are parametric and, while they can be fitted to any scattering curve, give no information on the structure; other models have no parameters and are incompatible with experimental data. Furthermore, the inverse scattering problem is non-unique, and one can ask whether any scattering experiment can be decisive.

We define a microemulsion as a thermodynamically stable (e.g. reversible to temperature cycles), fluid, clear solution containing bulk oil, bulk water and at least one surface active molecule, but which does not exhibit long range order through the appearance of Bragg peaks in scattering. Experimental data allow the classification of microemulsion samples into four broad families:

(a) a surfactant-cosurfactant (short chain alcohol) couple, with a typical alcohol to surfactant ratio of about five, giving interfacial films of very low bending energy ($K_c \propto kT$), stabilised by entropy;

(b) ternary microelmulsions made with double chain surfactants insoluble in both oil and water

"phase", for which a geometric model is very satisfactory, since the dominant constraint is the constant curvature of the water/oil interface at imposed volume fraction;

(c) a very peculiar branched short chain surfactant (AOT), for which any size of water droplets can be formed, which means that there is no preferred curvature radius for the interface; and

(d) ternary systems made with nonionic surfactants, for which there is not yet a clear picture of microstructure, although the phase behaviour can probably be unified by inversion of the temperature axis between ionics and nonionics.

Our purpose here is to compare currently used models and to test them with some simple experiments using typical ternary microemulsions.

## Currently used models of microstructure

We find reference to seven differential families of structural models used in the literature and we will first recall their main features.

### Interacting spherical droplets

In this model, one simply supposes that one of the two volumes — polar or apolar — is "internal" and is distributed in a dispersion of interacting monodisperse spherical droplets. The total interface per unit volume $\Sigma$ is set by the concentration of surfactant molecules $c$ and the molecular head-group area $a$ by the relation $\Sigma = ca$. The specific interface $\Sigma$ can be measured by the high-$q$ limit of the scattering, where the intensity decreases as $q^{-4}$ [1]. Since the polar volume fraction $\phi$ can be calculated precisely from the composition by adding the water and ionic headgroup volumes, the density $n$ (cm$^{-3}$) and the radius $R$ of the droplets is known a priori at any composition. The scattering $I(q)$ of interacting droplets can therefore be calculated on a absolute scale, without any parameter by using the factorisation [2]:

$$I(q) = P(q)S(q)$$

where the form factor $P(q)$ reflects the scattering of one sphere and the structure factor $S(q)$ the interference between adjacent particles. $S(q)$ can be calculated analytically for repulsive interactions using the simple expression given by Hayter and Penfold [2] and for attractive step-shaped function using the expression developed by Sharma and Sharma [3].

For instance, in the AOT/water/octane system, the simple assumption of a spherical droplet structure allows the prediction of the observed scattering at any concentration except close to the critical point, where the model fails due to large density fluctuations. The slight excess of the observed scattering at low $q$ can be assigned to polydispersity or attraction between droplets [4].

But often the assumption of spherical droplets fails completely to predict the observed scattering with a dispersion of droplets of known radius $R$ [5]. In this situation, the temptation to fit the data to another droplet radius is high since this allows reconciliation of the observed scattering peak with a simple structural picture, but then the interface $\Sigma$ is no longer compatible with the measured Porod limit.

For common quaternary systems, the simple picture of interacting droplets is incompatible with the electrical properties [6] as well as the isoviscosity lines of the phase diagram [7]. A review on the subject of the inconsistencies obtained from the droplet model is available [8].

Nethertheless, this picture of interacting droplets has been proven useful in the exploration of micellar structure and interaction in a number of surfactant systems [9].

### The Talmon-Prager model

This was the first model to offer a continuous transformation from water droplets to oil droplets via a bicontinuous network. The structure is bicontinuous over macroscopic distances for polar volume fractions between 18% and 82%.

The construction of a Talmon-Prager structure is as follows [10]:

(1) choose of a set of random (Poisson) points, density $n$ per cm$^3$ of sample and then construct the associated Voronoï tesselation of space using the bisector planes;

(2) fill the different Voronoï cells with water or oil according to the available volume fractions $\phi$ and $(l - \phi)$ of water and oil; then

(3) set the interface $\Sigma$ between adjacent cells with different contents.

The scattering of this structure can be calculated analytically. In any case, there is no peak in the scattering. Since nearly all microemulsions exhibit a scattering peak, no scattering experiment yet made is compatible with the Talmon-Prager model. Also, electrical conductivity behavior is much more complicated than predicted by Talmon-Prager model: a monotonic increase with polar volume fraction, with a percolation threshold at 18%.

### "Bicontinuous" models

These models derive from propositions originally made by Scriven [11]: the idea is to reconcile the high translation diffusion coefficients measured both for water and oil by using the structures of liquid crystalline cubic phases which consist of a minimal surface lying on a periodic lattice. Since no Bragg peaks are observed with microemulsions, one must then "melt" the underlying lattice in order to obtain a disordered structure [12, 13, 24].

There are actually two different families of bicontinuous structures proposed here:

— The first is a monolayer lying on a surface of constant but not necessarily zero mean curvature which separates the oil and the water domains. While this has since been found to be an unlikely structure for cubic phases [29], "melting" of the underlying lattice gives rise to the type of structures produced by the DOC cylinders model at values of the connectivity greater than about four; and

— The second is a bilayer centered on a minimal (zero mean curvature) surface separating two distinct "external" subvolumes. With an underlying crystalline lattice this is believed to be the structure of a number of known cubic phases [29]; when "melted" it gives rise to the structure of the so-called anomalous isotropic or $L_3$ phase [14].

Unfortunately, there is no known algorithm for generating a zero or other constant average curvature

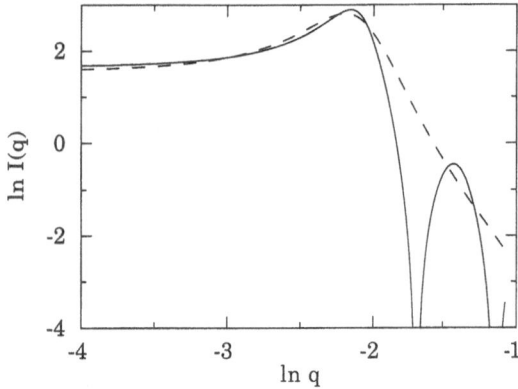

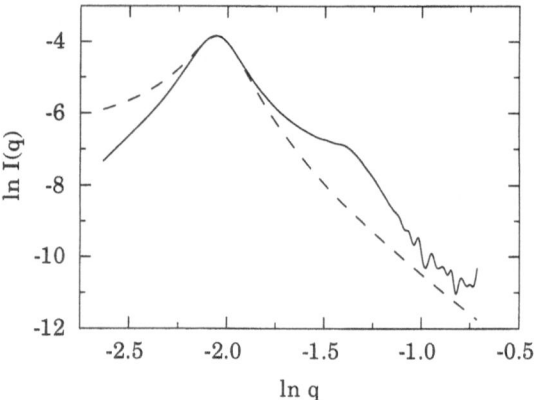

Fig. 1. (——) Calculated scattering of a set of hard spheres (radius 25 Å, volume fraction 34%), compared to the best fit obtained via the Teubner-Strey parametric expression (— — —), yielding a persistence length of $\xi = 32.1$ Å and a peak position $D^* = 53.2$ Å. The fit is acceptable in linear scale, but is shown here in log scale. The oscillations observed in the calculated $I(q)$ are due only to the monodispersity

Fig. 2. One of the rare experimental curves which could not be fitted by Teubner-Strey expression. Small angle X-ray scattering of 70% C12E8 in $D_2O$ at 60C compared to the best fit $\xi = 60.3$ Å, $D^* = 48.6$ Å

surface on a random network. The "bicontinuous model" proposed by Scriven is therefore a qualitative intuitive way of defining a structure which cannot yet be tested by experiment, since $I(q)$ cannot be calculated without a crystalline array.

### The cubic random cell model (CRC)

First introduced by De Gennes as a decorated Talmon-Prager model, this model assumes that the microstructure can be approximated by a set of cubes of size $\xi$ filled at random with water or oil, according to the available volume fraction. The persistence length $\xi$ is fixed by the available surfactant. The stability of the microemulsion is then related to the spontaneous persistence length, which is related to the surfactant parameter $v/al$. This model has only one distance built in. The persistence length $\xi$ is given by [15]:

$$\xi = \frac{6\varphi(1-\varphi)}{\Sigma}$$

and the peak position is predicted to be at $D^* = 2\pi/q_{max} = \xi$. Explicit calculations show that this model gives no peak at all in scattering [21]: the zero of the form factor of the cubes exactly cancels the peak due to the underlying lattice. This is an artefact of the exact calculation and the peak will reappear when a slight variation in the cube size — coming from the decoration by a surfactant monolayer — is taken into account.

However, as already noted by Auvray et al. [16], the peak is usually observed at about $D^* = 2\xi$: twice that real space distance. This has been taken as a proof of the validity of the CRC model, but this missing factor of two in the observed periodicity is *proof* of the existence of a local microstructure. This can be introduced within the framework of the DOC model. Alternatively, a microstructure — albeit less intuitive and predictive — can be introduced in terms of local correlations imposed by steric constraints such as bending energy [17]. Without such local microstructure, i.e. if the cells are filled at random, the predicted peak calculated in the tables is given by $D^* = 2\xi$ for the CRC model.

Calculation of the mean and Gaussian curvatures in a recent paper [18] shows that this is more complicated than previously thought. In particular, unless the volume fractions are equal, the mean curvature is not zero.

### Parametric models

On the basis of thermodynamic arguments, or assuming a locally lamellar model, two very similar expressions have been proposed to fit the scattering of a microemulsion. The first, proposed by Teubner and Strey [19], gives

$$I(q) = \frac{(8\pi/\xi)\bar{\eta}^2 c_2}{a_2 + c_1 q^2 + c_2 q^4}$$

from which the correlation function (or Patterson function) $\gamma(r)$ is

Table 1. Comparison of the predictions of the six non-parametric models for the topology, the peak position $D^*$, the average curvature $H$, the Gaussian curvature $K$ and the surfactant parameter $v/al$

| | Talmon-Prager | Cubic random cell | Spheres: water in oil | Spheres: oil in water | DOC cylinders | DOC lamellae |
|---|---|---|---|---|---|---|
| topology | Bicontinuous for $0.18 < \phi < 0.82$, oil continuous for $\phi < 0.18$ (water droplets in oil), water continuous for $\phi > 0.82$ (oil droplets in water) | Bicontinuous for $0.2 < \phi < 0.8$, oil continuous for $\phi < 0.2$ (water droplets in oil), water continuous for $\phi > 0.8$ (oil droplets in water) | Always oil-continuous | Always water-continuous | Bicontinuous for $Z > 1.2$ Always oil-continuous | Bicontinuous for $\psi > 0.18$ Always oil-continuous |
| position of scattering peak | no peak | $D^*\Sigma = 6\phi(1-\phi)$ | $D^*\Sigma = 4.84\,\phi^{1/3}$ $R = 3\phi/\Sigma$ | $D^*\Sigma = 4.84\,\phi^{2/3}$ $R = 3(1-\phi)/\Sigma$ | $D^*\Sigma = f(\phi,\Sigma,Z)$ | $D^*\Sigma = 11.64\,\psi(1-\psi)$ $t = 2\phi/\Sigma$ |
| $H$ | $1.57\,(1-2\Phi)n^{4/3}$ | $\approx \dfrac{(1-2\phi)}{12\xi}$ | $1/R$ | $1/R$ | | irrelevant |
| $K$ | $(3.65\,(2\phi-1)^2 - 1.49)n^{2/3}$ | $\approx \dfrac{8}{5\xi^2}\left[(1-2\phi)^2 - \dfrac{1}{4}\right]$ | $1/R^2$ | $1/R^2$ | | $(3.65\,(2\psi-1)^2 - 1.49)n^{2/3}$ |
| $v/al$ | $1 + Hl + Kl^2/3 \approx 1$ | $1 + Hl + Kl^2/3$ | $1 - Hl + Kl^2/3 > 1$ | $1 - Hl + Kl^2/3 < 1$ | $1 + Hl + Kl^2/3 > 1$ | $1 + \dfrac{K(tl + l^2/3)}{(1 + Kt)} \approx 1$ |
| what works | – curvature<br>– bicontinuity | – curvature<br>– bicontinuity | – peak position | NOTHING! | – peak position<br>– bicontinuity | – peak position<br>– bicontinuity<br>– curvature<br>– $v/al$ |
| what doesn't work | – no scattering peak | – peak at twice the correct angle | – not water-continuous<br>– $v/al$ too large | – not oil-continuous<br>– $v/al$ too small<br>– wrong peak position | – $v/al$ too large<br>– $v/al$ must change to give the right peak position | NOTHING! |

$$\gamma(r) = \gamma_0(r)e^{-r/\xi}$$

where $\gamma_0(r)$ is given by a periodic structure:

$$\gamma_0(r) = \frac{D^*}{2\pi r}\sin\left(\frac{2\pi r}{D^*}\right).$$

This expression allows one to extract the peak position $D^*$ and the peak width expressed in terms of a persistence length $\xi$ from any scattering curve by fitting the three parameters $a_2$, $c_1$ and $c_2$. Fitting an observed scattering $I(q)$ to these expressions succeeds as long as there is only one peak in the scattering.

Therefore no test of these models can be made – they have parameters instead of predictive power.

For example, as is shown in Fig. 1, one can calculate using the Hayter-Penfold procedure the intensity scattered by a system of hard spheres of radius 25 Å with an internal volume fraction of 34%. This theoretical curve can then be fitted by the Teubner-Strey expression to give values for the two parameters $D^*$ and $\xi$, as for any experimental curve. The discrepancy at high-$q$ between an oscillating form factor and a smooth behaviour of the parametric expression can be removed by any form of shape or mass polydispersity of the spheres [4].

Table 2. Composition of the samples, quantities measured from the scattering curves and values of the geometrical quantities given by the different models

| Sample | | | Tetra 1a | Tetra 3a | Tetra 5a |
|---|---|---|---|---|---|
| Composition | mass % | $D_2O$ | 27.0 | 42.0 | 60.0 |
| | | DDAB | 36.5 | 29.0 | 20.0 |
| | | Oil | 36.5 | 29.0 | 20.0 |
| | | $D_2O$ | 22.4 | 36.1 | 53.9 |
| | volume % | DDAB | 33.6 | 27.7 | 20.0 |
| | | Oil | 44.0 | 36.3 | 26.2 |
| | | $C_s(M)$ | 0.73 | 0.60 | 0.43 |
| | moles $D_2O$/DDAB | | 17.1 | 33.5 | 69.4 |
| | moles Oil/DDAB | | 2.3 | 2.3 | 2.3 |
| | $\Phi_1$ | | 27.9 | 40.6 | 57.2 |
| | $\Sigma$ (Å²/Å³) | | 0.030 | 0.025 | 0.018 |
| Measured quantities | $D^*$ (Å) | { neutron (Å) | 80 | 114 | 190 |
| | | { x-ray (Å) | 84 | 120 | 190 |
| | $l_c$ (average chord) | | 25 | 37 | 53 |
| | $\Sigma$ | { neutron | 0.031 | 0.027 | 0.019 |
| | $D^*\Sigma$ | { x-ray | 0.031 | 0.029 | 0.023 |
| | | | 2.5 | 3.0 | 3.5 |
| Models tested | spheres | | 28 | 49 | 95 |
| | w/o | { $R$ (Å) | 70 | 106 | 185 |
| | | { $D^*$ (Å) | 2.1 | 2.65 | 3.3 |
| | | { $D^*\Sigma$ | 1.5 | 1.26 | 1.13 |
| | | { $v/al$ | | | |
| | spheres | | 71 | 71 | 71 |
| | o/w | { $R$ | 130 | 140 | 150 |
| | | { $D^*$ | 3.9 | 3.4 | 2.8 |
| | | { $D^*\Sigma$ | 0.84 | 0.84 | 0.84 |
| | | { $v/al$ | | | |
| | CRC | { $D^*$ | 40 | 58 | 79 |
| | | { $D^*\Sigma$ | 1.21 | 1.45 | 1.47 |
| | DOC cylinders | { $D^*$ | 83 | 118 | 190 |
| | | { $D^*\Sigma$ | 2.5 | 2.95 | 3.4 |
| | | { $Z$ | 4.5 | 5 | 6 |
| | | { $R_s$ | 29 | 48 | 87 |
| | | { $v/al$ | 1.22 | 1.11 | 1.04 |
| | DOC lamellae | { $D^*$ | 82 | 116 | 162 |
| | | { $D^*\Sigma$ | 2.5 | 2.9 | 2.9 |
| | | { $t$ | 9.3 | 16.2 | 31.8 |
| | | { $\Psi$ | 0.3 | 0.5 | 0.5 |
| | | { $v/al$ | 0.98 | 0.97 | 0.97 |

One of the rare experiments which could not be fitted to the Teubner-Strey expression is shown in Fig. 2. In this case, the strong geometric constraints induce the formation of a second oscillation at higher $q$-values, as was also observed in DDAB/octane/water systems [20], and the Teubner-Strey model can no longer work.

It is worth noting here that for any scattering curve fitted up to now, a ratio $D^*/\xi$ around 2 was observed.

This suggests that the Voronoï tesselation is an appropriate underlying lattice for the construction of microemulsion microstructure, since for a Voronoï lattice, the order is lost over about one cell size. Therefore, the use of a Voronoï tesselation as a lattice in a DOC model will lead to right scattering peak witdth [21].

A general expression, similar to the Teubner-Strey expression has been derived by Kaler and Vonk: they

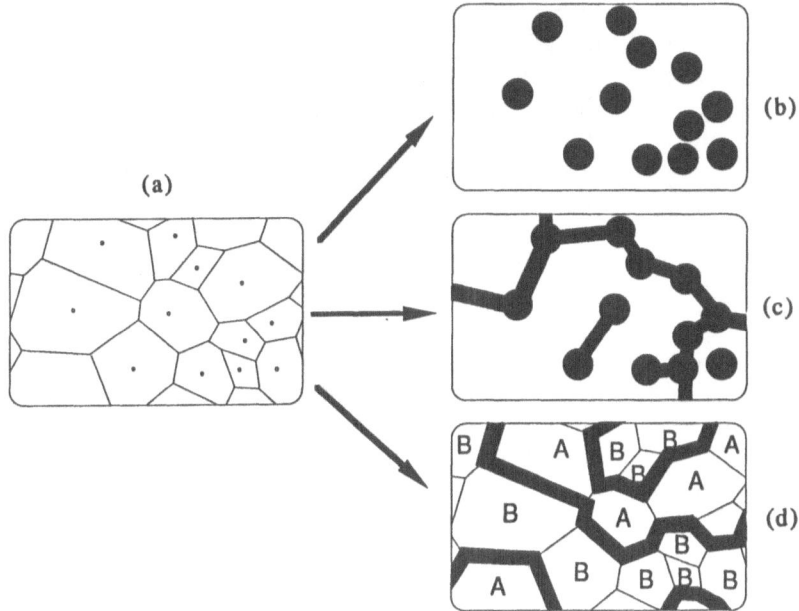

Fig. 3. Schematic 2-D view of the different textures as classified by the DOC model: a set of points, centers of imaginary repulsive spheres, allows construction of a Voronoï space tesselation (a). This space tesselation is then used as a random lattice to place spheres (b), cylinders (c), or bilayers (d). One of the solutions (b), (c) or (d) is acceptable if the "internal" volume fraction $\phi$ and interfacial area $\Sigma$ are imposed. All three structures have the same peak position

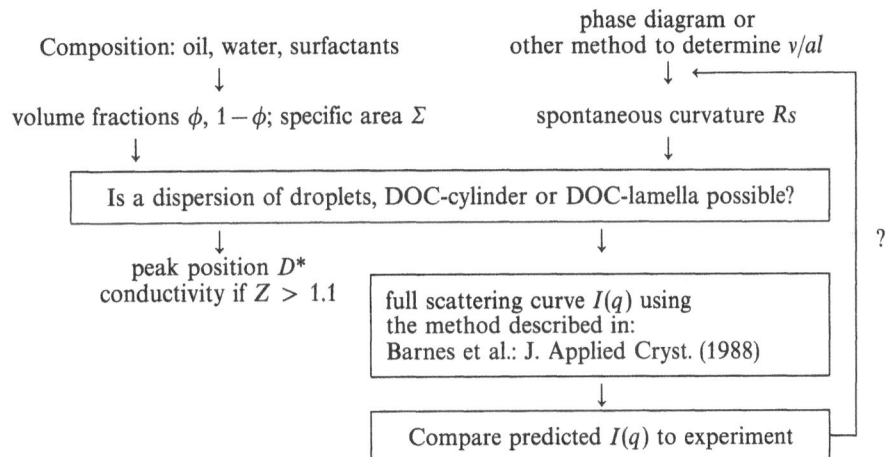

Fig. 4. Flow chart of the use of DOC model in order to interpret the scattering of a microemulsion. Composition and known molecular volumes allow determination of $\phi$ and $\Sigma$. The surfactant parameter $v/al$ must be known from phase diagram or be considered as a parameter. Solution of the geometrical constraints allows in favorable cases a numerical calculation of the peak position and the connectivity $Z$ for the three possible textures. Otherwise, full calculation of the scattering curve is required. This procedure must be done for a whole dilution line, where $a$ and $v/al$ have to remain constant

assume that any microemulsion structure is locally lamellar of periodicity $D^*$, with a persistence length $\xi$. The Patterson function of the structure is now given by [22]:

$$\gamma(r) = \gamma_0(r)e^{-\frac{2r}{\xi}}.$$

The third parameter $c_2$, introduced as the Porod limit by Teubner and Strey, is replaced in the model Kaler and Vonk by a lamellar thickness polydispersity $\sigma$. Unlike other parametric models, this even allows one to fit curves exhibiting no peak in the scattering. However, there is no guarantee that a given set of $D^*$, $\xi$ and $\sigma$ lead to a geometrically possible microstructure.

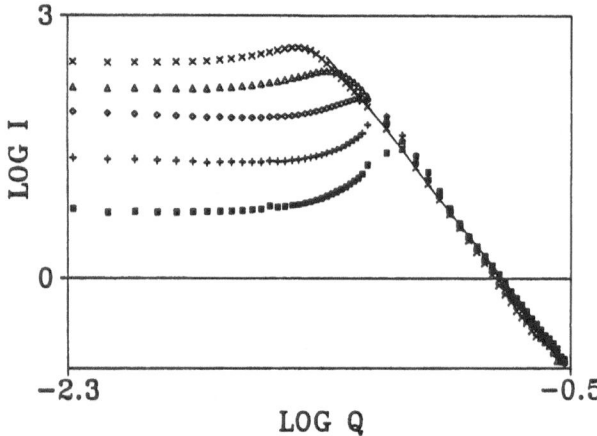

Fig. 5. Typical $\mathrm{Log}\,I$ versus $\mathrm{Log}\,q$ plot for a dilution line in the DDAB/tetradecane/water system. Two decades in $q$ are necessary for proper determination of $\Sigma$. The slope of the high-$q$ asymptotic limit is $-4$. The peak position moves with concentration. Composition of middle and extreme points are given in Table 1

Also, there is no spontaneous curvature embedded in these models, neither towards water, nor towards oil.

### The disordered open connected model [23]

We use a Voronoï space tesselation and construct a model for which the two important distances are embedded: the spontaneous curvature $R_S$ of the interface fixed by the nature of the surfactant film, and the average lattice spacing $D^*$ of the microstructure.

At the resolution of the scattering experiment, we say that the actual microstructure, a surface of constant curvature separating water and oil domains, can be approximated by sets of cylinders, spheres or bilayers.

Three constraints have to be satisfied [23]:

(A) the water $\phi$ and oil volume $(1 \cdot \phi)$ are set by the composition;
(B) the interfacial curvature must agree with the spontaneous curvature of the surfactant film: these are related by [21] in the case of a monolayer:

$$v/al = 1 + Hl + \frac{1}{3}Kl^2$$

where $H$ is the mean curvature and $K$ the Gaussian curvature; and
(C) the water/oil interfacial area per unit volume $\Sigma$ is set by the surfactant concentration.

Depending on the composition, these constraints can be satisfied by constructing a dispersion of spherical droplets, connected cylinders or random bilayers on the Voronoï tesselation (Fig. 3). In ternary systems, this framework allows the calculation of approximate phase boundaries and electrical conductivity [24] and of the full scattering curve [25] using the algorithm described in Fig. 4.

This model has until now only failed in two situations: the AOT/water/oil system, in which $v/al$ is not fixed by composition, and which in all experiments reported to date always keeps a droplet structure; and in systems such as SDS/pentanol/water/oil where the interface is mainly a short chain alcohol, in which case large bending radius variations are allowed. Both these failures are due to violation of constraint (B).

### The "random wave" model

This model was proposed by Berk using the algorithm of Cahn [26]. The idea is to produce a random structure of typical size $D^*$ by superposing waves of wavelength $D^*$ with random phases and random directions. The structure produced is random over long distances and the scattering has a peak at the desired position. No bending constraint or total interface is introduced: at a given water/oil ratio the total interface is fixed and cannot be varied. This model is useful in the case of large spontaneous radius of curvature ($v/al$ close to 1) and low bending energy ($K_c < kT$). In order to use this model, the position of the scattering peak has to be known. It is worth noticing that this model allows the calculation of the interfacial scattering spectrum from the bulk contrast one.

### Comparison with experiment

Our aim is now to compare the predictions of these different models on a typical three component microemulsion: the DDAB/tetradecane/water system. This system has some interesting features: the isotropic, flowing microemulsion region cannot be diluted with water nor with oil, without breaking into two phases [27]. This must correspond to a peculiar microstructure.

Whatever the water content, the conductivity for this system is always high: there is no inverse conductivity percolation threshold as for short chain alkanes or alkenes. According to the oil penetration theory, the surfactant parameter is expected to be close to one, since the tetradecane, unlike the shorter chain

TETRADECANE

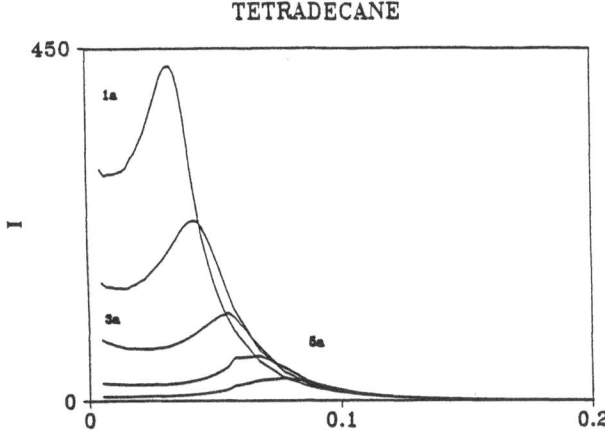

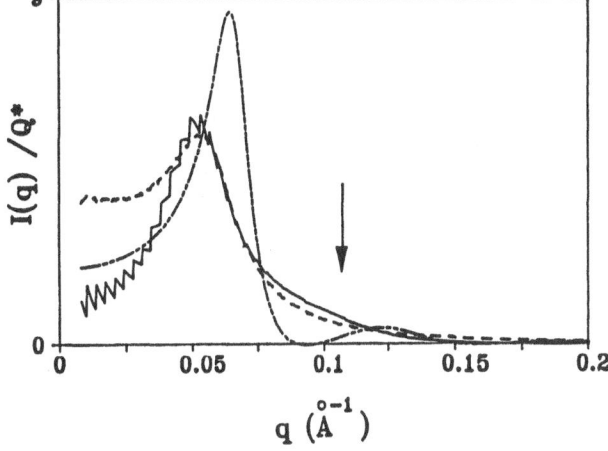

Fig. 7. Observed X-ray scattering (full line) of sample 3a of Table 1, as compared to the predictions of different models. The arrow indicates the peak position for randomly filled cubic cells (CRC); (–·–·–) is the prediction for water spheres in oil; (– – –) is the prediction for the DOC lamellae model. The intensities are significant and have only been rescaled by dividing through by the invariant $Q^*$

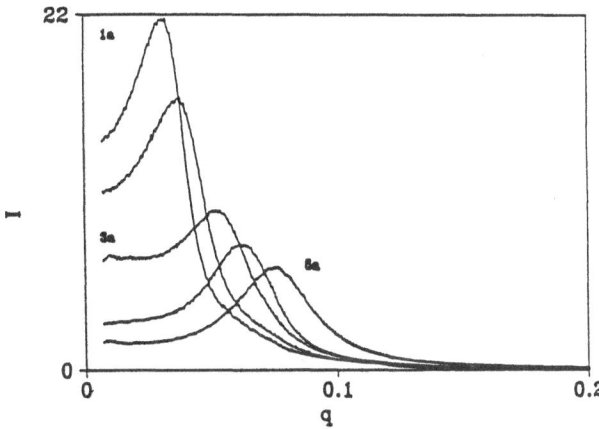

Fig. 6. X-ray and neutron scattering spectra for this dilution line are very similar: this shows that the underlying lattice is the origin of the peak, not the local form factor

## Test of the different models

The predictions of the non-parametric models are summarized in Table 1, where the volume fractions at which the structures are bicontinuous are given, as are approximate expressions for the scattering peak position $D^* = 2\pi/q_{max}$. Expressions for average and Gaussian curvature are also given, together with the constraint on formation of the interface by a surfactant of given $v/al$.

Table 2 indicates the compositions and molar ratios of the three components, together with the expected interfacial area, assuming that the area per headgroup is around 60 Å$^2$ per molecule as in the lamellar phase. The second part of the Table indicates the measured values: peak position $D^*$, interfacial area per unit volume $\Sigma$ (in Å$^2$/Å$^3$) measured by absolutely scaled neutron or X-ray scattering and the average chord length $l_c$ [23].

The seven available models are then tested. The peak positions for the interacting sphere model are derived from a Hayter-Penfold RMSA procedure, setting the droplet charge to zero. The peak positions for the DOC cylinder and DOC lamellae models are calculated by solving for the three steric constraints numerically. Only three models are compatible with the

oils, does not penetrate the surfactant hydrophobic chains [27]. Admixture of hexane, which does penetrate the chains, induces a percolation threshold and gives a powerful argument to support this assumption.

Small angle X-ray scattering experiments were done on the high resolution setup D22 at LURE (Orsay, France). Neutron scattering experiments on the same samples (all with D$_2$O) were done at the PACE facility at Orphee (Saclay, France). Typical spectra along a water dilution line are shown in Figs. 5 and 6. The peak position shifts with water content, but a sharply defined Porod decay allows determination of the interfacial area per surfactant head group. Neutron and X-ray spectra are very similar, hence the peak is due to the underlying lattice, not to microstructure.

We now try each available non-parametric model in turn.

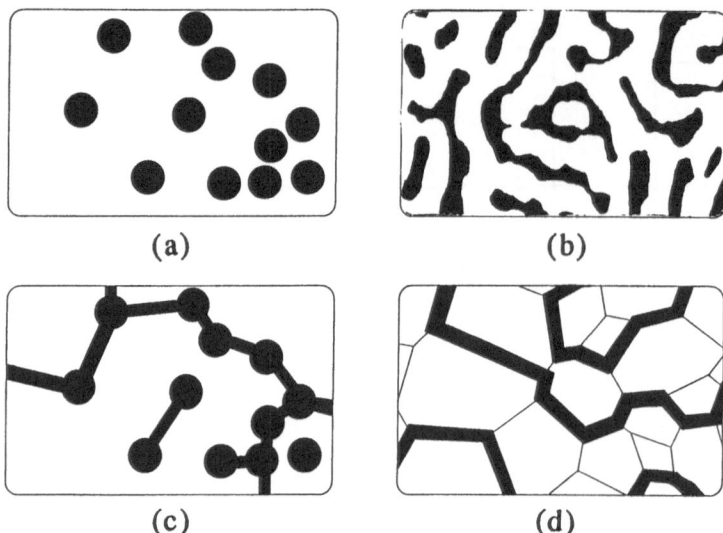

Fig. 8. The four different microstructures we have observed in microemulsion systems: (a) droplets; (b) the random wave model proposed by Berk, with local correlation as introduced by Welberry [28]; (c) connected cylinders; (d) random connected lamellae

measured peak position within 10%: water spheres in oil, DOC cylinders and DOC lamellae.

However, we have to reject the hard sphere droplet model for three main reasons: there is always high conductivity, samples are flow birefringent and the $v/al$ parameter has to be adjusted for each sample, because the radius is varying. In other oils, we have never observed such a strong variation of $v/al$ with water content.

Ruling out DOC-cylinder model is more difficult. The proposed structure has connectivity $Z$ about five throughout the stability range. However the packing constraints would have to be relaxed and $v/al$ would have to vary by 20% for this to work, which is probably not possible in a ternary system with a non-penetrating oil.

The only model of the seven proposed which is compatible with the packing constraints and the phase diagram is therefore the DOC lamellae model [20]. The microstructure is therefore a randomly folded infinite bilayer with no edges or seams, including water and separating two roughly equal oil subvolumes [20]. This structure also explains why the phase cannot be diluted with water or oil, and its high conductivity. Full scattering spectra predicted by CRC, water in oil spheres and DOC lamellae model are shown compared to the observed scattering in Fig. 7.

### Conclusion

We have shown here that instead of the usual fitting procedures used to fit a set of spectra to polydisperse ellipsoids or other sphapes, available models can be tested explicitly against a series of scattering experiments on an absolute scale.

In the DDAB/tetradecane/water system examined, all the standard models run into problems when compared to scattering: CRC predicts the peak in the wrong position, unless local correlations are introduced; droplets or DOC cylinder models would need unlikely *ad hoc* variation of the surfactant parameter $v/al$ with water content. Only the DOC lamellar model, describing a random bilayer, is compatible with the observed scattering. We therefore suggest that the observed microemulsion region is a DOC-lamellar (or $L_3$) phase. For surfactant systems with lower $k_c$, this structure is restricted to a very narrow channel along the edge of the $L_\alpha$ phase. In this system, due to the higher bending constant of the surfactant bilayer, this structure extends over a broader range of composition.

The four microstructures which have been observed so far in microemulsion systems are drawn schematically in Fig. 8. The DDAB/tetradecane/water microstructure is shown in Fig. 8d.

*Acknowledgements*

We thank Loic Auvray at Orphée and Claudine Williams at LURE for their assistance with the scattering measurements, and Stephen Hyde for extremely valuable discussions.

### References

1. Ciccariello S, Goodisman J, Brumberger H (1988) J Appl Cryst 21:117–128
2. Hayter JB, Penfold J (1985) In: Safran SA, Clark NA (eds) "Physics of Complex and Supermolecular Fluids", p 21

3. Sharma RV, Sharma KC (1977) Physica 89A:213–218
4. Chen S-H, Sheu EY (1988) Makromol Chem Macromol Symp 15:275–294
5. Ninham BW, Barnes IS, Hyde ST, Derian P-J, Zemb TN (1987) Europhys Lett 4:561–568
6. Singh HN, Shanti S, Singh RP, Daleem SM (1983) Ber Bunsenges Phys Chem 87:1115–1120
7. Rushforth DS et al. (1986) J Phys Chem 90:6668–6673
8. Kaler EW, Davies HT, Scriven LE (1983) J Chem Phys 79:5685–5692
9. Chevalier Y, Zemb TN (1990) Rep Prog Phys in press; Huang JS, Kotlarchyck M (1986) Phys Rev Lett 57:2587–2589
10. Talmon Y, Prager S (1982) J Chem Phys 76:1535
11. Scriven LE (1977) In: Mittal KL (ed) "Micellization, Solubilisation and Microemulsions", Plenum Press, Volume 2, p 877
12. Boned C, Clausse M, Lagourette B, Peyrelasse J, McClean VER, Sheppard RJ (1980) J Phys Chem 84: 1520–1525
13. Clausse M, Peyrelasse J, Heil J, Boned C, Lagourette B (1981) Nature 293:636–639
14. Porte G, Marignan J, Bassereau P, May R (1988) J Physique 49:511–519
15. Jouffroy J, Levinson P, de Gennes PG (1982) J Physique 43:1241
16. Auvray L, Coton J-P, Ober R, Taupin C (1985) In: Safran SA, Clark NA (eds) "Physics of Complex and Supermolecular Fluids", p 449
17. Andelman D, Cates ME, Roux D, Safran SA (1987) J Chem Phys 87:7229·7241
18. Hyde ST, Barnes IS, Ninham BW, J Chem (in press)
19. Teubner M, Strey R (1987) J Chem Phys 87:3195–3200
20. Warr GG, Zemb TN, Barnes IS (1988) Proceedings of the World Surfactants Congress, CESIO editor, Paris
21. Zemb TN, Hyde ST, Derian P-J, Barnes IS, Ninham BW (1987) J Phys Chem 91:3814–3820
22. Vonk CG, Billman JF, Kaler EW (1988) J Chem Phys 88:3970–3975
23. Barnes IS, Hyde ST, Ninham BW, Derian P-J, Drifford M, Warr GG, Zemb TN (1988) Progr Coll Polym Sci 76:90–95; Barnes IS, Hyde ST, Ninham BW, Derian P-J, Zemb TN (1988) J Phys Chem 92:2286–2293
24. Hyde ST, Ninham BW, Zemb TN (1989) J Phys Chem 93:1464–1468
25. Barnes IS, Zemb TN (1988) J Appl Cryst 21:373–379
26. Berk NF (1987) Phys Rev Lett 58:2718–2721
27. Evans DF, Mitchell DJ, Ninham BW (1984) J Phys Chem 88:6344; (1986) J Phys Chem 90:2817
28. Welberry TR, Zemb TN (1988) J Colloid Interf Sci 123:413
29. Hyde ST (1989) J Phys Chem 93:1458–1464

Authors' address:

Dr. Thomas Zemb
Dépt. Lasers et Physico-Chemie
Bat. 123
CBN-Saclay
F-91191 Gif-sur-Yvette
France
Fax (1) 69087963

**Progress in Colloid & Polymer Science**          Progr Colloid Polym Sci 81:30−35 (1990)

# On the interpretation of scattering peaks from bicontinuous microemulsions

S. H. Chen[1]), S. L. Chang[1]), and R. Strey[2])

[1]) Nuclear Engineering Department and Center for Materials Science and Engineering,
   Massachusetts Institute of Technology, Cambridge, MA USA
[2]) Max Planck Institut für Biophysikalische Chemie, Göttingen, FRG

*Abstract:* In several recent small angle neutron scattering (SANS) experiments made with three-component microemulsions in the one-phase region having oil-water contrast, one often finds in the SANS intensity distribution function a single prominent peak occuring at some finite Bragg wave number $Q_m$. This peak can be due either to spatial correlations between closely spaced microemulsion droplets when the droplet picture is valid, or to the length scale $\xi$ which is known to exist in bicontinuous microemulsions, when this type of description is appropriate. In this paper we examine a series of scattering patterns taken in the one-phase channel of AOT/water (0.6% NaCl)/decane ionic microemulsions. We propose a criterion for distinguishing between the bicontinuous and the droplet type microemulsions, based on analyses of the peak position and the peak height. We conclude that a bicontinuous structure can best be described as having a mean characteristic length $d$ with a certain amount of dispersion around it.

*Key words:* Bicontinuous; microemulsions; small angle neutron; scattering; Teubner Strey model; Berk model; Debye correlation function

## Introduction

In a pioneering small angle neutron scattering (SANS) experiment done with an oil-water contrast by Auvray, Cotton, Ober and Taupin [1], these authors found that for mono-phasic quaternary microemulsions consisting of water (6.5% NaCl added), toluene (oil), sodium dodecyl sulfate (surfactant) and 1-butanol (cosurfactant), having equal volumes of water and oil, the scattering intensity distribution showed a peak located at a characteristic wave number $Q_m$ given by

$$Q_m = \frac{0.5}{\Delta} \cdot \frac{\varphi_s}{\varphi_w \varphi_0}. \tag{1}$$

In Eq. (1), $\Delta = v_s/a_H$, where $v_s$ is the steric volume of the surfactant molecule and $a_H$ the head area subtended by each surfactant molecule at the oil-water interface, and $\varphi_s$, $\varphi_w$ and $\varphi_0$ are respectively volume fractions of the surfactant, water and oil. This scattering peak was thought to arise from the existence of a characteristic length scale, $\xi_K$, existing in a bicon-

tinuous microemulsion [2], derived from a finite curvature elastic constant of the surfactant film [3]. Subsequently, in a separate series of SANS experiments on the same microemulsion system, DeGeyer and Tabony [4] also found similar peaks from the middle-phase microemulsions, at an intermediate range of salinity (Windsor III microemulsions), in which water and oil volume fractions were nearly equal. Analyses of these scattering patterns were presented which showed evidence that the microstructure of these middle-phase microemulsions were indeed bicontinuous.

A theoretical connection between the persistence length $\xi_K$ and the characteristic wave number $Q_m$ was made by Jouffroy et al. [5] and by Milner et al. [6]. These authors divided the space into adjacent cubes of size $\xi_K$ (volume $\xi_K^3$) and each cube was filled either with oil or with water. Firstly, by imposing a condition of surfactant head area conservation, they obtained a relation [5]:

$$\xi_K = 6 \cdot \Delta \cdot \frac{\varphi_w \varphi_0}{\varphi_s}. \tag{2}$$

Secondly, by investigating the correlations between the water-filled regions in the cubic lattice, they obtained an approximate peak position in the structure factor occurring at [6]:

$$Q_m \doteq \frac{\pi}{\xi_K}. \tag{3}$$

Combining Eq. (2) and Eq. (3) one immediately obtains Eq. (1).

Lichterfeld et al. [7] have investigated, using small angle X-ray scattering (SAXS), a ternary nonionic microemulsion system consisting of $C_{12}E_5$-$H_2O$-$n$-tetradecane in the one phase region by changing the volume fraction of oil $\varphi_0$. The relation Eq. (1) was again well verified with the coefficient 0.5.

In this paper we shall rederive a relation similar to the Eq. (1) and also some other scaling relations by considering the Debye correlation function [8] for a bicontinuous microemulsion. We shall use both Teubner-Strey model [9] and Berk model [10] for the construction of the Debye correlation function. We analysed SANS data from a ternary microemulsion system: an ionic system, AOT/water (0.6% NaCl)/decane [11], in the one-phase region adjacent to the three-phase body [12, 13]. These analyses showed that even for microemulsions having equal oil and water volume fractions, the bicontinuous structure can only be realized at relatively low surfactent weight fractions, adjacent to the three-phase body.

**Theory**

For the purpose of analysing SANS data from three-component one-phase microemulsions, it is most appropriate to use Debye's theory [14] of two-component porous media since the scattering length densities of the head group and of the tail of the surfactant molecule can be taken to be approximately equal to that of water and oil respectively. We have, for the SANS intensity distribution function $I(Q)$ (in $cm^{-1}$),

$$I(Q) = \langle \eta^2 \rangle \cdot \int_0^\infty dr \cdot 4\pi r^2 \cdot \frac{\sin Qr}{Qr} \cdot \Gamma(r) \tag{4}$$

where $\eta(r) = \varrho(r) - \bar{\varrho}$ is the local deviation of the scattering length density from the average value $\bar{\varrho} = \varphi_1 \varrho_w + \varphi_2 \varrho_0$. From its definition one easily obtains $\langle \eta^2 \rangle = \varphi_1 \varphi_2 (\varrho_w - \varrho_0)^2$. We take $\varphi_1 = \varphi_w + \alpha \varphi_s$, $\varphi_2 = \varphi_0 + (1 - \alpha) \varphi_s$ where $\alpha$ is the fractional volume of the head group of the surfactant molecule. For AOT

$\alpha = 0.106$. The Debye correlation function in Eq. (4) is defined as $\Gamma(r) = \langle \eta(0)\eta(r) \rangle / \langle \eta^2 \rangle$.

In the model of Teubner and Strey [9] the Debye function is given as

$$\Gamma_{TS}(r) = e^{-r/\xi} \cdot j_0(kr) \tag{5}$$

where $d = 2\pi/k$ is the average interdomain (between water or between oil) distance and $\xi/d = k\xi/2\pi$ is a measure of polydispersity of the domain size. Inserting Eq. (5) into Eq. (4), one obtains

$$\frac{I(Q)}{\langle \eta^2 \rangle} = \frac{8\pi/\xi}{a^2 - 2Q_m^2 Q^2 + Q^4} \tag{6}$$

where $a^2 = k^2 + 1/\xi^2$ and $Q_m^2 = k^2 - 1/\xi^2$. The $I(Q)$ vs. $Q$ plot exhibits a peak at $Q = Q_m$ when $k\xi > 1$.

From the general property of the Debye function one obtains:

$$\Gamma'(r = 0) = -\frac{1}{\xi} = -\frac{1}{4\varphi_1\varphi_2} \cdot \frac{S}{V}. \tag{7}$$

Assuming all the surfactant molecules are on the oil-water interfaces, each subtending an area $a_H$, we have $\frac{S}{V} = \varphi_s/\Delta$, and thus combined with Eq. (7) to obtain a relation between $\xi$ and the system composition:

$$\xi = 4\Delta \cdot \frac{\varphi_1 \varphi_2}{\varphi_s}. \tag{8}$$

Using above relations, we can finally derive the following three useful relations:

$$Q_m \Delta = \frac{(k^2\xi^2 - 1)^{1/2}}{4} \cdot \frac{\varphi_s}{\varphi_1\varphi_2} = C_1 \frac{\varphi_s}{\varphi_1\varphi_2} \tag{9}$$

$$\frac{I(Q_m)}{I(0)} = \frac{1}{4} \cdot \frac{(k^2\xi^2 - 1)^2}{k^2\xi^2} = C_2 \tag{10}$$

$$\frac{Q_m^3 I(Q_m)}{\langle \eta^2 \rangle} = 2\pi \cdot \frac{(k^2\xi^2 - 1)^{3/2}}{k^2\xi^2} = C_3. \tag{11}$$

If we now, for an argument sake, impose the condition, $C_1 = 1/2$, in order that the empirical relation Eq. (1) be valid, then, from Eq. (9), we obtain $k\xi = 2.24$. We also get $C_2 = 1.8$ and $C_3 = 10.05$. We note that the theoretical result of Milner et al. [6] predicts that $C_1 = 0.52$ and $C_2 = 2.0$. We also remark that $k\xi = 2.24$ corresponds to the ratio of two parameter $\xi/d =$

0.36. It is interesting to note that a recent mean field solution of a Hamiltonian model of microemulsion due to Widom [15] gives $C_2 = 1.1$ and $\xi/d = 0.22$ in the isotropic phase.

In the model of Berk [10], one has, for the case of oil-water isometry, (i.e. $\varphi_1 = \varphi_2$),

$$\Gamma_B(r) = \frac{2}{\pi} \cdot \sin^{-1}[j_0(kr)] \tag{12}$$

where $k$ defines a length scale $d = 2\pi/k$. This model is not realistic in its original form because a sharp $k$ value would give rise to a very sharp peak in $I(Q)$ at $Q_m = k$ which is not experimentally observed. We therefore broaden the $k$ values by a Schultz distribution function:

$$f(k) = \left(\frac{Z+1}{\bar{k}}\right)^{Z+1} \cdot \frac{1}{\Gamma(Z+1)} \cdot k^Z$$

$$\cdot \exp\left[-\left(\frac{Z+1}{\bar{k}}k\right)\right]. \tag{13}$$

This amounts to replace $j_0(kr)$ in Eq. (12) by a function $\tau(r)$

$$\tau(r) = \int_0^\infty dk \cdot f(k) \cdot j_0(kr)$$

$$= \frac{Z+1}{\bar{k}r} \cdot (\cos\varphi)^Z \cdot \left(\frac{\sin Z\varphi}{Z}\right) \tag{14}$$

where $\tan\varphi = \bar{k}r/(Z+1)$ in Eq. (14). In this picture the dispersion of the length scale is given by

$$\frac{\Delta k}{\bar{k}} = \frac{\Delta d}{\bar{d}} = \frac{1}{\sqrt{Z+1}}. \tag{15}$$

In this modified Berk model, the boundary condition, Eq. (7), gives rise to a relation:

$$\bar{k} = \frac{\pi\sqrt{3}}{8\Delta} \cdot \frac{Z+1}{\sqrt{(Z+1)(Z+2)}} \cdot \frac{\varphi_s}{\varphi_1\varphi_2}. \tag{16}$$

We are able to find an additional, accurate empirical relation:

$$Q_m = \bar{k} \cdot \left(1 - \frac{3.01}{Z} + \frac{2.83}{Z^2} - \frac{1.64}{Z^3}\right). \tag{17}$$

Upon substituting Eq. (16) into Eq. (17), we get a relation

$$Q_m\Delta = \frac{\pi\sqrt{3}}{8\Delta} \cdot \frac{Z+1}{\sqrt{(Z+1)(Z+2)}}$$

$$\cdot \left(1 - \frac{3.01}{Z} + \frac{2.83}{Z^2} - \frac{1.64}{Z^3}\right) \cdot \frac{\varphi_s}{\varphi_1\varphi_2}. \tag{18}$$

For $Z = 8$, which has a polydispersity $\Delta k/\bar{k} = 0.33$, we have from Eq. (18), $C_1 = 0.429$. For $Z = 7$, for which $\Delta k/\bar{k} = 0.35$, $C_1 = 0.405$.

Because of the linear relationship between $Q_m$ and $\bar{k}$ given in Eq. (17), one can also derive a scaling relationship analogous to Eq. (11), i.e.

$$\frac{Q_m^3 I(Q_m)}{\langle \eta^2 \rangle} = G(Z) \tag{19}$$

where $G(Z)$ is a complicated function of $Z$ and can be evaluated only numerically. Lastly, one can recast Eq. (18) into a more interesting form

$$Q_m = \frac{\pi^2\sqrt{3}}{8} \cdot F(Z) \cdot \frac{Q^4 I(Q \to \infty)}{\int_0^\infty dQ \cdot Q^2 I(Q)} \tag{20}$$

where

$$F(z) = \frac{Z+1}{\sqrt{(Z+1)(Z+2)}}$$

$$\cdot \left(1 - \frac{3.01}{Z} + \frac{2.83}{Z^2} - \frac{1.64}{Z^3}\right). \tag{21}$$

For example for $Z = 8$, $F(Z) = 0.631$ and the prefactor in Eq. (20) has a value 1.35. This relation is interesting because it relates the peak position $Q_m$ to the asymptotic behavior of the $I(Q)$ function.

## Experiment

SANS data on AOT/water (0.6% NaCl)/decane system is taken from Ref. [11], where a detailed phase diagram of the system was given. In general the intensity $I(Q)$ is given in an absolute unit, but when in doubt, we use the invariant relation to normalize it. The invariant relation is

$$2\pi^2 \cdot (\varrho_w - \varrho_0)^2 \cdot \varphi_1\varphi_2 = \int_0^\infty dQ \cdot Q^2 I(Q). \tag{22}$$

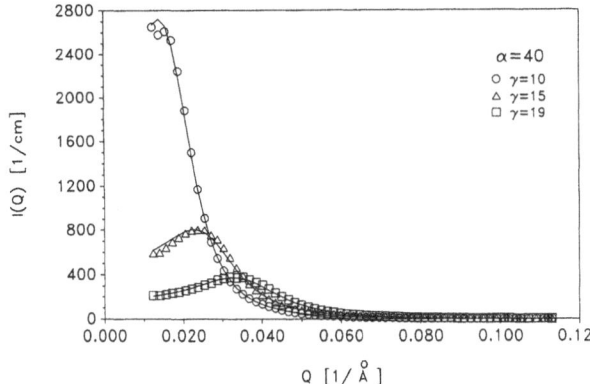

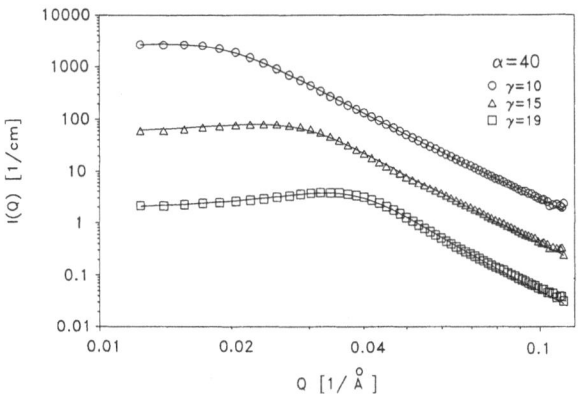

Fig. 1. (a) $I(Q)$ vs. $Q$ plots for AOT microemulsions with $\alpha = 40$ and $\gamma = 10, 15$ and $19$. Solid lines are the T-S model fits. (b) The same data and fits as in (a) but in log-log plots. Intensities for $\gamma = 15$ and $\gamma = 19$ cases have been divided by factors 10 and 100 respectively for the display purpose

Throughtout this paper, the important quantity, the effective surfactant length $\Delta$ is calculated from the data via a relation:

$$\frac{1}{\Delta} = \pi \cdot \frac{\varphi_1 \varphi_2}{\varphi_s} \cdot \frac{Q^4 I(Q \to \infty)}{\int_0^\infty dQ \cdot Q^2 I(Q)} \qquad (23)$$

which is independent of the intensity calibration.

## Results

In this paper we shall confine ourselves to those cases which have equal or approximately equal oil and water volume fractions. If we denote by $\alpha =$ wt. oil/ (wt. oil + wt. water), which is called the weight fraction of oil in oil plus water and by $\gamma =$ wt. surfactant/ (wt. oil + wt. water + wt. surfactant), which is called

the weight fraction of surfactant in solution, then $\alpha = 40\%$ would correspond to the equal volumes of $D_2O$ and oil case. At $\alpha = 40\%$, the reference point in the phase diagram where the one-phase region begins, is: $\gamma_0(\text{AOT}) = 6\%$. The corresponding temperature is: $T_0(\text{AOT}) = 40.5\,^\circ\text{C}$.

We shall begin by showing that Teubner-Strey (T-S) model fits all SANS data having $\gamma > \gamma_0$ (one-phase region). Figs. 1a and 1b show a pair of plots, one in linear and the other in log scale, for the AOT one-phase microemulsions, for the cases $\alpha = 40$, $\gamma = 10$, 15 and 19. Agreement between the model and the data are quite satisfactory. Parameters obtained are listed in Table 1.

Figs. 2a and 2b give the comparison of Berk model analysis and the AOT microemulsion data already shown in Fig. 1. We observe that the agreement is reasonable only for the case of $\gamma = 10$. Thus for AOT microemulsions Berk model works well up to $\gamma = 12$ (analysis not shown) but begins to fail at $\gamma = 15$. All the extracted model parameters from the analyses are listed in Table 2.

Fig. 3 present an experimental test of various equivalent scaling relations, for bicontinuous microemulsions, as given by Eq. (1), Eq. (9) and Eq. (18). Since T-S model fits all SANS data for $\alpha = 40$ microemulsions (see Fig. 1 and Ref. [21]), we can use it to accurately locate $Q_m$ of each SANS intensity distribution. We can also use the large $Q$ data of the distribution to determine the parameter $\Delta$ (see Eq. (23)). So the dimensionless quantity $C_1 = (Q_m \Delta)/(\varphi_s/\varphi_1 \varphi_2)$ is a model independent, experimentally measurable quantity. We plot in Fig. 3 $C_1$ vs. $\gamma$ for all data we have for AOT microemulsions at $\alpha = 40$. At each $\gamma$ value there are more than one data point because we took SANS spectra at different temperatures within the one-phase channel. Generally, the points with lower $C_1$ values correspond to data at lower temperatures. One sees that for all microemulsions having $\gamma = 10$ to 48, $C_1$ values lie between 0.26 and 0.50. We also observe that for all cases which are known to agree with Berk's model of bicontinuous microemulsions, value of $C_1$ is below 0.43. This corresponds to $Z < 8$ in Eq. (18).

Equations (11) and (19) are more sensitive method of testing whether a microemulsion is bicontinuous in Berk's sense. In Fig. 4 we plot the dimensionless quantity $Q_m^3 I(Q_m)/\langle \eta^2 \rangle$, which is again a model independent quantity, against $\gamma$, for all $\alpha = 40$ AOT microemulsions. We see that this quantity, what we called $C_3$, increases rapidly with $\gamma$ values. For the region of $\gamma$ and temperature where we have indication [11] that the microemulsions are bicontinuous, the value of $C_3$

Table 1. Parameters from Teubner-Strey models fits

AOT/water (0.6% NaCl)/decane ionic microemulsions at $\alpha = 40$

| $\gamma$ | $T$ (°C) | $\varphi_1$ | $\xi$ (Å) | $k\xi$ | $\xi/d$ | $Q_m$ (Å$^{-1}$) | $\Delta$ (Å) |
|---|---|---|---|---|---|---|---|
| 10 | 46 | 0.4671 | 107 | 1.80 | 0.29 | 0.0141 | 8.1 |
| 15 | 46 | 0.4512 | 78 | 2.04 | 0.33 | 0.0228 | 8.0 |
| 19 | 52 | 0.4383 | 70 | 2.50 | 0.40 | 0.0328 | 8.2 |

$$(\varrho_w - \varrho_{C10}) = 4.69 \times 10^{21} \text{ cm}^{-4}$$

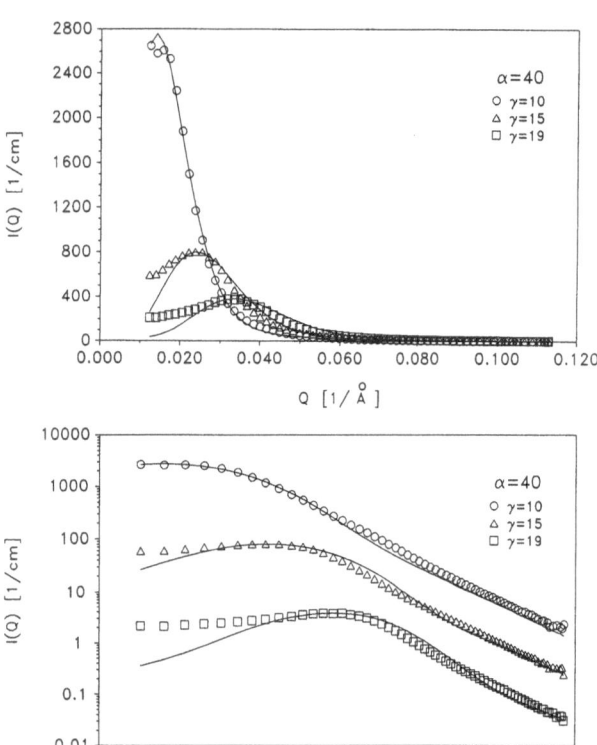

Fig. 2. (a) $I(Q)$ vs. $Q$ plots for AOT microemulsions with $\alpha = 40$ and $\gamma = 10$, 15 and 19. Solid lines are the fits by Berk model. (b) Same as in (a) but presented in log-log plots. Intensities for $\gamma = 15$ and $\gamma = 19$ cases have been divided by factors 10 and 100 respectively for the display purpose

is less than 10. We may therefore called this a <u>rule of ten</u>, a very useful rule indeed for judging whether a microemulsion is bicontinuous or not.

## Conclusions

The above analyses lead us to the following observations and conclusions: (1) Although T-S model gives satisfactory fits to all SANS data for microemulsions with equal oil and water volume fractions, it is not capable of distinguishing between microemulsions with a bicontinuous structure and those with some other disordered structures. The ratio $\xi/d$ of the two extracted parameters is a measure of the domain size polydispersity, the smaller the ratio, the larger the polydispersity. (2) Berk model fits SANS spectra for $\alpha = 40$ microemulsions with relatively low $\gamma$ values, near the reference point $\gamma_0$, in the one-phase channel. These microemulsions can be expected to be bicontinuous in its microstructure. The polydispersity index obtained for these cases is generally larger than 33% (i.e. $Z < 8$). As $\gamma$ and temperature increase the $Z$ value increases, indicating the decrease in polydispersity, and deviation from bicontinuous structure. (3) The well known scaling relation Eq. (1) supposedly valid for a bicontinuous microemulsion is shown to be insensitive to the deviation from bicontinuity. Instead, the newly proposed alternate scaling relation Eq. (11) or Eq. (19) is a more sensitive indicator for the bicontinuous structure. One may be certain that when $C_3 < 10$ the structure is what one normally imagines, a mutually inter-penetrating network of branched tubular domains of water and oil. (4) The scattering peak position $Q_m$ is intimately related to the asymptotic behavior of $I(Q)$ through a scaling relation Eq. (20). (5) The bicontinuous structure as we dipict in this article is better described as a disordered structure formed by oil and water domains in which one can identify an average length scale given by $\bar{d} = 2\pi/\bar{k}$, having a polydispersity $\Delta k/\bar{k} > 33\%$.

It should be remarked here that Berk model is not the only possible approach for the description of bicontinuous structure. There is another recent successful model due to Zemb et al. [16]. The future task will be to deduce these length scale and polydispersity from microscopic theories such as given in Ref. [6], [15] and [17].

Table 2. Parameters from Berk model fits

AOT/water (0.6% NaCl)/decane ionic microemulsions at $\alpha = 40$

| $\gamma$ | $T$ (°C) | $\varphi_1$ | $\bar{k}$ (Å$^{-1}$) | $Z$ | $\Delta k/\bar{k}$ | $Q_m$ (Å$^{-1}$) | $\Delta$ (Å) |
|---|---|---|---|---|---|---|---|
| 10 | 46 | 0.4671 | 0.0221 | 7.2 | 0.349 | 0.0141 | 8.1 |
| 15 | 46 | 0.4512 | 0.0342 | 9.1 | 0.315 | 0.0228 | 8.0 |
| 19 | 52 | 0.4383 | 0.0444 | 11.4 | 0.284 | 0.0328 | 8.2 |
| 12 | 40 | 0.4608 | 0.0261 | 7.2 | 0.349 | 0.0140 | 8.0 |
| 12 | 46 | 0.4608 | 0.0275 | 8.0 | 0.333 | 0.0172 | 8.0 |
| 12 | 52 | 0.4608 | 0.0280 | 8.4 | 0.326 | 0.0185 | 8.1 |

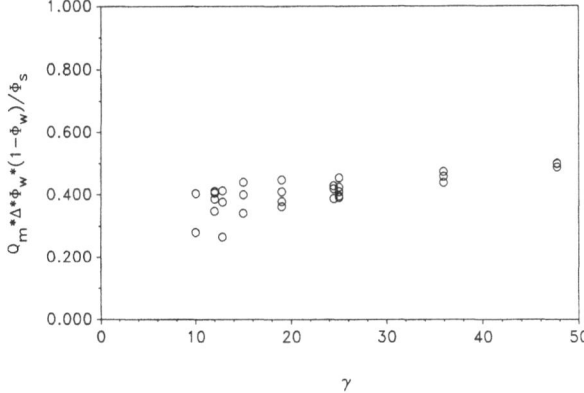

Fig. 3. The coefficient $C_1$ defined in Eq. (9) plotted as a function of $\gamma$ for all AOT microemulsions we studied at $\alpha = 40$

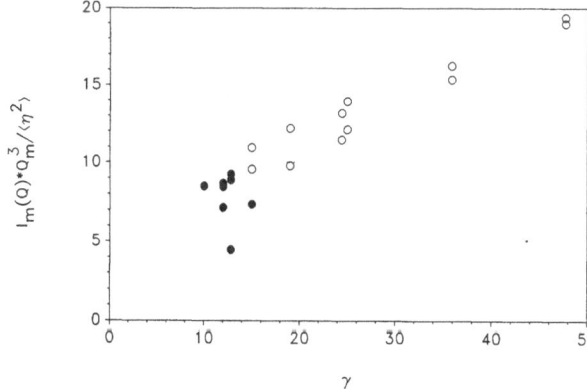

Fig. 4. The coefficient $C_3$ as defined in Eq. (11) plotted as a function of $\gamma$ for all AOT microemulsions we studied at $\alpha = 40$

*Acknowledgements*

This research is supported by grants from National Science Foundation. We are indebted to Prof. M. Kahlweit for his support.

# References

1. Auvray L, Cotton JP, Ober R, Taupin C (1984) J Phys Chem 88:4586
2. Scriven LE (1977) In: Mittal K (ed) Micellization, Solubilization and Microemulsions, Plenum Press 2:877 Talmon Y, Prager S (1978) J Chem Phys 69: 2984
3. DeGennes PG, Taupin C (1982) J Phys Chem 86:2294
4. DeGeyer A, Tabony J (1985) Chem Phys Lett 113:83; ibid (1986) 124:357
5. Jouffroy J, Levinson P, DeGennes PG (1982) J Physique 43:1241
6. Milner ST, Safran SA, Andelman D, Cates ME, Roux D (1988) J Physique 49:1065
7. Lichterfeld F, Schmeling T, Strey R (1986) J Phys Chem 90:5762
8. Debye P, Bueche AMJ (1949) Appl Phys 20:518
9. Teubner M, Strey R (1987) J Chem Phys 87:3195
10. Berk NF (1987) Phys Rev Lett 58:2718
11. Chen SH, Chang SL, Strey R, Structural Evolution Within one-phase Region of A Three-Component Microemulsion System: water-n-decane-AOT (submitted to J Chem Phys)
12. Kahlweit M, Strey R, Firman P, Hasse D, Jen J, Schomaker R (1988) Langmuir 4:499
13. Kahlweit M, Strey R, Schomacker R, Hasse D (1989) Langmuir 5:305
14. Debye P, Anderson Jr. HR, Brumberger H (1957) J Appl Phys 28:679
15. Widom B (1989) J Chem Phys 90:2437
16. Zemb T et al. (1987) J phys Chem 91:3814; Ninham BW et al. (1987) Europhys Lett 4:561
17. Gompper G, Schick M (1989) Phys Rev Lett 62:1647

Authors' address:

Prof. S.-H. Chen
Mit 24 – 209
Cambridge, Ma 02139, USA
Fax: 61 72 58 74 37

**Progress in Colloid & Polymer Science**  Progr Colloid Polym Sci 81:36−40 (1990)

# Mixing of oils with surfactant monolayers

R. Aveyard, B. P. Binks, P. Cooper and P. D. I. Fletcher

School of Chemistry, University of Hull, Hull, England

*Abstract:* In systems containing oil, water and surfactant, the penetration of oil into the chain region of the surfactant monolayer governs, in part, the system behaviour. For example, the limiting area per molecule for Aerosol-OT in saturated monolayers at the oil-water interface depends markedly on the alkane chain length. This, and other results, are discussed in connection with the occurrence of low oil-water interfacial tensions and microemulsion inversion from Winsor I-III-II systems. We also present new data for the effects of aliphatic remove oils on the cloud points of micellar solutions of nonionic surfactants, and discuss the probable locus of oil solubilisation. In connection with an attempt to determine alkane penetration into surfactant monolayers quantitatively, we report preliminary data for mixed films of oil and cationic surfactant at the air-water interface.

*Key words:* Microemulsion; oil penetration; effective surfactant molecular geometry, mixed films

## Introduction

It has long been recognised that hydrocarbons have a tendency to penetrate into monolayers of amphipathic materials. For example, volatile hydrocarbons adsorb from the vapour phase into insoluble monolayers at the air-water interface giving substantial changes in the surface pressure-surface area isotherms [1]. Equally, it has been explicitly recognised that e.g. alkane molecules mix with the chains of monolayers of surface active materials at the alkane-water interface [2] and that such monolayers are two-dimensional solutions. We will be concerned here with the penetration of (liquid) hydrocarbons into concentrated surfactant layers at oil-water interfaces, in microemulsions and micelles, and at air-water surfaces. Such penetration may well be responsible for the occurrence of ultralow oil-water interfacial tensions, microemulsion inversion in Winsor systems, and changes in cloud points of aqueous solutions of nonionic surfactants containing solubilised oils. We report here also on some preliminary work directed towards the quantitative determination of alkane penetration into surfactant monolayers at the air-water surface.

## Microemulsion inversion and ultralow tensions

Based on old ideas, which have been revived by Mitchell and Ninham [3], one may ascribe cross-sectional areas $a_c$ and $a_h$ to the chain and headgroup regions respectively of a surfactant molecule; a packing factor $P$ may then be defined as $a_c/a_h$. The nature and size of the surfactant aggregates which form above the critical micelle (or microemulsion) concentration, cmc (or cμc), in systems containing both oil and water phases depend, in this simple picture, on the value of $P$. For $P < 1$, aggregates form in the aqueous phase (positive curvature). As $P$ increases towards unity and the curvature falls, the aggregates solubilise more oil, and oil-in-water microemulsion droplets of increasing size result (Winsor I systems). When $P$ exceeds unity, aggregated surfactant transfers to the oil phase and water-in-oil microemulsion droplets (negative curvature) are produced (Winsor II systems). As $P$ increases further the size of the droplets falls. For $P \sim 1$, when the curvature of the surfactant monolayers is close to zero on average, a surfactant rich phase is formed in equilibrium with both excess oil and aqueous phases (Winsor III systems). These ideas are conceptually attractive and useful, but unfortunately the value of $P$ cannot be determined numerically. A major reason for

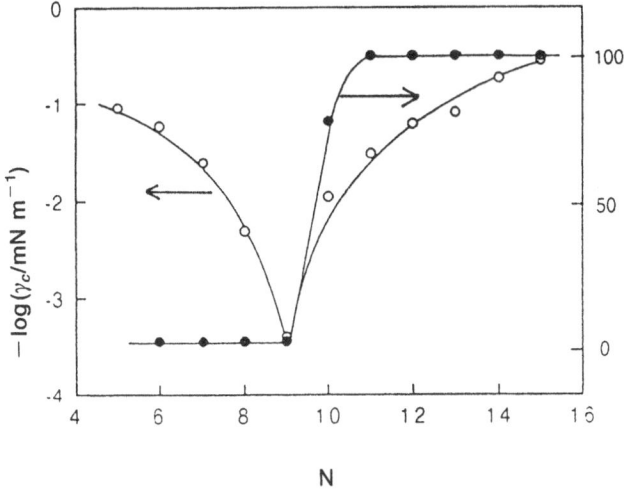

Fig. 1. Oil-water interfacial tensions ($\gamma_c$) and AOT content in aqueous phases versus alkane chain length ($N$) in systems containing 0.0684 M NaCl at 25°C

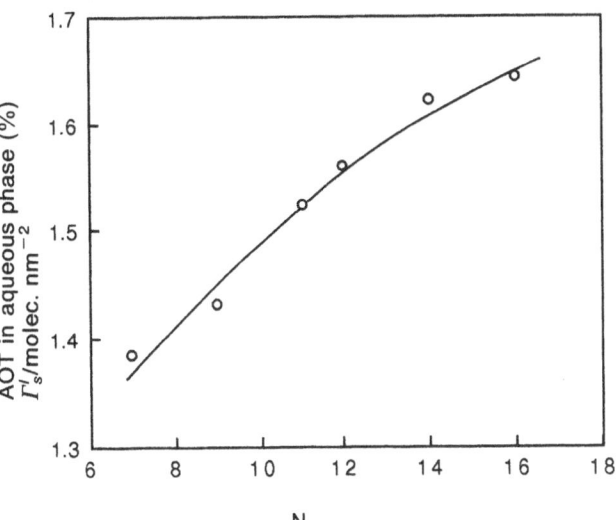

Fig. 2. Variation of $\Gamma_s^l$ with alkane chain length for AOT adsorbed at the 0.1 M aq. NaCl-alkane interface at 25°C

this is that $a_c$ and $a_h$ are not related simply to the molecular structure of the surfactant. They are *effective* values and include, in the case of $a_h$, effects of hydration of, and electrostatic repulsion between, head groups. In the case of $a_c$, the solvation of surfactant chains, i.e. the extent of penetration of oil molecules, into the chain region of the surfactant monolayer must be included.

If we take a system containing an alkane and an aqueous phase for which $P$ is close to unity, it is possible to effect a Winsor transition by scanning through the alkane chain length. In Fig. 1 we show how surfactant (AOT) distributes in systems containing 0.068 M aqueous NaCl and alkane, chain length $N$ [4]. For $N$ up to 9 the aqueous phase surfactant concentration is low, close to the expected cmc in the absence of oil, and this phase contains no surfactant aggregates. For hexane all the aggregated surfactant resides in the oil as w/o microemulsion droplets, and for heptane to decane third surfactant-rich phases are formed. For undecane upwards, the aqueous phases are o/w microemulsions, and no third phase is formed. The tensions between oil and aqueous (outer) phases reflect the Winsor transition (Fig. 1), passing through an ultralow minimum around nonane, where the aqueous phase surfactant concentration starts to rise rapidly.

If transitions of the kind described are accompanied by changes in $P$, this should be reflected in changes in the surface concentration $\Gamma_s$ of surfactant molecules in saturated monolayers at the oil-water interface. In

regimes where $P < 1$, $\Gamma_s$ should be determined by the head group size since $a_h > a_c$. The values of $a_h$ can be reduced by addition of inorganic electrolyte and it is well known that $\Gamma_s$ can be increased in this way. In systems which exhibit Winsor transitions however, at a certain salt concentration, where $a_h = a_c$ and $P = 1$, $\Gamma_s$ should attain a limiting value ($\Gamma_s^l$) which is determined by $a_c$ and which is independent of salt concentration. This indeed has been shown to be the case in appropriate systems containing AOT [4]. In this region then it should be possible to detect the effects of alkane penetration on $\Gamma_s^l$. Values of $\Gamma_s^l$ for AOT at alkane-aqueous NaCl interfaces are shown in Fig. 2 as a function of $N$. It is clear that the monolayers are more expanded ($\Gamma_s^l$ smaller) the lower the value of $N$, indicating the stronger alkane penetration for the lower homologues. This is entirely consistent with the nature of the observed Winsor systems, i.e. for lower alkanes, $a_c$ is large and $P > 1$, giving w/o microemulsions, whereas for the larger alkanes $a_c$ is reduced and $P < 1$, resulting in the formation of o/w microemulsions.

Although the simple model discussed appears to be satisfactory for the anionic surfactant AOT, the picture is probably more complicated in the case of nonionic surfactants of the polyoxyethylene (EO) type. Attempts have been made by us to determine $\Gamma_s$ as a function of temperature, passing through $T$ corresponding to that giving minimum $\gamma_c$ (the PIT). From the above arguments it was expected that $\Gamma_s$ should increase with increasing $T$ (due probably to the de-

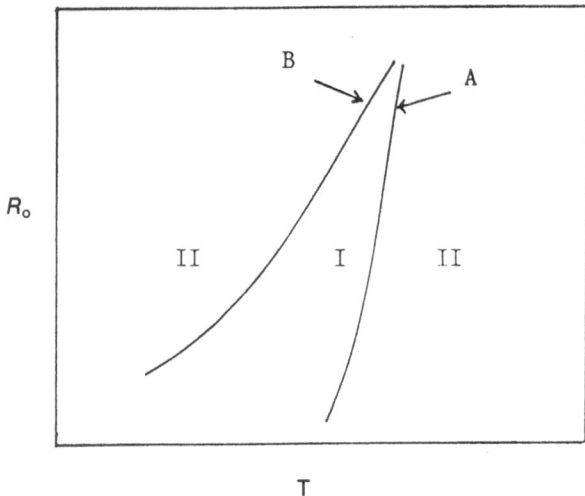

Fig. 3. Schematic representation of one phase o/w microemulsion domain formed in nonionic surfactant-oil-water systems

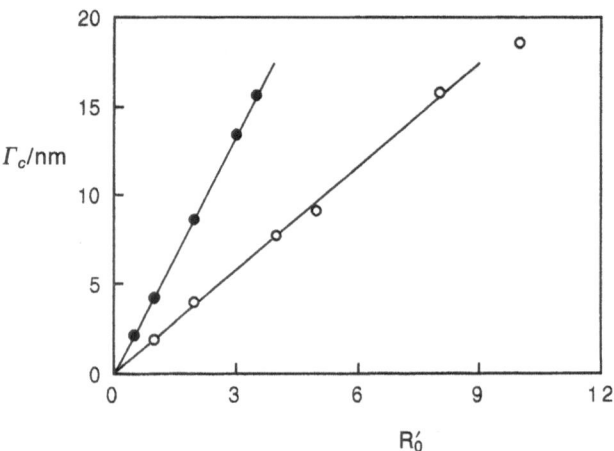

Fig. 4. Variation of $r_c$ with $R_0'$ (at the solubilisation boundary) for o/w microemulsions stabilised by $C_{12}E_5$ with heptane ($\bigcirc$) or tetradecane ($\bullet$)

hydration of the EO headgroups) until the PIT is attained, and thereafter attain values dependent on $a_c$. In this case however, $a_c$ may well be $T$-dependent also, in some unknown way, so a limiting value of $\Gamma_s$ may not be attained in the range of $T$ studied. In any event, for experimental reasons it is difficult in such systems to determine $\Gamma_s$ precisely, and understanding here must await further developments.

## Alkane penetration into curved monolayers

We have discussed above the effect of alkane penetration on $\Gamma_s$, in planar monolayers, where $\Gamma_s$ is readily determined from interfacial tensions obtained as a function of surfactant concentration below the cμc. For curved (droplet) interfaces $\Gamma_s$ cannot be determined so directly, and in any case at curved interfaces the area per surfactant molecule is different in the head and chain regions. However, useful information can be obtained by measurement of droplet sizes in microemulsions.

We consider here o/w microemulsions in systems containing the surfactant $C_{12}H_{25}(OCH_2CH_2)_5OH$, designated $C_{12}E_5$, and either heptane or tetradecane. Uptake of alkane by surfactant as a function of temperature to give o/w microemulsions is shown schematically in Fig. 3, where the ordinate $R_0$ is the molar ratio of oil to surfactant in the system. The single phase microemulsion region (I) is defined by the two phase boundaries $A$ and $B$. The upper $T$ boundary ($A$)

is determined mainly by interactions between droplets whereas the lower boundary $B$ (termed the solubilisation curve) is governed primarily by the spontaneous curvature of the surfactant layers. Hydrodynamic radii, $r_h$, of droplets in one phase systems at the solubilisation boundary have been determined by dynamic light scattering in systems containing heptane and tetradecane [5]. Assuming droplets to be spherical and monodisperse, simple geometry gives $r_h$ as

$$r_h = 3R_0' V_{mol} \Gamma_{s,d} + t = r_c + t \qquad (1)$$

where $R_0'$ is the molar ratio of oil to surfactant in the *droplets* (i.e. total surfactant concentration $-$ cμc) [5] and $V_{mol}$ is the molecular volume of the alkane. The quantity $t$ is the thickness of the (solvated) surfactant layer and $\Gamma_{s,d}$ is the surface concentration of surfactant at the interface between the droplet cores (radius $r_c$) and the (chain side) of the surfactant layers coating the droplets.

On this rather crude model, $\Gamma_{s,d}$ can (if it is constant) be obtained from the slope of a plot of $r_h$ against $R_0'$, the $r_h$ and $R_0'$ corresponding to a series of points along the solubilisation curve. The $r_h$ versus $R_0'$ plots for both heptane and tetradecane are within the errors linear, both giving an intercept ($t$) of 3.2 nm, a physically reasonable value for $C_{12}E_5$. We show plots in Fig. 4 of core radii against $R_0'$ for the two alkanes. It is clear that the slopes, and hence the $\Gamma_{s,d}$, differ greatly for the two alkanes. The $\Gamma_{s,d}$ values are, for the heptane and tetradecane systems respectively, 2.6 and 3.5 molecule nm$^{-2}$. For close packed alkyl chains, $\Gamma_{s,d}$ would

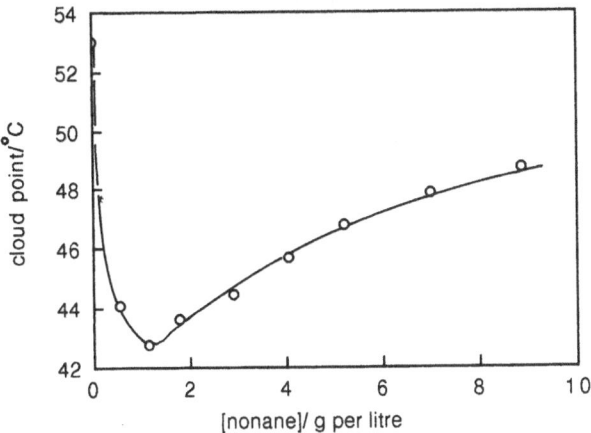

Fig. 5. Cloud points of 0.5 wt. % aqueous solutions of $C_{12}E_6$ as a function of the amount of added nonane

should result in increased inter-micellar attraction and hence reduce the CP of the solution. Conversely, incorporation of alkane in the micellar core will produce spherical aggregates and, if the micelles are initially rather asymmetrical, give rise to an increase in CP.

On the basis of these arguments, it appears that nonane first penetrates into the surfactant chains, giving a sharp drop in CP, but as more is added it begins to form a core, and the aggregates are transformed into spherical microemulsion droplets with an accompanying increase in the CP. It is known that longer chain alkanes tend to give only an increase in CP whereas shorter alkanes can give very strong reductions [7]. It appears then that CP data again support the view that alkane penetration in aggregates is stronger the shorter the alkane.

be about 4.8 molecule $nm^{-2}$, indicating that both alkanes penetrate into the surfactant tail region, with heptane penetration being significantly greater than that for tetradecane, in line with the results for planar monolayers discussed earlier.

## Possible role of oil penetration in changes in cloud points

Aqueous solutions of nonionic surfactants above the cmc phase separate (at the cloud point, CP) as $T$ is increased as a result of inter-micellar attractions and possibly micellar growth. Cloud points are concentration dependent and solutions exhibit a lower consolute temperature. For a fixed surfactant concentration cloud points are known to be affected in a complex fashion by the presence of increasing amounts of solubilised alkane [6, 7]. The cloud point at saturation with alkane is $2-3$ K below the temperature at which minimum oil-water tension ($\gamma_c$) is observed, and corresponds closely to the temperature at which a third phase forms in systems containing both alkane and aqueous phases.

We show in Fig. 5 cloud points of 0.5 wt % aqueous solutions of $C_{12}E_6$ as a function of the amount of nonane added to the solution. The CP first drop sharply and then rise as more nonane is added, finally levelling off as saturation is approached. Solubilised alkane can be located either within the surfactant chain region of the micelle, or can preferentially form an oil core [8]. In the former case, the value of $P$ (see earlier) will be increased towards unity and the micelles will be expected to become more asymmetrical. This in turn

## Quantitative determination of alkane penetration at the air-water surface

As seen, alkane mixing with surfactant chains can have drastic effects on phase behaviour in systems containing surfactant aggregates. It is however difficult to determine alkane uptake by surfactant chains quantitatively in the case of surfactant aggregates and planar surfactant layers at oil-water interfaces. We have however, made preliminary quantitative measurements of alkane uptake by surfactant layers at air-water interfaces [9].

Longer chain liquid alkanes e.g. dodecane do not spread macroscopically on a water surface, nor do they cause a detectable lowering of the surface tension i.e. they do not adsorb as a monolayer either. We have found however, that although such alkanes do not spread for example on a solution of cetyltrimethylammonium bromide (CTAB) above its cmc, they do lower the surface tension by an amount $\Delta\gamma$. This lowering is presumably due to the formation of a mixed alkane/surfactant monolayer. Further, the long chain liquid hydrocarbon squalane neither spreads on, or significantly mixes with, surfactant monolayers. Values of $\Delta\gamma$ obtained using mixtures of squalane + dodecane of varying mole fraction are linearly related to the mole fraction $x$ of dodecane. The surface concentration of the shorter alkane, $\Gamma_0$, can be obtained (for constant surfactant concentration) using the Gibbs equation in the form

$$\Gamma_0 = (x/kT)(d\Delta\gamma/dx) \qquad (2)$$

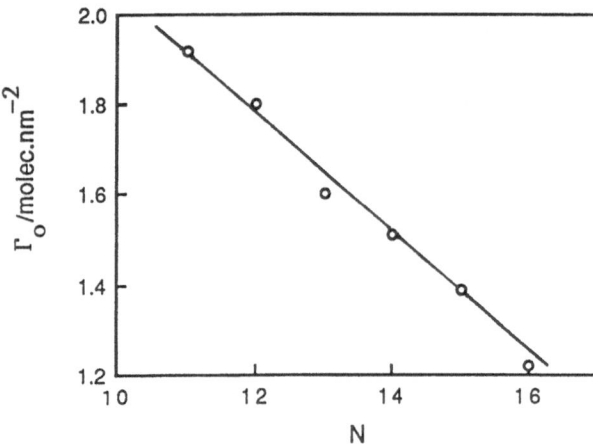

Fig. 6. Surface concentration of alkane, $\Gamma_0$, versus alkane chain length in mixed monolayers with CTAB (bulk concentration = 1.56 mM) at 25 °C

on the reasonable assumption that the alkane mixtures are almost ideal. Consequently, for any pure alkane, we may assume $\Gamma_0 \sim \Delta \gamma / kT$.

We show in Fig. 6 surface concentrations of alkanes in mixed monolayers with CTAB, in equilibrium with lenses of excess pure alkanes. It is clear that the shorter chain alkanes are present in higher concentrations, as expected from all our other findings. From measurement of tensions just below the cmc, we know that the surface concentration of surfactant, $\Gamma_s$, is not much affected (slightly lowered) by incorporation of alkane. If we suppose then that $\Gamma_s$ in mixed films is the same as that in the pure films at the cmc, we can estimate the surface mole fractions $x_0^\sigma$ of the various alkanes in CTAB + alkane monolayers. Values of $x_0^\sigma$ range from 0.57 in the case of decane to 0.42 for hexadecane.

Work is in progress looking at the effects of both surfactant and oil (polar and nonpolar) molecular structure on mixing in monolayers at air-water surfaces. Neutron reflection studies are also planned to determine film compositions by an alternative route.

*Acknowledgements*

We are grateful to BP Research for the provision of an Extramural Research Award and to Bevaloid for a Research Studentship (for PC).

## References

1. Hayes KE, Dean RB (1953) J Phys Chem 57:80
2. Fowkes FM (1962) J Phys Chem 66:1863
3. Mitchell DJ, Ninham BW (1981) J Chem Soc Faraday Trans II 77:601
4. Aveyard R, Binks BP, Mead J (1986) J Chem Soc Faraday Trans I 82:1755
5. Aveyard R, Binks BP, Fletcher PDI (1989) Langmuir 5:1210
6. Saito H, Shinoda K (1967) J Colloid Interface Sci 24:10
7. Aveyard R, Lawless TA (1986) J Chem Soc Faraday Trans I 82:2951
8. Aveyard R, Binks BP, Clark S, Fletcher PDI, J Chem Tech Biotech, in press
9. Aveyard R, Cooper P, Fletcher PDI (1990) J Chem Soc Faraday Trans 86:211

Authors' address:

Dr. B. P. Binks
School of Chemistry
University of Hull
Hull, HU6 7RX
England

**Progress in Colloid & Polymer Science**                    Progr Colloid Polym Sci 81:41−53 (1990)

# The multiple chemical equilibrium approach to the theory of droplet microemulsions

J. C. Eriksson and S. Ljunggren

Department of Physical Chemistry, Royal Institute of Technology, Stockholm, Sweden

*Abstract:* A novel theory of droplet microemulsions is presented that is based on surface and small systems thermodynamics, and is formulated in terms of a large set of chemical complex equilibria which account for the size, shape and composition fluctuations. The fluctuation-caused polydispersity is shown to be of a rather limited relative magnitude but, nevertheless, it is of paramount importance when calculating the total volume fraction of microemulsion droplets.

In the Winsor II (w/o) two phase regime the fluctuations take place about an equilibrium droplet for which the Laplace pressure $2\gamma/R$ is exactly counterbalanced by the curvature pressure $\pi_c = -(\partial\gamma/\partial R)_{\text{state}}$. Hence, a sufficiently large curvature dependence of the interfacial tension $\gamma$ is crucial for attaining thermodynamically stable droplet microemulsions.

*Key words:* Droplet microemulsions; surface thermodynamics; small systems thermodynamics; multiple chemical equilibria; fluctuations; polydispersity

## Introduction

About thirty years ago T. L. Hill devised a framework of general thermodynamic relations which is adequate for describing small, open systems with sizeable fluctuations [1]. In rather detailed ways he even considered the formation of micelles in a dilute soap solution (loc. cit., Vol. II, p. 120) and the growth of linear aggregates (loc. cit., Vol. II, p. 77). For surfactant aggregates of various shapes, this kind of approach was followed-up, later on, by the present authors [2−5] who combined it with modelling the main physical factors that govern surfactant association in aqueous systems, following and extending the lines already established by Tanford [6] and Israelachvili et al. [7]. In addition, we developed alternative, surface-thermodynamic methods for treating spherical [2], rod-shaped [3] and disc-shaped [4] micelles.

In Hill's small systems thermodynamics the interest is primarily focused on an aggregate with ensemble-averaged intrinsic properties. The entropy-dominated free energy that is associated with the fluctuation-caused polydispersity is then added afterwards so as to yield the overall free energy. For the case of micelle formation, it was explicitly shown by Hill (loc. cit., Vol. II, p. 131) that this approach is equivalent to the conventional procedure based on a set of chemical equilibria, i.e.

$$N A_1 \rightarrow A_N \quad (N = N_{\min}, \cdots, N_{\max}). \tag{1}$$

Here $N_{\min}$ and $N_{\max}$ refer to the lowest and highest aggregation numbers, respectively, which are compatible with the notion of a micellar aggregate of a certain kind. Denoting the intrinsic free energy per monomer in a $N$-micelle by $\hat{\mu}_N$ (adopting Hill's notation) and assuming the intramicellar interactions to be negligible, we have

$$N(\hat{\mu}_N - \mu_1) + kT\ln\phi_N = 0$$

$$(N = N_{\min}, \cdots, N_{\max}) \tag{2}$$

as the condition for aggregation equilibrium. In the above equations, $\mu_1$ is the monomer chemical potential prevailing in the surrounding solution and $\phi_N$ the volume fraction of micelles with aggregation number $N$. According to Eq. (2), for each $N$-micelle, the excess free energy $\varepsilon_N \equiv N(\hat{\mu}_N - \mu_1)$ is counterbalanced by the corresponding (partial) free energy of dispersion in the solution.

In addition, a "phase" equilibrium condition is implied in the Hill treatment. The pseudo-phase considered in this context is composed of $N$ identical (average) micelles dispersed in the solution. Their size is determined from the relation

$$\langle \mu_N \rangle = \left\langle \frac{\partial(N\hat{\mu}_N)}{\partial N} \right\rangle = \mu_1 \qquad (3)$$

which implies that the ensemble average of the chemical potential of a surfactant monomer in a micelle equals the monomer chemical potential $\mu_1$ in the adjacent solution. The condition (3) is satisfied when the excess free energy $\varepsilon_N = N(\hat{\mu}_N - \mu_1)$ has a minimum for a certain aggregation number $N$, i.e. when $\varepsilon_N$ has attractive (for large $N$) as well as repulsive (for small $N$) components [2]. Hence, it provides a rational for the "principle of opposing forces" that is commonly invoked when discussing surfactant aggregation phenomena.

It is fairly obvious that the same general thermodynamic rules that govern micelle formation in a $\mu$, $p$, $T$-environment must also hold for the formation of microemulsion droplets in a surfactant-containing solution. In other words, it should constitute a valid theoretical approach to consider microemulsion droplets similarly as molecular complexes of variable stoichiometry. Alternatively, one has to explicitly treat the fluctuations in size, shape and composition of the droplets. Most previous theories [8–16] have ignored some, if not all, of the thermodynamically important fluctuations in their theoretical descriptions of microemulsion droplets. In the end, this has caused difficulties with the quantitative aspects. After this work was completed, however, Borcovec et al. [17, 18] published a theoretical treatment which is in principle correctly based on assuming a set of chemical equilibria to account for the formation of microemulsion droplets of various sizes. Nevertheless, the microemulsion complexes considered by these last-mentioned authors were not exactly the same as in our theory to be presented below, resulting in some rather important differences to which we shall return in the discussion section. It is also worth nothing that mixed surfactant micelles have been considered previously by employing a largely equivalent approach [19, 20].

The purpose of the present paper is twofold. Firstly, we describe a novel method centered on using the Jacobian of a variable transformation to estimate the enormously large influence, even of rather moderate fluctuations, when it comes to quantifying the total volume fraction of microemulsion droplets. Secondly,

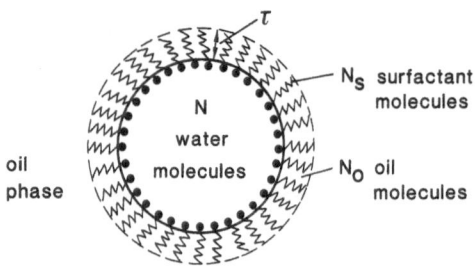

Fig. 1. Sketch of a spheroidal w/o microemulsion droplet. The interior surfactant/water solution is supposed to be very dilute and is considered approximately as pure water. The size of the droplet, its shape and the composition of the surfactant film may vary because of fluctuations which are constantly occurring due to collisions and molecular transport processes

we address the surface thermodynamics of spherical microemulsion droplets and show the crucial importance of having a significant curvature-dependence of the interfacial tension $\gamma$ in order to ever attain thermodynamic stability. When developing the latter part we have been guided by our own previous works on surfactant micelles [2, 3] and, in particular, by the thorough works of Overbeek et al. [14] on w/o microemulsions that, in our opinion, are properly based on surface thermodynamics. We begin with considering the nature of the fluctuations which have to be taken into account in the present context for droplets of a three-component microemulsion.

## Fluctuations of microemulsion droplets

To gain simplicity we shall restrict ourselves to considering three-component w/o microemulsions with an oil ($o$) as the continuous phase. The additional components are water ($w$) and a surfactant ($s$) which may be either of the ionic or the nonionic type. The interior of a droplet contains a dilute water solution of surfactant monomers. For the most part, however, we assume that the water concentration at the centre is approximately the same as for pure water (Fig. 1). This restriction can easily be removed but at the price of a somewhat more complex thermodynamic treatment.

A multitude of experimental investigations indicate that microemulsion droplets are dynamic entities in the size range between about 2 to 100 nm in radius, which may form in regimes of low or ultralow interfacial tensions, $\gamma = 1$ to $10^{-4}$ mNm$^{-1}$ (cf. Refs. [21, 22]). A state of complete equilibrium is reached for one particular size of spherical droplets only. Still, we have to anticipate that droplets of variable shape, size and composition are present in the solution as a result of fluctuations about the strict equilibrium conditions.

For a spherical droplet, mechanical equilibrium demands that the generalized Young-Laplace equation,

$$p_i - p_e = 2\gamma/R - \pi_c \qquad (4)$$

is fulfilled where $p_i$ is the internal pressure, $p_e$ the external pressure and $\pi_c$ the curvature-related pressure [2] (to be discussed below). Note that Eq. (4) holds independently of our choice of position for the dividing surface. In this context, it turns out to be convenient to determine $R$ from the simple relation

$$4\pi R^3/3 = N\bar{V} \qquad (5)$$

where $\bar{V}$ is the (constant) molecular volume of pure water and $N$ the number of water molecules contained inside the droplet. However, for any given volume of the water core there will be deviations from the strictly spherical shape, associated, e.g., with the collision processes which are constantly occurring. Hence, these shape variations constitute dynamic fluctuations about the mechanical equilibrium condition given by Eq. (4). Here, we are not in the first place referring to vibrations in the ordinary sense but rather to shape variations accompanied by transfer of surfactant molecules to and from the droplet surface. The freeze fracture electron micrographs recorded by Jahn and Strey [23] and the NMR relaxation studies of Carlström and Halle [24] indicate that it might be sufficient to consider spheroidal shapes only corresponding to spherical harmonics with $l = 2$.

By the same token there will be fluctuations (that leave the spherical shape unaffected) about the equilibrium size with $R = R_{eq}$. A droplet with this radius for which the physico-chemical equilibrium condition $\mu_w(\text{drop}) = \mu_w(\text{oil})$ is fulfilled, subscript $w$ denoting the water component, we call an equilibrium droplet. We have to anticipate, of course, that both the shape and size fluctuations can occur quite readily (though normally, not very fast), i.e. in shallow free energy minima, because the interfacial tension which should be the main controlling factor here is so much lower than in the analogue case of surfactant micelles [25, 26].

In addition, we can have variations of the composition of the interior water solution (that we presently ignore) and of the surfactant film at the water-oil interface (Fig. 1). This is to say that for a given surface area of the water core, the number of surfactant mol-

ecules will be variable. On the other hand, it should, as a rule, be a reasonable approximation to assume incompressibility, i.e. to neglect the density fluctuations in the water core and correspondingly so for the surfactant film part.

Toward this background we find that in order to describe any one of the various microemulsion droplet complexes we need to specify the following set of independent variables: $N, N_s, N_0$. The number of water molecules, $N$, determines the radius of the corresponding spherical droplet (Eq. (5)). The $N_s$ surfactant molecules in the surfactant film yield the interfacial area $A$ in accordance with the relationship

$$A = N_s a \qquad (6)$$

where $a$ denotes the surface area per surfactant head group. Moreover, $N_s$ and $N_0$, the number of oil molecules, determine together the volume of the surfactant film:

$$A\tau = N_s \bar{V}_s + N_0 \bar{V}_0 \qquad (7)$$

where $\tau$ is the surfactant film thickness which is assumed to remain constant. This equation should be sufficiently accurate, at least for shape variations of a limited amplitude.

### The total volume fraction of disperse water phase

In order to compute the overall volume fraction of dispersed microemulsion droplets we have to add up the volume fractions of all the different aggregates which are specified by the aggregation numbers $N, N_s$ and $N_0$. In the dilute regime these are given by equations similar to Eq. (2), i.e.

$$\varepsilon(N, N_s, N_0) + kT\ln\phi(N, N_s, N_0) = 0 \qquad (8)$$

where we have introduced the excess free energy of an aggregate, $\varepsilon(N, N_s, N_0)$, corresponding to the work of formation of the droplet at constant chemical potentials, $\mu_w, \mu_s, \mu_0$. The resulting sum of Boltzmann factors

$$\phi - \sum_{N,N_s,N_0} e^{-\varepsilon(N_s,N_0)/kT} \qquad (9)$$

has the character of a partition function for an open system. This analogy which is, of course, more than accidental, was noted already by Hill (Ref. [1], p. 133).

Replacing the sum by its corresponding integral we have

$$\phi = \int_0^\infty \int_0^\infty \int_0^\infty e^{-\varepsilon(N,N_s,N_0)/kT} dN \, dN_s \, dN_0 . \tag{10}$$

In order to solve this integral it is convenient to change to a new set of independent variables: the radius $R$ of the undistorted, spherical water core, the area per surfactant head group, $a$, and the shape factor, $\chi = A/4\pi R^2$. The latter accounts for the increase in droplet surface area associated with a fluctuation away from the spherical shape at a fixed $N$. Thus, Eq. (10) may be rewritten as follows:

$$\phi = \int_0^\infty \int_{\alpha\bar{a}}^\infty \int_1^\infty e^{-\varepsilon(R,a,\chi)/kT} |J| \, dR \, da \, d\chi \tag{11}$$

where $J$ denotes the Jacobian

$$J = \frac{\partial(N,N_s,N_0)}{\partial(R,a,\chi)} \tag{12}$$

and $\alpha$ is a numerical factor somewhat less than 1 (cf. the Appendix).

The matrix elements can easily be evaluated by employing Eqs. (5)−(7), resulting in

$$J = -\frac{64\pi^3 R^6 \chi \tau}{\bar{V}\,\bar{V}_0} \tag{13}$$

and, hence, we finally obtain

$$\phi = \frac{64\pi^3 \tau}{\bar{V}\bar{V}_0} \int_0^\infty \int_{\alpha\bar{a}}^\infty \int_1^\infty \frac{\chi R^6}{a^2} e^{-\varepsilon(R,a,\chi)/kT} dR \, da \, d\chi . \tag{14}$$

We reiterate that this expression for the total volume fraction of microemulsion droplets is based on the notion that each combination of the aggregation numbers $N, N_s, N_0$ corresponds to one particular molecular complex. A great number of different complexes will, of course, be present simultaneously in the solution because of the fluctuations in size ($R$), composition ($a$) and shape ($\chi$).

We are justified in restricting the possible shapes to $l = 2$ spheroids by reference to the well-known circumstance that surfactant films in microemulsion systems possess a significant degree of rigidity or bending elasticity. This implies that droplet shapes with strongly wrinkled parts are effectively prohibited. In addition, at least insofar as $\chi = A/4\pi R^2$ is not very

much larger than unity, it should, as a rule, constitute a rather satisfactory approximation to assume that $\varepsilon(R,a,\chi)$ is independent of the exact shape. This is so because the curvature effects generally tend to cancel in a similar manner as for ordinary spherical surfactant micelles [26], the main reason being that parts with lower *and* higher curvature than $2/R$ are formed simultaneously at constant $V$.

In order to estimate the free energy function $\varepsilon(R,a,\chi)$, we next need to consider the thermodynamics of spherical w/o microemulsion droplets.

## Thermodynamics of a size-fluctuating microemulsion droplet

In this section we apply surface and small systems thermodynamics to a spherical microemulsion droplet of radius $R$ which consists of $N$ water molecules, $\bar{N}_s$ surfactant molecules that are located in the interfacial film, and $\bar{N}_0$ oil molecules. The bar superscript denotes an average value, i.e. we are in fact treating a spherical droplet core with a certain (variable) number of water molecules, $N$, that is coated with an ensemble-averaged surfactant/oil film where the packing density ($\bar{a}$) is such that physico-chemical equilibrium is established with the surrounding oil phase:

$$\mu_s(\text{drop}) = \mu_s(\text{oil}) \tag{15}$$

$$\mu_0(\text{drop}) = \mu_0(\text{oil}). \tag{16}$$

Toward this background we may write down the Helmholtz free energy differential, $dF$, for a microemulsion droplet by employing, in essence, the approach which we developed a few years ago for an ordinary surfactant micelle [2].

$$dF = -S dT - p_i dV + \gamma dA - \pi_c A dR$$
$$+ \mu_N dN + \mu_s d\bar{N}_s + \mu_0 d\bar{N}_0 . \tag{17}$$

Note that all of the extensive thermodynamic properties involved here: $F$, $S$, $\bar{N}_s$, $\bar{N}_0$, etc. are strictly defined by means of the dividing surface located at the radius $R = (3N\bar{V}/4\pi)^{1/3}$ from the centre of the droplet. The pressure $p_i$ is either the actual, isotropic pressure at the droplet centre (for large enough droplets) or, alternatively (for very small droplets), the pressure that would be required to realize the prevailing droplet water chemical potential, $\mu_N$, for a pure bulk phase of water that serves as a reference phase. The latter approach is originally due to Gibbs [27].

Let us now, as the next step, introduce the Legendre transformation

$$\hat{G} = F + p_e V - \mu_s \bar{N}_s - \mu_0 \bar{N}_0 \qquad (18)$$

by means of which we effect a change of independent variables to the solution state variables $p_e$ (external pressure), $\mu_s$, $\mu_0$. Thus we get from Eq. (17)

$$d\hat{G} = -S\,dT + V\,dp_e - (p_i - p_e)dV$$
$$+ \gamma\,dA - \pi_c A\,dR$$
$$+ \mu_N dN - \bar{N}_s d\mu_s - \bar{N}_0 d\mu_0. \qquad (19)$$

However, by taking into account the Young-Laplace condition, Eq. (4), and the simple geometrical relations $dV = 4\pi R^2 dR$, $dA = 8\pi R\,dR$ for a sphere, we find that Eq. (19) can be simplified to a considerable extent. Hence, for the droplet we obtain

$$d\hat{G} = -S\,dT + V\,dp_e + \mu_N dN$$
$$- \bar{N}_s d\mu_s - \bar{N}_0 d\mu_0. \qquad (20)$$

From this differential of $\hat{G}$ it appears that the chemical potential of the water inside the droplet may be defined by the partial derivative

$$\mu_N = \left(\frac{\partial \hat{G}}{\partial N}\right)_{T,p_e,\mu_s,\mu_0}. \qquad (21)$$

Consequently, when the quotient $\hat{G}/N = \hat{\mu}_N$ varies irregularly with $N$ way to a significant degree, as might be the case if the interfacial tension $\gamma$ has a small enough value which depends on the droplet radius $R$, there is a chance that for some particular aggregation number $N = N_{eq}$, $\mu_N$ may attain the same value as in the external oil phase, $\mu_w$, implying that complete physico-chemical equilibrium is established (cf. Fig. 2).

We now introduce the full $\Omega$-potential of a microemulsion droplet through the relationship

$$\varepsilon = \hat{G} - N\mu_w. \qquad (22)$$

Obviously $\varepsilon$ stands for the overall excess free energy of the microemulsion droplet that corresponds to the work of formation at constant state of the surrounding solution. By making use of Eq. (20) we arrive at the following $\varepsilon$-differential

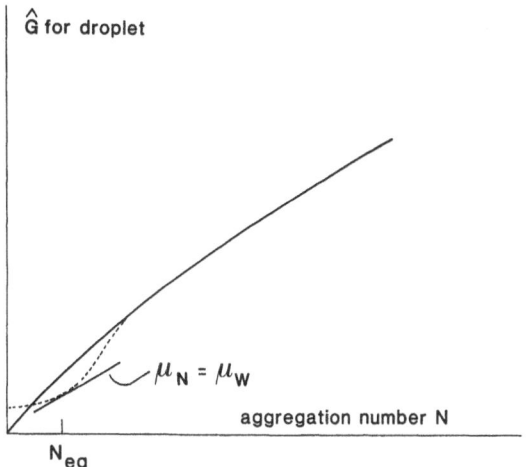

Fig. 2. Plot of $\hat{G}$-functions vs. the aggregation number, $N$, for small w/o droplets. The full-drawn line refers to the normal case where the interfacial tension $\gamma$ is independent of curvature whereas the dotted curve exemplifies a more unusual case where $\gamma$ depends on the radius $R$ in such a way that for $N = N_{eq}$ the derivative $\partial \hat{G}/\partial N = \mu_N$ is equal to $\mu_\infty = \mu_w$ implying that this particular droplet, the equilibrium droplet, can coexist with an excess water phase (Winsor II case)

$$d\varepsilon = -S\,dT + V\,dp_e + (\mu_N - \mu_w)dN$$
$$- N\,d\mu_w - \bar{N}_s d\mu_s - \bar{N}_0 d\mu_0 \qquad (23)$$

which includes the illuminating partial derivative

$$\left(\frac{\partial \varepsilon}{\partial N}\right)_{\text{state}} - \mu_N - \mu_w. \qquad (24)$$

Complete droplet/oil phase equilibrium is characterized, of course, by the condition

$$\mu_N - \mu_w = 0 \qquad (25)$$

which evidently implies that $\varepsilon = \varepsilon(N)$ must have a minimum for some particular, finite droplet size, $N = N_{eq}$. If this is not the case, $\langle\mu_N\rangle$ will be appreciably different from $\mu_w$ and a thermodynamically stable microemulsion will not exist. Hence, we conclude that it is of great importance in the present context to explore what general conditions and molecular mechanisms that may give rise to a descending $\varepsilon(N)$-branch for small $N$-values and an ascending branch for large $N$, i.e. how we can realize a range of droplet states which are properly described in terms of "opposing forces".

By applying Eq. (19) for a conical sector of a droplet, integrating over the solid angle from 0 to $4\pi$ at constant $T, p_e, \mu_s, \mu_0$ and making use of Eq. (22) we can write down the work of formation of a small spherical droplet in the following, alternative way

$$\varepsilon = -(p_i - p_e)V + \gamma A + N(\mu_N - \mu_w) \tag{26}$$

or, inserting the Young-Laplace condition (4),

$$\varepsilon = (\gamma + \pi_s R)A/3 + N(\mu_N - \mu_w). \tag{27}$$

However, with pure water as the incompressible reference phase we can express $\mu_N$ and $\mu_w$ in terms of pressure differences:

$$\mu_N = \mu_w(p_e) + (p_i - p_e)\bar{V} \tag{28}$$

$$\mu_w = \mu_w(p_e) + (p_r - p_e)\bar{V} \tag{29}$$

where $p_r$ is the hypothetical pressure that would be required to realize the solution chemical potential $\mu_w$ for a bulk water phase. In the two-phase, Winsor II regime, when a nearly pure water phase is actually present in the system, $p_r$ equals $p_e$. Inserting the above relations in Eq. (26) yields

$$\varepsilon = \varepsilon(R) = \gamma A + (p_e - p_r)V \tag{30}$$

where the second term vanishes for Winsor II microemulsions. Hence, when the interfacial tension $\gamma$ is known as a function of the droplet radius $R$, and the pressure difference $p_e - p_r$ is likewise a known quantity (e.g. from vapour pressure measurements), the excess free energy $\varepsilon(R)$ can readily be obtained by means of Eq. (30).

By differentiating Eq. (26), combining with Eq. (23) and invoking the Gibbs-Duhem condition for the pure water reference phase, it is an easy matter to reassure the validity of the Gibbs surface tension equation. Accordingly,

$$A\,d\gamma = -(S - N\bar{S}_w)dT - \pi_c A\,dR$$
$$- \bar{N}_s d\mu_s - \bar{N}_0 d\mu_0 \tag{31}$$

where the partial derivative

$$\left(\frac{\partial\gamma}{\partial R}\right)_{T,\mu_s,\mu_0} = \left(\frac{\partial\gamma}{\partial R}\right)_{\text{state}} = -\pi_c \tag{32}$$

is included, showing that, in a straightforward fashion, the curvature pressure $\pi_c$ accounts for the curvature dependence of the interfacial tension $\gamma$. At the radius

$R_0$, corresponding to the *spontaneous curvature*, $\pi_c$ is thus equal to zero.

In particular, for a Winsor II microemulsion with an excess water bulk phase we find that

$$\frac{\partial\varepsilon}{\partial N} = \left(\gamma\frac{\partial A}{\partial V} + \frac{\partial\gamma}{\partial R}\right)\frac{\partial V}{\partial N}$$
$$= (2\gamma/R - \pi_c)\bar{V} = \mu_N - \mu_w. \tag{33}$$

Consequently, for an equilibrium microemulsion droplet of this kind it must hold true that

$$p_i - p_e = 2\gamma/R - \pi_c = 0. \tag{34}$$

In other words, the curvature pressure plays a crucial role for microemulsion droplets in counterbalancing the Laplace pressure $2\gamma/R$. Note that since $\gamma/R$ is always $>0$ it is a necessary condition that $\pi_c > 0$, which means that the equilibrium size is always found somewhere on the descending, (repulsive) $\gamma(R)$ branch, below the radius of spontaneous curvature, $R_0$, where $\pi_c = 0$.

It is, after all quite astonishing that some of the above, rather fundamental thermodynamic relations for microemulsions have not hitherto been fully recognized. Eqs. (33) and (34), for instance, supply a very rapid and clarifying answer to the old, but still intriguing question why microemulsions, as opposed to ordinary emulsions, are thermodynamically stable. Hence, one is prone to speculate about what the reason for this state of affairs might be.

It is evident that the very existence of thermodynamically stable microemulsions is closely related to the curvature dependence of the interfacial tension $\gamma$. By and large, the actual importance of such a dependence was underrated by Guggenheim [28] who, for decades, was a leading authority on surface thermodynamics. Further back, Gibbs [27] developed his brilliant and famous "Theory of Capillarity" on basis of the surface tension concept employing the surface of tension (sot) as the dividing surface (especially for curved interfaces). A few years ago, however, it was pointed out by Mitchell and Hall [29] that this particular choice of dividing surface is highly inconvenient when it comes to treating microemulsion droplets. In fact, it can be shown (Ref. [2], Eq. (8)) that the radius, $R_{\text{sot}}$, of the surface of tension is given by the expression

$$R_{\text{sot}}^3 = \frac{6}{(p_i - p_e)}\int_0^\infty (p_e - p_T)r^2\,dr \tag{35}$$

where $p_T$ is the tangential pressure tensor component and $r$ is the distance from the centre of the droplet. Thus, $R_{sol} \to \infty$ when the equilibrium-sized microemulsion droplet is approached. For this reason, one has to resort to choosing some other dividing surface which preferably, like the one defined by Eq. (5) which obviously corresponds to the equimolar dividing surface for the water component, is always positioned in the interfacial region. This means that it becomes improper to invoke the conventional Young-Laplace equation (without the $\pi_c$-term) when considering a microemulsion droplet and, furthermore, that it is also inappropriate to employ the standard version of the Gibbs surface tension equation (without the $\pi_c dR$-term). In short, the Gibbs approach is definitely not very well suited for developing the thermodynamics of a microemulsion droplet. Instead, one has to start out from some of the more general formulations of the thermodynamics of curved interfaces, such that an arbitrary criterion can be employed to fix the position of the dividing surface. Classical treatments of this kind were worked out long time ago by Hill [30], Ono and Kondo [31] and Rusanov [32] and, hence, these authors have provided the foundation of the thermodynamics of microemulsion droplets presented here.

**The excess free energy $\varepsilon(R, a, \chi)$ of a microemulsion droplet**

In order to solve the multiple integral of Eq. (14) it is necessary to make some more specific *ansatz* about the excess free energy of a microemulsion droplet, $\varepsilon(R, a, \chi)$. With this ultimate goal in mind we refer to Huh [9] and de Gennes and Taupin [11] and assume the following expansion in $1/R$ for $\gamma(R)$ at a certain solution state of a spherical droplet with an equilibrated surfactant film ($a = \bar{a}$):

$$\gamma = \gamma_\infty + k_1/R + k_2/R^2 \qquad (36)$$

where, obviously, $\gamma_\infty$ denotes the value of the interfacial tension $\gamma$ for very large $R$. In the present context, $\gamma_\infty$ is primarily considered as an experimentally accessible parameter. A negative constant $k_1$ is anticipated in conjunction with a positive $k_2 > k_1^2/4\gamma_\infty$. The $\gamma(R)$ function generated in accordance with Eq. (36) will then have a minimum for $R_0 = -2k_2/k_1$, i.e. the spontaneous mean curvature equals $-k_1/2k_2$ (Fig. 3). The minimum value of $\gamma(R)$ is related to $\gamma_\infty$ by the expression

$$\gamma_0 = \gamma_\infty - k_1^2/4k_2. \qquad (37)$$

Moreover, we can easily show that Eq. (36) can be written in the alternative form

$$\gamma = \gamma_0 + k_2\left(\frac{1}{R} - \frac{1}{R_0}\right)^2. \qquad (38)$$

Thus, our constant $k_2$ is essentially the same as the well-known bending elasticity (rigidity) constant $k_c$ which was introduced for amphiphilic bilayers several years ago by Helfrich [33]. In fact by comparing with a simple Helfrich expression, similar to Eq. (38) above, without the saddle splay term, we may identify $k_2$ with $2k_c$ and $k_1$ with $-4k_c/R_0$. However, comparing instead with the full Helfrich expression (spherical geometry), i.e.

$$\gamma = \text{const} + 2k_c\left(\frac{1}{R} - \frac{1}{R_0}\right)^2 + \bar{k}_c/R^2 \qquad (39)$$

we find that $k_2 = 2k_c + \bar{k}_c$, $k_1 = -4k_c/R_0$ and $\gamma_\infty = \text{const} + 2k_c/R_0^2$.

The bending stress coefficient used by Overbeek et al. [14] is closely related to the curvature pressure $\pi_c$. As a matter of fact, it may be defined as

$$c = \frac{1}{2}\pi_c R^2 \qquad (40)$$

and it increases linearly with $1/R$ insofar as Eq. (36) holds because from this equation we find the expression

$$c = k_1/2 + k_2/R. \qquad (41)$$

In Fig. 3 we have reproduced an example of how $\gamma$ as given by Eq. (36) may vary with the droplet radius $R$ for a typical microemulsion case. The elaborate investigations of Overbeek et al. [14] where direct and detailed comparison between theory and experiments are invoked, yield support of the quadratic form of Eq. (36). Additional support is provided by our recent model calculations on interfacial tensions of curved w/o interfaces [34] carried through by extending the corresponding calculations made for planar surfactant-loaded interfaces [35].

Combining Eqs. (30) and (36) we get for the Winsor II case (spherical droplets only):

$$\varepsilon(R) = 4\pi(\gamma_\infty R^2 + k_1 R + k_2). \qquad (42)$$

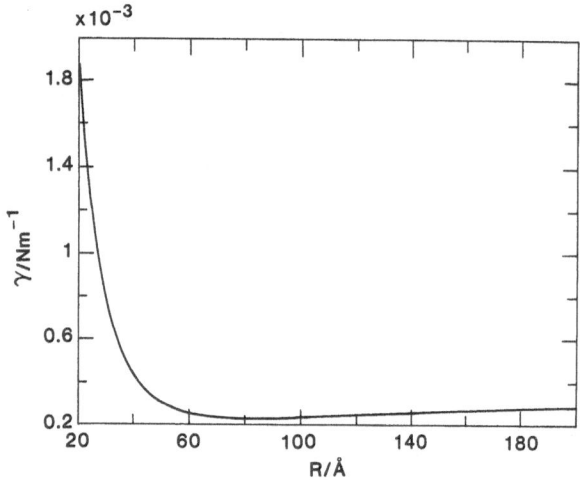

Fig. 3. Plot of the interfacial tension $\gamma$ vs. the droplet radius $R$ for a case yielding comparatively small microemulsion droplets. The constants in Eq. (36) were chosen as follows: $\gamma_\infty = 4.0 \times 10^{-4}$ Nm$^{-1}$, $k_1 = -2.87 \times 10^{-12}$ N, $k_2 = 1.2 \times 10^{-20}$ J. The radius $R_0$ corresponding to the spontaneous curvature (where the curvature pressure $\pi_c$ equals zero) is equal to 83.62 Å

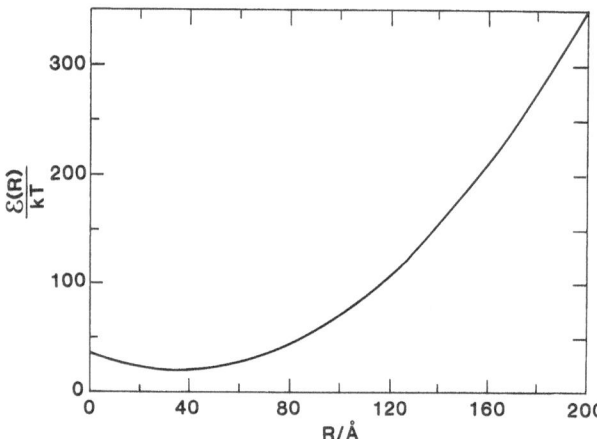

Fig. 4. Plot of excess free energy function $\varepsilon(R)$ according to Eq. (42) with the same choice of parameters as in Fig. 3. The equilibrium droplet has the radius $R_{cq} = 35.88$ Å

This excess free energy function has a minimum equal to $4\pi(k_2 - k_1^2/4\gamma_\infty)$ for $R_{eq} = -k_1/2\gamma_\infty$ (cf. Fig. 4). It is easily verified that $R_{eq} < R_0$ which immediately implies that the curvature pressure $\pi_c$ is $> 0$ at $R = R_{eq}$, as is required for thermodynamic stability. In addition, it is worth observing that since $|k_1|$ is of the order of $10^{-12}$ $N$, one must typically reach $\gamma_\infty$-values as low as about 1 mNm$^{-1}$ or less to be able to form a stable microemulsion.

To allow for the shape fluctuations (which for the most part involve diffusive surfactant transport and occur on a rather slow time-scale) and the concentration fluctuations in the surfactant film we now modify $\varepsilon(R)$ of Eq. (42) in the following way,

$$\varepsilon(R,a,\chi) = 4\pi\chi[\gamma_\infty R^2 + k_1 R + k_2 + \varkappa(a/\bar{a} - 1)^2 R^2] + (4\pi R^3/3)(p_e - p_r) \tag{43}$$

where, as before, the shape factor $\chi$ is defined as $A/4\pi R^2$, and the last term has been added in accordance with Eq. (30). The new quadratic term involving the deviation of the surface area per surfactant molecule, $a - \bar{a}$, accounts for the variation of $\varepsilon$ associated with a concentration change in the surfactant film. We anticipate that the constant $\varkappa$ can be estimated from theoretical calculations on planar films [35].

From Eq. (43) it is seen that we have really not included any particular curvature effects connected with the spheroidal shapes that we consider. In other words, the important shape fluctuations in this context are generally supposed to be surface-tension rather than curvature-limited. As already state above this should constitute a reasonable approximation, at least for $1 < \chi < 1.1$, because portions of higher and lower curvature than $2/R$ are formed simultaneously at constant $V$ and, furthermore, the predominant droplet sizes are in the range of $R_{eq} < R_0$ where $\pi_c$ is always $<0$, resulting in a tendency to cancellation of the various free energy changes which are associated with a certain shape deformation. Moreover, the much distorted shapes have lesser weight, anyhow, due to the comparatively larger increase of the interfacial area.

### The droplet size distribution function $\phi(R)$

We are now prepared to return to Eq. (14) and insert the excess free energy function $\varepsilon(R,a,\chi)$ furnished by Eq. (43), yielding the expression

$$\phi = \frac{64\pi^3\tau}{\bar{V}\bar{V}_0} \int_0^\infty e^{-4\pi R^3(p_e - p_r)/3kT}$$

$$\times \, dR \int_{\alpha\bar{a}}^\infty da \int_1^\infty \frac{\chi R^6}{a^2}$$

$$\times \, e^{-4\pi[\gamma(R) + \varkappa(a/\bar{a} - 1)^2]/kT} d\chi \tag{44}$$

where $\gamma(R)$ is given by Eq. (36). In the Appendix we

show how this integral can be solved, approximately, resulting in

$$\phi = C \int_0^\infty (R^3/\gamma) e^{-4\pi R^2 \gamma / kT} e^{-4\pi R^3 (p_e - p_r)/kT} dR \qquad (45)$$

where the factor $C$ is

$$C = \frac{8\pi^2 \tau (kT)^{3/2}}{\bar{V} \bar{V}_0 \bar{a} \sqrt{\varkappa}}. \qquad (46)$$

Hence, in case of a Winsor II microemulsion, we obtain the following size distribution function

$$\phi(R) = (CR^3/\gamma) c^{-4\pi R^2 \gamma / kT} \qquad (47)$$

where the pre-exponential factor normally is very large, typically in the range $10^{15}-10^{23}$ m$^{-1}$. In Figs. 7 and 8 we show two examples of size distributions computed by means of Eq. (47) and inserting the following numerical values to obtain the constant factor $C$:, $\tau = 17$ Å, $kT = 0.412 \times 10^{-20}$ J, $\bar{V} = 30$ Å$^3$, $\bar{V}_0 = 200$ Å$^3$, $\bar{a} = 40$ Å$^2$, $\varkappa = 0.05$ Nm$^{-1}$. The corresponding $\gamma(R)$- and $\varepsilon(R)$-functions are reproduced in Figs. 3 to 6. The distribution curves are approximately Gaussian and peak at about 1.15 $R_{eq}$. This upward shift is caused by the "degeneracy factor" $CR^3/\gamma$ that favours the larger droplets. The standard deviation of the droplet radius, $\sigma_R/R$, is estimated to be about 0.15 in both cases which is in broad agreement with currently available experimental information [36].

From Figs. 4 and 6 it is seen that the minimum values of $\varepsilon(R)$ (those corresponding to the respective equilibrium droplets) are about 20.9 $kT$ and 37.9 $kT$ in thew two examples. This means that the volume fractions of equilibrium droplets are $8.4 \times 10^{-10}$ and $3.5 \times 10^{-17}$, respectively. Nevertheless, the corresponding total volume fractions of microemulsion droplets are many orders of magnitude larger, $1.00 \times 10^{-2}$ and $0.95 \times 10^{-2}$. Hence, we conclude that it is of a major importance to correctly account for the fluctuations in size, shape and composition when calculating the total volume fraction $\phi$ of microemulsion droplets. These fluctuations bring about an enormous number of molecular complexes with slightly differing stoichiometry as compared with the true equilibrium droplet, each of them obeying the aggregation equilibrium condition given by Eq. (8).

The whole problem is extremely sensitive to the exact numerical values we insert for the parameters $\gamma_\infty$, $k_1$ and $k_2$ which all have to be considered as functions of the solution state. Our present choice of parameter

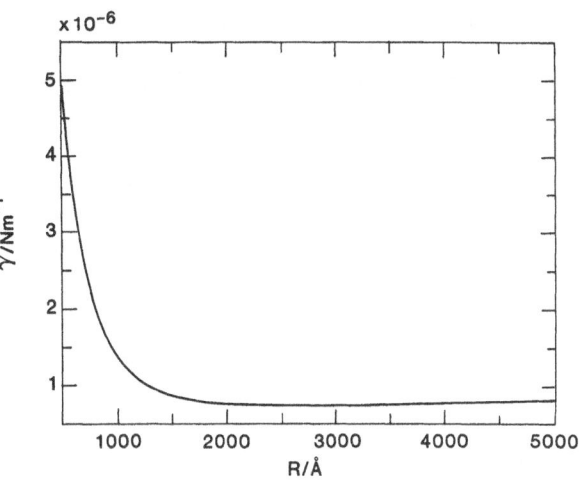

Fig. 5. Plot of the interfacial tension $\gamma$ vs. the droplet radius $R$ for a case yielding relatively large microemulsion droplets. The constants in Eq. (36) were taken to be $\gamma_\infty = 1.0 \times 10^{-6}$ Nm$^{-1}$, $k_1 = -1.28 \times 10^{-13}$ N, $k_2 = 1.65 \times 10^{-20}$ J. The radius $R_0$ corresponding to the spontaneous curvature (where $\pi_c = 0$) is 2578 Å

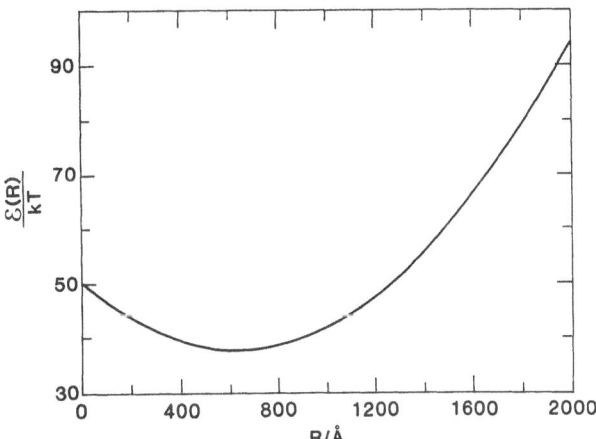

Fig. 6. Plot of the excess free energy function $\varepsilon(R)$ in accordance with Eq. (42) with the same choice of parameters as in Fig. 5. The equilibrium droplet has the radius $R_{eq} = 640$ Å

values is presumably rather realistic, however, since it is well-known that Winsor II microemulsions tend to form in regimes where $\gamma_\infty$ ranges between 1 and $10^{-3}$ mNm$^{-1}$, $k_2$ amounts to a few $kT$-units and $|k_1|$ is about $2k_2/R_0$. Such values of $k_1$ and $k_2$ correspond, by the way, to bending stress coefficients $c$ between $1 \times 10^{-13}$ and $5 \times 10^{-12}$ N. Moreover, the well-known empirical relation $\gamma_\infty = kT/\bar{R}^2$ [21] is approximately fulfilled in both of the above examples where $\bar{R}$ is the average droplet radius. Our recent theoretical calculations which were based on extending previous calculations of the surface tension of surfactant solutions [35] by invoking the effects of curvature on the electrostatic and hydrocarbon/oil-mixing free energies, also yield $k_1$- and $k_2$-values of the above order for a $C_{12}NH_3^+/C_{12}OH$/salt w/o microemulsion [34].

It is worth noting that if we were to assume the equation $\gamma_\infty = kT/R_{eq}^2$ to hold exactly, we could easily derive the relation $k_1 = -2\sqrt{kT\gamma_\infty}$, and, furthermore, that the constant $k_2(= 2k_c + \bar{k}_c)$ should have a value between about 2.5 and 4 $kT$ (depending on the droplet size) in order to generate a microemulsion with $\phi = 0.1$.

## The shape polydispersity

Our theory also permits an approximate estimation of the shape polydispersity of the microemulsion droplets. As discussed above, in this context we generally neglect those bending free energy contributions which are related to the deformations at constant volume of the spherical water core of radius $R$, i.e. the shape fluctuations considered are supposed to be surface-tension-limited. As shown in the Appendix, the average value of the shape factor $\chi$ can be computed by means of the formula ($\lambda = \varepsilon(R)/kT$):

$$\langle\chi\rangle \approx 1 + 2/(2\lambda - 1) \approx 1 + 1/\lambda \qquad (48)$$

with a standard deviation of about $1/\lambda$.

Accordingly, for the most abundant droplets belonging to the distributions of Figs. 7 and 8 we find $\langle\chi\rangle = 1.049$ and $\langle\chi\rangle = 1.027$, respectively. This indicates that even the shape deformations are, after all, of a limited magnitude, relatively speaking. Hence, we are to some extent justified *à posteriori* in neglecting the curvature-related deformation free energy. Nevertheless, since the various shapes are, to a good approximation, exponentially distributed over the $\chi$-variable, there will always be a certain fraction of substantially deformed droplets. Considering oblate ellipsoidal shapes only we find that about 15% of the droplets actually have an axial ratio larger than 2 for $\lambda = 20$. This is in reasonable agreement with the interpretation favoured by Carlström and Halle [24] of their $^2H$ and $^{17}O$ spin relaxation data for AOT/D$_2$O/ isooctane microemulsions according to which a significant fraction of the droplets are distinctly nonspherical with axial ratios between 2 and 3.

On the other hand, Milner and Safran [37] have considered the comparatively rapid (curvature-limi-

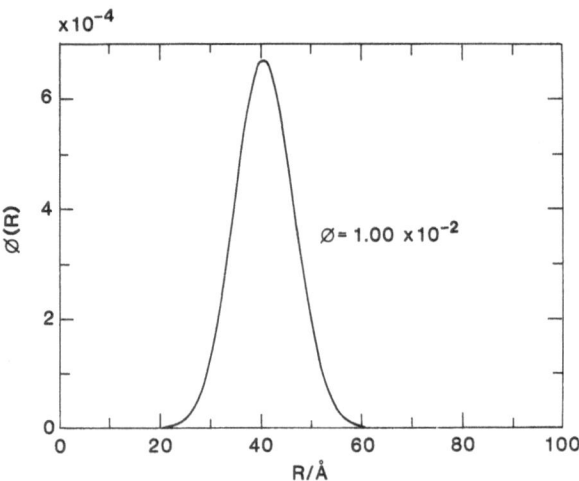

Fig. 7. Example of a size distribution function, $\phi(R)$, for small microemulsion droplets. The parameters were chosen to be the same as in Figs. 3 and 4. The average radius is 40.9 Å and $\sigma_R/R = 0.14$

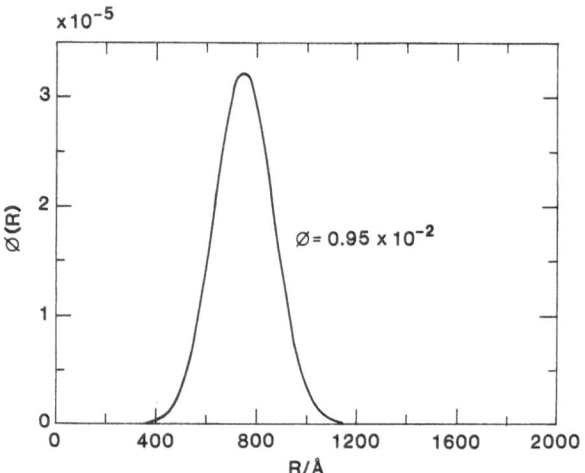

Fig. 8. Example of a size distribution function, $\phi(R)$, for large microemulsion droplets. The parameter values are the same as in Fig. 5 and 6. The average radius is 742 Å and $\sigma_R/R = 0.16$

ted) vibrations which may occur under constraints of constant $V$ (i.e. $R$), $\chi$ and $a = \bar{a}$ for thermodynamically *closed* droplets. In the present context we have tentatively assumed these fluctuations to yield a small, approximately constant, contribution to $\gamma$ independently of the exact values of $R$, $\chi$ and $\bar{a}$. In other words, their direct influence on the formation of microemulsions is supposedly rather minor compared to the enormous effects of the other fluctuations discussed above. In addition, we note that the rather slow, coupled size/shape fluctuations treated earlier by Safran [12] that are subject to imposed constraints of constant total volume and surface area of the entire droplet phase are automatically covered by our scheme.

## Discussion

In the present paper we have shown how the multiple chemical equilibrium approach to treating droplet microemulsions can be joined with the classical thermodynamic description of curved fluid interfaces. The general notions of the thermodynamics of small systems derived by Hill have been most useful in this context.

During recent years it is above all Overbeek and coworkers [14] who have developed and quantified the surface-thermodynamic approach, in particular to Winsor II microemulsions. At the outset they assumed monodisperse (equilibrium) droplets and did not invoke any fluctuations. By proceeding in this way and carrying through, with great care, comparisons with experimental data, they found support of the idea put forward by Reiss [38] that the dispersion of small droplets in a solution is associated with an extraordinary entropy-caused free energy change, in the present case equal to

$$\Delta\varepsilon(\text{Reiss}) = -1.5\,kT\ln(16R^3/\bar{V}) \tag{49}$$

per droplet that, supposedly, would be due to the greater freedom of motion of the center of mass in the disperse state. This term would accordingly amount to about $-15.77\,kT$ for the smaller droplets in Fig. 7 and to about $-28.82\,kT$ for the larger droplets of Fig. 8 and would raise the volume fractions by about $7.06 \times 10^6$ and $3.28 \times 10^{12}$, respectively. These figures correspond rather closely to the influence of the various fluctuations, in particular so for the smaller droplets where we got $\phi/\phi_{\text{eq}} = 1.19 \times 10^7$. However, since the statistical-mechanical basis of the Reiss term is obscure and controversial and has been severely criticized in the literature [39] whereas the concept of a fluctuating microemulsion droplet follows in a straightforward way from the conventional idea of a multiple chemical equilibrium by invoking small systems thermodynamics and is, furthermore, supported by several experimental findings as to the droplet polydispersity, we believe that the latter approach is the correct one. Incidentally, we note that prof. Overbeek has recently extended his theoretical description so as to encompass the size fluctuations [40].

The theory of Borcovec et al. [17, 18] which is justly founded on the multiple chemical equilibrium concept is different from our treatment in several respects: i) they invoke spherical $N$, $N_s$ complex only whereas we automatically include also the shape fluctuations as they will give rise to a large number of additional molecular complexes, ii) we cover the surface-thermodynamic aspects more completely and consider $\gamma_\infty$ primarily as an experimental parameter [41–43], i.e. in the present context we refrain from invoking any theory of $\gamma_\infty$ since this is an issue of considerable complexity that needs separate treatment [35], iii) we address the problem of calculating the total volume fraction of (non-interacting) droplets at a given state of the continuous oil phase following a thermodynamic route, whereas Borcovec et al. focused on deriving the size distribution. It is reassuring, however, that the calculations of these authors indicate a droplet size polydispersity of similar order, $\sigma_R/R \approx 0.25$.

## Conclusion

The theoretical picture of a microemulsion droplet revealing itself as a result of the present study is essentially as follows. Random molecular transport processes and collision events bring about fluctuations in size, shape and composition which are large on an absolute scale but of a rather limited magnitude, relatively speaking. Hence, for many practical purpose, we may consider a microemulsion droplet as being approximately spherical and of a certain size and composition. These fluctuations must be properly accounted for, however, when evaluating the thermodynamic properties, otherwise one is actually neglecting the corresponding entropies which yield sizeable free energy contributions for large, dynamically formed aggregates held together by weak forces. The conventional multiple chemical equilibrium approach provides a perfectly valid and rather convenient method to deal with these matters for microemulsions formed in water/oil/surfactant systems.

*Appendix*

In order to evaluate the integral

$$\phi = \frac{64\pi^3\tau}{\bar{V}\bar{V}_0} \int\limits_0^\infty \int\limits_{\alpha\bar{a}}^\infty \int\limits_1^\infty \frac{\chi R^6}{a^2} e^{-\varepsilon(R,a,\chi)/kT}\, dR\, da\, d\chi \tag{50}$$

where $\alpha$ is of the order of 0.9, $\chi = A/4\pi R^2$, and

$$\varepsilon(R,a,\chi) = 4\pi R^2\chi\tilde{\gamma} + 4\pi R^3\Delta p/3$$

with

$$\tilde{\gamma} = \gamma_\infty + c(a - \bar{a})^2 + k_1/R + k_2/R^2$$

where $k_1 < 0$, the following abbreviated notation is used

$$\gamma = \gamma_\infty + k_1/R + k_2/R^2$$

$$\lambda = 4\pi R^2 \gamma/kT$$

and

$$\eta = 4\pi R^2 c/kT.$$

Then, it follows that

$$4\pi R^2 \tilde{\gamma}/kT = \lambda + \eta(a - \bar{a})^2$$

and

$$\varepsilon/kT = [\lambda + \eta(a - \bar{a})^2]\chi + 4\pi R^3 \Delta p/3kT$$

and we have

$$\phi = \frac{64\pi^3 \tau}{\bar{V}\bar{V}_0} \int_0^\infty dR\, R^6 e^{-4\pi R^3 \Delta p/3kT}$$
$$\times \int_1^\infty \int_{\alpha\bar{a}}^\infty da\, d\chi \frac{\chi}{a^2} e^{-[\lambda + \eta(a-\bar{a})^2]} \quad\quad (51)$$

where we denote the last double integral by $I$. Thus,

$$I = \int_1^\infty \chi e^{-\lambda\chi} d\chi \int_{\alpha\bar{a}}^\infty \frac{e^{-\eta\chi(a-\bar{a})^2}}{a^2} da. \quad\quad (52)$$

Since the integrand in the last integral falls off very rapidly with increasing $|a - \bar{a}|$ we may replace $a^2$ in the denominator by $\bar{a}^2$ without introducing an appreciable error, and at the same time we can replace the lower limit of the integral with $-\infty$. Then,

$$I \approx \int_1^\infty d\chi\, \chi e^{-\lambda\chi} \bar{a}^{-2} \int_{\alpha\bar{a}}^\infty da\, e^{-\eta\chi(a-\bar{a})^2}$$

$$\approx \int_1^\infty d\chi\, \chi e^{-\lambda\chi} \bar{a}^{-2} \int_{-\infty}^{+\infty} da\, e^{-\eta\chi(a-\bar{a})^2}$$

$$= \frac{1}{\bar{a}^2} \sqrt{\frac{\pi}{\eta}} \int_1^\infty d\chi\, \chi^{1/2} e^{-\lambda\chi}$$

$$= \frac{1}{\bar{a}^2} \sqrt{\frac{\pi}{\eta}} \lambda^{-3/2} \Gamma(3/2, \lambda). \quad\quad (53)$$

Since $\lambda$ is usually of the order of 10 or larger for all relevant $R$-values we may replace $\lambda^{-3/2} \Gamma(3/2, \lambda)$ by its asymptotic expansion

$$\lambda^{-3/2} \Gamma(3/2, \lambda) \approx (e^{-\lambda}/\lambda)(1 + 1/2\lambda - 1/4\lambda^2 + \ldots)$$

where the second term may be necglected with a relative error of $1/2\lambda$.

If this result is inserted into Eq. (2), we conclude that

$$\phi = \int_0^\infty \phi(R) dR \quad\quad (54)$$

where

$$\phi(R) = \frac{8\pi^2 \tau (kT)^{3/2} R^3}{\bar{V}\bar{V}_0 \bar{a} \sqrt{\varkappa \tilde{\gamma}(R)}}$$
$$\times \exp[-4\pi R^2 \gamma(R) + R\Delta p/3)/kT]_2 \quad\quad (55)$$

where we have introduced the quantity $\varkappa = c\bar{a}$, which has the same dimension as $\gamma$.

The above formalism also allows us to calculate the mean value of $\chi$. We obtain

$$\langle\chi\rangle = \frac{\displaystyle\int_1^\infty d\chi\, \chi e^{-\lambda\chi} \sqrt{\pi/\eta\chi}}{\displaystyle\int_1^\infty d\chi\, e^{-\lambda\chi} \sqrt{\pi/\eta\chi}} \quad\quad (56)$$

or

$$\langle\chi\rangle = \frac{\displaystyle\int_1^\infty d\chi\, \chi^{1/2} e^{-\lambda\chi}}{\displaystyle\int_1^\infty d\chi\, \chi^{-1/2} e^{-\lambda\chi}}. \quad\quad (57)$$

Now,

$$\int_1^\infty d\chi\, \chi^{-1/2} e^{-\lambda\chi} = \lambda^{-1/2} \Gamma(1/2, \lambda)$$
$$= (e^{-\lambda}/\lambda)(1 - 1/2\lambda + 3/4\lambda^2 + \ldots).$$

Thus,

$$\langle\chi\rangle = \frac{(e^{-\lambda}/\lambda)(1 + 1/2\lambda - 1/4\lambda^2 + \ldots)}{(e^{-\lambda}/\lambda)(1 - 1/2\lambda + 3/4\lambda^2 + \ldots)} \quad\quad (58)$$

or

$$\langle\chi\rangle \approx 1 + 2/(2\lambda - 1). \quad\quad (59)$$

In addition we can calculate the variance, $\langle(\chi - \langle\chi\rangle)^2\rangle$. We have

$$\langle\chi^2\rangle = \frac{\displaystyle\int_1^\infty d\chi\, \chi^{3/2} e^{-\lambda\chi}}{\displaystyle\int_1^\infty d\chi\, \chi^{-1/2} e^{-\lambda\chi}} = \frac{\lambda^{-5/2} \Gamma(5/2, \lambda)}{\lambda^{-1/2} \Gamma(1/2, \lambda)}$$

$$= \frac{(e^{-\lambda}/\lambda)(1 + 3/2\lambda + 3/4\lambda + \ldots)}{(e^{-\lambda}/\lambda)(1 - 1/2\lambda + 3/4\lambda + \ldots)}$$

$$\approx 1 + \frac{4}{2\lambda - 1}. \quad\quad (60)$$

Simple calculations show that the variance is approximately $1/\lambda^2$, i.e. quite small, and the standard deviation approximately $1/\lambda$.

## References

1. Hill TL (1963, 1964) Thermodynamics of Small Systems. Benjamin, New York, Vol. I and II
2. Eriksson JC, Ljunggren S, Henriksson U (1985) J Chem Soc, Faraday Trans 2, 81:833
3. Eriksson JC, Ljunggren S (1985) J Chem Soc, Faraday Trans 2, 81:1209
4. Ljunggren S, Eriksson JC (1986) J Chem Soc, Faraday Trans 2, 82:913
5. Ljunggren S, Eriksson JC (1988) J Chem Soc, Faraday Trans 2, 84:329
6. Tanford C (1974) J Phys Chem 78:2469
7. Israelachvili JN, Mitchell DJ, Ninham BW (1976) J Chem Soc, Faraday Trans 2, 72:1525
8. Robbins ML (1977) In: Mittal KL (ed) Micellization, Solubilization and Microemulsions, Plenum, New York, Vol 2, p 713
9. Huh C (1983) Soc Pet Eng J 23:829; (1984) J Colloid Interface Sci 97:201
10. Ruckenstein E, Chi JC (1975) J Chem Soc, Faraday Trans 2, 71:1690; (1978) J Colloid Interface Sci 66:369
11. De Gennes PG, Taupin C (1982) J Phys Chem 86:2294
12. Safran SA (1983) J Chem Phys 78:2073; Safran SA, Turkevitch LA (1983) Phys Rev Lett 50:1930
13. Grimson JM, Honary F (1984) Phys Lett 102 A:241
14. Overbeek JTG (1986) In: Debye Symposium, Proc Kon Ned Akad Wetenschap B, 89:61; Overbeek JTG, Verhoeckx GJ, De Bruyn PL, Lekkerkerker HNW (1987) J Colloid Interface Sci 119:422
15. Mitchell DJ, Ninham BW (1981) J Chem Soc, Faraday Trans 2, 77:601
16. Israelachvili JN (1987) In: Mittal KL, Bothorel P (eds) Plenum, New York, Vol. 4, p 3
17. Borcovec M, Eicke H-F, Ricka J (1989) J Colloid Interface Sci 131:366
18. Borcovec M (1989) J Chem Phys 91:6268
19. Rao IV, Ruckenstein E (1987) J Colloid Interface Sci 119:211
20. Ljunggren S, Eriksson JC (1987) Progr Colloid Polymer Sci 74:38
21. Langevin D (1988) Acc Chem Res 21:255
22. Shinoda K, Lindman B (1987) Langmuir 3:135
23. Jahn W, Strey R (1988) J Phys Chem 92:2294
24. Carlström G, Halle B (1989) J Phys Chem 93:3287
25. Ljunggren S, Eriksson JC (1984) J Chem Soc, Trans Faraday Soc 2, 80:489
26. Ljunggren S, Eriksson JC (1989) J Chem Soc, Trans Faraday Soc 2, 85:1553
27. Gibbs JW (1878) In: The Collected Works of Joshia Willard Gibbs, Longmans, Green & Co, London, 1928, Vol. 1
28. Guggenheim EA (1977) Thermodynamics, North-Holland, Amsterdam, p 55
29. Hall DG, Mitchell DJ (1983) J Chem Soc, Faraday Trans 2, 79:185
30. Hill TL (1952) J Phys Chem 56:526
31. Ono S, Kondo S (1960) Flügge S (ed) Handbuch der Physik, Springer, Berlin, Vol X, p 1934
32. Rusanov AI (1978) Phasengleichgewichte und Grenzflächenerscheinungen, Akademie-Verlag, Berlin
33. Helfrich W (1973) Z Naturforsch C 28:693
34. Eriksson JC, Ljunggren S (1990) To be published
35. Eriksson JC, Ljunggren S (1989) Colloids Surf 38:179
36. Chen SH (1986) Annu Rev Phys Chem 37:351
37. Milner ST, Safran SA (1987) Phys Rev A, 36:4371
38. Reiss H (1975) J Colloid Interface Sci 53:61
39. Nishioka K, Pound GM (1977) Adv Coll Interf Sci 7:205
40. Overbeek JTG (1989) Progr Colloid Polymer Sci, in press
41. Verhoeckx GJ, De Bruyn PL, Overbeek JTG (1987) J Colloid Interface Sci 119:409
42. Aveyard R, Binks BP, Fletcher PDI (1989) Langmuir 5:1210
43. Binks BP, Meunier J, Abillon O, Langevin D (1989) Langmuir 5:415

Authors' address:

Prof. J. C. Eriksson
Department of Physical Chemistry
Royal Institute of Technology
S-10044 Stockholm, Sweden

**Progress in Colloid & Polymer Science**

Progr Colloid Polym Sci 81:54−59 (1990)

# SANS study of polymer-containing microemulsions

S. Radiman[1], L. E. Fountain[1], C. Toprakcioglu[1,2], A. de Vallera[3] and P. Chieux[4]

[1] Cavendish Laboratory, University of Cambridge, Cambridge, UK
[2] Institute of Food Research (AFRC), Norwich, UK
[3] Department of Physics, University of Lisbon, Lisbon, Portugal
[4] Institut Laue-Langevin, Grenoble, France

*Abstract:* The water soluble polymer polyethylene oxide (PEO) was introduced into dilute water-in-oil microemulsions stabilised by the surfactant Aerosol-OT (AOT). Since PEO is insoluble in the oil (n-heptane), the polymer is confined within the aqueous microphase. The phase stability of the resulting microemulsions is significantly affected by the concentration and molecular weight of the polymer. At fixed mass concentration of the polymer relative to the aqueous phase, any structural or phase behaviour changes as a function of molecular weight are primarily entropic. We have used small-angle neutron scattering (SANS) and conductivity measurements to show that the polymer (PEO) can induce droplet aggregation, with increased polydispersity at higher volume fractions of the aqueous microphase, and critical-like scattering on approaching the upper-temperature phase boundary. At $R = $ [Water]/[AOT] $= 40$, the polymer-free microemulsion shows critical scattering, with the correlation length, $\xi$, diverging as the cloudpoint, $Tc$, is approached. It appears that when the polymer radius of gyration, $Rg$, is large compared to the water droplet radius, $r$, the polymer chains are only solubilised when the temperature is sufficiently high so that $\xi \approx Rg$. The critical exponents extracted from the data ($\nu = 0.67 \pm 0.08$ and $\gamma = 1.25 \pm 0.16$) are consistent with the 3-dimensional Ising model. Thus, the presence of polymer in these microemulsions does not affect their universality class.

*Key words:* Polyethylene oxide; aerosol-OT; small-angle neutron scattering; critical phenomena; polymer-containing microemulsions; conductivity

## Introduction

Polymer-microemulsion interactions are important because of various industrial applications, for example in enhanced oil recovery where the polymers are added to reduce drag. However, despite the fairly advanced state of the theory of polymer chains, and recent theoretical advances in the field of microemulsions, polymer-microemulsion interactions in which the polymer chains are confined within the dispersed phase have received relatively little theoretical attention. On the experimental side most of the work on polymer-microemulsion systems was mainly for the application in enhanced oil recovery (EOR) [1]. There have been several investigations of protein containing systems notably that of Luisi and co-workers and Chen and co-workers [2]. Even though with addition

of proteins the phase diagrams change, there is as yet no adequate theoretical understanding of these shifts. In fact, the focus was mainly on the solubilised proteins themselves i.e. their conformations and activities. There are a number of complexities which any theoretical treatment of polymer-microemulsion interactions has to address, some of which are:

a) in the random mixing approximation of microemulsions, the entropy of mixing is considered to be that of the Flory-Huggins type. With addition of polymers confined to the droplet phase, the entropy of polymer is reduced but also the entropy of mixing of the microemulsion is affected. This is because a polymer that is too large for a single droplet may induce coagulation among the droplets so as to relax its own configuration.

b) The polymer can, in addition, affect the bending energy and preferred curvature of the interface through specific interactions or by adsorption at the oil/surfactant/water interface.

To treat two differing species, one of which performs a random walk under confinement and the other with complicated self-assembling behaviour requires a detailed understanding of both the specific interactions and self-assembling behaviour in the presence of a "field". Work has already been done on this for polymers outside the droplets [3, 4].

## Experimental

### Sample preparation

Polyethylene oxide, PEO (with Mw/Mn = 1.05) was obtained from Polymer Laboratories Ltd, Church Stretton, U.K. Haptane was Analar grade purchased from BDH Chemicals Ltd., Poole, England. Deuterated water was 99.8% atom $D$ bought from Aldrich Chemical Co., Gillingham, England. The surfactant AOT (di-2-ethylhexyl sodium sulphosuccinate) was obtained from Sigma Chem. Co. Ltd, Poole, England. Known amounts of PEO were first weighed to 0.1 mg accuracy. To the PEO was then added the appropriate amount of $D_2O$ and the polymer allowed to dissolve. A known volume of a stock solution of 0.2 mol $dm^{-3}$ AOT was added, and finally heptane to fill up to 10 ml total volume. The phase diagrams were determined using the same sample employed in the SANS measurements with two temperature scans only to determine the cloud point, preventing possible decomposition at high temperature. Repeat of the phase diagram measurements using the same stock solution showed the cloud point to be accurate to within 0.5 °C.

## Instruments

The SANS experiments were conducted mainly on D17 at the ILL, Grenoble. The $Q$ range was obtained using $\lambda = 15$ Å with $\Delta\lambda/\lambda \approx 10\%$, with a sample to detector distance of 2.8 m. The sample changer was thermostated with a water bath (Haake) available at the D17 facility. 1 mm path length quartz cells (Hellma (England) Ltd., Essex) were used for the SANS measurements and the sample temperature was kept constant to within 0.2 °C. Conductivity measurements used stainless steel electrodes at a spacing of about 3 mm and the cell was a calibrated 10 ml flask. The bridge used was a Model 5007 Kent Industrial Measurements Portable Conductivity meter with 2 frequencies — 70 Hz was used for low conduction and 1 kHz for higher conductions. Both frequencies are sufficiently low and the conductivity may be regarded as frequency independent. The cell constant was determined using salt solutions of known conductivity. Triply distilled water was used for these measurements.

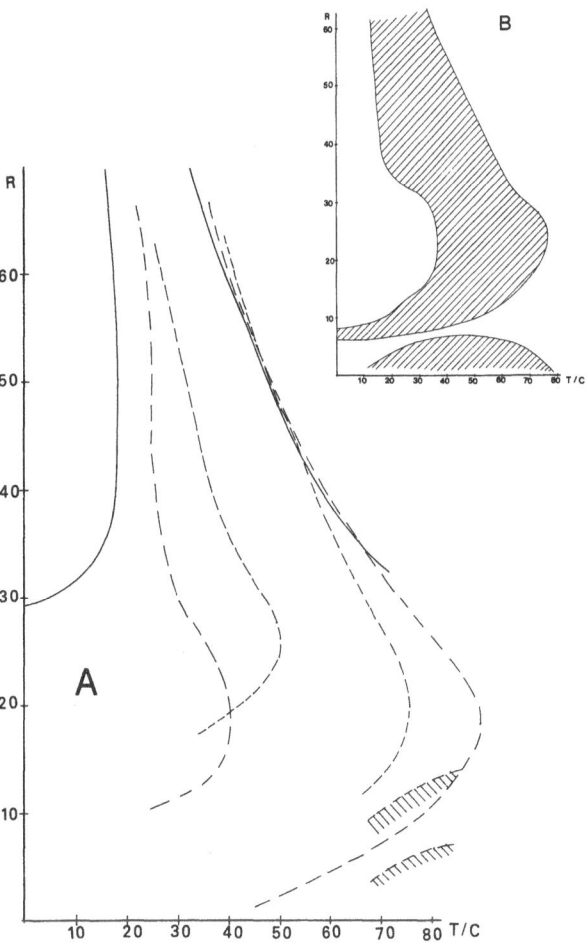

Fig. 1. A) Phase diagram of $D_2O$/AOT/n-heptane at 0.1 mol $dm^{-3}$ AOT where $R = [D_2O]/[AOT]$. a) Polymer-free system (——). b) $D_2O$/AOT/n-heptane system containing PEO of M 7.1 × 10³ at a concentration of 2% in the aqueous phase (—·—·—). c) $D_2O$/AOT/n-heptane system containing PEO of M 105 × 10³ at a concentration of 2% in the aqueous phase (— — —). The microemulsion region is the area between the two curves in each case. The shaded area in systems (b) and (c) denotes the re-entrant microemulsion phase.
B) Inset: Phase diagram of $H_2O$/AOT/n-heptane at 0.1 mol $dm^{-3}$ AOT with 1.5% PEO (M = 7.1 × 10³) in the aqueous phase. The shaded area denotes the microemulsion region. The re-entrant microemulsion phase can be seen at low values of $R$

## Results and discussion

### Phase behaviour

For a ternary system such as water/AOT/alkane it is convenient to construct a pseudo-binary phase diagram by fixing the concentration of the surfactant and varying the temperature and molar ratio, $R$, of

Table 1

Fits to Ornstein-Zernicke structure function

| Samples | Radius (Å) | $\xi$ (Å) | $\varkappa$ | Temperature (°C) |
|---|---|---|---|---|
| *R = 40 systems* | | | | |
| R = 40 | 56 ± 1 | 189 ± 11 | 6.3 ± 0.7 | 55.0 |
| | 56 ± 2 | 246 ± 20 | 10.4 ± 1.5 | 56.4 |
| | 55 ± 3 | 372 ± 66 | 24.7 ± 7.8 | 58.6 |
| PEO 1.08 × 10³ | 59 ± 1 | 122 ± 5 | 2.5 ± 0.2 | 49.4 |
| | 57 ± 1 | 183 ± 13 | 6.7 ± 0.9 | 52.9 |
| | 56 ± 2 | 232 ± 32 | 11.8 ± 3 | 55.0 |
| | 56 ± 3 | 302 ± 108 | 17.1 ± 10 | 56.4 |
| PEO 12.6 × 10³ | 59 ± 2 | 118 ± 5 | 3.0 ± 0.2 | 52.9 |
| | 58 ± 2 | 148 ± 7 | 5.3 ± 0.5 | 55.0 |
| | 58 ± 2 | 179 ± 12 | 7.5 ± 0.9 | 56.4 |
| | 57 ± 3 | 267 ± 20 | 17.9 ± 4 | 58.6 |
| | 56 ± 4 | 287 ± 116 | 17 ± 11 | 60.4 |
| PEO 56.3 × 10³ | 58 ± 2 | 98.6 ± 4 | 3.3 ± 0.2 | 52.9 |
| | 57 ± 2 | 121 ± 6 | 4.9 ± 0.4 | 55.0 |
| | 56 ± 3 | 144 ± 9 | 6.5 ± 0.8 | 56.4 |
| | 55 ± 3 | 196 ± 21 | 12.2 ± 2.4 | 58.6 |
| | 54 ± 3 | 296 ± 79 | 26.7 ± 12 | 60.4 |
| PEO 105 × 10³ | 57 ± 1 | 107 ± 4 | 4.3 ± 0.3 | 52.9 |
| | 56 ± 2 | 128 ± 7 | 6.0 ± 0.6 | 55.0 |
| | 55 ± 2 | 151 ± 10 | 8.2 ± 1 | 56.4 |
| | 54 ± 3 | 225 ± 17 | 17 ± 4 | 58.6 |
| | 54 ± 4 | 318 ± 129 | 29.2 ± 19 | 60.4 |
| *R = 60 systems* | | | | |
| PEO 56.3 × 10³ | 65 ± 2 | 61 ± 3 | 4.0 ± 0.4 | 40.9 |
| | 66 ± 3 | 78 ± 7 | 4.1 ± 5 | 43.2 |
| | 65 ± 2 | 127 ± 34 | 6.2 ± 2.8 | 45.7 |
| PEO 105 × 10³ | 65 ± 2 | 64 ± 4 | 4.3 ± 0.5 | 40.9 |
| | 68 ± 2 | 103 ± 10 | 3.8 ± 0.7 | 43.2 |
| | 65 ± 2 | 124 ± 58 | 6.5 ± 4.6 | 45.7 |
| PEO 145 × 10³ | 67 ± 2 | 101 ± 10 | 3.7 ± 0.4 | 43.2 |
| | 65 ± 2 | 123 ± 69 | 6.2 ± 5 | 45.7 |

Fits to Gaussian distribution of spheres

| | Mean radius (Å) | $\sigma$ (Å) | Temperature (°C) |
|---|---|---|---|
| R = 60 | 84 ± 2 | 16 ± 1 | 22.4 |
| | 79 ± 2 | 11 ± 1 | 28.1 |
| | 81 ± 2 | 14 ± 1 | 29.9 |
| | 83 ± 3 | 19 ± 2 | 32.0 |
| | 90 ± 4 | 31 ± 3 | 34.7 |

the water to surfactant. $R$ is a useful structural parameter as it is, to a first approximation, linearly related to the droplet radius, $r$. In polymer containing microemulsions, similar phase diagrams can be constructed at fixed polymer/water ratio. The effect of adding PEO into the water/AOT/heptane microemulsion at 0.1 mol dm$^{-3}$ AOT is shown in Fig. 1 as a function of polymer molecular weight ($M$). The area between the two boundaries represents the isotropic one-phase microemulsion domain. Beyond these boundaries the system separates to two or three phases. Below the lower phase boundary a small water-rich lower phase exists in equilibrium with a droplet upper phase. Above the higher phase boundary, two or three phases are observed depending on the composition and temperature. At $R = 40$, on phase separation two clear isotropic phases are obtained with roughly equal volumes. While the upper phase boundary changes little when polymers are added, the lower-temperature phase boundary shifts systematically to higher temperatures with increasing $M$, and the microemulsion "window" is thus significantly reduced for a given $R$. A further feature of the phase behaviour is that for certain molecular weights of PEO, and in particular $M = 7 \times 10^3$, there appears to be a "re-entrant phase", i.e. an additional clear microemulsion phase at low $R$ whose nature is not adequately understood. When water is replaced with D$_2$O, there is a boundary shift to higher temperature as reported before in the absence of polymer [6]. This shift is particularly pronounced in the case of the re-entrant region.

*Small-angle neutron scattering (SANS) and conductivity*

SANS measurements were carried out at $R = 40$ and $R = 60$ both with and without PEO in the aqueous phase and always in the one-phase region of the phase diagram. The polymer-free microemulsion at $R = 40$ shows critical scattering at low $Q$ as the upper cloud point is approached. This behaviour is consistent with the results of previous investigations on the AOT system [5, 6]. The data are well described by an expression for the scattered intensity, $I(Q)$, of the form,

$$I(Q) = F(Q,r)S(Q) \qquad (1)$$

where $F(Q,r)$ is the single-particle form factor for a sphere of radius $r$ and $S(Q)$ is the Ornstein-Zernicke function $S(Q) = 1 + \dfrac{\varkappa}{1 + Q^2 \xi^2}$ where $\xi$ is the correlation length. Very good fits can be obtained with Eq. (1) for all measurements at $R = 40$ (see Fig. 2A). In the polymer-containing systems, in particular, it is also possible to obtain equally good fits using Texeira's [7] model of fractal aggregation of spheres (see

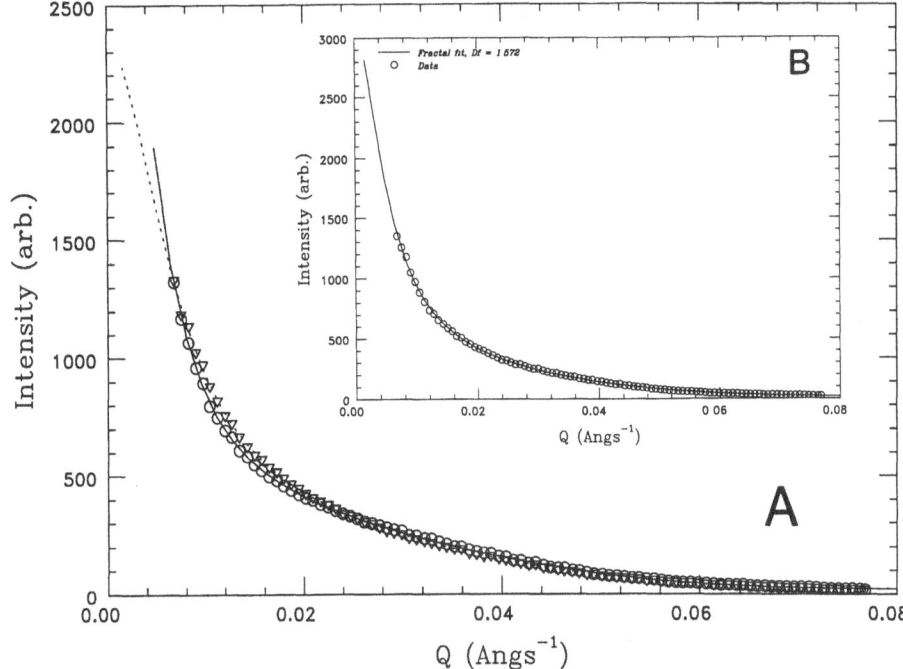

Fig. 2. A) SANS profiles of $D_2O$/AOT/n-heptane microemulsions with 0.1 mol $dm^{-3}$ AOT a) $R = 40$ at 56.4 °C with fitted parameters $r = 56 \pm 2$ Å, $\xi = 246 \pm 20$ Å and $\varkappa = 10.4 \pm 1.5$ (○)and b) $R = 40$ with 2% PEO 56300 in the aqueous phase at the same temperature as (a) giving $r = 56 \pm 2$ Å, $\xi = 144 \pm 9$ Å and $\varkappa = 6.5 \pm 0.8$. (▽) B) Inset: Fractal fit on sample $R = 40$ with 2% PEO 105000 in the aqueous phase at 56.4 °C giving $r = 50 \pm 2$ Å, $\xi = 233 \pm 10$ Å and $D_f = 1.57 \pm 0.05$

Fig. 2B). Then the structure factor in Eq. (1) takes the form,

$$S(Q) = 1 + \frac{1}{(Qr)^D}\frac{D\,\Gamma(D-1)}{(1 + (1/Q^2\xi^2))^{(D-1)/2}}$$
$$\times \sin[(D-1)\arctan(Q\xi)] \qquad (2)$$

where $r$ is the radius of the spherical droplets, $D$ is the fractal dimension of the aggregates and $\Gamma$ is the Gamma function.

The polymer free microemulsions at $R = 60$ on the other hand, do not show a sharp increase in intensity at low $Q$ (Fig. 3A), and no satisfactory fits can be obtained with Eqs. (1) or (2). The SANS data are best described by a model of polydisperse spheres, i.e. $I(Q,r) = \int F(Q,r)p(r)dr$, where $p(r)$ is a Guassian distribution function with the polydispersity increasing with temperature. This is due to the fact that unlike the $R = 40$ system which appears to be close to a critical composition, at $R = 60$ the microemulsion is far from the critical point. Thus significant aggregation seems to occur only very close to the cloud point in this system. Addition of PEO to the $R = 60$ microemulsion however, induces substantial aggregation, so that at the same reduced temperature, the polymer-containing microemulsion shows critical scattering unlike its polymer-free counterpart (Fig. 3B). When the polymer is added at $R = 40$, there is no substantial qualitative change in the scattering profile, but the correlation lengths extracted from the data are reduced (Fig. 2A).

The conductivity data similarly demonstrate a marked difference between the $R = 40$ and $R = 60$ systems (Fig. 4). At $R = 40$ the conductivity, $K$, is highest in the absence of polymer, although the polymer-containing microemulsions follow the same trend as the polymer-free system (Fig. 4A). At $R = 60$, however, $K$ is significantly higher in the polymer containing systems, while the polymer-free microemulsion shows a much reduced conductivity.

While it is difficult to describe polymer-containing microemulsions in detail without an adequate theory, our experiments suggest at least some qualitative structural features. The problem of confining polymer chains inside aqueous droplets is essentially an entropic one. At fixed mass concentration of PEO relative to the aqueous phase, increasing the molecular weight of the polymer is bound to cause pronounced structural changes in the microemulsion when the polymer radius of gyration, $R_g$, becomes comparable to the droplet radius, $r$. The situation might be improved by clustering of the droplets which produces much larger aggregates. This seems to be borne out by the phase behaviour. On adding PEO, large shifts are observed in the lower phase boundary for low values of $R$ (20 to 40). It appears that at these lower values of $R$, the polymer chains are only solubilised when the cluster size of the polymer free microemulsion is suf-

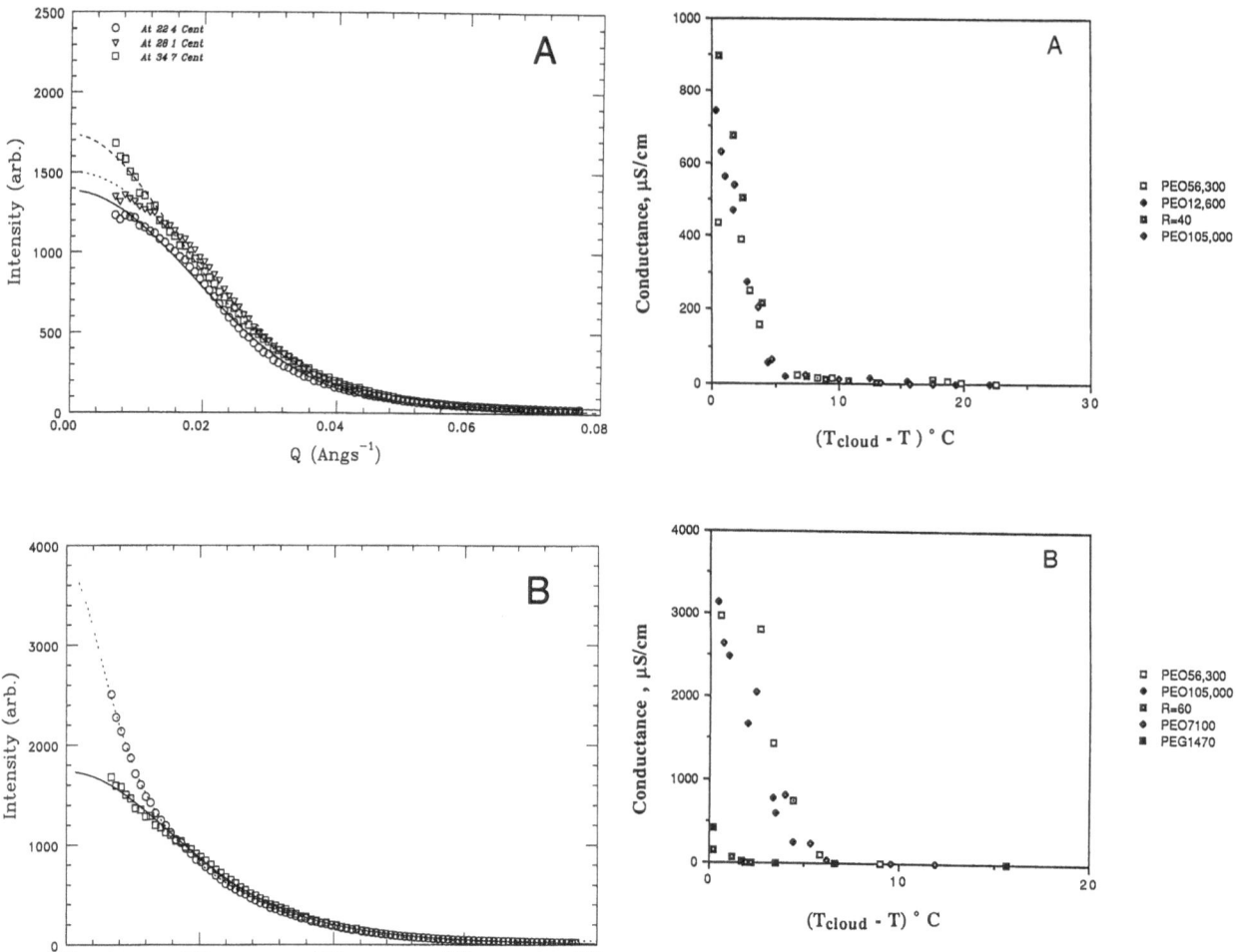

Fig. 3. A) SANS profile of $D_2O$/AOT/n-heptane microemulsion with 0.1 mol dm$^{-3}$ AOT and $R = 60$ at a) 22 °C (◯), b) 28.1 °C (▽), and c) 34.7 °C (□). The curves through the experimental points are least squares fits to a Gaussian distribution of spheres.
B) SANS profile of $D_2O$/AOT/n-heptane microemulsion with 0.1 mol dm$^{-3}$ AOT and $R = 60$ a) without polymer (□) and b) with PEO of M 12.6 × 10$^3$ at a concentration of 2% in the aqueous phase and at the same reduced temperature $\varepsilon = (1.45 \pm 0.2) \times 10^{-3}$ (◯). The curves through the points are least squares fits to Eq. (1) in (a) and a Gaussian distribution of spheres in (b)

Fig. 4. Conductivity ($\mu S$/cm) of $D_2O$/AOT/n-heptane microemulsion with 0.1 mol dm$^{-3}$ AOT as a function of temperature for different molecular weights of PEO. The concentration of the polymer in the aqueous phase is 2% in each case. A) At $R = 40$ and B) at $R = 60$

ficiently large to accommodate them. Since the droplets are smaller at low values of $R$, large temperature shifts are required to increase $\xi$ sufficiently (so that $\xi \approx R_g$). At high values of $R$ (60 and above) the increased mean droplet size and polydispersity facilitate the solubilisation of the chains so the temperature shifts in the phase diagram are less pronounced. The chains, however, may still cause droplet aggregation since their confinement is much less severe in a cluster than in a single droplet. When polymer chains are added to an already aggregating system (as in the case of $R = 40$) they reduce the cluster size and increase the apparent fractal dimension of the aggregate. It appears the polymer chains impose a more compact structure on the aggregate, and this effect is more pronounced at higher molecular weights.

Since the $R = 40$ system appears to be close to a critical composition it may be appropriate to attempt an evaluation of critical exponents from the SANS data. If we assume that the cloud point, $T_c$, of this system is close to the critical point we may use the familiar expressions for the critical parameters as a function of reduced temperature, $\varepsilon$, i.e.

$$\xi = \xi_0 \varepsilon^{-\nu} \quad \text{and} \quad \varkappa = \varkappa_0 \varepsilon^{-\gamma}$$

$$\text{where} \quad \varepsilon = (T_c - T)/T_c.$$

R=40 with 2 % polyethylene oxide

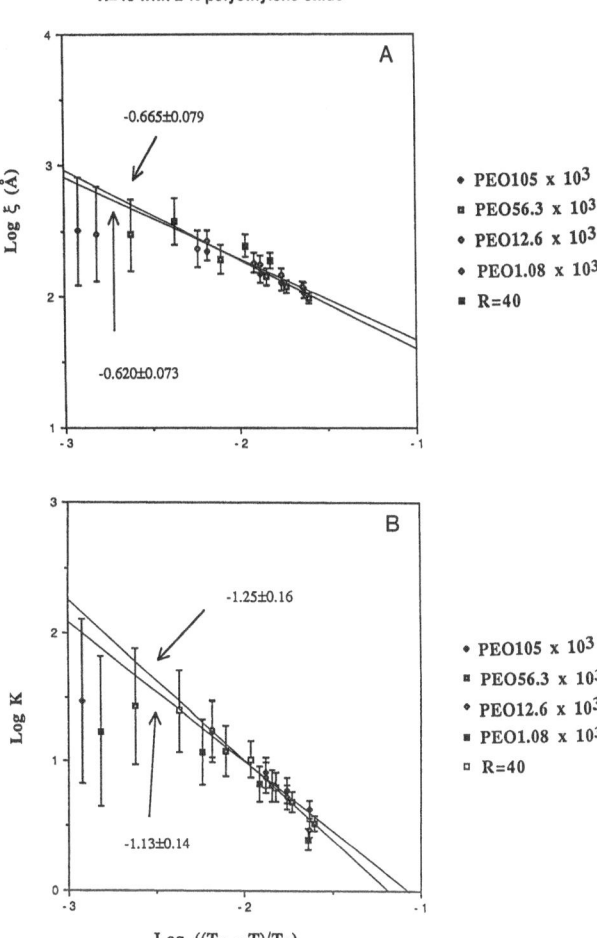

Fig. 5. Plot of $\log \xi$ (A) and $\log \varkappa$ (B) against $\log \varepsilon$ where $\varepsilon = (T_c - T)/T_c$. The gradients of the straight lines, obtained with a least squares fit, yield $v$ and $\gamma$ respectively. Two slopes are given, one which includes all the points and another which ignores the first 3 points

The results shown in Fig. 5 appear to justify this approach for the $R = 40$ system, with the exponents extracted from the data ($v = 0.62 \pm 0.073$, $\xi_0 = 11 \pm 3.5$ Å, and $\gamma = 1.13 \pm 0.14$) consistent with the Ising model. The samples associated with the first three points are very close to $T_c$ (within 1 °C) and consequently carry very large errors under the temperature control conditions of our experiments. Further-

more, the $\xi$ values of these samples probably tend to be underestimated due to the lack of data at very low $Q$. Neglecting the first 3 points the fitted values are $v = 0.665 \pm 0.079$, $\xi_0 = 9.3 \pm 1.4$ Å, and $\gamma = 1.25 \pm 0.16$. Ising exponents have already been reported in the literature [5, 6] for AOT microemulsions. It appears, therefore, that the presence of polymer in these microemulsions does not affect their universality class. Although the SANS data of the polymer-containing microemulsions at $R = 60$ can be fitted to Eq. (1) using the Ornstein-Zernicke function for $S(Q)$, these systems do not yield any meaningful exponents as they are far from the critical composition. The data are summarised in Table 1. We have also considered the possibility of the formation of interconnected cylindrical aqueous conduits induced by the solubilisation of the PEO chains. This type of microemulsion structure could be consistent with our conductivity data. The SANS data appear to rule out such a model, however, as a cylindrical form factor could not be fitted to the experimental points within our $Q$-range for any of the systems reported here. Finally, at very low values of $R$ (less than 10) there is a "re-entrant" microemulsion phase in some systems (see Fig. 1B). The structure of this phase is not well-understood.

### References

1. Shah DO, Schechter RS (1977) (eds, 1977) Improved Oil Recovery by Surfactant and Polymer Flooding, Academic Press
2. Caselli M, Luisi PL, Maestro M, Roselli R (1988) J Phys Chem 92:3899−3905; Battistel E, Luisi PL, Rialdi G (1988) J Phys Chem 92:6680−6685; Bratko D, Luzar A, Chen SH (1988) J Chem Phys 89(1):545
3. Nagarajan R (1989) J Chem Phys 90(3):1980
4. Siano DB, Bock J (1982) J Colloid Int Sci 90(2):477
5. Kotlarchyk M, Chen SH, Huang JS (1983) Phys Rev A 28:508
6. Toprakcioglu C, Dore JC, Robinson BH, Howe A, Chieux P (1984) J Chem Soc Faraday Trans I 80:413
7. Texeira J (1986) In: Stanley HE, Ostrowsky N (eds) On Growth and Form, Martinus Nijhoff Publishers

Authors' address:

Dr. C. Toprakcioglu
Cavendish Laboratory
Madingley Rd.
Cambridge, CB3 0HE, UK

Progress in Colloid & Polymer Science                    Progr Colloid Polym Sci 81:60—63 (1990)

# Microemulsion shape fluctuation measured by neutron spin echo

B. Farago[1], J. S. Huang[2], D. Richter[3], S. A. Safran[2] and S. T. Milner[2]

[1]) Institute Laue Langevin, Grenoble, France
[2]) Exxon Research and Engineering, Annandale, New Jersey, USA
[3]) KFA Julich, Julich, FRG

*Abstract:* Neutron spin echo (NSE) and small angle neutron scattering (SANS) technique were employed to study the shape fluctuations of di-2-ethylhexyl sulfosuccinate (AOT)/water/decane microemulsions. We show that NSE can directly measure the shape fluctuations of these water-in-oil droplets. A comparison of the dynamics and static measurements yields both the splay modulus $K$, as well as the saddle-splay modulus $\bar{K}$. With the addition of butanol as cosurfactant, softening of the interface was observed.

*Key words:* Microemulsions; being elasticity; small angle neutron scattering; neutron spin echo

## Introduction

Microemulsions can form a great variety of structures: droplets, tubes, lamellar or more complicated bicontinuous structures. Although there are many experimental probes for the static structure, the dynamics of the system is less well understood. The basic structure can be studied using X-ray, synchroton radiation, SANS, but to access at the same time the dynamics we need the highest resolution neutron spectroscopy available: neutron spin echo.

## Experimental

The AOT-water-decane model system was chosen for several reasons:

1. It is well established that a water droplet structure exists as a large single phase region in the phase diagram [3, 4] without any cosurfactant.
2. The droplets are reasonably monodisperse, with a polydispersity $\approx 20\%$.
3. The interaction between the droplets, due to the nonpolar oil continuous phase, is mainly hard sphere repulsion plus a short ranged attractive interaction [5]. At low ($\approx 5\%$) volume fraction of the dispersed phase (water + surfactant) the scattering is only weakly affected by the structure factor ($S(q)$). Indeed at this concentration $S(q)$ is simply a smooth increasing function toward $q = 0$ and can be approximated as 1 for $q > 0.03$ Å$^{-1}$.

Most of the samples were composed of deuterated decane, normal (protonated) AOT and D$_2$O. Since the scattering length density of deuterated decane and D$_2$O are nearly the same, only the surfactant shell pariciptes in the scattering. In this way most of the scattering comes from the volume, which undergoes the shape fluctuation. To make sure that there is no spurious $S(q)$ effect, one sample was prepared with H$_2$O too. Since this case the scattering comes from the full volume (sphere) the fluctuations should be less prominent and should show up at in a different range of wavevector. On the other hand $S(q)$ is the same for both sample as it is not influenced by the contrast.

The spin echo measurements were performed in the ILL on the IN11 spectrometer with wavelengths of $\lambda = 11.5$ Å and $\lambda = 8$ Å, $\Delta\lambda/\lambda = 18\%$, and the SANS spectra on the D11 instrument with $\lambda = 8$ Å $\Delta\lambda/\lambda = 10\%$.

## Results and dicussion

With NSE the normalized intermediate scattering function $I(q,t)$ is measured directly [6]. For a fluctuating droplet of radius $R$ this can be written as:

$$I(q,t) = \exp(-D_{tr}q^2 t) V_s^2 (\Delta\varrho)^2$$
$$\times \Bigg[ f_0(qR)$$
$$+ \sum_{l>2} \frac{2l+1}{4\pi R^2} f_l(qR) \langle |u_l|^2 \rangle \exp(-t/\tau_l) \Bigg] \quad (1)$$

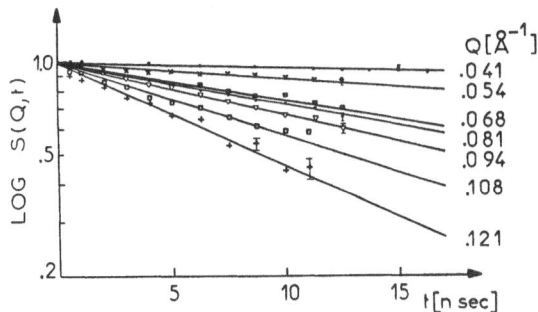

Fig. 1. NSE spectra $[I(q,t)/I(q,0)]$ as measured for the S33 sample (see Table 1). There is no strong visible deviation from exponential decay

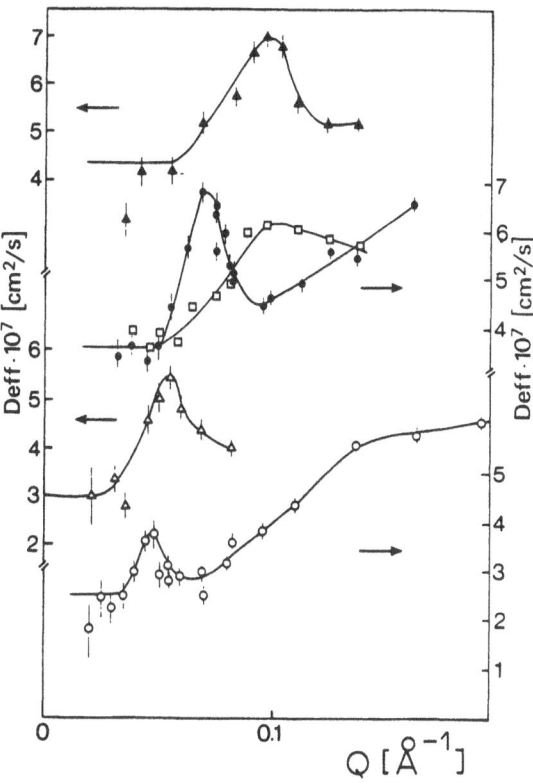

Fig. 2. $D_{eff}(q)$ for the different size shell contrast samples and for one sphere contrast (shell samples: S32 (▲), S33 (●), S34 (△), S35 (○); and in sphere contrast the S33 (□) sample). The lines are only guides for the eye

The scattering wavevector is expressed in terms of the neutron wavelength $\lambda$ and scattering angle $\theta$ as $q = 4\pi/\lambda \sin(\theta/2)$. $D_{tr}$ is the tracer diffusion coefficient, $V_s$ the scattering volume (proportional to $R^2$ for the shell and $R^3$ for the sphere), $\Delta\varrho$ the scattering length density contrast.

Equation (1) can be derived by expanding the shape fluctuations into spherical harmonics [7, 8] with dimensionless amplitudes $u_l$. Thereby $f_0(qR)$ is the static form factor of the droplet: $f_0(qR) = \langle[3j_1(qR)/(qr)]^2\rangle$ for the sphere and $f_0(qR) = \langle[\sin(qR)/qR]^2\rangle$ for the shell; the inelastic form factors are $f_l(qR) = [3j_l(qR)]^2$ for the sphere and $f_l(qR) = [(l + 2)j_l(qR) - (qR)j_{l+1}(qR)]^2$ for the shell. $j_l(x)$ stands for the spherical Bessel function of order $l$, the brackets $\langle\rangle$ means ensemble average. The expressions for the shell suppose an infinitely thin layer thickness. For the fitting of both the SANS and NSE data a more accurate expression was used taking into account the finite thickness of the interface $\approx 12$ Å. Due to the viscosity of the inner and outer liquid the shape fluctuation will show up as overdamped motion, the time constant of mode $l$ being $\tau_l$.

Although $I(q,t)$ is a sum of exponentials, it did not show a very strong deviation from a single exponential decay, so to visualize the NSE data we fit $\exp(-D_{eff}q^2t)$ to the measured points (typically $10 - 12$ points in the $0 - 18$ ns range) thereby defining $D_{eff}(q)$ (Fig. 1). In the absence of shape fluctuation $D_{eff}(q)$ is equal to $\langle D_{tr}\rangle$, the average taken over the size distribution.

The pronounced peak structure in Fig. 2 can be understood as follows: at small $q$ ($qR \ll 1$) we have $f_0(qR) \gg f_l(qR)(l > 1)$ and only the translational diffusion is observed. At $qR \approx \pi$, in the case of shell contrast, $f_0(qR)$ has a minimum and the lowest $l = 2$ mode form factor has a maximum. At $qR > \pi$ $f_0(qR)$

increases again, while $f_2(qR)$ decreases, and this interplay between the form factors results in the observed maximum in $D_{eff}(q)$. The same effect takes place but at higher $q$ for the sphere contrast too. The experimental findings (Fig. 2) correspond to these expectations. The position of the maximum for the different size shell samples scales with the radius and for the sphere contrast sample it shows up at higher $q$. From this we see that the peak in $D_{eff}(q)$ is most sensitive to the shape fluctuations, which come essentially from the $l = 2$ mode.

In our previous publication [9] we showed that the ratio of the peak height to the value of $D_{eff}(q = 0)$ scales with $R$, consistent with a restoring force which originates from the bending elasticity and not the surface tension. Milner and Safran [7] calculated the fluctuation amplitudes $u_l$ and the decay times $\tau_l$ with the assumption of: 1. fixed total surface; 2. fixed total volume of the droplets; 3. a given spontaneous curvature ($R_s$); 4. the splay bending energy as the main contribution to the free energy. The theory also related the width of the size distribution being the $l = 0$ mode,

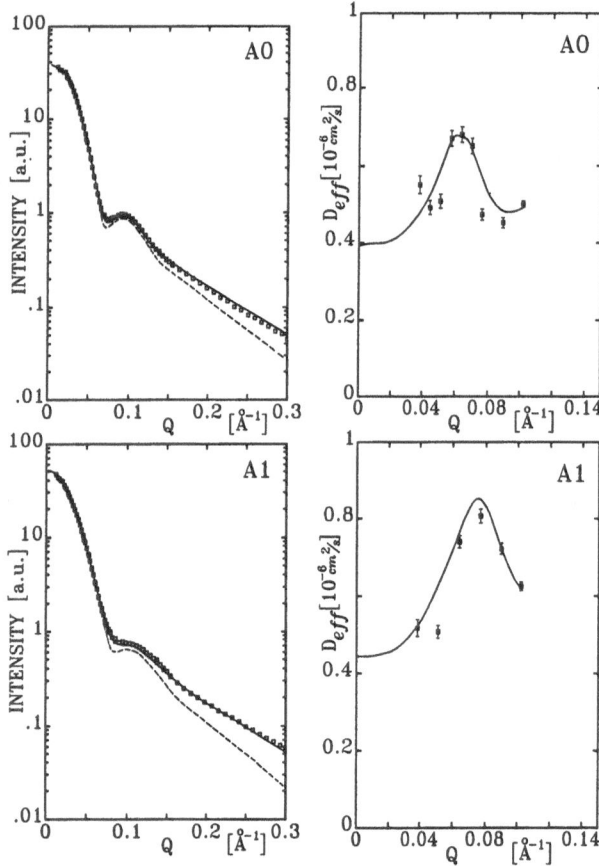

Table 1. The composition of the samples on Fig. 1 and the parameters determined from SANS. The NSE data could be fitted with a common $K = 2kT$ and $\bar{K}/K = -1.88$

|  | $R_{in}$ [Å] | $Z$ | [D$_2$O/AOT] molar ratio |
|---|---|---|---|
| S 32 | 24.7 | 18 | 16.3 |
| S 33 | 38 | 18.7 | 24.5 |
| S 34 | 49.8 | 16 | 32.6 |
| S 35 | 62 | 16.2 | 40.8 |

Fig. 3. SANS and NSE spectra of the alcohol free and butanol containing spectra. The compositions are given in Table 2. The increase in polydispersity with butanol is clearly visible, with a simultaneous increase of floppiness

$u_0$. The observed maximum drop size is 100 Å; this implies the spontaneous curvature $R_s \approx 100$ Å (for $R > R_s$ there is an emulsification failure instability [10]). With these inputs we fit with a single parameter, the splay bending constant ($K$), the measured polydispersity of the SANS and the $l = 2$ mode dynamical fluctuation seen by NSE. It turns out that to describe the polydispersity we had to take $K = 0.5\, kT$ which contradicted the NSE fit which gave $K \approx 5\, kT$.

The results can be reconciled if one includes the saddle-splay bending energy. The full expression for the bending elastic energy is:

$$E_{bend} = \frac{K}{2} \int dS \left( \frac{1}{R_1} + \frac{1}{R_2} - \frac{2}{R_s} \right) + \bar{K} \int dS \frac{1}{R_1 R_2}. \tag{2}$$

Where $R_1$ and $R_2$ are the two radii of curvature, $R_s$

is the spontaneous or preferred curvature. The second term, correspondingly to the Gaussian curvature or saddle-splay elasticity, was previously neglected since it gives a constant energy contribution when the number of the droplets is fix (Gauss-Bonett theorem). However in this self-assembling system, the number of drops is not fixed; only the total surface and volume are constrained by the conservation of water and surfactant. Consequently the contribution of the Gaussian curvature is not constant. Details of the modified theory will be given elsewhere [11], here we summarize the results for the $l = 0$ and $l = 2$ mode:

$$\langle |u_0|^2 \rangle = \frac{kT}{2K(6 - A)} \tag{3}$$

$$\langle |u_2| \rangle = \frac{kT}{4KA} \qquad \tau_2 = \frac{\eta R^3}{K} \frac{2.29}{A} \tag{4a,b}$$

where

$$A = 4\frac{R}{R_s} - 3\frac{\bar{K}}{K}.$$

The $l = 0$ mode, as measured by SANS, is the most conveniently parametrized by the Schultz distribution [12] where the width parameter $Z$ is equivalent to

$$Z = \frac{4\pi}{u_0^2} - 1. \tag{5}$$

Fitting the SANS data for the different size shell samples, we found that the polydispersity depends very weakly, if any, on the droplet size (Table 1). That implies that $A$ in Eqs. (3) and (4a,b) depends only weakly on $R$, consequently $R_s \gg R$ must hold. To describe simultaneously the relatively broad size distribution and the $l = 2$ mode the only possibility is if $K \approx 2kT$ and $A$ is close to 6, which implies $\bar{K}/K$ close to $-2$ (see Table 1). These values are in reasonable agreement with other measurement [13].

Table 2. The compositions and the results of the SANS and NSE fit for the buthanol free [A0] and the butanol containing sample [A1]

|     | $R_{in}$ [Å] | $Z$ | $K$ [$kT$] | $\bar{K}/K$ | [$D_2O$/AOT] molar ratio | $D_2O$ [cm³] | butanol [cm³] | d-decane [cm³] |
|-----|------|------|------|-------|------|------|------|------|
| A0  | 37.5 | 18.7 | 3.0  | −1.88 | 24.5 | 0.06 | 0    | 1.92 |
| A1  | 30.7 | 12.7 | 1.6  | −1.86 | 28.6 | 0.07 | 0.02 | 2.00 |

In a second series of experiment [11] we investigated the influence of butanol as cosurfactant on the interface rigidity. Fig. 3 compares the SANS and NSE data for an alcohol free (A0), and a buthanol containing (A1) sample. Our first observation was that to obtain drops with approximately the same radius we had to increase the water/AOT ratio. This means that the available surface in the system increased which is a direct indication that the butanol molecules go into the interface. Secondly there is a strong increase of polydispersity with the addition of alcohol. With no change of the $l = 2$ mode amplitude this would smear out the peak structure of the NSE data. However this is not what we observe; instead the peak height increases (Fig. 3). Thus independently of any model we can conclude that the rigidity is substantially reduced with the addition of the alcohol. If we go through the fitting procedure [11] we find from sample A0 to A1 (Table 2) the reduction of $K$ from $3kT$ to $1.6kT$ with only a slight change in the $\bar{K}/K$ ratio.

Finally we want to remark that for the butanol free sample (A0) we have found a somewhat bigger $K$ ($3kT$) than in our size series experiment ($2kT$). As in this case the AOT we used, was from a different batch, and also we could make under identical condition slightly bigger spheres (90 Å compared to 75 Å), we think that this discrepancy must have its origin in the purity of the samples.

## References

1. Helfrich W (1978) Z Naturforsch 33a:305
2. Safinya CR, Roux D, Smith GS, Sinha SK, Dimon P, Clark NA, Bellocq AM (1986) Phys Rev Lett 57:2718
3. Kotlarchyk M, Stephens RB, Huang JS (1988) J Phys Chem 92:1533
4. Kotlarchyk M, Chen SH, Huang JS, Kim MW (1984) Phys Rev A29:2054
5. Huang JS, Safran SA, Kim MW, Grest GS, Kotralchyk M, Quirke N (1984) Phys Rev Lett 53:592
6. Neutron Spin Echo edited by F. Mezei, Lecture Notes in Physics Vol. 128. Springer-Verlag, Heidelberg
7. Milner ST, Safran SA (1987) Phys Rev A36:4371
8. Safran SA (1983) J Chem Phys 78:2073
9. Huang JS, Milner ST, Farago B, Richter D (1987) Phys Rev Lett 59:2600
10. Safran SA, Turkevich LA (1983) Phys Rev Lett 50:1930
11. Farago B, Richter D, Huang JS, Safran SA to be published
12. Kotlarchyk M, Chen SH (1983) J Chem Phys 79:2461
13. Binks BP, Meunier J, Abillon O, Langevin D (1989) Langmuir 5:415

Authors' address:

Bela Farago
Institut Laue Langevin
156 X Cedex
38042 Grenoble, France

# Control of curvature in microemulsions: The AOT/Krypotofix couple

L. J. Magid, R. Weber, M. E. Leser and B. Farago

Department of Chemistry, University of Tennessee, Knoxville, USA and
Institut Laue-Langevin, Grenoble, France

*Abstract:* Aerosol OT (AOT) exhibits a pronounced tendency towards reverse curvature, forming w/o droplets in the $L_2$ region of the phase diagram in the three-component water/cyclohexane/AOT system. The net curvature of the interfacial layer can be tuned by complexing AOT's sodium ion with the cryptand Kryptofix 222 (Kr 222). With increasing Kryptofix, a structural evolution occurs from w/o droplets through a bicontinuous system to o/w droplets. This contribution focuses on the structural changes which occur in the $L_2$ phase upon addition of 0.25 equivalent of Kr 222. Results from PGSE $^1$H-FTNMR and static SANS measurements lead to the conclusion that the AOT droplets decrease in size and exhibit a significant increase in the droplet-droplet interaction when the Kr 222 is added.

*Key words:* Reverse micelles; AOT; cryptand; SANS; NMR

## Introduction

Recently Evans [1] and Magid [2] and coworkers studied the structural changes for certain anionic normal micelles in aqueous solutions, induced by the complexation of the counterions by the macrocyclic multidentate ligand 4,7,13,16,21,24-hexaoxa-1,10-diazabicyclo[8.8.8]hexacosane (Kryptofix 222, abreviated with Kr 222 or just KR). The observed micellar reoragnizations are rationalized mainly in terms of changes in head group repulsion which is manifested in an increased curvature of the interfacial layer and in a decrease in micellar aggregation numbers.

It is also of interest to ask how the addition of Kr 222 alters the curvature of the interfacial surfactant film in ternary systems such as sodium bis(2-ethylhexyl)-sulfosuccinate (AOT)/water/cyclohexane. Studies of the phase behavior with different amounts of Kr 222 suggest that the macrocycle acts as a "curvature adjuster", i.e., changes the net curvature of the interfacial layer dramatically. With increasing Kryptofix, which has a complexation constant of 5000 for $Na^+$ [3], a small $L_1$ region appears at the water corner of the pseudoternary phase diagram (see Fig. 1). The $L_1$ region continues to grow at the expense of the $L_2$; the $L_2$ disappears entirely at a ratio of 1.0. At that point there is a large $L_1$ region extending to high AOT con-

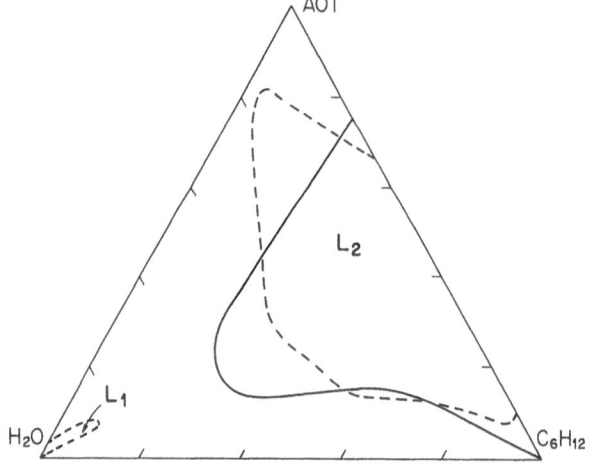

Fig. 1. Phase diagram at 25 °C of the system water/cyclohexane/AOT without (solid line) and with 0.25 equivalents of Kr 222 (dashed line). $L_2$ stands for w/o microemulsion, $L_1$ for o/w microemulsion

tent along the water-AOT edge of the pseudoternary phase diagram [4]. Such an evolution in phase behavior with increasing Kr 222 is certainly suggestive of a corresponding evolution in microstructure: w/o droplets to bicontinuity to o/w droplets.

In this work we will focus on the description of the $L_2$ region of the pseudoternary phase diagram when 0.25 equivalents of Kr 222 are present. The structural and dynamic changes in the microemulsion, induced by the added macrocycle, are studied by means of PGSE $^1$H-FTNMR and static SANS.

## Experimental and methods

*Phase behavior:* Phase boundaries were determined in tightly closed glass tubes. The solutions were mixed by vortex mixer and allowed to equilibrate in a water bath at 25°C for 30 minutes to 1 hour. The changes in phases (appearance or disappearance of turbidity) were followed by eye. $W_0$ is defined as the molar ratio of water to surfactant ($w_0 = [H_2O]/[AOT]$).

*PGSE $^1$H-FTNMR:* Pulsed-gradient spin-echo $^1$H Fourier transform NMR measurements were made on a Bruker WM-250 spectrometer at 25°C. The pulsed field gradient system used has been described in detail elsewhere [5]. The solvent used in the experiments was a 1/1 (v/v) mixture of $C_6H_{12}$ and $C_6D_{12}$.

*SANS:* Static Small-angle neutron scattering measurements were made at the low-angle spectrometer at the High Flux Beam Reactor of Brookhaven National Laboratory (BNL) and at the D11 spectrometer of the Institut Laue-Langevin (ILL), Grenoble. The $Q$ ranges employed ($Q$ is defined as $(4\pi/\lambda)\sin\theta$, where $2\theta$ is the scattering angle) were $0.009-0.32$ Å$^{-1}$ at BNL and $0.01-0.29$ Å$^{-1}$ at ILL. Each two-dimensional set of raw scattering data was corrected in a manner described elsewhere [6].

*Fitting of the scattering curves:* The SANS intensity from a polydispersed population of globular droplets (here modeled as spheres) can be given by:

$$I(Q) = N_p \bar{P}(Q)\bar{S}(Q). \tag{1}$$

A Shulz distribution $f(R)$ was used for the droplet radii [7]. $\bar{P}(Q)$ was obtained by integration of $|F(Q,R)|^2 f(R)$ over $R$. The two main contrasts discussed in this work for water/AOT/oil are h/h/d (sphere) and d/h/d (shell). A few spectra were also obtained using d/h/h contrast ($D_2O$, $C_6H_{12}$), in order to size directly the water pools of the droplets. In cases where all three contrasts were used, a self-consistent description was obtained: $R_{wp}$ from the d/h/h contrast plus d (thickness of the interfacial film) from d/h/d spectra agreed with the overall sphere radius from d/h/d and h/h/d spectra. This held true both with and without Kr 222.

The form factor for the spheres is given by:

$$F_{sph}(Q,R) = (4\pi/3)R^3 \Delta\varrho [3j_1(QR)/QR] \tag{2}$$

and for the shells by:

$$F_{sh}(Q,R) = (4\pi d)R^2 \Delta\varrho [\sin(QR)/QR]. \tag{3}$$

The shell form factor has a first minimum at $Q*R = \pi$. The depth of the minimum decreases with decreasing $Z$ (increasing polydispersity). Typical $Z$ values in the Shulz distribution were 20 (no Kr 222) and 16 (with Kr 222 present).

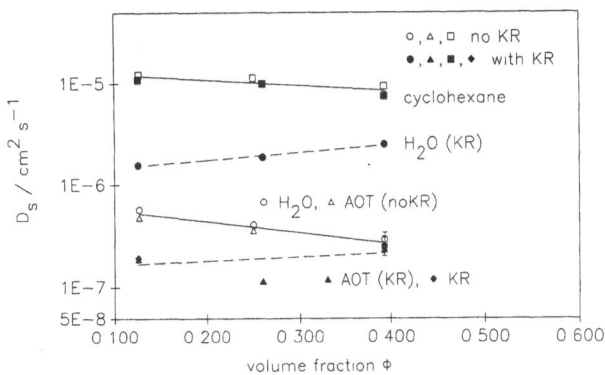

Fig. 2. Self-diffusion coefficients in the system water/cyclohexane/AOT without and with the addition of 0.25 equivalent Kr 222 as a function of the volume fraction of the dispersed phased; $w_0 = 17.5$

At the lowest volume fractions studied ($<0.1$), we expressed $S(Q)$ according to the decoupling approximation [8]. At higher volume fractions, the microemulsions with and without Kr 222 have h/h/d SANS curves with interaction peaks. The decoupling approximation does not adequately reproduce the peak positions. The shift in $Q_{max}$ to lower $Q$ with increasing polydispersity is not pronounced enough. For these samples we used instead the approach of Vrij [9, 10] to compute $S(Q)$.

*Materials:* AOT, $D_2O$ and $C_6D_{12}$ were obtained from Aldrich, Kr 222 from Sigma, $C_6H_{12}$ (highest purity available) from Brand.

## Results

### Phase behaviour

The extent of the $L_2$-phase (w/o microemulsion) is not drastically modified by the presence of 0.25 equivalents of Kr 222 (see Fig. 1). However, at relatively low AOT/Kryptofix content (up to a volume fraction of ca. $0.06-0.08$) no one phase system is observable at 25°C. Above this volume fraction one obtains a clear and stable one-phase w/o microemulsion. In the water-rich corner of the phase diagram a small clear area is observed which is thought to be an $L_1$-phase (w/o microemulsion).

### PGSE $^1$H-FTNMR

Measurement of the self-diffusion coefficients $D_s$ for water, cyclohexane, AOT and Kr 222 which were accomplished with the PGSE FTNMR method are exemplified in Fig. 2. In the system without Kr 222 the AOT and $H_2O$ have the same $D_s$ values, which strongly suggests that they are diffusing together in

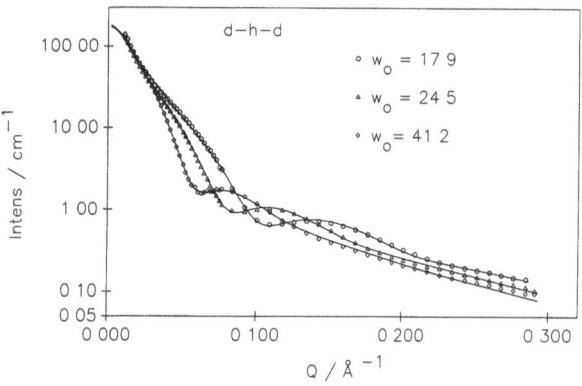

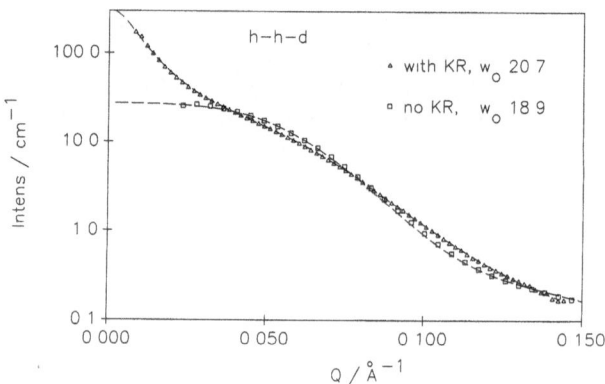

Fig. 3. SANS data (shell contrast d-h-d) of $D_2O$/AOT/0.25 eq of Kr 222/$C_6D_{12}$ microemulsions at different $w_0$ values. The volume fractions $\phi$ are 0.09 – 0.1. The solid lines represent the fits of the curves (see experimental)

Fig. 4. SANS data (sphere contrast h-h-d) and corresponding fits (solid lines) of $H_2O$/AOT/$C_6D_{12}$ without and with the addition of 0.25 equivalents of Kr 222 at a volume fraction of 0.088; estimated water pool radius plus interfacial film: without Kr 222: 36 Å; with Kr 222: 31 Å

the same aggregate. The observed diffusion coefficients are therefore a measure of the diffusion of the entire aggregate [11]. The observed values are in agreement with the result obtained by Stilbs and Lindman for a similar AOT system [12].

The addition of 0.25 equivalents of Kr 222, however, changes the situation dramatically: the AOT and the water no longer have the same $D_s$ values, and the added Kr 222 seems to diffuse with the AOT. This change in the microemulsion dynamics can be rationalized either with a formation of a bicontinuous structure at the expense of the spherical structures or with a strongly interacting droplet system. In this context we should also note that the conductivity in the microemulsion solution is increased by an order of about 3 magnitudes after the addition of 0.25 equivalents of Kr 222 (data not shown).

*Static small-angle neutron scattering (SANS)*

We present here the results of experiments carried out with two different scattering contrasts, which under the assumption of droplets correspond to shell contrast ($D_2O$/AOT/$C_6D_{12}$) and sphere contrast ($H_2O$/AOT/$C_6D_{12}$).

Inspection of the scattering patterns arising from $L_2$-region solutions of $D_2O$/AOT/$C_6D_{12}$ (shell contrast at various $w_0$ values) when Kr 222 is present reveals that spherical aggregates are present (see Fig. 3). The minima observed correspond to the first minimum in the form factor for a hollow shell, which occurs at $Q^*R = \pi$. $Q_{min}$ shifts to lower $Q$ as $w_0$ increases, consistent with the growth of the droplets. This could also be observed in the scattering patterns seen with the

second contrast. The fits obtained, shown as solid lines in Fig. 4, employed an interfacial film thickness of 8 Å.

The SANS data in Fig. 4 demonstrate that without Kr 222 the droplet interactions are those of hard spheres. With Kr 222 present, the system exhibits at low volume fraction and low $Q$ values a significant increase in the scattered intensity, consistent with substantial attractive interactions between the droplets. The estimated $S(0)$ values for a Kr 222 containing sample at the lowest volume fraction is $8-10$ times higher than the estimated $S(0)$ value for the system without Kryptofix. At a volume fraction of 0.4, the SANS curves with and without Kr 222 can both be fit using only the $S(Q)$ for polydispersed spheres.

At constant $w_0$, the addition of Kr 222 causes the microemulsion droplets to decrease in size. Figure 4 demonstrates this clearly for a solution with a volume fraction of 0.08 and a $w_0 = 18.0$: the radii of the water pool plus interfacial film are respectively 36 Å and 31 Å. This decrease can be understood by considering the fact that because of the complexation of the $Na^+$ ions by the macrocycle the effective area per surfactant headgroup in the interface increases. Equation (4) shows that the droplet radii should decrease when this area is increased [13]:

$$R_{wp} = (3\,\phi_d)/(c_s A_{HG}) \qquad (4)$$

where $\phi_d$ is the volume fraction of the dispersed phase (water, surfactant headgroups and counterions), $c_s$ is the number of surfactant molecules per unit volume, and $A_{HG}$ is the area per headgroup. For the system without Kr 222 $A_{HG}$ is 65 Å$^2$ [14], whereas after the

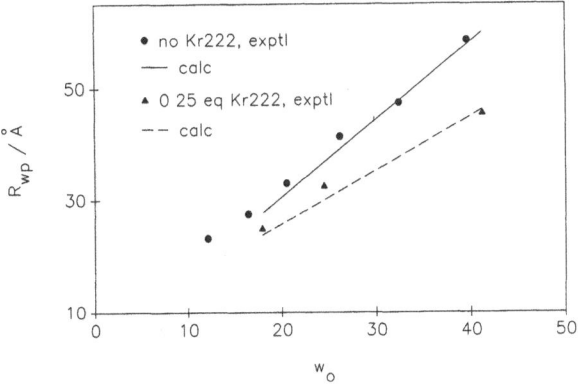

Fig. 5. Experimental (from SANS fits) and calculated (according to Eq. (4)) monodisperse water pool radii as a function of $w_0$ with and without Kr 222. The slopes are respectively 0.97 and 1.38

addition of 0.25 equivalents of Kr 222 $A_{HG}$ is estimated to be 93.8 Å$^2$, assuming that each Kr 222 requires 115 Å$^2$. The high $Q$ portion of the SANS curves (the Porod limit) give $A_{HG}$'s of ca. 80 Å$^2$ when KR 222 is present. Figure 5 compares values of the water pool radii obtained from fitting of the SANS curves to those calculated using Eq. (4). The agreement is excellent. The increased $A_{HG}$'s which result when Kr 222 is present lead to a decreased interfacial film density. This effect, as shown previously by Bellocq and coworkers [15], can lead to increased attractive interactions between the droplets.

## Conclusions

After the introduction of 0.25 mole equivalent of Kr 222 into AOT/water/cyclohexane solutions, w/o droplets in the $L_2$ region of the phase diagram are still present. The mean droplet size decreases at constant $w_0$, because of an increase in the surfactant's area per head group. Furthermore, Kr 222 produces a large increase in the attractive forces between the droplets. Although the NMR data are also consistent with a

bicontinuous structure where a fast exchange of the water pools is possible, we believe that the SANS data establishes that the picture of strongly interacting droplets is the more correct description of the system in the presence of 0.25 equivalents of Kr 222. Further studies of this microemulsions using dynamic light-scattering, time-resolved fluorescence quenching, conductance and viscosity measurements are in progress in order to clarify this point.

*Acknowledgements*

Financial support of the National Science Foundation (CHE86-11586) and ACS-PRF (18483-AC7) is gratefully acknowledged. We thank D. Schneider at BNL and Dr. C. S. Johnson, Jr. and Q. He at the University of North Carolina in Chapel Hill for the assistance with the NMR work.

## References

1. Evans DF, Sen R, Warr GG (1986) J Phys Chem 90:5500
2. Payne KA, Magid LJ, Evans DF (1987) Progr Colloid & Polymer Sci 73:10
3. Sauvage JP, Lehn JM (1975) J Am Chem Soc 97:6700
4. to be published
5. Saarinen TR, Woodward WS (1988) Rev Sci Instrum 59:761
6. Magid LJ (1986) Colloids and Surfaces 19:129
7. Shulz GV (1939) Z Phys Chem 43:25
8. Kotlarchyk M, Chen S-H (1983) J Chem Phys 79:2461
9. van Beurten P, Vrij A (1981) J Chem Phys 74:2744
10. Vrij A, de Kruif CG (1985) In: The Effects of Polydispersity on the Scattering Spectrum of Interacting Colloidal Particles, ACS National Meeting
11. Lindman B, Stilbs P, Moseley ME (1981) J Colloid Interface Sci 83:569
12. Stilbs P, Lindman B (1984) J Colloid Interface Sci 99:290
13. Langevin D (1988) Acc Chem Res 21:255
14. Kotlarchyk M, Chen S-H (1984) Phys Rev A 29:2054
15. Dichristina T, Roux D, Bellocq AM, Bothorel P (1985) J Phys Chem 89:1433
16. Nicholson JD, Clarke JHR (1984) In: Mittal KL, Lindman B (eds) Surfactants in Solution. Plenum Press, New York and London, Vol. 3, pp 1663–1674

Authors' address:

Dr. Linda J. Magid
Department of Chemistry, University of Tennessee
Knoxville, TN 37996-1600 USA

# Bending elasticity and dynamical fluctuations of giant lipidic vesicles

P. Méléard, M. D. Mitov, J. F. Faucon, P. Bothorel

Centre de Recherche Paul Pascal, C.N.R.S., Pessac, France

Bending elasticity is a fundamental property of flexible membranes. In physical systems, it controls the stability of diluted phases (microemulsions, swelling lamellar phases). In living cells, many natural movements of cytoplasmic membranes are dependent on their elastic behaviour (endo-, exocytosis, fusion and membrane-membrane interactions). Since the early theoretical work of Helfrich [1] in 1973, different attempts have been made to measure the elastic modulus $k_c$ of phospholipidic membranes. This was generally done by analysing the fluctuations of giant tubular [2–4] or quasi-spherical vesicles [5, 6] and the different published values ranged from 0.4 to 2.4 · $10^{-19}$ J.

So, our aim was to understand the reasons for such a scattering of $k_c$ values and to work out a method allowing accurate and reproducible measurements of this modulus. We showed that this can be actually done by analysing the thermal fluctuations of giant vesicles, using optical microscopy, video-enhanced contrast technics and image analysis [7, 8]. However, some requirements have to be fulfilled, both at the experimental and theoretical level:

1. introduction in the data analysis of the tension of the vesicle membrane as a fitting parameter to take into account the conservation of the membrane area [9] and of the vesicle volume,
2. analysis of a large number (several hundred) of contours to improve the accuracy of the analysis,
3. decomposition of the autocorrelation function of the radius fluctuations into a series of Legendre polynomials to decrease the noise contribution,
4. limitation of the analysis to modes having long correlation times compared to the video scan time (40 ms) or use of a correction factor to avoid any effect due to the integration time of the video camera.

The effect of these experimental or theoretical requirements on the measurement of the bending elastic stiffness has been illustrated using simulations [8] or experimental data [10]. The static analysis of a large number of egg-yolk phosphatidylcholine vesicles led to a bending elastic modulus $k_c = (0.4-0.5) \times 10^{-19}$ J, the corresponding membrane tensions being very low, in every case smaller than $15 \times 10^{-5}$ mN/m.

We also studied the dynamics of the thermal fluctuations of giant vesicles, following earlier studies [5, 11] and showed [12] that a rigorous and complete analysis of the dynamics of the fluctuations can be performed. Indeed, the space-time autocorrelation function can be expressed as a series of Legendre polynomials whose coefficients are directly related to the time correlation functions of the spherical harmonics derived by Milner and Safran [13]. The experimental space-time autocorrelation function was calculated using about ten thousand contours of a single vesicle. Owing to this very large number of contours which considerably increases the accuracy, the time dependence of the Legendre polynomials coefficients could be evaluated up to the $8^{th}$ mode. Furthermore, a mono-exponential decrease was observed versus time with correlation times ranging from 2 to 0.18 s. Finally, these data led to values of the bending modulus and of the membrane tension practically identical to those derived from the static analysis.

The time correlation functions of the Fourier amplitudes of the contours were also determined. In this case, as expected, the time dependence was not truly mono-exponential. However, the correlation times of a given mode $q$ could also be extracted, by using a linear combination of the correlation functions of the different Fourier amplitudes. The obtained values were then in very good agreement with those derived from the space-time autocorrelation function.

These results clearly demonstrate the validity of the theoretical model of Milner and Safran [13] for the dynamics of fluctuations, which strongly supports their hypothesis that the effect of convective and inertial terms can be neglected to describe the hydrodynamic behaviour of the lipid bilayer.

## References

1. Helfrich W (1973) Z Naturforsch 28 c:693 − 703
2. Deuling HJ, Helfrich W (1976) J Physique 37:1335 − 1345
3. Servuss RM, Harbich W, Helfrich W (1976) Biochim Biophys Acta 436:900 − 903
4. Schneider MB, Jenkins JT, Webb WW (1984) Biophys J 45:891 − 899
5. Schneider MB, Jenkins JT, Webb WW (1984) J Physique 45:1457 − 1472
6. Engelhardt H, Duwe HP, Sackmann E (1985) J Physique Lett 46:L395 − L400
7. Faucon JF, Méléard P, Mitov MD, Bivas I, Bothorel P (1989) Prog Colloid Polym Sci 79:11 − 17
8. Faucon JF, Mitov MD, Méléard P, Bivas I, Bothorel P (1989) J Physique 50:2389 − 2414
9. Kwok R, Evans E (1981) Biophys J 35:637 − 652
10. Mitov MD, Faucon JF, Méléard P, Bothorel P, Advances in Organized Media Ed Fendler JH, Jai Press Inc, to be published
11. Duwe HP, Zeman K, Sackmann E (1989) Prog Colloid Polym Sci 79:6 − 10
12. Méléard P, Mitov MD, Faucon JF, Bothorel P (1990) Europhys Lett 11:355 − 360
13. Milner ST, Safran SA (1987) Phys Rev A 36:4371 − 4379

Authors' address:

Dr. P. Méléard
Centre de Recherche Paul Pascal, C.N.R.S.
Château Brivazac, Avenue A. Schweitzer
33600 Pessac, France

**Progress in Colloid & Polymer Science**        Progr Colloid Polym Sci 81:70 – 75 (1990)

# Dynamic rigidity percolation of inverted AOT micellar solutions

J. S. Huang, L. Ye, D. A. Weitz, Ping Sheng, S. Bhattacharya and M. J. Higgins

Exxon Research and Engineering Company, Annandale, New Jersey, USA

*Abstract:* We use ultrasonic techniques and Brillouin scattering to study the elastic response of AOT surfactant solutions. This micellar solution features a short-range attractive interaction between the droplets. We find behavior consistent with a dynamic rigidity percolation wherein clusters that span the system can form in the solution at surfactant volume fractions above $\Phi \approx 0.16$. The percolation clusters contribute a real shear modulus causing an increase in the sound velocity if the frequency is higher than the characteristic relaxation rate of the cluster ($\sim 10^8$ Hz). By contrast, at low frequencies the solution behaves as an effective medium with isolated micelle aggregates imbedded in the oil continuum, and the anomalous contribution of the shear modulus disappears. This experiment provides a unique measurement of the scaling of the elastic properties for a percolating system. In particular, the rigidity exponent is found to be $\tau' \simeq 2.5$, consistent with the theoretical predictions.

*Key words:* Dynamic percolation; percolation; rigidity; micelles; AOT

## Introduction

When surfactants are dissolved in hydrocarbon media, they often form stable spherical aggregates called inverted micelles [1 – 6]. It is known, for example, that AOT surfactant (sodium di ethylhexyl sulfosuccinate) forms near-spherical micellar aggregates in decane over a wide range of concentrations [5], with each micelle containing 20 – 22 AOT molecules. Furthermore, it is reported that there exists a short range attractive interaction between the aggregates [7]. This interaction, characterized by a square-well potential with a width of roughly 3 Å, is thought to arise from the overlapping of the surfactant tails when two micelles are sufficiently close to each other [8, 9]. This attractive interaction promotes the formation of micellar clusters composed of the basic spherical units. At room temperature, the interaction is not strong enough to cause phase separation, but the average size of this dynamic cluster grows quickly as the concentration of the surfactant increases. If a small amount of water is incorporated into the micelles, the surfactant aggregates swell into minute water droplets [10], coated with a monolayer of the ionic surfactants. Because of the transient nature of these aggregates, the surfactants as well as the interior aqueous phase are exchanged freely between the interacting micelles [11, 12]. Thus when the average dynamic cluster increases in size, the ac electric conductivity increases accordingly. At the point when the average cluster grows to be macroscopic in size and begins to span the system, the normally non-conducting hydrocarbon solution loses much of its resistance to the flow of electric current. The critical volume fraction of the surfactant in solution that marks the insulator-conductor transition is called the (electric) percolation threshold, $\Phi_p$.

There have been several reports of investigations of the percolation behavior in AOT-water-oil systems [13 – 16]. In particular, Peyrelasse, et al. [15] have measured of $\Phi_p$ for an AOT-water-undecane system as a function of the [water]/[AOT] molar ratio, $X$. They found that the percolation threshold decreases from 0.6 to roughly 0.2 as $X$ decreases from 30 to 5. This suggests that percolating clusters may form in dry micelles (corresponding to $X = 0$) at concentrations below 20%. However, due to the lack of free water in these micellar clusters, electric conductivity measurements lose the sensivitiy required to detect the existence of percolation clusters.

The exact mechanism responsible for the occurrence of electrical percolation is still somewhat uncertain at the moment. There is a widely help belief that the percolation of a water internal microemulsion is due to a structural transition, from a water-droplet phase to a bicontinuous phase [17, 18]. But the evidence obtained from small angle neutron scattering [10, 19] and other measurements [4, 20, 21] do not support the occurrence of any structural transition at $\Phi_p$. One alternative explanation for the conductivity percolation is based on a surfactant hopping model [13, 22], wherein charge is carried by either the surfactants on the surface layer or the counterion in the water cores within the percolation cluster, which is itself a dynamic aggregate of the individual micelles. Unlike electric conductivity, measurements of the mechanical properties of the micellar solution may serve to resolve this dilemma and help unravel the structure of the percolation cluster. In this paper, we study these mechanical properties using ultrasonic techniques and Brillouin scattering to probe the propagation and the attenuation of in micellar solutions. This probes the complex elastic moduli of the micellar system, which depend in turn on the structure of the micelles in solution.

## Experimental

The ultrasonic measurements are carried out in the frequency range $2 < f < 45 \times 10^6$ Hz (MHz), employing an interferometric time-of-flight technique. The sound velocity, $v$, and attenuation coefficient, $\alpha$, are directly measured from the echo trains. The same quantities are obtained at higher frequencies using Brillouin scattering [23], the inelastic light scattering by the thermal sound wave fluctuations in the solution. After suitable calibration and deconvolution of the instrumental linewidth, we obtain the sound velocity from the peak position and the attenuation coefficient from the full width at half maximum, FWHM, of the Brillouin peaks. We use a five-pass Fabry-Perot interferometer with a finesse of 65. Both a $Kr^+$ ion laser (operating at 6471 Å) and an $Ar^+$ ion laser (operating at 5145 Å) are used as light sources. Scattering measurements are performed at several angles, providing measurements in the range of $0.6 < f < 5 \times 10^9$ Hz.

AOT surfactants are obtained from Fluka, and purified according to a published procedure [24]. The $n$-decane is a 99% + Gold Label product purchased from Aldrich Chemical Co. It is used without further purification. For the few microemulsion samples used in this study, a small amount of double distilled, deionized water is added to the micellar solutions. The added water swells the aggregates into reasonably mono-disperse droplets whose mean radius is given by the [water]/[AOT] molar ratio. Both the structure and the interaction potential for this 3-component system is known [5, 7, 8].

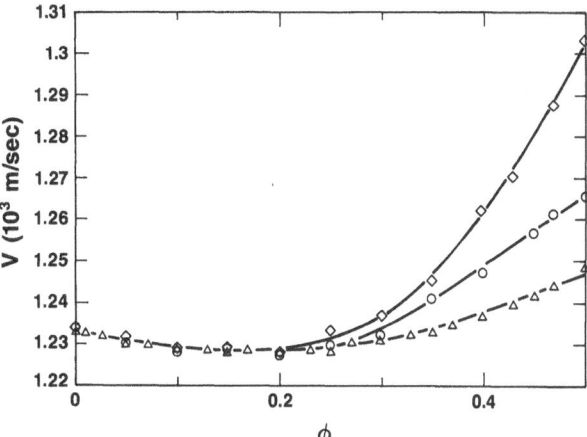

Fig. 1. Sound velocity plotted versus surfactant volume fraction $\Phi$ for three different frequencies (*) 2 MHz, (○) 15 MHz, and (+) 5 GHz. Solid curves are calculated using effective medium theory

## Results and discussion

A typical Brillouin spectrum is shown in Fig. 1 for a 3% micellar solution. The solid curve is a fit to the data using the line-shape proposed by Mountain [23], after convolution with the measured resolution function. This expression for the Brillouin line-shape explicitly accounts for the coupling of the sound waves to internal degrees of freedom of the medium, which leads to an additional damping of the sound. This results in the increased intensity between the Rayleigh peak and the Brillouin doublet. We find very good agreement between the predicted form of the Brillouin spectra and that measured, as shown in Fig. 1. By contrast, other possible descriptions of the shape of the spectra, such as three Lorenzations, do not fit the data as well. The additional damping due to the internal degrees of freedom is characterized by a relaxation time $\tau$. The best fit value for $\tau$ appears to be independent of the surfactant concentration and is of the order of $10^{-8}$ sec. Finally, we also note that the values obtained for the apparent Landau-Placzek ratio (the ratio of the integrated intensity in the central Rayleigh peak to that in the Brillouin doublets) is about 4. This is much higher than the typical values obtained in simple fluids but is not uncommon for complex fluids.

The adiabatic sound velocity, $v$, obtained from both Brillouin scattering and ultrasonic measurements, is plotted as a function of $\phi$ in Fig. 2 for three different frequencies. The data exhibit an initial drop in $v$ with increasing $\Phi$, followed by a rise at still higher volume

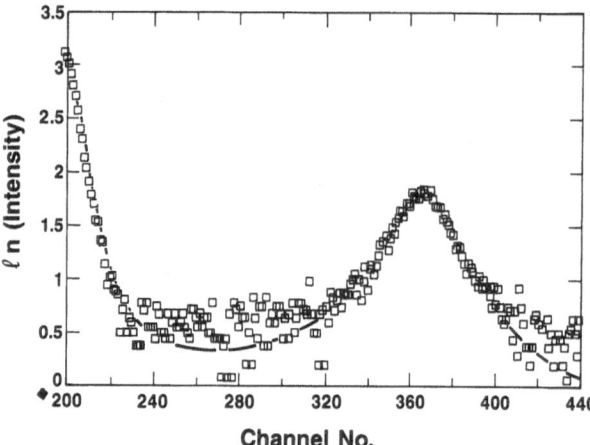

Fig. 2. Typical Brillouin Spectrum obtained for a 3% micellar solution. Solid curve is the theoretical fit

fractions. For $\Phi > 0.2$, there is a pronounced frequency dependence for $v$, with a significantly more rapid rise at higher frequencies. At low frequencies the dependence of $v$ can be modelled very well using an effective medium theory appropriate to isolated independent micellar droplets embedded in the oil. The velocity in the effective medium is given by $\langle v \rangle = (\langle \beta' \rangle / \langle \varrho \rangle)^{1/2}$, where the angular brackets represent the average values of the appropriate quantities for the effective medium and where:

$$1/\langle \beta \rangle = \Phi/\beta_2 - (1 - \Phi)/\beta_1 \qquad (2)$$

$$\langle \varrho \rangle = \Phi \varrho_2 + (1 - \Phi)\varrho]. \qquad (3)$$

Here $\beta$ is the complex elastic constant $\beta = \beta' + i\beta''$, and $\varrho$ is the density [25]. A volume fraction, $\phi$, of phase 2 is embedded in phase 1, with the subscripts denoting the appropriate phases. For the low frequency (2 MHz) measurements, the dependence of the sound velocity on volume fraction is very well described by Eqs. (2) and (3), for all $\phi$ up to 0.5, as shown by the solid line through the data in Fig. 2. In obtaining this fit, we use values for $\beta_1'$ and $\beta_1''$ obtained directly from experiment from the velocity and damping of the sound measured in pure decane. We cannot measure the values of $\beta_2$ for a pure phase of surfactant droplets, and thus take them as fitting parameters in obtaining the results shown in Fig. 2.

Unlike the low frequency data, the simple, but elegant, Wood's formula (Eqs. (2), (3)) cannot account for dispersion of the sound velocity at higher frequencies. The simple model of isolated micelles imbedded

in the oil medium is no longer valid, particularly at higher volume fractions. Since there is good evidence [15] to think that the clusters of micellar aggregates can percolate at volume fractions above 0.2, we employ an effective medium theory that treats both phases as interconnected continua [26, 27]. In this model, the complex elastic constant of the effective medium is related to the components in the following way:

$$\frac{1}{\langle \beta \rangle} = \frac{1 - \Phi}{\beta_1 + \frac{4}{3}(\langle \mu \rangle - \mu_1)}$$
$$+ \frac{\Phi}{\beta_2 + \frac{4}{3}(\langle \mu \rangle - \mu_2)} \qquad (4)$$

$$\frac{1}{\langle \mu \rangle + H} = \frac{1 - \Phi}{\mu_1 + H} + \frac{\Phi}{\mu_2 + H} \qquad (5)$$

where

$$H = \langle \mu \rangle \cdot \frac{9\langle \beta \rangle - 4\langle \mu \rangle}{6\langle \beta \rangle + 4\langle \mu \rangle} \qquad (6)$$

where $\mu$ is the complex shear modulus given by

$$\mu = \mu' + i\mu''$$
$$= \mu' + i2\pi f\eta \qquad (7)$$

where $\eta$ is the shear viscosity.

We propose that when the probable size of the dynamic clusters reaches the scale of the wavelength of the ultrasound (about a micron), it will appear to be stiffer at frequencies higher that the inverse relaxation time of clusters. This additional stiffness arises from the ability of the percolating clusters in the surfactant phase to support shear at these high frequencies. This will lead to the observed frequency dispersion in the ultrasonic velocity at micellar volume fractions sufficiently high to produce these large dynamic clusters.

To quantitatively account for the dependence of $v$, we self-consistently solve Eqs. (4) – (6) using the known values of $\mu_1$ and $\beta_1$ for decane and using $\mu_2$ and $\beta_2$ for the AOT phase as fitting parameters at each frequency. Excellent fits are obtained in each ase, as shown by the solid lines in Fig. 2. Furthermore, in Fig. 3 we plot the real part of the shear modulus of the AOT micelles, $\mu_2'$, obtained from the fits, as a function of the frequency. The observed increase in the shear modulus at higher frequencies leads directly to

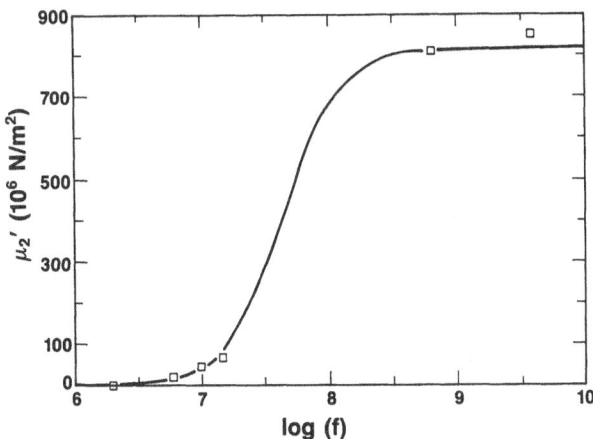

Fig. 3. The real part of the shear modulus for the surfactant phase obtained from fits to the effective medium theory, plotted against frequency. The solid curve represents the form of a single-relaxation-time Debye model

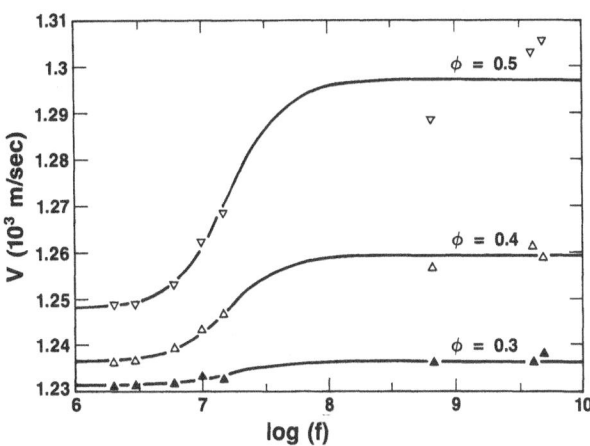

Fig. 4. Frequency dispersion of the sound velocity for several volume fractions. Solid lines again represent the form of the single-relaxation-time Debye model

the increased rigidity of the micellar phase at these frequencies. This suggest that the AOT micelles are locked in position at high frequency, and thus support shear. By contrast, at low frequency, their diffusion results in the destruction of the dynamic clusters on time scales shorter than the measuring frequency, and thus they cannot support shear. The solid curve in Fig. 3 is a fit to the data using a Debye model with single relaxation constant $\tau$:

$$\mu' = \mu'_0 + (\mu'_\infty - \mu'_0)(\omega^2\tau^2)/(1 + \omega^2\tau^2) \qquad (8)$$

where the subscripts represent the zero and high frequency limits. The characteristic frequency $1/\tau \simeq 10^8$ Hz, is presumably related to the relaxation frequency of the percolation cluster structure. This single relaxation model also works reasonably well for the observed frequency dependence of the velocity at volume fractions up to 0.5, as shown in Fig. 4. The three curves shown are fitted with a similar form,

$$v' = v_0 + (v_\infty - v_0)(\omega^2\tau^2)/(1 + \omega^2\tau^2). \qquad (9)$$

A relaxation time of $\tau \simeq 10^{-8}$ sec is obtained here also, while the value of $v_\infty$ is obtained from the fits to the Brillouin spectra. The consistency of the behavior of the frequency dependence of the velocity and $\mu'$ suggests that the solution properties properly reflect the rigidity percolation of the surfactant phase.

To further establish rigidity percolation, it would be useful to show that a percolation threshold exists. The threshold implies that at a certain volume fraction

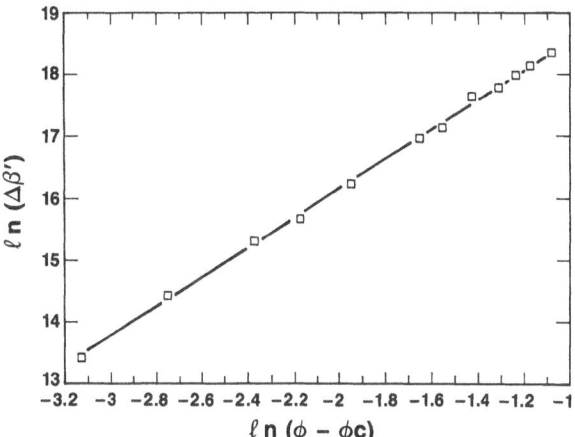

Fig. 5. The critical behavior of the shear modulus contribution of the surfactant phase at volume fractions $\Phi > \Phi_c$. Here $\Phi_c = 0.16$; The best straight-line fit represents the power-law $\Delta\beta\alpha(\Phi - \Phi_c)^\tau$ where $\tau' \simeq 2.5$

of the micelles, $\Phi_c$, there is a singularity in the elastic property. This singularity can often be observed experimentally in term of a scaling behavior as the singularity is approached. The fact that an electric percolation threshold exists for a similar system, containing a small amount of water in the neighborhood of the volume fraction where the frequency dispersion in $v$ is first observed is very suggestive. The rigidity of the micellar phase is given by the difference between the high and low frequency shear moduli, $\Delta \langle \beta' \rangle$. Fig. 5 shows a log-log plot of the $\Delta \langle \beta' \rangle$ versus $\Phi - \Phi_c$ for $\phi > 0.16$. A straight line fit is obtained:

$$\Delta \langle \beta' \rangle \alpha (\Phi - \Phi_c) \tau' \qquad (10)$$

where $\tau' \simeq 2.5$ and $\phi_c = 0.16$. This scaling behavior is consistent with theoretical expectations [28]. Elastic percolation with central forces have be shown to be mathematically equivalent to an electric percolation [28]. It is known that for the conductivity, $\sigma \propto (\Phi - \Phi_e)^t$ where $\Phi_e$ is the electric percolation threshold, and $t \simeq 2$ [29]. For the AOT microemulsions, $t$ has been measured by several groups recently [14, 15], with the observed values roughly about 2.0, as expected theoretically [29]. However, elastic percolation with only central forces is probably not realistic for the present case [30], as we might expect bond bending (noncentral) forces to exist in the micellar clusters. In this case, the exponent $\tau'$ is no longer the same as $t$. Kantor and Webman [31] obtained a value for singly connected network $\tau' \simeq 3.5$, which is higher than the value observed in micellar systems. This value might be expected to increase for the multiply connected clusters above percolation. However, the influence of the surrounding medium, the oil continuous phase, might in turn reduce the exponent. Clearly more theoretical work is required to account for the observed value of the exponent.

There are some significant differences between a static percolation cluster and that of the micelles, which exists only dynamically. On time scales short compare with some cluster relaxation time $\tau_c$, which might be of the order of $10^{-8}$ seconds as suggested by the observed Debye relaxation time, the cluster is a mechanically linked entity. At times longer than $\tau_c$, the clusters fall apart, and while some new linkages will form, all the elastic strain of the old cluster is released. This dynamic nature of the percolating elastic cluster could lead to a different rigidity percolation threshold, and perhaps to a different scaling behavior in the volume fraction. This is, of course, highly speculative and must remain, for the moment, a challenge for further theoretical treatment.

In the case of the 3-component microemulsion with swollen micelles, the velocity dispersion with $\Phi$ is not nearly as pronounced as that observed for the micellar solution, while the dispersion with frequency is suppressed below $\Phi \approx 0.5$. There are at two possible factors that may account for these observations. First, the electric percolation limit for the larger droplets is higher [15] depending on the water-to-surfactant ratio. For example, if 25 water molecules are added to every AOT molecule in solution, the droplet size is roughly 45 Å, and the electric percolation threshold is measured as nearly 0.6. If one assumes that the rigidity percolation threshold is near that of the electric percolation threshold, then we do not expect to see a significant velocity dispersion below $\Phi = 0.6$. The second factor is that, unlike micelles which are themselves quite rigid, the swollen microemulsion droplets contain relatively larger amounts of water, and thus the droplets are not as stiff as the micelles, thereby supporting less shear stress. For these reasons, it is harder to ascertain whether the microemulsion droplets also form percolating clusters at either $\Phi_c$ or $\Phi_e$. However, from evidence gathered from electron microscopic measurements, SANS measurements, NMR measurements [32], as well as these measurements, we believe that the electric percolation cluster in AOT microemulsion system is likely to consist of individual closed droplets.

## Summary

In conclusion, we find a dynamic rigidity percolation behavior, with $\phi \approx 0.16$, for the AOT micellar solution wherein a short-range attractive interaction between the droplets is prominent. When a sound wave is propagated through such a system, the percolation cluster possesses a shear modulus which leads to an increase in sound velocity in the solution, provided the frequency is higher than the characteristic relaxation rate of the cluster (about $10^8$ Hz). By contrast, for lower frequency sound, the dynamic nature of the percolating clusters causes the solution to behave as an effective medium with isolated micelles imbedded in the oil continuum, so that the additional contribution of the shear modulus disappears. This results in the pronounced velocity dispersion with frequency observed at higher volume fractions. This experiment provides one of the very rare measurements of the elastic properties of a percolation cluster. The exponent for the rigidity percolation is found to be $\tau' \simeq 2.5$.

## References

1. Eckwell P, Mandell L, Fontell K (1970) J Colloid Interface Sci 33:215
2. Jean YC, Aacke HJ (1978) J Am Chem Soc 100:6320
3. Zulauf M, Eicke HF (1979) J Phys Chem 83:480
4. Assih T, Larche F, Delford P (1982) J Colloid Interface Sci 89:35
5. Kotlarchyk M, Huang JS, Chen SH (1985) J Phys Chem 89:4382
6. Winsor PA (1968) Chem Rev 68:1
7. Huang JS, Safran SA, Kim MW, Grest GS, Kotlarchyk M, Quirke N (1984) Phs Rev Lett 53:592

8. Huang JS (1985) J Chem Phys 82:480
9. Lemaire B, Bothorel P, Roux D (1983) J Chem Phys 89:1023
10. Kotlarchyk M, Chen SH, Huang JS, Kim MW (1984) Phys Rev A 29:2054
11. Lang J, Jada A, Malliaris A (1988) J Phys Chem 92:1946
12. Eicke HF, Shepherd JCW, Steinemann A (1976) I Coll Interface Sci 56:168
13. Kim MW, Huang JS (1986) Phys Rev A 34:719
14. Bhattacharya S, Stokes JP, Kim MW, Huang JS (1985) Phys Rev Lett 55:1884
15. Peyrelasse J, Moha-Ouchane M, Bonded C (1988) Phys Rev A 38:904
16. Mathew C, Patanjali PK, Nabi A, Maitra A (1988) Coll Surf 30:253
17. deGennes PG, Taupin C (1982) J Phys Chem 86:2294
18. Borkovec M, Eicke HF, Hammerich H, Gupta BD (1984) J Phys Chem 92:206
19. Huang JS, Kotlarchyk M (1986) Phys Rev Lett 57:2587
20. Fletcher JD, Galai MF, Robinson DH (1984) J Chem Soc Faraday Trans I 80:3307
21. Jahn W, Strey R (1988) J Phys Chem 92:2294
22. Van Dijk MA (1985) Phys Rev Lett 55:1003
23. See, for instance, Berne BJ, Pecora R (1976) In: Dynamic Light Scattering, John Wiley, New York, p 247
24. Williams EF, Woodberry NJ, Dixon JK (1957) J Coll Sci 12:452
25. Hashin Z (1962) J Appl Mech 29:143
26. Sheng P (1986) In: Ericksen JL, Kinderlehrer D, Kohn R, Lions JL (eds) Homogenization and Effective Moduli of Materials and Media. Springer-Verlag, New York, p 196
27. Berryman JG (1980) J Acoust Soc Am 68:1809
28. deGennes PG (1976) J Phys Lett 37:L1
29. Stauffer D (1985) In: Introduction to Percolation Theory. Taylor and Francis, London, p 52
30. For the case of central force clusters, the percolation threshold $\Phi_c = 1$
31. Kantor Y, Webman I (1984) Phys Rev Lett 52:1891
32. Maitra AN (1989) preprint

Authors' address:

Dr. J. S. Huang
Room LA-352
Exxon Research and Engineering Company
Annandale, New Jersey, 08801
USA

**Progress in Colloid & Polymer Science**                    Progr Colloid Polym Sci 81:76–80 (1990)

# Dynamic light scattering on liquid-like polyelectrolyte solutions: correlation spectroscopy on dilute solutions of virus particles at very low ionic strength

S. F. Schulz*), E. E. Maier, R. Krause and R. Weber

Universität Konstanz, Fakultät für Physik, Konstanz, FRG

*Abstract:* Dilute aqueous solutions of semiflexible and rodlike polyelectrolytes (fd-virus particles, $L = 880$ nm, $d = 6$ nm, and Tobacco-Mosaic-virus, $L = 300$ nm, $d = 18$ nm) at very low ionic strength show a liquid-like structure that was observed by static and dynamic light scattering. Dynamic light scattering is strongly influenced by the Coulomb forces between the particles that arise from dissociated acid groups on the macroion surface. Effective short time diffusion coefficients extracted from field correlation functions are given by those of freely diffusing particles divided by the structure factor $S(q)$. The long time diffusion coefficient of a single particle in solutions with structure is significantly smaller than that of a particle without interaction and scales roughly with $D_0/S(q_{max})$. At low $q$, in the region that precedes the first structure peak, a slow mode component of the field correlation function is observed. Long time diffusion coefficients related to this region are smaller than single particle diffusion by more than one order of magnitude.

*Key words:* Light scattering; liquid-like structure; diffusion; rod-like particles; polyelectrolytes

## Introduction

Polyelectrolytes in dilute aqueous solution exhibit a varietry of effects of Coulomb-interaction, if the ionic strength is low and screening takes place only at the length scale of interparticle distances. To investigate the properties of these solutions visible laser light scattering is a suitable method since the wavelength $\lambda_0$ of the laser light is in the region of the interparticle distances of a few hundred nm. The wavevector of the scattered light

$$q = 4\pi n/\lambda_0 \cdot \sin(\vartheta/2) \qquad (1)$$

ranges from $3 \times 10^{-3}$ nm$^{-1}$ to $3.3 \times 10^{-2}$ nm$^{-1}$ with an incident light beam of $\lambda_0 = 488$ nm, a refractive index in the aqueous solution of $n = 1.33$ and scattering angles $\vartheta$ from $10°$ to $150°$.

There have been static and dynamic light scattering experiments with solutions of spherical polyelectrolytes that are well understood in theory [1–5]. Less is known in the case of linear and rodlike macroions, if the length of these particles is comparable to the interparticle distance and must not be regarded as small.

We have used two types of virus particles to study the properties of solutions of linear macroions. The Tobacco Mosaik Virus (TMV) is a rigid rod-like particle with a length of $L = 300$ nm, a diameter of $d = 18$ nm and about 2000 ionizible acid groups on its surface. The fd-virus is a filamenteous bacteriophage of $L = 880$ nm, $d = 6$ nm and about 10000 ionizible acid groups. It can be regarded as a semiflexible rod [6].

In purely aqueous solution of deionized water the acid groups cause a negative charge of a few hundered $e^-$ on the protein surface layer of each virus particle, which gives reason for a long range Coulomb-interaction. To maintain a minimum ionic strength in the solution we added a mixed bed ion exchanger to the

---

*) Adress after March 1, 1990: Universität Bayreuth, Physikalische Chemie I, Bayreuth, FRG

samples, which removes all other small ions than $H^+$-counterions and remains in the sealed sample cuvette during the measurements. In samples without ionexchanger the ionic strength was increased by carbonic acid that comes from the absorption of $CO_2$ during preparation. Although ionic strength is still very low, Coulomb-interaction is partly screened in these solutions.

The virus concentration $c$ ranges from 0.16 to 0.38 mg/ml in the case of fd-virus (MW = $16.4 \times 10^6$ g/mol) and from 0.7 to 1.55 mg/ml in the case of TMV (MW = $40 \times 10^6$ g/mol). In units of $cL^3$ there is less than 1 virus per $L^3$ in TMV-solutions, but 4 to 10 viruses per $L^3$ in the fd-virus solutions.

Static light scattering measurements [7, 8] prove that an isotropic, liquid-like order occurs in these solutions that do not show any birefringence at rest. A static structure factor $S(q)$ is derived from the intensity measurements by normalization with a standard (toluene) and division by the isotropic one particle formfactor $F(q)$ for rods of $L = 300$ nm and $L = 880$ nm respectively. This structure factor is very similar to the structure factor of solutions of spherical macroions, but it includes additional information on mutual orientation of the rods. The scattering vector $q_{max}$ of the structure peak maximum that corresponds to a prefered mean interparticle distance (not center-of-mass distance) varies with $c^{1/3}$ only for $cL^3 < 1$ and with $c^{1/2}$ for $cL^3 > 1$. Further minima and maxima are observed for $q > q_{max}$ and $S(q)$ approaches 1 in the high $q$-region.

To study the dynamics of these virus solutions the intensity correlation functions $\langle I(q,0)I(q,t)\rangle$ are measured in the same $q$-region as for static light scattering with a digital correlator (ALV, FRG) in up to 1023 channels for delay time $t$ and with the true background information of $\langle I(q,0)\rangle$. The intensity correlation function is converted to the normalized field correlation function by

$$g_E(q,t) = \sqrt{\frac{\langle I(q,0)I(q,t)\rangle}{\langle I(q,0)\rangle^2} - 1}. \qquad (2)$$

The field correlation function reflects the different types of motion that take place within the scattering volume. The Brownian motion results in a center-of-mass diffusion of the viruses and also in a diffusive rotation and bending motion of the rod-like and semiflexible particles. All three types of single particle motion are expected to be influenced by the structure and the Coulomb-interaction between the charged ma-

croions. The correlated motion may lead to density fluctuations that exceed the fluctuations caused by free diffusion within the scattering volume.

Till now there is no theory that predicts correlation functions that include all these effects of this type of "isotropic anisotropic particle solution" with particle interaction. The structure of such a correlation function may be expected to look like

$$g_E(q,t) \sim \frac{\left\langle \sum_{i=1}^{N}\sum_{j=1}^{N} f_i(0)f_j(t)e^{iq(R_i(0)-R_j(t))}\right\rangle}{\langle I(q,0)\rangle}$$

$$= \frac{1}{NF(q)S(q)}\left[\sum_{i=j}^{N} ... (\text{self part})\right.$$

$$\left. + \sum_{i \ne j}^{N} ... (\text{distinct part})\right] \qquad (3)$$

if there is a constant number of the $N$ viruses in the scattering volume and density-density correlations are neglected. $R_i(t)$ describe the center-of-mass positions and $f_i(t)$ the time dependent formamplitudes of the particles.

Dynamic light scattering studies have been made with virus solutions to determine diffusion constants and the effect of rotation and bending in the absence of particle interaction [9,10,6]. The virus charge was screened by the high ionic stength of the buffer solution. Assumptions had to be made about coupling of length- and sidewise motion, rotation and flexibility to fit the experimental light scattering data. In our experiments only experimental correlation functions of samples with screened Coulomb-interaction were used as reference for noninteracting particles.

## Experimental and data analysis

Sample preparation and light scattering apparatus are described in our previous papers [7, 8, 11]. All intensity correlation functions were converted to normalized field correlation functions using [2]. Short time and long time behavior of the measured correlation functions were analysed in different ways. The first cumulant

$$K_1(q) = -d/dt\,[\ln g_E(q,t)]_{t \to 0} \qquad (4)$$

was extracted in a 2-cumulant fit from the datapoints of the first 100−200 µsec. From this we calculate an effective diffusion coefficient

$$D_{eff}(q) = K_1(q)/q^2 \qquad (5)$$

that depends on the scattering vector and is generally higher than the center-of-mass diffusion coefficient $D_0$, because it includes the effects of the other types of motion [6, 9, 10].

The long time behavior was analysed starting by the assumption that $g_E(q,t)$ roughly follows an exponential form of

$$g_E(q,t) = A e^{-q^2 W(q,t)} \qquad (6)$$

with an amplitude factor $A$ that has to be adjusted for each measurement. The time dependence of $g_E(q,t)$ is transfered to a function $W(q,t)$ that has the dimension of a mean square displacement. In case of diffusing spherical macromolecules $W(q,t)$ is related to the mean square displacements of the center-of-mass positions $R(t)$

$$\langle \delta R^2(t) \rangle_{sph} = 6 D_{sph}\, t = 6 W(q,t). \qquad (7)$$

$W(q,t)$ of spheres without particle interaction does not depend on $q$, whereas $W(q,t)$ of anisotropic particles is $q$-dependent as long as internal motions (rotation, bending) contribute to the correlation function.

The time derivative of $W(q,t)$ gives $q$-dependent values of diffusion coefficients that are also time dependent, if $W(q,t)$ is not linear in $t$. However, a linear behavior of $W(q,t)$ is often observed in the long time tail that is defined as

$$t > (q^2 D_0)^{-1}. \qquad (8)$$

A long time diffusion coefficient is then given by

$$D_L(q) = \frac{d}{dt}[W(q,t)]_{t > (q^2 D_0)^{-1}}. \qquad (9)$$

Two regions of the scattering vector $q$ must be distinguished. In the high $q$-region mainly single particle scattering is observed, because the typical distance of $2\pi/q$ is smaller than the mean interparticle distance. $D_L(q)$ is then refered as $D_{L,single}(q)$. In the low $q$-region many particles contribute to the scattering within $2\pi/q$ and $D_L(q)$ is refered as $D_{L,coll}(q)$. The crossover between high $q$- and low $q$-region depends on particle length and concentration.

## Results and discussion

Dynamic light scattering measurements on liquid-like virus solutions show a different behavior in $g_E(q,t)$ in these two $q$-regions. In both regions $g_E(q,t)$ decays slower than in a measurement of an equivalent sample with completely screened interaction. But in the low $q$-region an extra slow mode component occures in the correlation function that often exceeds the time scale of the correlator. This slow mode component is most dominant in virus samples with ionexchanger and lowest ionic strength, but it is also present in samples without ionexchanger and low structure peak. Comparable measurements in solutions of interacting latex spheres do not show such a slow mode component, but it was observed by Pusey and van Megen [12] in latex samples with glass-like structure.

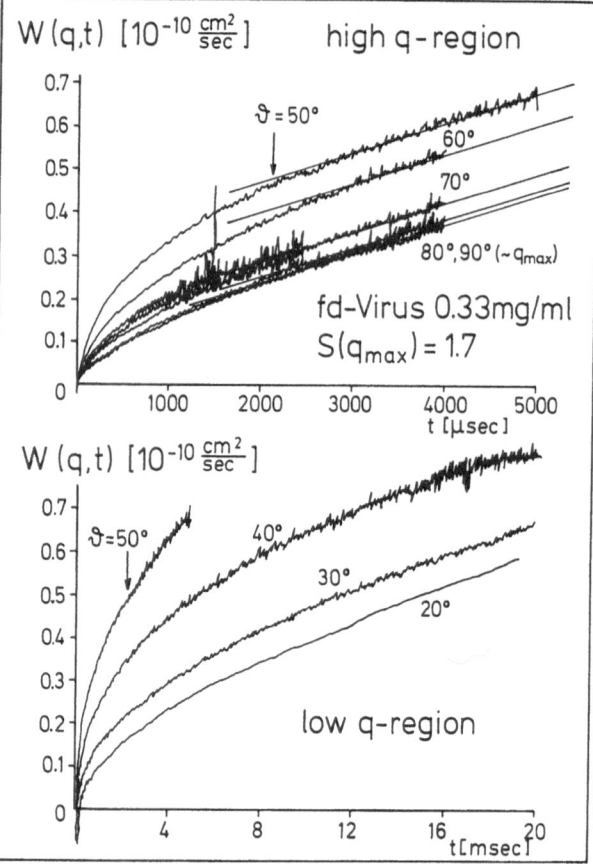

Fig. 1. $W(q,t)$ from correlation functions of an fd-virus sample with liquid like structure and a structure peak height of about 1.7. The peak position $q_{max}$ corresponds to a scattering angle of $\vartheta = 95°$. In the high $q$-region ($50°-150°$/not labeled) the initial slope is proportional to $1/S(q)$, and the long time slope reaches a constant independent of $q$ (indicated by parallel lines). For reference the $50°$-curve is also plotted in the low $q$-region, but only $40°$ and lower scattering angles belong to this region, where the slow mode component occurs in the correlation function and the long time slope is much smaller than in the high $q$-region

To study the short time region of the correlation function the influence of the structure was examined by a comparison to well known relationships for samples of interacting spheres. There $D_{eff.}(q)$ is given by

$$D_{eff.}(q) = D_0/S(q) \qquad (10)$$

respectively the first cumulant $K_{1,i}(q)$ measured in a sample with interaction

$$K_{1,i}(q) = K_{1,0}(q)/S(q) \qquad (11)$$

where $K_{1,0}(q)$ is the first cumulant measured in an

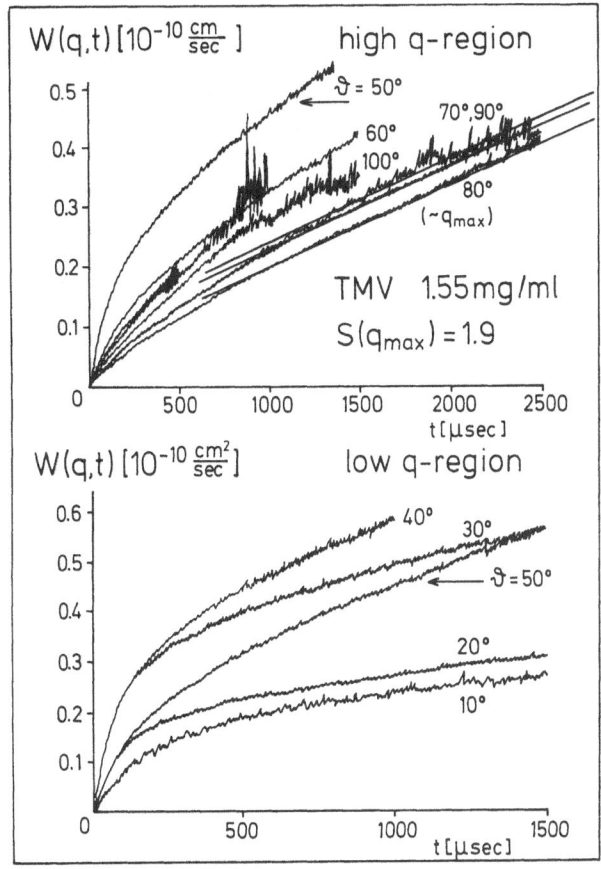

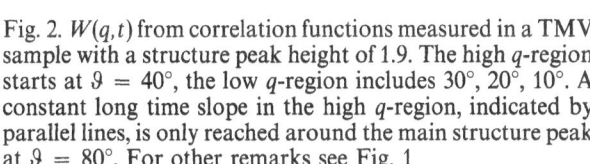

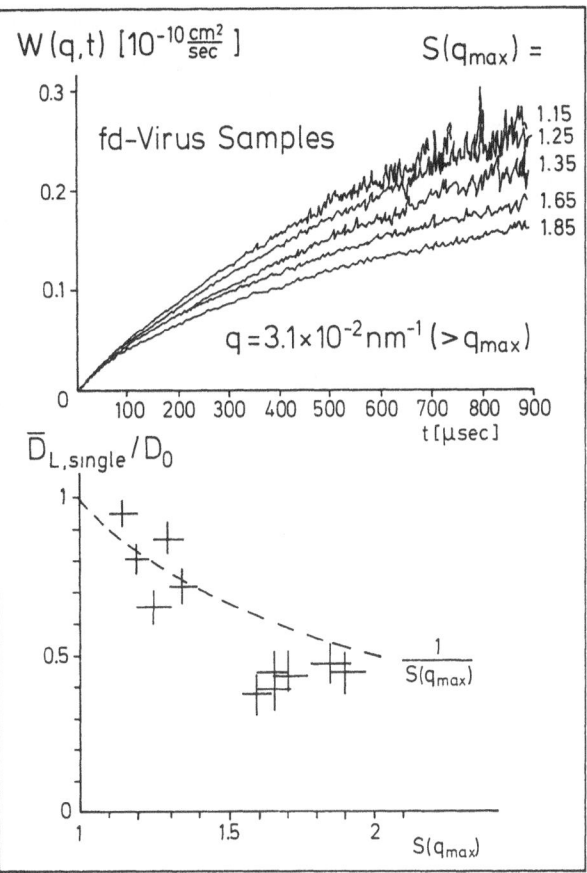

Fig. 2. $W(q,t)$ from correlation functions measured in a TMV sample with a structure peak height of 1.9. The high $q$-region starts at $\vartheta = 40°$, the low $q$-region includes 30°, 20°, 10°. A constant long time slope in the high $q$-region, indicated by parallel lines, is only reached around the main structure peak at $\vartheta = 80°$. For other remarks see Fig. 1

Fig. 3. $W(q,t)$ of various fd-virus samples with different structure peak heights measured at the same scattering angle of $\vartheta = 130°$ that is behind the main structure peak for all samples. The long time slope clearly decreases with increasing structure peak height. In the lower part $\bar{D}_{L,single}/D_0$ is plotted versus structure peak height for fd- and TMV-samples. For comparison $1/S(q_{max})$ is drawn as broken line

equivalent sample without or with screened particle interaction.

If we use (11) to determine $S(q)$ from the dynamic light scattering measurements, we find a good agreement to the $S(q)$-values derived from static light scattering measurements [8, 11], only with significant deviations in the lowest $q$-region, where the slow mode component is observed. In the $q$-region around the main structure peak and at higher $q$ short time $g_E(q,t)$ behaves the same manner as in solutions of interacting polyballs, only that in (10) $D_0$ must be replaced by $q$-dependent $D_{eff,0}(q)$.

The occurance of the slow mode component is clearly seen in the low $q$-region of $W(q,t)$ (lower parts of Fig. 1 and Fig. 2 for fd and TMV resp.). The initial slope does not hold relation (10) in strength, and there is not a well defined linear time dependence in $W(q,t)$ in the long time region.

$D_{L,coll.}(q)$ values derived by (9) are much smaller than $D_0$ values ($1.8 \times 10^{-8}$ cm$^2$/sec for fd, $3 \times 10^{-8}$ cm$^2$/sec for TMV) and the slope of $W(q,t)$ seems to decrease further and further with increasing $t$ (see also [11]). In this $q$-region the long time regime (8) belongs to times that are comparible to diffusing times on center-of-mass distances between the viruses.

If $W(q,t)$ is plotted for a series of $q$-values in the high $q$-region, the initial slope varies according to (10). The lowest slope corresponds to the maximum in $S(q)$, the highest to $q < q_{max}$ where $S(q) \ll 1$. The long time slopes of many curves reach a constant value, which is very similar for all $W(q,t)$, so that the $W(q,t)$-curves run parallel in the long time region (upper parts of Fig. 1 and Fig. 2 for fd and TMV resp.). The $D_{L,single}(q)$ values extracted by (9) do not show a significant $q$-dependence so that we find a mean value $\bar{D}_{L,single}$ that is typical for the high $q$ region.

It is useful to compare the $\bar{D}_{L,single}$-values to the mean center-of-mass diffusion coefficient $D_0$ of the viruses for free diffusion and to the structure peak height $S(q_{max})$ in the solution that reflects the strength of interaction between the particles. $\bar{D}_{L,single}$ values are always found to be smaller than $D_0$ and decrease with increasing structure peak height. The evolution of long time slopes of $W(q,t)$ in dependence of $S(q_{max})$ is shown in the upper part of Fig. 3 for five fd-samples. A concentration dependence of $\bar{D}_{L,single}$ is not found with our concentrations in different samples of the same species variing by a factor of two. The values of $\bar{D}_{L,single}/D_0$ seem to decrease stronger than $D_0/S(q_{max})$, as demonstrated in the lower part of Fig. 3, but the error bars in detecting diffusion coefficients and peak heights in our measurements are still to large to find a scaling law.

## Concluding remarks

We could show that dynamic light scattering measurements on dilute solutions of rod-like polyelectrolytes reflect the static interparticle structure as well as diffusion properties with the influence of Coulomb interaction. For low scattering vectors the correlation functions of scattered light from samples with structure are dominated by a slow mode component, which is not present in samples with screened Coulomb interaction or in dilute solutions of spheres. High density fluctuations in the scattering volume as an origin of this slow mode would indicate that the solution is not in thermal equilibrium, but more in a glass-like state. Particle rearrangement on interparticle distances is hindered by strong Coulomb repulsion. Long time diffusion coefficients extracted from field correlation functions at high $q$-vectors are strongly correlated with structure peak heights as a measure of strength of interaction.

*Acknowledgements*

Virus material was kindly supplied by Prof. Rasched, Konstanz, and by Prof. Wetter, Saarbrücken. This work was supported by the Deutsche Forschungsgemeinschaft (SFB 306).

## References

1. Brown CJ, Pusey PN, Goodwin JW, Ottewill RH (1975) J Phys A: Math Gen 8:664—682
2. Grüner F, Lehmann W (1979) J Phys A Math Gen 12:L303—L307
3. Grüner F, Lehmann W (1982) J Phys A Math Gen 15:2847—2863
4. Hess W, Klein R (1983) Adv Phys 32:173—283
5. Krause R, Nägele G, Karrer D, Schneider J, Klein R (1988) Physica A 153:400—419
6. Fujime S, Tagasaki-Ohsita M, Maeda T (1987) Macromolecules 20:1292—1295
7. Maier EE, Schulz SF, Weber R (1988) Macromolecules 21:1544—1546
8. Schulz SF, Maier EE, Weber R (1989) J Chem Phys 90:7—10
9. Schaefer DW, Benedek GB, Schofield P, Bradford E (1971) J Chem Phys 55:3884—3895
10. Maeda T, Fujime S (1985) Macromolecules 18:2430—2437
11. Schulz SF, Maier EE, Krause R, Hagenbüchle M, Deggelmann M, Weber R (1989) submitted to J Chem Phys
12. Pusey PN, van Megen W (1987) Phys Rev Lett 59:2083—2086

Authors' address:

Susanne F. Schulz
Universität Bayreuth
Physikalische Chemie I
Postfach 101251
D-8580 Bayreuth, FRG

**Progress in Colloid & Polymer Science**                    Progr Colloid Polym Sci 81:81 – 86 (1990)

# A small-angle neutron scattering study on AOT/toluene/(water + acrylamide) micellar solutions

C. Holtzscherer[1]), F. Candau[2]) and R. H. Ottewill[3])

[1]) Faculté de Pharmacie, Laboratoire de Pharmacie Galénique, Chatenay-Malabry, France
[2]) Institut Charles Sadron (CRM-EAHP), Strasbourg, France
[3]) School of Chemistry, University of Bristol, Bristol, England

*Abstract:* The effect of the presence of acrylamide inside water-swollen micelles of Aerosol OT dispersed in toluene on the interparticular interactions and micellar size has been investigated by small angle neutron scattering. The structure factor obtained for $D_2O$/AOT/toluene systems (no acrylamide) can be fitted by a hard sphere model. The incorporation of acrylamide induces attractive interparticular forces. The micellar sizes obtained from the form factor are in good agreement with previous findings by quasi-elastic light scattering.

*Key words:* Acrylamide; aerosol AOT micellar systems; neutron scattering

## Introduction

Polymerization of various monomers in microemulsions is a new and rapidly developing field and several studies have been reported in the past years [1 – 7]. Recently, one of us has described a new method of preparation of inverse microlatices by free radical polymerization of acrylamide inside water-swollen micelles of Aerosol OT (AOT) dispersed in toluene [8]. A thorough study by light scattering, viscometry and ultracentrifugation of the micellar solutions prior to polymerization has shown that the addition of acrylamide had a strong influence on the structural properties of the systems. More specifically, it was found that:

i) The $L_2$ inverse micellar domain was notably extended in the phase diagram, indicating that acrylamide acted as a cosurfactant.

ii) There was a strong enhancement of the interparticular attractive forces as indicated by the negative values of the second virial coefficient of the osmotic pressure [8].

iii) The conductivity of the systems was several orders of magnitude larger than in the absence of acrylamide [9].

A qualitative explanation of the observed effects was based on the assumption that acrylamide was partially located at the interface in between the AOT molecules.

This led to the possibility of an increase of the attractive interactions as a result of some disorganization of the interfacial film and consequently rapid exchanges between droplets [10].

The purpose of the present paper was to obtain confirmatory evidence of these interactions by an examination of the small angle neutron scattering (SANS) from the microemulsions. SANS is a powerful tool to determine the structural properties of colloids and reverse AOT micelles have already been the object of several studies [11 – 15]. Moreover, this technique allows one to measure the particle size as a function of concentration. These data can be compared to previous measurements using quasi-elastic light scattering in the dilute regime [8].

## Experimental

### Materials

Water was deionized and doubly distilled. Deuterium oxide (99%) was obtained from Aldrich. $h_8$-toluene (Prolabo) and $d_8$-toluene (Aldrich, 99% deuterated material) were used as supplied.

Aerosol OT or AOT (sodium bis-2-ethylhexylsulfosuccinate) was purchased from Fluka and purified as described elsewhere [8].

Acrylamide (AM) supplied by Merck was recrystallized twice from chloroform and dried under vacuum.

*Preparation of microemulsions*

Microemulsions were prepared just before the scattering measurements by adding the acrylamide dissolved in water to a solution of AOT in toluene. A clear microemulsion was obtained almost instantaneously with gentle agitation.

The compositions of the microemulsions investigated in the present work are gathered in Table 1. The coherent scattering lengths of the materials used are given in Table 2.

*Quasi-elastic light scattering (QUELS)*

The determination of the hydrodynamic diameter of the micelles was obtained according to a procedure described previously [8]. Measurements were made at a scattering angle of 90°. The light scattering vector was defined by

$$Q = \frac{4\pi}{\lambda} \sin \frac{\theta}{2} \tag{1}$$

where $\lambda$ is the wavelength of light in the medium, i.e. $\lambda_0/n$ with $n$ the refractive index of the dispersion, and $\theta$ the scattering angle.

*Small angle neutron scattering (SANS)*

The SANS experiments were performed at the Institut Laue Langevin (I.L.L.) Grenoble, France, using the neutron diffractometer D17 [16]. The microemulsions were contained in optical-quality quartz cells having a path length of 1 mm. The sample to detector distance was 1.41 m or 3.4 m with incident wavelengths of 10.0 Å or 12.0 Å respectively: the full width of half height of the distribution of wavelength $\delta\lambda/\lambda$ was 10%. For elastic scattering of neutrons, the scattering vector is also defined by Eq. (1) with $\lambda =$ the wavelength of the neutron beam. Examinations were carried out over the $Q$ range 0.01 to 0.22 Å$^{-1}$.

The basic data were processed to give the absolute intensity of scattering at a particular $Q$, $I(Q)$, relative to water.

**Theory**

The theory of SANS from colloidal systems has been described in detail in several papers [17–19] and we will only recall here the main theoretical predictions which are relevant for the discussion of our results.

The observed coherent scattered intensity of a neutron beam $I(Q)$ is given for a monodisperse system by

$$I(Q) = (\varrho_p - \varrho_m)^2 V^2 N S(Q) P(Q) \tag{2}$$

where $\varrho_p$ is the mean scattering-length density of the particle of volume $V$, $\varrho_m$ the scattering length density of the medium, and $N$ the number of particles. $P(Q)$ is the scattering form factor for a single micelle and $S(Q)$ the structure factor for the correlation between particles.

*Dilute colloidal systems*

For very dilute systems corresponding to non interacting particles, $S(Q) \simeq 1$ and $I(Q) \propto P(Q)$. $I(Q)$ reflects the size, shape and internal features of the particles and for spherical particles of radius $R$ is given by

$$I(Q) = 9(\varrho_p - \varrho_m)^2 \Phi V$$
$$\times \left[ \frac{\sin QR - QR \cdot \cos QR}{Q^3 R^3} \right]^2 \tag{3}$$

where $\Phi$ is the volume fraction of the dispersed phase ($\Phi = NV$).

At zero scattering angle ($Q = 0$), the particle form factor $P(Q)$ becomes unity and hence,

$$I(0) = (\varrho_p - \varrho_m)^2 \Phi V. \tag{4}$$

For small values of $QR(QR < 1)$, one can use the Guinier approximation, which gives

$$I(Q) = I(0) \exp\left( -\frac{Q^2 R_g^2}{3} \right) \tag{5}$$

where $R_g$ is the radius of gyration of the scattering particle.

Table 1. Composition of the systems

| Sample | Composition (wt %) | | | | | | |
| --- | --- | --- | --- | --- | --- | --- | --- |
| | D$_2$O | H$_2$O | Acrylamide | AOT | d$_8$ toluene | h$_8$ toluene | $\Phi$ % |
| 1 | 8.92 | — | — | 17.01 | — | 74.07 | 21.78 |
| 2 | — | 8.95 | — | 17.00 | 74.05 | — | 23.27 |
| 3 | 5.32 | — | 3.99 | 16.81 | — | 73.88 | 21.48 |
| 6 | — | 5.15 | 4.00 | 16.96 | 73.89 | — | 23.10 |
| 8 | 5.04 | — | — | 16.94 | — | 78.02 | 17.89 |

Table 2. Coherent neutron scattering length densities

| Molecules | $\varrho/10^{10}$ cm$^{-2}$ |
|---|---|
| Water, $H_2O$ | $-0.56$ |
| Heavy water, $D_2O$ | 6.4 |
| $h_8$-toluene | 0.94 |
| $d_8$-toluene | 5.63 |
| Acrylamide | 1.56 |
| AOT | 0.615 |

Plotting $\ln I(Q)$ versus $Q^2$ gives in the limit $Q_{\to 0}$ a straight line of slope $(-R_g^2/3)$ for spherical particles.

*Concentrated colloidal systems*

In the case of interacting particles, the shape of $I(Q)$ depends through $S(Q)$ on the pair correlation function $g(r)$ with $r$ the distance of interparticle centre-centre separation, according to

$$S(Q) = 1 + \frac{4\pi N}{Q} \int_0^\infty [g(r) - 1] r \sin Qr \, dr. \qquad (6)$$

A useful expression for $S(Q)$ for hard spheres is obtained from the Percus-Yevick approximation [20, 21], the expression of $S(Q)$ being in the form

$$S(Q) = \frac{1}{[1 - NC(2QR_{HS})]} \qquad (7)$$

with

$$C(2QR_{HS}) = -32\pi R_{HS}^3 \int_0^1 \left( \frac{\sin(2sQR_{HS})}{2sQR_{HS}} \right)$$
$$\times (\alpha_0 + \beta s + \gamma s^3) s^2 ds \qquad (8)$$

where $R_{HS}$ is the radius of the hard-sphere such that the hard-sphere volume fraction is defined by

$$\Phi_{HS} = 4\pi R_{HS}^3 N/3 \qquad (9)$$

and the coefficients $\alpha_0, \beta$ and $\gamma$ are defined by

$$\alpha_0 = (1 + 2\Phi_{HS})^2/(1 - \Phi_{HS})^4 \qquad (10)$$

$$\beta = -6\Phi_{HS}(1 + 0.5\Phi_{HS})^2/(1 - \Phi_{HS})^4 \qquad (11)$$

$$\gamma = 0.5\Phi_{HS}(1 + 2\Phi_{HS})^2/(1 - \Phi_{HS})^4. \qquad (12)$$

In order to apply this relationship, it can be assumed that

$$R_{HS} = R_c + t \qquad (13)$$

where $R_c$ is the radius of the water core and $t$ the thickness of the hydrocarbon adsorbed layer [22].

The shape of $I(Q)$ versus $Q$ provides a straightforward qualitative information on the interaction potential. For hard spheres suspensions, the presence of oscillations in the curves of $S(Q)$ versus $Q$ are characteristic of repulsive interactions. In the case of microemulsions, models based on different interaction potentials have been proposed to account for the presence of attractive interactions between droplets. Vrij et al. have introduced an attractive perturbation term in addition to the hard sphere repulsive forces [23]. Other models [24, 25] proposed that the interaction potential is proportional to the volume of penetration of the droplets; this would account for the existence of attractive forces in microemulsions containing alcohols and oils of different lengths. In the case of AOT systems, an attraction also appears to exist which tends to damp out the oscillations [26].

Other representations of SANS data are also useful. For dispersions characterized by a sharp interface, the Porod limit [27] shows that $I(Q)$ decreases as $Q^{-4}$ when $Q_{\to\infty}$, so that

$$I(Q) - B \simeq AQ^{-4} \qquad (14)$$

where $B$ represents the signal due to background incoherent scattering etc. ... A plot $(I(Q) - B)Q^4$ versus $Q$ shows oscillations at large angles which contain information at high resolution; more specifically, for hard spheres, the values of the first minimum $Q_{min}$, attributed to the extinction of the scattering of a sphere, provides an estimation of the particle radius [28] through the equation

$$Q_{min} R = 4.5. \qquad (15)$$

Also, the absence of oscillations of higher orders is indicative of possible contributions from shape fluctuation, polydispersity or strong attractive forces.

**Results**

*Effect of the composition of the droplet core on the interparticular interactions*

In this set of experiments, the interior of the micelles was $D_2O$ which allowed one to probe directly

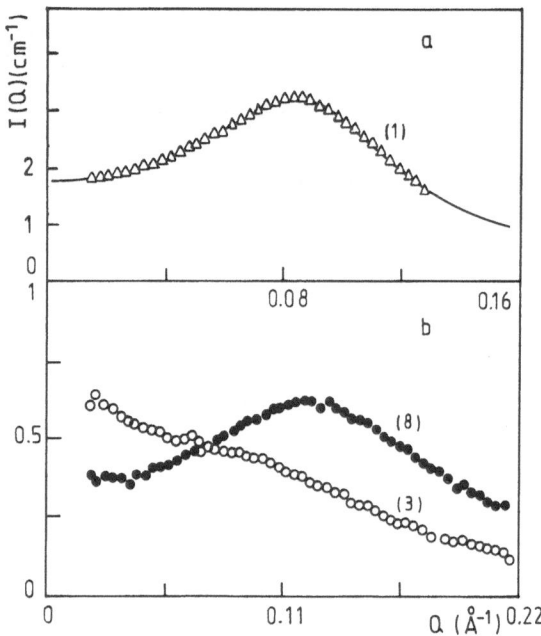

Fig. 1. Intensity spectra of samples 1 ($\triangle$), 8 ($\bullet$) and 3 ($\circ$). The solid line of sample 1 is the calculated curve for hard spheres using Eq. (7) (see text)

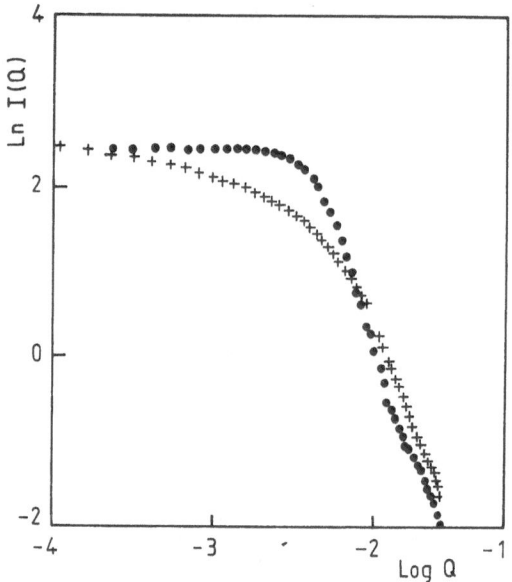

Fig. 2. $Ln I(Q)$ against $\log Q$ for samples 2 ($\bullet$) and 6 ($+$)

the micellar core. Figure 1 shows typical examples of the curves of $I(Q)$ versus $Q$ for systems with acrylamide (3) and without (1 and 8). Sample 1 shows a well-defined first order peak (Fig. 1a). The solid line which is drawn was obtained using Eq. (7) for $S(Q)$ with the following parameters: $R_c = 17$ Å, $R_{HS} = 30$ Å and the scattering length densities shown in Table 2.

One can remark that the thickness of the shell is of the same order of magnitude as the length of the hydrophobic tail of the AOT molecules ($\simeq 12.5$ Å) [14]. Contrary to the case of other investigated microemulsions [24, 26], there is no evidence of attractive interactions in this area of the phase diagram, if one considers the good agreement between experimental data and the hard sphere model.

Sample 8 (Fig. 1b) shows also an interaction peak which is shifted to higher $Q$ values; this indicates that, as expected, the core of the micelles becomes smaller upon decreasing the water content.

The shape of the curve of sample 3 in which part of the deuterium oxide was replaced by acrylamide is strikingly different from those observed for microemulsions without acrylamide. In particular, no repulsive interaction peak is observable. Such a behavior confirms previous studies which have shown that incorporation of acrylamide induces attractive interparticular forces resulting in a percolative behavior [8, 9].

*Effect of addition of acrylamide on the structural properties of the dispersed particles*

In order to gain information on the structural properties of the whole particle (core + shell), we have carried out measurements on microemulsions containing deuterated toluene as the continuous medium. Microemulsions 2 and 6 contain $\simeq 9\%$ of $H_2O$ and ($\simeq 5\%$ $H_2O$ + $\simeq 4\%$ AM) respectively (see Table 1).

*Concentrated systems:* Fig. 2 shows the log-log plots of $I(Q)$ versus $Q$ for samples 2 and 6. The effect of attractive forces is again evident in the intermediate $Q$ range and in the higher $Q$ range there are still some differences. In the latter range of $Q$, the resolution of the experiment is of the order of 30 Å ($2\pi/Q_{max}$) that is of the order of the tail length of the surfactant ($\simeq 12$ Å) or of the water core ($\simeq 17$ Å). The best least square fits to the data yield straight lines whose slopes are $-4.0$ for sample 6 and $-5.8$ for sample 2. Experiments performed on dilute samples led to slopes ranging between $-4.6$ and $-7$. With the available experimental set up, the highest attainable value of $Q_{max}$ was $\simeq 0.22$ Å$^{-1}$ which corresponds to $Q_{max}R \simeq 6$. The latter value was to low to reach the Porod limit.

*Effect of dilution:* Figs. 3 and 4 show another representations of the scattering data in the form $I(Q)Q^4$ versus $Q$ for samples 2 and 6, at various volume fractions. One observes a minimum which can be asso-

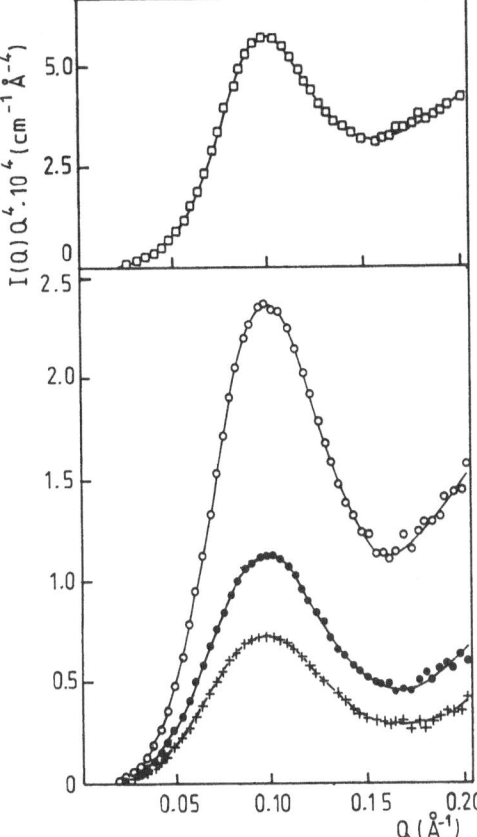

Fig. 3. $I(Q)^4$ versus $Q$ plots for sample 2: $\Phi = 0.233$ □; $\Phi = 0.093$ ○; $\Phi = 0.052$ ●; $\Phi = 0.031$ +

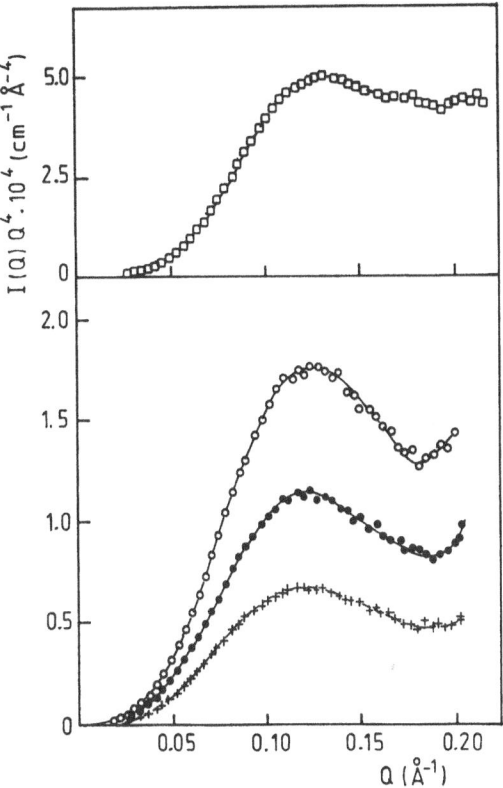

Fig. 4. $I(Q)Q^4$ versus $Q$ plots for sample 6. $\Phi = 0.231$ □; $\Phi = 0.0924$ ○; $\Phi = 0.0513$ ●; $\Phi = 0.0308$ +

ciated with the first minimum of the scattering pattern of dense particles. The radii of the particles have been calculated using Eq. (15) and are reported in Table 3.

Inspection of Figs. 3 and 4 and Table 3 show that, for both systems, the structural properties are not affected by dilution: the positions of the maximum and of the minimum of $I(Q)Q^4 = f(Q)$ do not significantly vary with volume fraction. An exception is sample 6 which contains the largest amount of AM (Fig. 4, top); in this case, the minimum is rather indistinct. This would be in agreement with previous measurements which indicated that this volume fraction was above the percolation threshold [9].

To obtain the particle size in an alternative way, we have performed the Guinier plots as represented in Fig. 5. From the initial slopes of $Ln I(Q)$, versus $Q^2$, we have deduced the values of the radii of gyration listed in Table 3.

In the same Table are reported the values of the equivalent hard sphere particles, $R_p = (5/3)^{1/2} R_g$ together with the hydrodynamic radii determined from

Table 3. Characteristics of the systems

| Sample | $\Phi$ (%) | $R_g$ (Å) | $R_p$ (Å) | $R_{Q\,min}$ (Å) | $R_{QELS}$ (Å) | $R_{HS}$ (Å) |
|---|---|---|---|---|---|---|
| 2 | 23.17 | 22.8 | 29.4 | 28.6 | 30 | 30 |
| 6 | 23.10 | 18.8 | 24.3 | 24.5 | 25.5 | — |

QELS experiments. There is an excellent agreement between $R_{Q\min}$ (from the minimum of $I(Q)Q^4$), $R_{HS}$ (from the hard sphere fit), $R_p$ (from the Guinier plot) and $R_H$ (from QELS). Systems 6 containing AM are characterized by lower particle size. This may be correlated to the partial localization of the monomer at the w/o interface, resulting in a decrease of the interfacial tension as shown elsewhere [29].

A more detailed analysis of the structural and interfacial changes caused by the incorporation of AM in AOT systems will be provided by the use of the variable contrast method. This study is under investigation together with the results on polymerized systems and will be presented in a further report.

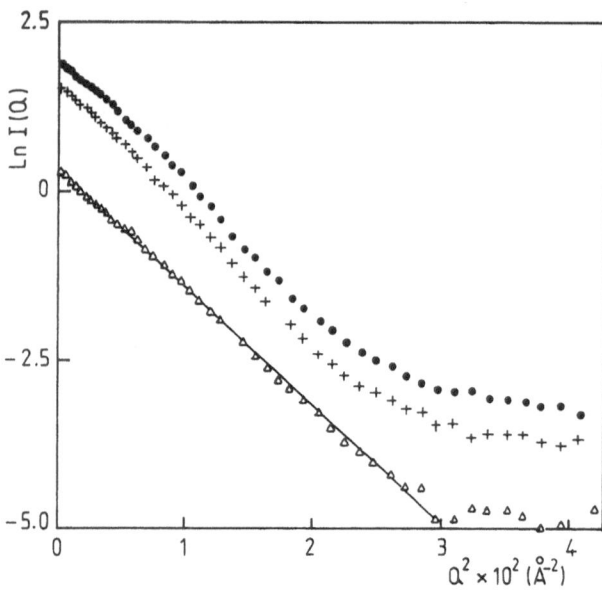

Fig. 5. Guinier plots of sample 2. $\Phi = 0.052$ ●; $\Phi = 0.031$ +; $\Phi = 0.010$ △

*Acknowledgements*

We wish to acknowledge support from NATO in the form of a Research Grant and the Institut Laue-Langevin for neutron beam facilities. Our thanks are also due to I. Markovic for her help in early experiments and Dr. A. Rennie and Dr. T. Zemb for many useful discussions.

# References

1. Atik SS, Thomas KJ (1982) J Am Chem Soc 104:5868
2. Jayakrishnan A, Shah DO (1984) J Polym Sci, Polym Lett Ed 22:31
3. Tang HI, Johnson PL, Gulari E (1984) Polymer 25:1357
4. Schauber C, Riess G (1989) Makromol Chem 190:725
5. Leong YS, Candau F (1982) J Phys Chem 86:2269
6. Kuo PL, Turro NJ, Tseng CM, El Aasser M, Vanderhoff JW (1987) Macromolecules 20:1216
7. Stoffer JO, Bone T (1980) J Polym Sci 18:2641
8. Candau F, Leong YS, Pouyet G, Candau SJ (1984) J Coll Interf Sci 101:167
9. Carver MT, Hirsch E, Wittmann JC, Fitch RM, Candau F (1989) J Phys Chem 93:4867
10. Leong YS, Candau SJ, Candau F (1984) In: Mittal K, Lindmann B (eds) Surfactants in Solution. Plenum Press, New York, 3:1987
11. Cabos C, Delord P (1979) J Apply Cryst 12:502
12. Cabos C, Delord P (1980) J Phys (Letters) 41:455
13. Robertus C, Joosten JGH, Levine YK (1988) Progr Coll & Polym Sci 77:115
14. Sheu E, Chen SH, Huang JS (1987) J Phys Chem 91:3306
15. Brochette P, Zemb T, Mathis P, Pileni MP (1987) J Phys Chem 91:1444
16. Neutron Beam Facilities at the H.F.R. available for Users (Institut Laue-Langevin, Grenoble, 1977)
17. Ottewill RH (1982) In: Goodwin W (ed) Colloidal Dispersions. RSC, London, Chap 7 p 143 and Chap 9 p 197
18. Cabane B (1987) In: Zana R (ed) Surfactant Solutions New Methods of Investigation. Marcel Dekker, New York, Chap 2 pp 57–139
19. Hayter JB (1985) In: Degiorgio V, Corti M (eds) Physics of Amphiphiles: Micelles, Vesicles and Microemulsions. North Holland, Amsterdam, p 59
20. Percus JK, Yevick GJ (1958) Phys Rev 110:1
21. Ashcroft NW, Lekner J (1966) Phys Rev 45:33
22. Cebula DJ, Ottewill RH, Ralston J, Pusey PN (1981) J Chem Soc, Faraday Trans 77:2585
23. Agterof WGM, Van Zomeren JAJ, Vrij A (1976) Chem Phys Lett 43:363
24. Lemaire B, Bothorel P, Roux D (1983) J Phys Chem 87:1023
25. Brunetti S, Roux D, Bellocq AM, Fourche G, Bothorel P (1983) J Phys Chem 87:1028
26. Huang JS, Safran S, Kim MW, Grest G, Kotlarchyk M, Quirke N (1984) Phys Rev Lett 53:592
27. Porod G (1951) Kolloid Z 124:83
28. Auvray L, Cotton JP, Ober R, Taupin C (1984) J Phys 45:913
29. Pichot C, Graillat C, Revillon A (1987) Proceedings of the XVII Congress AFTPV, Nice, p 270

Authors' address:

Dr. Françoise Candau
Institut Charles Sadron (CRM-EAHP)
6, rue Boussingault
67083 Strasbourg Cedex, France

**Progress in Colloid & Polymer Science**                    Progr Colloid Polym Sci 81:87–88 (1990)

# Diffusion of latex particles in viscoelastic surfactant solutions

S. Bucci, H. Hoffmann and G. Platz

Lehrstuhl für Physikalische Chemie I, Universität Bayreuth, Bayreuth

*Key words:* Diffusion coefficients; latex particles; viscoelastic surfactants; networks

The viscoelastic properties of surfactants solutions are due to a three-dimensional network of rodlike micelles. The network does not contribute to the scattering process because the refractive index of the perfluorosystems is matched to the refractive index of water. The scattering of the solutions is therefore coming from the latex-particles and their self diffusion coefficient is measured. The perfluoro surfactants $C_9F_{19}CO_2N(CH_3)_4$ and $C_8F_{17}CO_3N(C_2H_5)_4$ were used for the preparation of the systems.

The zero shear viscosity in these systems passed over a maximum with increasing concentration. At the maximum the viscosity was five orders of magnitude higher than the one of water. The selfdiffusion coefficient of the latex particles is effected by the temporary network of the viscoelastic surfactant. However the diffusion coefficient of the latex particles does not monitor the makroscopic viscosity of the system. The selfdiffusion coefficient did not pass through a minimum with increasing surfactant concentration but decreased monotonically with surfactant concentration.

We obtain the microviscosity of the network solution from the measured latex self diffusion coefficients using the Stokes-Einstein-equation (Fig. 1).

In particular, in the system with excess salt we find that the value of the diffusion coefficient and therefore the microviscosity are substantially not effected by the presence of the network even in the gel-like solutions, and this result does not depend on the diameter of the latex particles.

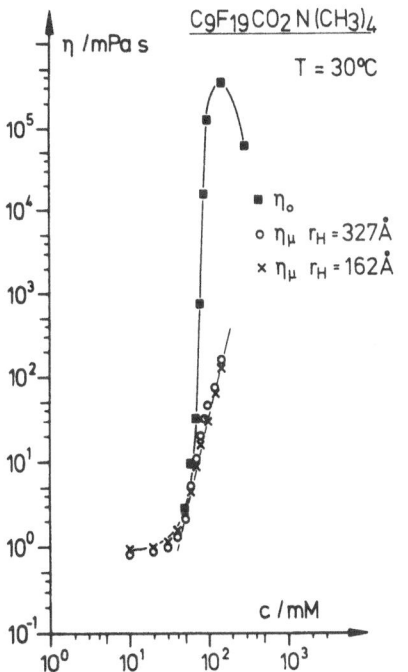

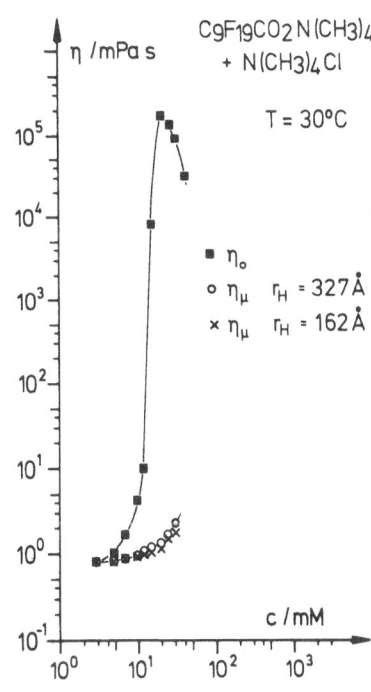

Fig. 1. Double log plot of the microviscosity $\eta_\mu$ which is calculated from the correlation times or self diffusion coefficient in comparison to the macroscopic viscosity $\eta_0$. 1a no excess salt, 1b $10^{-2}$ m $N(CH_3)_4Cl$ present

The meshsize of the network can be obtained from SANS-experiments [1] or from rheological measurements. [2] The storage modulus $G_0$ give the size $l_0$ by the following equation:

$$G_0 = v k_s T \quad \text{and} \quad l_0 = v^{-1/3}.$$

The simple picture that the particles should become trapped when the meshsize of the network becomes smaller then the dimension of the diffusing particles was found to be invalid. For conditions at which the dimensions of the particles were considerably larger than the meshsize which can be calculated from the amount of surfactant material or from the shear modulus of the network the particles were not trapped but could diffuse. Even in these situations there was little dependence of the microviscosity which is monitored by the diffusing particle on the diameter of the particle. The microviscosity which was calculated from the diffusion coefficient was up to four orders of magnitude smaller than the makroscopic shear viscosity.

Under the experimental conditions the correlation time as measured from the dynamic light scattering method was always considerably shorter than the structural relaxation time of the network. We can assume therefore that the network does not relax during the experiment. It follows than that the meshsize have a large size distribution and that any given mesh in the system fluctuates rapidly in size at a time scale which is shorter than the correlation time. In this way even particles with a dimension which is larger than the average meshsize can diffuse.

### References

1. Hoffmann H, Kalus J, Schwandner B (1987) Ber Bunsenges Phys Chem 91:99–106
2. De Gennes PG (1976) Macromolecules 9:587

Authors' address:

Prof. H. Hoffmann
Lehrstuhl für Physikalische Chemie I
Universität Bayreuth
Universitätsstraße 30
D-8580 Bayreuth

**Progress in Colloid & Polymer Science**

Progr Colloid Polym Sci 81:89–94 (1990)

# Light-scattering experiment on anisotropic spherical particles

R. Piazza, J. Stavans*), T. Bellini, D. Lenti, M. Visca and V. Degiorgio

Dipartimento di Elettronica-Sezione di Fisica Applicata, Università di Pavia, Pavia, Italy
Montefluos, Spinetta Marengo, Allesandria, Italy

*Abstract:* Static and dynamic light scattering measurements are performed on dilute aqueous dispersions of spherical latex particles made of a polytetrafluoroethylene copolymer. The particles are found to possess an intrinsic anisotropy, and to be likely polycrystalline. A correlation is found between the value of the measured anisotropy and the value of the melt flow index (MFI). Because of the existence of the intrinsic anisotropy, the dynamic light scattering data allow to derive not only the translational diffusion coefficient, but also the rotational diffusion coefficient, notwithstanding the fact that the particles are spherical.

*Key words:* TFE latex; crystalline particles; depolarized scattering; rotational diffusion

## Introduction

We present in this paper static and dynamic light scattering measurements performed on a dispersion of spherical latex particles made of polytetrafluoroethylene (PTFE) with an added copolymer. We have recently studied by light scattering [1] and electric birefringence [2] rod-like PTFE latex particles which are crystalline [3] and optically anisotropic. We show in this study that even the spherical latex prepared by adding an appropriate copolymer to PTFE is crystalline to some degree.

We have measured the polarization of scattered light, the average scattered intensity and the intensity correlation function versus the scattering angle $\theta$. The most striking property of the dispersion is that it is impossible to suppress scattering even by index-matching the particles with an appropriate solvent (a mixture of water and glycerol, in our case). In the index-matching situation, the scattered light is strongly depolarized, so that the intensity correlation measurements allow to easily derive both the translational and the rotational diffusion coefficients of the latex particles.

## Experimental

The used samples were colloidal spherical particles of a melt processable polytetrafluoroethylene copolymer, a commercial product by Montefluos S.p.A., Milano, Italy, obtained by emulsion polymerization. As observed with the electron microscope (see Fig. 1), the particles are fairly monodisperse, and have a spherical shape. The original latex was purified by dialysis which is very efficient in removing weakly adsorbed molecules and ions. It should be noted that the dialysis process only partially removes the strongly adsorbed fluorinated surfactant used in the polymerization procedure. Dialysis was carried out for over 12 weeks, until a stable conductivity value of about 100 µS was reached. The light scattering samples were prepared by diluting the latex solution with BDH AnalaR water, with a small amount of nonionic surfactant added to prevent slow coagulation. All the used solutions contained about 10 mM NaCl, which was added to screen electrostatic interactions. For the intensity measurements we used two different kinds of spheres, with the same copolymer composition, but differing for the "melt flow index" (MFI), the higher

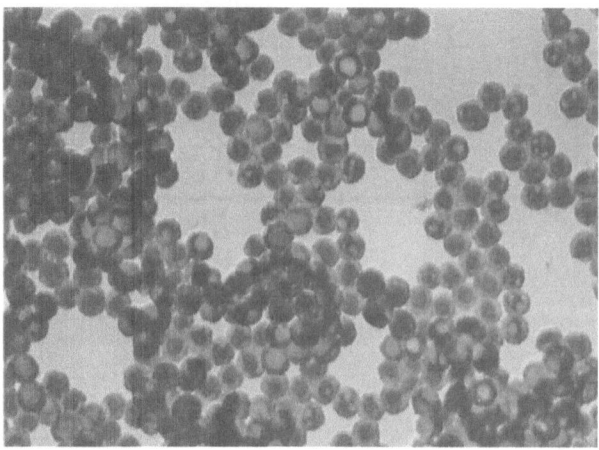

Fig. 1. Transmission electron microscope photograph of spherical latex particles. The average particle radius is 85 nm

MFI corresponding to a lower molecular weight of the polymeric chains. The average radius is 70 nm for the spheres having MFI = 0.6 (henceforth called copolymer 1), and 90 nm for the spheres having MFI = 5.5 (copolymer 2). Spheres coming from a preparation similar to that of copolymer 2 were used for the intensity correlation measurements.

The light scattering apparatus includes an argon laser operating at $\lambda = 514.5$ nm, a cylindrical scattering cell, a photomultiplier tube mounted on a rotating arm, and a digital correlator for the measurement of the intensity correlation function of scattered light. All measurements were performed at room temperature. The index of refraction of the solvent, $n_s$, was varied by mixing water with glycerol. In order to keep the turbidity of the sample approximately constant, when the optical mismatch between particle and solvent was varied the particle volume fraction was changed accordingly. The used volume fractions were in the range $2.5 \cdot 10^{-3} - 4 \cdot 10^{-2}$. The values of $n_s$ reported on the abscissa axis of Figs. 2 and 3 are relative to $\lambda = 589$ nm (sodium line).

By measuring the ratio $P_i/P_t$ between the incident and transmitted power of the laser beam as a function of $n_s$, we derive the turbidity $T$ as $T = (1/L)\ln(P_i/P_t)$, where $L$ is the pathlength of the beam in the scattering cell. The experimental data for copolymer 1 are shown in Fig. 2, where we have reported the specific turbidity $T/\phi$, $\phi$ being the volume fraction occupied by the latex particles. We see that the turbidity never goes to 0, that is, it is impossible to optically match particles and

solvent. The minimum turbidity corresponds to a weight fraction of glycerol of 20.5%.

The intensity and polarization of scattered light were measured at $\theta = 90°$ as functions of the refractive index of the solvent for both copolymers 1 and 2. The geometry of the experiment is the following: the incident beam is linearly polarized in the vertical direction, the scattering plane is horizontal. We call $I_{VV}(I_{VH})$ the intensity of scattered light with vertical (horizontal) polarization. The results obtained with copolymer 1 are shown in Fig. 2. We see that the behavior of $I_{VV}$ is similar to that of the turbidity, whereas $I_{VH}$ is independent of $n_s$. The minimum of $I_{VV}$ occurs in the same position as the minimum of $T$. At the minimum, the ratio $I_{VV}/I_{VH}$ is equal to 1.85. Figure 3 presents the comparison between the $I_{VV}$ curves obtained with copolymer 1 and 2. We see that both have a parabolic shape, but the two minima are in different positions (the minimum turbidity for copolymer 2 corresponds to a glycerol weight fraction of 17.5%), and the widths of the two parabolas are also different. The ratio $I_{VV}/I_{VH}$ for copolymer 2 is the same as that of copolymer 1.

Intensity correlation function measurements were performed at various scattering angles for horizontally polarized scattered light. We have chosen to measure $G_{VH}(\tau)$ because the results are simpler to interpret. The obtained correlation functions show a nearly exponential behavior. We have analyzed the data with the usual cumulant fit. We report in Fig. 4 the behavior of the first cumulant $\Gamma$ versus $k^2$, where $k$ is the scattering vector given by $(4\pi n_s/\lambda)\sin(\theta/2)$, $\lambda$ being the wavelength of incident light. The measured $\Gamma$ behaves as a linear function of $k^2$, and shows a nonzero intercept when extrapolated at $k = 0$.

## Discussion

The general treatment of static and dynamic light scattering from large anisotropic particles is rather complex [4–7]. In our case we can take advantage of the fact that the latex particles are suspended in a nearly index-matched solvent, so that we can interpret the experimental results by using the Rayleigh-Gans approximation [4], even if the particle size is comparable to $\lambda$.

We first recall the expression of the field scattered by a small anisotropic particle characterized by a polarizability tensor $\boldsymbol{\alpha}$. If the incident field is given as

$$\vec{E}_i(\vec{r},t) = \vec{n}_i E_0 \exp[i(\vec{k}_i \cdot \vec{r} - \omega t)] \qquad (1)$$

where $\vec{n}_i$ is a unit vector in the direction of the incident field, the scattered field $\vec{E}_s$ observed at distance $R$ in the far field is given by

$$\vec{E}_s(R,t) = (E_0/4\pi\in R)\vec{k}_s \times [\vec{k}_s \times (\boldsymbol{\alpha}\cdot\vec{n}_i)]$$
$$\times \exp[i(k_s R + \vec{k}\cdot\vec{r} - \omega t)] \qquad (2)$$

where $\vec{r}$ is the particle position with respect to the origin of the coordinate system, $\vec{k}_s$ is the wave vector of the scattered field, and $\vec{k} = \vec{k}_s - \vec{k}_i$. Assuming that the incident field is linearly polarized in the vertical direction and that the scattered field is observed in the horizontal plane ($\vec{k}_s \perp \vec{n}_i$), we can write the following expressions for the vertical and horizontal components of $\vec{E}_s$, $E_{VV}$ and $E_{VH}$:

$$E_{VV} = -(E_0 k_s^2/4\pi\in R)\alpha_{VV}$$
$$\times \exp[i(k_s R + \vec{k}\cdot\vec{r} - \omega t)] \qquad (3)$$

$$E_{VH} = -(E_0 k_s^2/4\pi\in R)\alpha_{VH}$$
$$\times \exp[i(k_s R + \vec{k}\cdot\vec{r} - \omega t)] \qquad (4)$$

where

$$\alpha_{VV} = \vec{n}_V\cdot\boldsymbol{\alpha}\cdot\vec{n}_V, \quad \text{and} \quad \alpha_{VH} = \vec{n}_H\cdot\boldsymbol{\alpha}\cdot\vec{n}_V. \qquad (5)$$

Note that $\alpha_{VV}$ and $\alpha_{VH}$ are calculated in the laboratory-fixed frame, and are, therefore, time-dependent because of the rotational motion of the particle.

Assuming cylindrical symmetry for $\boldsymbol{\alpha}$ along a main optical axis, calling $\alpha_\parallel$, $\alpha_\perp$ the diagonal components of the polarizability tensor in the particle-fixed frame, the laboratory-fixed components $\alpha_{VV}$ and $\alpha_{VH}$ are given by [6] and [7]

$$\alpha_{VV} = \alpha + (16\pi/45)\beta\, Y_{2,0}(\delta,\varphi) \qquad (6)$$

$$\alpha_{VH} = i(2\pi/15)\beta[Y_{2,1}(\delta,\varphi) + Y_{2,-1}(\delta,\varphi)] \qquad (7)$$

where $\alpha = (\alpha_\parallel + 2\alpha_\perp)/3 - V(n_s^2 - 1)$ is the average excess polarizability of the particle with respect to the solvent, $V$ is the particle volume, $\beta = \alpha_\parallel - \alpha_\perp$ is the anisotropy of the particle polarizability, $Y_{2,m}$ is the second order spherical harmonic of index $m$, and $\delta$ and $\varphi$ are the angles describing the optical axis direction relative to the laboratory frame. By decoupling rotations from translations, assuming that the solution is sufficiently dilute to make interactions negligible, and averaging over all particle orientations, we can calculate from Eqs. (3) and (4) the field correlation functions

$$|G_{VV}(\tau)| = C[\alpha^2\exp(-D_T k^2\tau)$$
$$+ (4/45)\beta^2\exp(-D_T k^2\tau - 6D_R\tau)] \qquad (8)$$

$$|G_{VH}(\tau)| = C(\beta^2/15)\exp(-D_T k^2\tau - 6D_R\tau) \qquad (9)$$

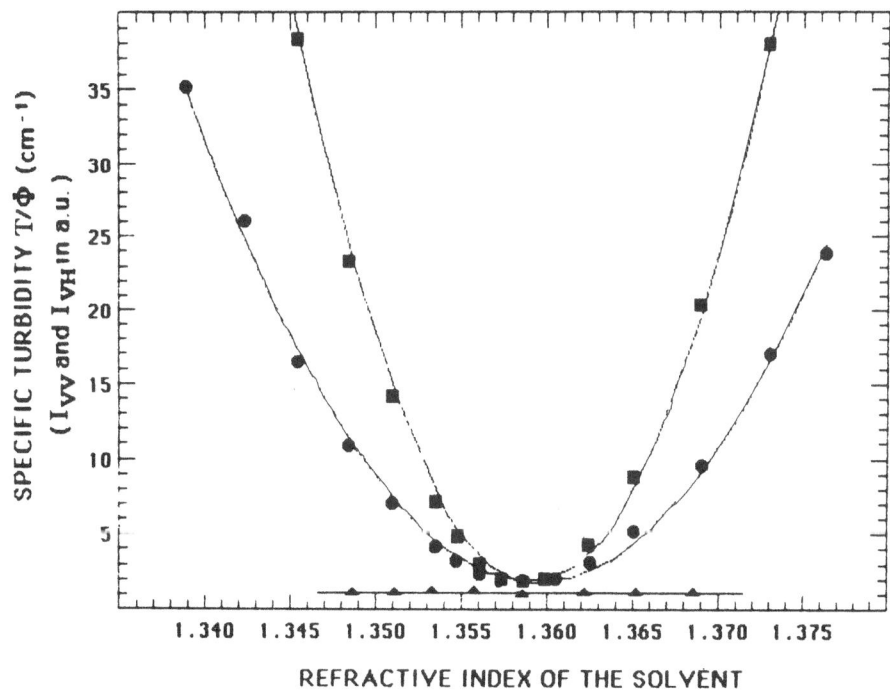

Fig. 2. Specific turbidity (●), and intensity of scattered light $I_{VV}$ (■) and $I_{VH}$ (▲) as a function of the refractive index of the solvent $n_s$

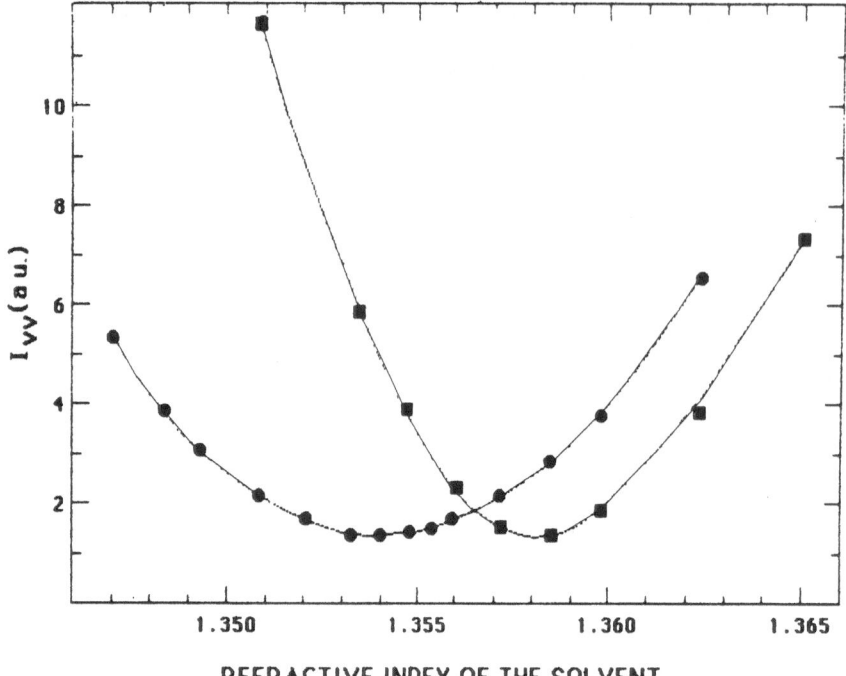

Fig. 3. $I_{VV}$ as a function of the refractive index of the solvent for copolymer 1 (■) and 2 (●)

where $D_T$ and $D_R$ are, respectively, the translational and rotational diffusion coefficients of the particle, and $C$ is a constant. Note that the average scattered intensities are simply derived from Eqs. (8) and (9) as: $I_{VV} = G_{VV}(0)$, $I_{VH} = G_{VH}(0)$.

The turbidity $T$ is given by

$$T = N\sigma \qquad (10)$$

where $N$ is the concentration of scatterers and $\sigma$ is the total scattering cross-section which can be expressed as [1, 4]:

$$\sigma = A[\alpha^2 + (2/9)\beta^2] \qquad (11)$$

where $A$ is a constant. Since, in our case, both $\beta$ and $\alpha$ are much smaller than $(\alpha_\parallel + 2\alpha_\perp)/3$, we can write approximately $\sigma$ and the scattered intensities as:

$$\sigma = A'[(\bar{n} - n_s)^2 + (2/9)(n_\parallel - n_\perp)^2] \qquad (12)$$

$$I = C'[(\bar{n} - n_s)^2 + (4/45)(n_\parallel - n_\perp)^2] \qquad (13)$$

$$I = C'(n_\parallel - n_\perp)^2/15 \qquad (14)$$

where $\bar{n} = [(n_\parallel^2 + 2n_\perp^2)/3]^{1/2}$ is the average index of refraction of the particles, and $A'$ and $C'$ are constants. As discussed previously, since the particles are almost

index-matched by the solvent, we can treat light scattering in the Rayleigh-Gans approximation: if the particle is homogeneous, this is equivalent to say that the field scattered by the particle is obtained by multiplying the small-particle results (Eqs. (3) and (4)) by the particle form factor. This will affect only the angular distribution of the scattered intensity, leaving unchanged the dependence on $n_s$, and the temporal behavior of the correlation functions. We can still use, therefore, Eqs. (12)–(14) and Eqs. (8)–(9), respectively, for the interpretation of the static and dynamic light scattering data.

We see from Fig. 2 that the measured turbidity is indeed a parabolic function of the refractive index of the solvent. The full line in the figure represents the fit with Eq. (12). The fit parameters are $n_\parallel - n_\perp = 0.0086 \pm 0.0008$, and $\bar{n} = 1.3581 \pm 0.0002$. A similar analysis is applied to the plot of $I_{VV}$ presented in Fig. 2. The fit with Eq. (13) gives: $n_\parallel - n_\perp = 0.0087 \pm 0.0008$ and $\bar{n} = 1.3581 \pm 0.0002$, in very good agreement with the values derived from the turbidity curve.

We note that the ratio $I_{VV}/I_{VH}$, as calculated from Eqs. (13) and (14) for $\bar{n} = n_s$, is equal to 4/3. The fact that the experimental ratio is larger than 4/3 indicates that the particle cannot be treated as a single crystal, but has a more complicated internal structure which includes, perhaps, both amorphous and crystalline regions. In order to understand the effect of a possible non-uniformity of the local polarizability, we have calculated the scattered intensity for polycrystalline spheres, by assuming a completely random distribution for the orientation of the optical axis of different crystalline regions. If we call $r$ the number of crystalline regions, we find that the scattered intensity can still be expressed by Eqs. (13) and (14) with the only difference that, instead of the anisotropy $n_{\parallel} - n_{\perp}$, we have an "effective" anisotropy $(n_{\parallel} - n_{\perp})/\sqrt{r}$. This suggests that the large difference in the measured anisotropy between the PTFE latex studied in Ref. [1] and the PTFE copolymers studied in this paper is mainly due to the fact that the pure PTFE particle is essentially a single crystal, whereas the latex particles obtained with the PTFE copolymer are polycrystalline. Furthermore, the difference in the measured anisotropy between copolymers 1 and 2 is probably due to the difference in the number $r$ of crystalline regions.

In the case of copolymer 2, the fit of the $I_{VV}$ curve of Fig. 3 gives $n_{\parallel} - n_{\perp} = 0.0160 \pm 0.0008$ and $\bar{n} = 1.3540 \pm 0.0005$. It is interesting to note that different values of the MFI correspond to different values of the optical anisotropy of the latex particles.

Finally, we discuss the dynamic light scattering data. As it is well-known, the measured quantity is the intensity correlation function which, under the hypothesis that the scattered field is a Gaussian process,

has a time-dependent part proportional to the square modulus of the field correlation function. According to Eq. (9), the first cumulant of $G_{VH}$ is given by

$$\Gamma = D_T k^2 + 6 D_R. \tag{15}$$

We see, indeed, from the data of Fig. 4 that the measured $\Gamma$ is a linear function of $k^2$. The fit with Eq. (15) gives the diffusion coefficients $D_T = (1.43 \pm 0.05) \times 10^{-8}$ cm$^2$/s, and $D_R = (135 \pm 5)$ s$^{-1}$. By taking $T = 22\,^{\circ}$C and a solvent viscosity of 1.73 cp, we derive from the experimental value of $D_T$ a particle radius of 87 nm, and from the experimental $D_R$ a particle radius of 88 nm. The internal agreement between the two data is excellent, and the consistency with the electron microscope evaluation ($R \approx 85$ nm) is satisfactory.

As a conclusion, we have given in this work a rather complete characterization of spherical latex particles made of a PTFE copolymer. In particular, we have discovered that the particles are partially crystalline, and that the degree of crystallinity shows a correlation with the melt flow index of the polymer. Our data indicate that the light scattering technique can easily be used as a method to detect the crystalline structure of colloids. From a more fundamental point of view, the latex we have studied represents a very interesting model system for optically anisotropic particles. As far as we know, the data reported in this paper represent

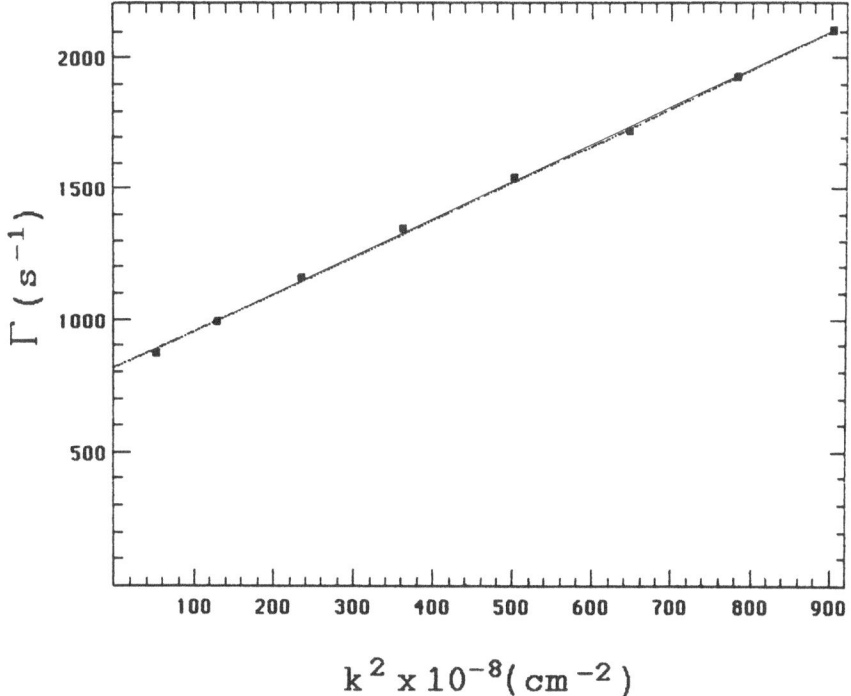

Fig. 4. Reciprocal of the decay time of the correlation function of $I_{VH}$ plotted as a function of $k^2$

the first light-scattering measurement of the rotational diffusion constant of a spherical colloid particle.

*Acknowledgements*

We thank M. Corti for useful discussions. J. Stavans was supported by a grant from the International Center for Theoretical Physics (Trieste, Italy) under the Training Program in Italian Laboratories. This work was supported by funds from the Italian Ministero della Pubblica Istruzione (MPI 40%).

## References

1. Piazza R, Stavans J, Bellini T, Degiorgio V (1989) Opt Commun 73:263
2. Bellini T, Piazza R, Sozzi C, Degiorgio V (1988) Europhys Lett 7:561
3. Ottewill RH, Rance DG (1986) Coll Polymer Sci 264:982
4. van de Hulst HC (1981) "Light Scattering by Small Particles", Dover, New York
5. Kerker M (1969) "The Scattering of Light and Other Electromagnetic Radiation", Academic Press, New York
6. Berne B, Pecora R (1975) "Dynamic Light Scattering", Wiley, New York
7. Aragon SR, Pecora R (1977) J Chem Phys 66:2506

Authors' address:

Prof. V. Degiorgio
Dipartimento di Elettronica, Università di Pavia
via Abbiategrasso 209
27100 Pavia, Italy

# Hydrodynamic interactions in suspensions

B. U. Felderhof

Institut für Theoretische Physik A, R.W.T.H. Aachen, Aachen, FRG

*Abstract:* We give a brief review of the concept of hydrodynamic interactions in fluid suspensions. We discuss their relevance for the calculation of effective viscosity, collective mobility and self-mobility.

*Key words:* Hydrodynamic interactions; diffusion; mobility; sedimentation; brownian motion

It is easily shown from statistical mechanics that hydrodynamic interactions do not influence the thermal equilibrium properties of a fluid suspension. In particular the static structure factor and the phase diagram of a suspension are independent of hydrodynamic interactions. In contrast, the dynamics of a suspension is dominated by hydrodynamic interactions, unless the suspended particles are charged and the Debye length is much larger than the actual particle radius. Especially in dense suspensions the transport coefficients are strongly influenced by hydrodynamic interactions. Peculiar features of these interactions are their extremely long range and their manybody character.

Hydrodynamic interactions between solute particles are due to the flow patterns caused by their motion in the ambient fluid. To elucidate the concept we consider first the hydrodynamic interactions between a pair of particles suspended in a uniform fluid of infinite extent. If the particles are much larger than the molecules of the fluid, then the latter may be described as a continuum. In good approximation the fluid may be taken to be incompressible. Provided the particle velocities are sufficiently small, the fluid motion is well described by the so-called creeping flow equations [1]

$$\eta \nabla^2 \vec{v} - \nabla p = 0, \quad \nabla \cdot \vec{v} = 0, \qquad (1)$$

where $\eta$ is the shear viscosity, $\vec{v}(\vec{r})$ the fluid flow velocity, and $p(\vec{r})$ the pressure, which is determined by the condition of incompressibility. If external forces $\vec{E}_1$, $\vec{E}_2$ act on the particles, then they acquire velocities $\vec{U}_1$, $\vec{U}_2$ which are linearly related to the forces. The linear equations are

$$\begin{aligned} \vec{U}_1 &= \vec{\mu}_{11}(\vec{R}) \cdot \vec{E}_1 + \vec{\mu}_{12}(\vec{R}) \cdot \vec{E}_2, \\ \vec{U}_2 &= \vec{\mu}_{21}(\vec{R}) \cdot \vec{E}_1 + \vec{\mu}_{22}(\vec{R}) \cdot \vec{E}_2. \end{aligned} \qquad (2)$$

For spherical particles the mobility tensors $\vec{\mu}_{ij}(\vec{R})$ depend only on the distance vector $\vec{R} = \vec{R}_2 - \vec{R}_1$ between the two centers. The tensors are found from the solution of Eqs. (1). For hard particles we must specify boundary conditions at the particle surface. Usually stick boundary conditions are appropriate. For microemulsion droplets we must specify the properties of the membrane and the viscosity of the inner fluid. On account of isotropy the mobility tensors take the form

$$\begin{aligned} \vec{\mu}_{11}(\vec{R}) &= \alpha_{11}(R) \hat{R}\hat{R} + \beta_{11}(R)(\vec{1} - \hat{R}\hat{R}), \\ \vec{\mu}_{12}(\vec{R}) &= \alpha_{12}(R) \hat{R}\hat{R} + \beta_{12}(R)(\vec{1} - \hat{R}\hat{R}), \end{aligned} \qquad (3)$$

with scalar functions $\alpha_{ij}(R)$, $\beta_{ij}(R)$. The mobility tensors incorporate the hydrodynamic interactions between the two particles.

For two particles the hydrodynamic interactions are known in full detail [2−4]. A method of reflections, in which successive flow patterns act between the two particles, naturally leads to a series expansion of the scalar functions in Eq. (3) in inverse powers of the distance $R$. Some years ago [5] I evaluated these expansions by hand up to the power $R^{-7}$. Later longer expansions were obtained [6, 7]. Recently we have developed the method into a powerful scheme which allows arbitrary numerical accuracy [4]. For two equal hard spheres we have determined up to 843 terms of the series expansions [8]. For most practical purposes a smaller number of terms is sufficient.

For two equal hard spheres of radius $a$ with stick boundary conditions the first few terms of the series expansions read explicitly

$$\alpha_{11}(R) = \frac{1}{6\pi\eta a} - \frac{7}{8\pi\eta a}\left(\frac{a}{R}\right)^4 + O(R^{-6}),$$

$$\beta_{11}(R) = \frac{1}{6\pi\eta a} + O(R^{-6}),$$

$$\alpha_{12}(R) = \frac{1}{4\pi\eta R} - \frac{1}{6\pi\eta a}\left(\frac{a}{R}\right)^3 + O(R^{-7}),$$

$$\beta_{12}(R) = \frac{1}{8\pi\eta R} + \frac{1}{12\pi\eta a}\left(\frac{a}{R}\right)^3 + O(R^{-11}).$$

(4)

The last two expressions exhibit the long range of the hydrodynamic interactions. This long range originates from the Stokes flow pattern generated by a single sphere acted on by an external force $\vec{E}$. The Stokes flow pattern reads

$$\vec{v}^{St}(\vec{r}) = \frac{1}{8\pi\eta}\left[\frac{\overset{\rightrightarrows}{1+\hat{r}\hat{r}}}{r} + \frac{1}{3}a^2\frac{\overset{\rightrightarrows}{1-3\hat{r}\hat{r}}}{r^3}\right] \cdot \vec{E}.$$

(5)

The sphere acquires the Stokes velocity $\vec{U}^{St} = \vec{E}/6\pi\eta a$.

For $N$ particles we must solve the complete flow problem. In principle the solution of Eqs. (1) exists for specified applied forces $\vec{E}_1,\ldots,\vec{E}_N$ acting on the particles and leads to a set of translational particle velocities $\vec{U}_1,\ldots,\vec{U}_N$ and rotational velocities $\vec{\Omega}_1,\ldots,\vec{\Omega}_N$. The solution is unique if we specify that the torques and the fluid velocity at infinity vanish. The linear relationship between translational velocities and forces is expressed by

$$\begin{pmatrix}\vec{U}_1\\\vdots\\\vec{U}_N\end{pmatrix} = \tilde{\tilde{\mu}}(\vec{R}_1,\ldots,\vec{R}_N)\begin{pmatrix}\vec{E}_1\\\vdots\\\vec{E}_N\end{pmatrix}$$

(6)

with a $3N \times 3N$ mobility matrix dependent on the particle configuration. This matrix cannot be expressed in terms of two-body mobility matrices. The more complicated dependence implies that the hydrodynamic interactions have many-body character. The reason for this complicated dependence is the propagation of flow disturbances due to reflection from the set of particles. The complete flow pattern takes the form

$$\vec{v}(\vec{r}) = \sum_{j=1}^{N} \vec{v}_j^{St}(\vec{r} - \vec{R}_j) + \sum_{j=1}^{N} \hat{\vec{v}}_j(\vec{r}; \vec{R}_1,\ldots,\vec{R}_N),$$

(7)

where the first term is the sum of initial Stokes flow patterns, generated by the applied forces according to Eq. (5), and the second term is the sum of flows caused by reflections against the freely moving particles. The latter may be expressed in terms of a multiple scattering expansion [9].

In suspensions we deal with a complicated many-body problem, but on the other hand we are not interested in the full details of the solution. We can make headway by use of statistical methods. Thus we average over a probability distribution $P(\vec{R}_1,\ldots,\vec{R}_N)$, which in first approximation may be assumed to be known. For example, we may employ an equilibrium distribution of hard spheres, or another equilibrium distribution which follows from the direct particle interactions. In this manner we obtain an equation for the average suspension flow velocity, which for slow spatial variations takes the form [10]

$$\eta_{eff}\nabla^2\langle\vec{v}\rangle - \nabla\langle p\rangle = -n\vec{E}(\vec{r}), \quad \nabla\cdot\langle\vec{v}\rangle = 0$$

(8)

with an effective viscosity $\eta_{eff}$ and mean particle number density $n$. We have assumed that the applied forces are derived from a field $\vec{E}(\vec{r})$ according to the rule $\vec{E}_j = \vec{E}(\vec{R}_j)$. We also find an equation for the average particle current density

$$\langle\vec{J}\rangle - n\langle\vec{v}\rangle = n\mu_C\vec{E}(\vec{r}),$$

(9)

where the current density $\vec{J}$ is defined by

$$\vec{J}(\vec{r}) = \sum_{j=1}^{N}\vec{U}_j\delta(\vec{r} - \vec{R}_j),$$

(10)

and $\mu_C$ is the collective mobility. It is the task of theory to evaluate the transport coefficients $\eta_{eff}$ and $\mu_C$. In addition one is interested in the self-mobility $\mu_S$, which gives the average velocity of a particle when a force is applied to that particle only. For these transport coefficients one may derive statistical expressions involving the many-body hydrodynamic interactions [11, 12].

For a low density suspension the task may be reduced to the solution of a two-body problem. Since the two-body hydrodynamic interactions are well known this allows an exact evaluation of the first few terms in a series expansion of the transport coefficients in powers of the volume fraction $\phi = (4\pi/3)na^3$. Thus one finds for hard spheres with stick boundary conditions

$$\eta_{eff} = \eta \left[ 1 + \frac{5}{2}\phi + k_H \left(\frac{5}{2}\phi\right)^2 + O(\phi^3) \right]. \qquad (11)$$

The term linear in $\phi$ was found by Einstein. The latest accurate value [13] of the Huggins coefficient $k_H$ is 0.800. Similarly one finds for the collective mobility

$$\mu_C = (6\pi\eta a)^{-1}[1 - k_C\phi + O(\phi^2)] \qquad (12)$$

with coefficient $k_C = 6.546$. For the self-mobility one finds

$$\mu_S = (6\pi\eta a)^{-1}[1 - k_S\phi + O(\phi^2)] \qquad (13)$$

with coefficient $k_S = 1.831$. The values for the coefficients [13] slightly improve earlier results of Batchelor [14, 15].

The expressions (11)–(13) provide values for the transport coefficients valid for volume fractions of at most a few percent. The higher order correction terms cannot be evaluated exactly and approximations need be made. At the present time such approximate calculations are still in a preliminary stage. It is hoped that a comparison of theoretical results with computer simulations will provide sufficient guidance to allow one to select a reliable and useful approximation scheme.

An exciting result of recent theory deserves to be mentioned. Cichocki and I have found [11] that an approximate calculation of the collective mobility $\mu_C$, taking into account higher order terms in $\phi$, yields a behavior which numerically differs hardly from the linear relation as given in Eq. (12). As a consequence we find that $\mu_C$ vanishes at the rather low volume fraction $\phi_C \approx 0.15$. This indicates a dynamical phase transition.

In deriving this result we have assumed an equilibrium distribution of hard spheres. In an actual steady state sedimentation experiment the distribution function will change. In particular the pair distribution will become anisotropic. However, if we start from an equilibrium distribution, then at short times or high frequency the distribution function has no time to change. Therefore we expect our result to apply to the short-time collective diffusion coefficient and the high-frequency collective mobility.

Closer analysis of the linear coefficient $k_C$ in Eq. (12) shows that it is a sum of two contributions and may be written

$$k_C = 5 + k'_C, \qquad (14)$$

where the first term is kinematic, i.e. it arises as a general property of the flow equations, and is independent of the structure of the (spherical) particles. On the other hand the second term $k'_C$ does depend on the nature of the particles. For hard spheres with stick boundary conditions it takes the value $k'_C = 1.546$, but for slip boundary conditions or for liquid droplets a different value is found. Some years ago I derived the kinematic contribution in Eq. (14) from a calculation similar to that of the Lorentz local field in dielectrics [16]. From this point of view the kinematic term is of mean field type and accounts for collective effects. The higher order terms, accounting for correlations, make a relatively small correction to the mean field behavior.

As mentioned above, it is desirable to test the theory with the aid of computer simulations. Such simulations have the advantage that the nature of the particles and their distribution function can be specified precisely. To avoid boundary effects due to the finite system size it is necessary to use periodic boundary conditions. The long range of the hydrodynamic propagator in combination with periodic boundary conditions gives rise to a conceptual difficulty [17]. A rather subtle analysis is needed to resolve the issue [18]. At present we are performing computer simulations using about six hundred particles per unit cell. Our preliminary results indicate that the dynamical phase transition in the collective mobility occurs.

### References

1. Happel J, Brenner H (1973) Low Reynolds Number Hydrodynamics, Noordhoff, Leiden
2. Jeffrey DJ, Onishi Y (1984) J Fluid Mech 139:261
3. Kim S, Mifflin RT (1985) Phys Fluids 28:2033
4. Cichocki B, Felderhof BU, Schmitz R (1988) Physico Chem Hyd 10:383
5. Felderhof BU (1977) Physica 89 A:373
6. Schmitz R, Felderhof BU (1982) Physica 116A:163
7. Jones RB, Schmitz R (1988) Physica 149A:373
8. Cichocki B, Felderhof BU (1988) J Chem Phys 89:3705
9. Felderhof BU (1988) Physica 151A:1
10. Felderhof BU (1988) Physica 153A:217
11. Cichocki B, Felderhof BU (1989) Physica 154A:213
12. Cichocki B, Felderhof BU, Schmitz R (1989) Physica 154A:233

13. Cichocki B, Felderhof BU (1988) J Chem Phys 89:1049
14. Batchelor GK, Green JT (1972) J Fluid Mech 56:401
15. Batchelor GK (1976) J Fluid Mech 74:1
16. Felderhof BU (1976) Physica 82A:611
17. Smith ER, Snook IK, van Megen W (1987) Physica 143A:441
18. Felderhof BU (1989) Physica 159A:1

Author's address:

B. U. Felderhof
Institut für Theoretische Physik A
R.W.T.H. Aachen
Templergraben 55
5100 Aachen, FRG

**Progress in Colloid & Polymer Science**                    Progr Colloid Polym Sci 81:99 (1990)

# Shear-flow-induced structural changes

S. Hess and W. Loose

Institut für Theoretische Physik, Technische Universität Berlin

*Key words:* Nonequilibrium molecular dynamics; pair-correlation function; kinetic theory; shear-induced order; static structure factor

Nonequilibrium molecular dynamics (NEMD) computer simulation studies are reviewed, some recent results are presented and their relevance for dense colloidal dispersions are discussed. The shear-induced anisotropy of the velocity distribution function [1] and of the short range structure described by the pair-correlation function [2] as observed by NEMD are in good agreement with kinetic theory [1, 2]. The NEMD data for the short range structure are qualitatively similar to the distortion of the structure found by light scattering from sheared colloidal solutions [3, 4]. The formation of a long range ordered state seen in NEMD at high shear rates [5] has been questioned [6]. Recent NEMD studies, however, confirm the existence of a long range anisotropy as well as the coexistence of ordered and amorphous states and the occurrence of a plug like flow [7]. This is also in accord with a stability analysis based on generalized hydrodynamics [8]. The direct calculation of the static structure factor of model fluids under shear [9] allows a qualitative comparison with recent small angle neutron scattering data from colloidal solutions of spherical particles [10].

# References

1. Loose W, Hess S (1987) Phys Rev Lett 58:2443 (1988) Phys Rev A 37:2099
2. Hanley HJM, Rainwater JC, Hess S (1987) Phys Rev A 36:1795
3. Ackerson BJ, Clark NA (1983) Physica 118 A:221
4. Hanley HJM, Rainwater JC, Clark NA, Ackerson BJ (1983) J Chem Phys 79:4448
   Rainwater JC, Hanley HJM, Hess S (1988) Phys Lett 126 A:450
5. Erpenbeck JP (1984) Phys Rev Lett 52:1333
   Hess S (1985) Int J Thermophys 6:657
   (1985) J Mec Theor Appliq, Numéro Special, p 1
6. Evans DJ, Morriss GP (1986) Phys Rev Lett 56:2176
7. Hess S, Loose W (1988) Molecular dynamics: Test of microscopic models for the material properties of matter. In: Axelrad DR, Muschik W (eds) Constitutive laws and microstructure, Springer, Berlin, p 93
8. Loose W, Hess S (1989) Rheol Acta 28:101
9. Hess O, Loose W, Weider T, Hess S (1989) Physica B 156 & 157:505
10. Lindner P et al. (1987) Annual Report of the ILL, Grenoble, p 76

Authors' address:

Prof. Dr. Siegfried Hess
Institut für Theoretische Physik
Technische Universität Berlin
Sekr. PN 7-1
Hardenbergstr. 36
D-1000 Berlin 12, West-Germany

**Progress in Colloid & Polymer Science**

Progr Colloid Polym Sci 81:100 – 106 (1990)

# Time-dependent smallangle neutron measurements of aligned micelles

J. Baumann[1]), G. Hertel[2]), H. Hoffmann[2]), K. Ibel[3]), V. Jindal[1]), J. Kalus[1]), P. Lindner[3]), G. Neubauer[1]), H. Pilsl[1]), W. Ulbricht[2]) and U. Schmelzer[1])

[1]) Experimentalphysik I, Universität Bayreuth, Bayreuth, FRG
[2]) Physikalische Chemie I, Universität Bayreuth, Bayreuth, FRG
[3]) Institut Laue-Langevin, Grenoble, France

*Abstract:* The results of time dependent small angle neutron scattering (SANS) experiments on the alignment of four samples are described. The relaxation times encountered vary in magnitude from 150 ms to 60 min. For the shorter relaxation times a sampling method was used. The four different processes of which results are reported are:

— the transient alignment of *p*-TFE fibrilles in an electric field
— the rotation of a prealigned nematic phase in a magnetic field
— the build up and decay of a shear induced micellar structure in a shear field
— the overshoot during the alignment of rodlike micelles in a shear field.

*Key words:* SANS; micelles; transient alignment

## Introduction

Surfactant molecules in aqueous solutions have the ability to form large aggregates of different shapes which are called micelles. In dilute solutions these micelles can almost behave as noninteracting entities, but at higher concentrations, interaction plays an essential role. At high concentration the micelles are expected to form liquid crystalline phases. The micelles can be globular, rodlike or disclike in shape. The shape of the micelles as well as the interaction is reflected in the intensity distribution of scattered neutrons or x-rays. A typical length scale of a micelle may be around 5 nm. Neutrons or x-rays having a wavelength of 0.1 to 1 nm show the diffraction pattern concentrated in the region of small angles. The micelles in solution can be randomly oriented, can be aligned by some external forces or, as mentioned above, can form liquid crystalline structures. Particular attention needs to be paid to rodlike micellar solutions as these can be highly viscoelastic, which means elastic like a solid and viscous like a liquid at the same time. A hypothesis for this behaviour assumes, that the micelles form entanglements.

Small angle neutron scattering (SANS) experiments are an obvious choice for an experimental study of liquid solutions of micelles since large two dimensional detectors are available. Furthermore the scattering power is large too, because of the good contrast between micelles and the solvent; for the latter preferentially $D_2O$ is used. The twodimensional detector of instrument D11 of the Institut Laue-Langevin has for example $64 \times 64$ elements, each element of 1 cm$^2$ size. The detector can share the counting events in 32 time channels of preselected width, one following the other with a time for data handling of less than a ms. In this mode of operation a time dependent SANS pattern can be followed easily. Examples are presented in the next chapters.

## The sampling method

Neutron sources like reactors are notoriously weak in intensity. If, as an example, one is interested in a time dependent measurement of a system with a characteristic relaxation time $\tau$ of say one second, a useful measurement has to be done in a time of several seconds. Normally in such a short time the scattering intensity in a SANS experiment is by far too low. Therefore, the experiment has to be repeated several times and the subsequent results have to be added to

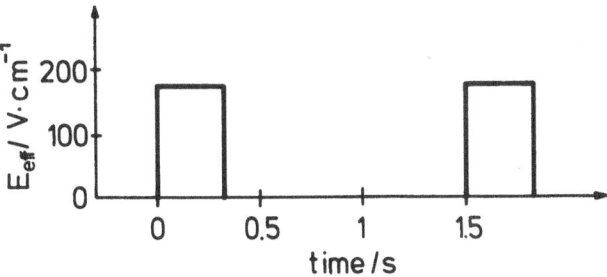

Fig. 1. The time dependence of an electric field as an example for the sampling method

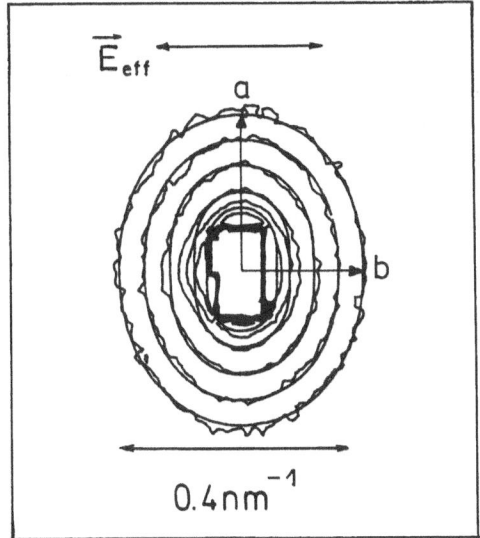

Fig. 2. Contour plot of the scattering intensities for the time interval $280 < t < 320$ ms for $p$-TFE. The solid smooth lines are due to a fit. The disturbance in the middle of the figure is due to the beam stop. The length of the arrow indicates a momentum transfer of $Q = 0.4$ nm$^{-1}$

get a useful result. This is illustrated in Fig. 1, where as an example, a certain time sequence of a disturbance, in this case an electric field, is applied to the sample, giving an alignment. The time dependence of the electric field can be chosen according to the physical details one is interested in. In our case a step function was chosen. The electric field was turned on for a certain time, say $t = 0.32$ s, and then suddenly dropped to zero. Measurements can be done, as mentioned above, for 32 time channels of a certain width, say 40 ms, to observe the build up and decay of the anisotropic scattering pattern induced by the electric field. Some time after the removal of the electric field the anisotropy has disappeared. Then a second experiment can be started in the same way as the previous one, as is indicated in Fig. 1 and so on. In this way hundreds or thousands of experiments can be performed and added, giving reasonable intensity.

## Alignment of micelles

The alignment we used was achieved by electric or magnetic fields and by shear. In the first two cases the anisotropy of the shapes of the micelles and the differences of the electric or magnetic susceptibilities between micelles and solution give rise to torques. Of course if permanent dipoles are present, torques are present anyway. For a single monomer the electrostatic or magnetostatic energy $\Delta E$ related to the torques is small compared to the thermal energy $kT$ and no alignment can be observed. Because of the fact that in a micelle $N$ monomers are organised ($N \gg 1$), the total energy becomes $\sim N \cdot \Delta E$ which may be comparable to or larger than $kT$. Then a substantial alignment is observed.

In the presence of a shear the axis of non spherical particles like platelets or rods are aligned more or less perpendicular or parallel to the velocity vector, respectively. Stochastic forces, described by diffusion and a rotational diffusion coefficient $D$ play an essential role for the degree of alignment.

In the next sections we give some results of relaxation measurements.

## Electric alignment of $p$-tetrafluorethylene ($p$-TFE)

The sample is a 0.4% solution by weight of $p$-TFE fibrilles dissolved in $D_2O$. To achieve a good solubility, 0.1% per weight of the ionic perfluorsurfactant $C_8F_{17}(CH_2)_2CO_2NH_4$ was added. Under these conditions most of the surfactant will be adsorbed on the surface of the fibrilles and the particles will be highly charged. The fibrilles had been examined by electron microscopy [1, 2] and the micrographs had indicated some polydispersity in the diameter and the length. Typically the diameter is around 6 nm [3], whereas the length is around 3 µm. An AC voltage was applied, exactly as shown in Fig. 1. The frequency was 16 kHz. The direction of the electric field was perpendicular to the direction of the neutrons hitting the sample. A typical contour plot of the anisotropic scattering intensities is shown in Fig. 2. In this experiment the

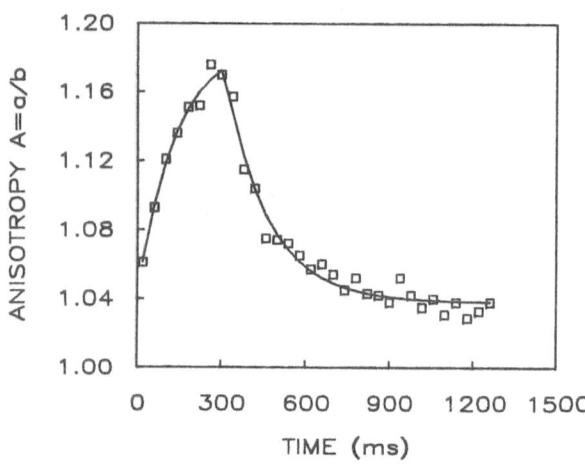

Fig. 3. The anisotropy $A = a/b$ ($a$ and $b$ are shown in Fig. 2) as a function of time for $p$-TFE. The smooth lines are due to a fit procedure

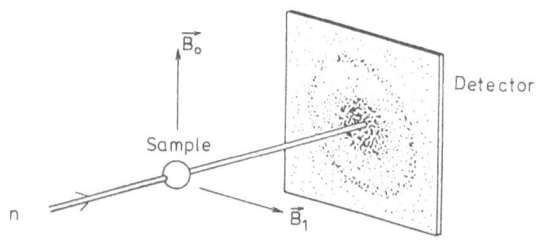

Fig. 4. The experimental set up for the measurement of alignment in magnetic fields $\vec{B}_0$ and $\vec{B}_1$

intensity of 8000 measurements were sampled. The width of the time channels were 40 ms. As a simple measure of the anisotropy we defined an anisotropy factor $A = a/b$ (see Fig. 2). The values of $A$ as a function of time are shown in Fig. 3. The anisotropy is not equal to 1 for long times, but levels off around $A = 1.038$. Presumably a long relaxation time of the order of several seconds or more is present, which could not be resolved in our experiment.

An equation of motion of non-interacting long rods in an electric field $E_z$ in $z$-direction is available [4, 5]. The time dependent distribution function $f$ of the rod axis is given by

$$f = K\{1 + B(g_1 + (g_3 - g_1) \cdot \cos^2\theta) \\ \times (1 - \exp(-6Dt))\}, \quad t > 0 \tag{1}$$

for a step function of the electric field with

$$E = 0 \quad \text{for} \quad -\infty < t < 0,$$

$$E = E_{eff} \quad \text{for} \quad 0 < t < \infty$$

and

$$f = K\{1 + B(g_1 + (g_3 - g_1) \cdot \cos^2\theta) \\ \times \exp(-6Dt))\}, \quad t > 0 \tag{2}$$

with

$$E = E_{eff} \quad \text{for} \quad -\infty < t < 0$$
$$\text{and} \quad E = 0 \quad \text{for} \quad 0 < t < \infty.$$

In Eqs. (1) and (2) small ripples stemming from the AC − voltage are neglected [5], and

$$g_1 = 2\varepsilon_1(\varepsilon_2 - \varepsilon_1)/(\varepsilon_1 + \varepsilon_2)$$

$$g_3 = \varepsilon_2 - \varepsilon_1$$

$$B = V\varepsilon_0 E_{eff}^2/(2kT)$$

$$1/K = 2\pi(2 + V\varepsilon_0 E_{eff}^2(g_3 + 2g_1)/3kT).$$

$\varepsilon_1$ and $\varepsilon_2$ are the dielectric constants of water ($\varepsilon_1 = 79$) and $p$-TFE ($\varepsilon_2 = 2.1$), respectively. $V$ is the volume of the rod and $\theta$ is the angle between the rod-axis and the electric field. $D$ is a rotational diffusion coefficient and is related to the relaxation time according to $D = 1/6\tau$. The solid line in Fig. 3 is the result of a fit, giving for $\tau$ a value of around 150 ms. More details are given in Ref. [3].

## Alignment in a magnetic field

A solution of 10.2% per weight CTAB + 9.9% CTA-Benzolsulfonate + 2.45% $n$-Dekanol + 77.45% $D_2O$ gives a nematic lyotropic liquid crystal which consists of dislike micelles. CTAB is $C_{16}H_{33}N(CH_3)_3Br$, and CTA-Benzolsulfonate is $C_{16}H_{13}(CH_3)_3C_6H_5SO_3$. The sample was prealigned in a magnetic field $\vec{B}_0$ of 7 T at a temperature of 28 °C. At this temperature the alignment stays constant for days. The orientation of $\vec{B}_0$ relative to the neutron beam is shown in Fig. 4. The anisotropic scattering intensity thus measured is shown in Fig. 5a. The field $\vec{B}_0$ was then switched to zero and a second magnetic field $\vec{B}_1$ of 1.4 T was applied perpendicular to the direction of $\vec{B}_0$ (see Fig. 4). The $\vec{B}_1$ field now tries to rearrange the huge texture seen in Fig. 5a. We again measured the anisotropic

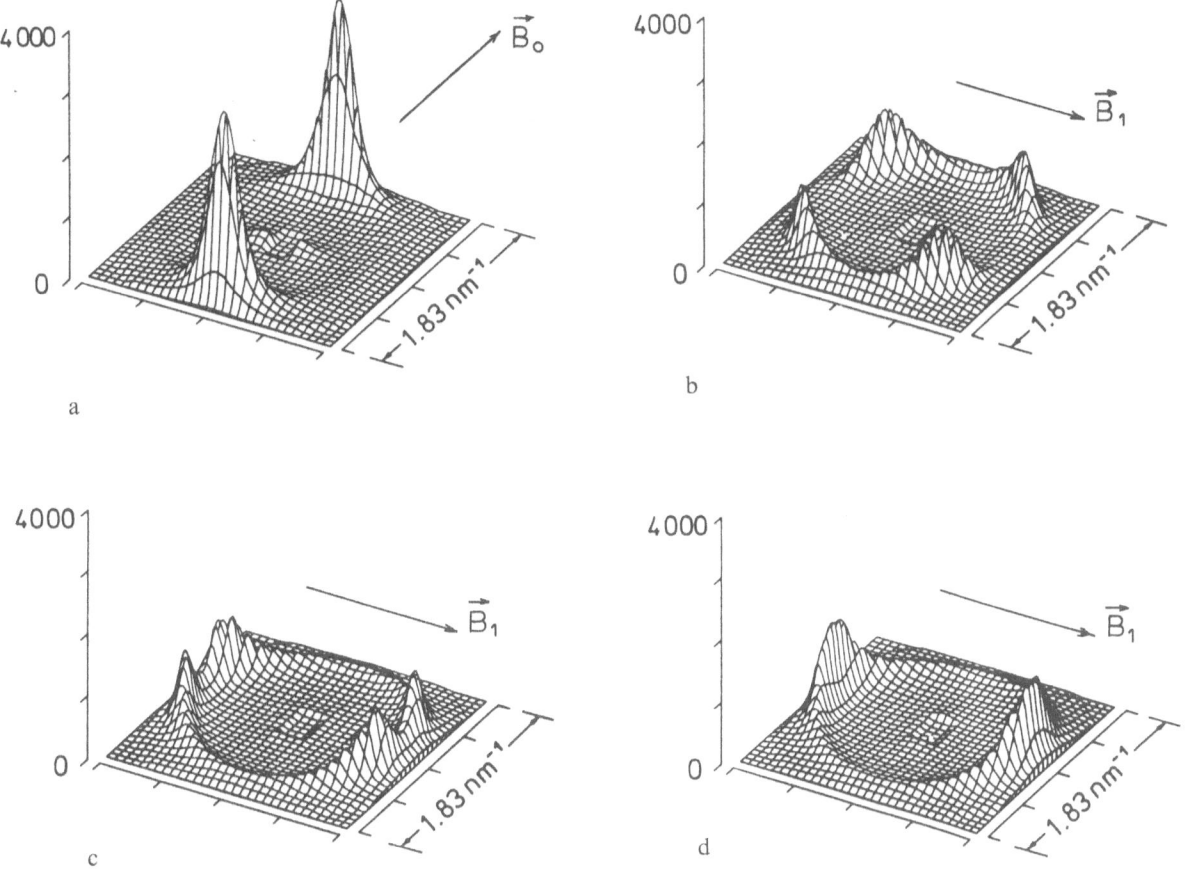

Fig. 5. The SANS scattering curve of a prealigned sample of CTAB. a) shows the results in a magnetic field $\vec{B}_0$ and b), c) and d) show the results in a magnetic field $\vec{B}_1$ after $t = 20$ min, 50 min and 90 min, respectively $\vec{B}_0 = 7$ T, $\vec{B}_1 = 1.4$ T. $\vec{B}_0 \perp \vec{B}_1$ (see Fig. 4)

scattering intensity as a function of time. This is shown in Fig. 5b−d. Finally a new equilibrium is obtained, in which the anisotropic intensity distribution has changed its orientation about an angle of 90°.

Technically each of the scattering pattern was measured for ∼200 s and no sampling was necessary for this experiment. The typical time for reorientation in this case is in the region of 60 min.

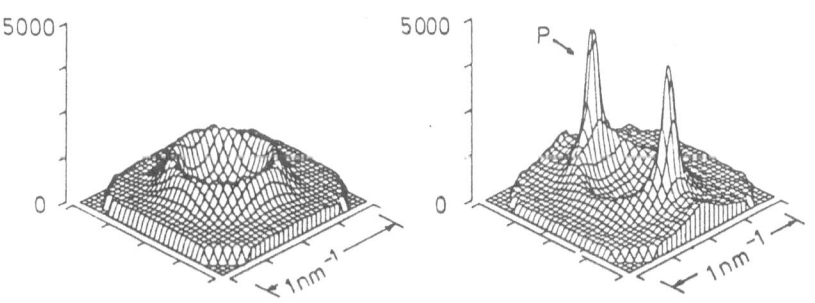

Fig. 6. Isometric intensity plot as observed on the two-dimensional detector, before (left) and 154 s after (right) the shear rate $\Gamma$ was increased stepwise from 0 to 600 s$^{-1}$. $T = 30$ °C. The sample is Hexadecyloctyldimethylammonium-bromide

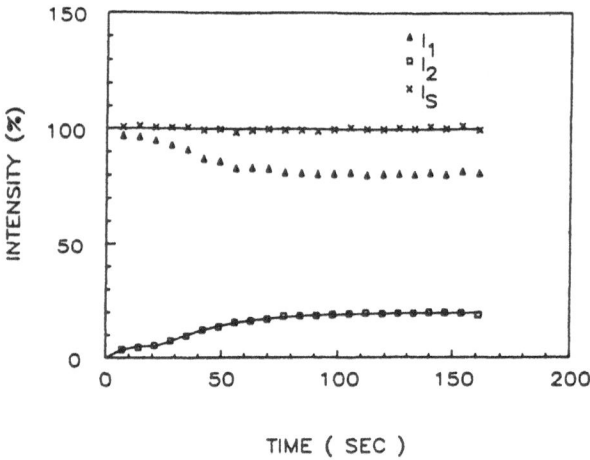

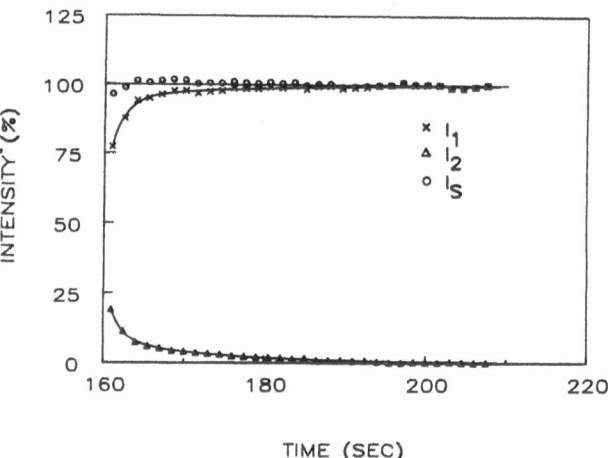

Fig. 7. Amount of type I ($I_1$) and type II ($I_2$) micelles as a function of time. At $t = 0$ shear $\Gamma$ was increased stepwise from 0 to 600 s$^{-1}$. $I_s$ is the sum of $I_1$ and $I_2$. The sample is Hexadecyloctyldimethylammoniumbromide

Fig. 8. Amount of type I ($I_1$) and type II ($I_2$) micelles as a function of time. At $t = 161$ s shear was reduced stepwise from 600 s$^{-1}$ to zero. $I_s$ is the sum of $I_1$ and $I_2$. The sample is Hexadecylotyldimethylammoniumbromide

### Transient alignment in a sheared solution

*Shear induced phase transition*

SANS-experiments have been performed on a 50 mM aqueous solution of Hexadecyloctyldimethyl-ammoniumbromide ($C_{16}H_{33}C_8H_{17}N(CH_3)_2Br$) under a shear $\Gamma$, changing its value suddenly from 0 to 600 s$^{-1}$ at time $t = 0$. At a time $t = 161$ s, $\Gamma$ was reduced to zero again. In fact a sampling program according to Fig. 1 was performed, beginning the next $\Gamma = 600$ s$^{-1}$ step at a time 231 s and so on. In this experiment the intensity of 23 measurements were sampled.

This shear is well above a threshold value [6] at which a shear induced phase transition begins. The intensity pattern shows a sharp peak $P$ superimposed over an ringlike anisotropic pattern as seen in Fig. 6.

It turned out, that the anisotropic pattern is caused by weakly aligned rodlike micelles (type I micelles), whereas the sharp peak $P$ is due to rodlike micelles ordered in a hexagonal liquid crystalline lattice (type II micelles), which starts forming when the shear is applied. The type II micelles are growing in number as time grows, till they reach an equilibrium after about 2 min, as shown Fig. 7.

The anisotropic ringlike structure was analysed by the evaluation of the distribution function $f$ of the

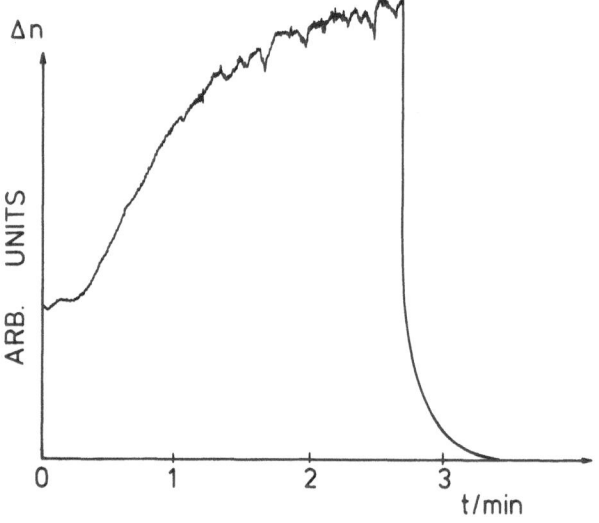

Fig. 9. The transient behaviour of the flow birefringence $\Delta n$ under the same conditions, as for the SANS measurements. The sample is Hexadecyloctyldimethylammoniumbromide

orientation of the rod axes [4], which is valid for very long rods.

$$\frac{\partial f}{\partial t} = D \cdot \left[ \frac{1}{\sin\theta} \frac{\partial}{\partial\theta} \left( \frac{\partial f}{\partial\theta} \sin\theta \right) + \frac{1}{\sin^2\theta} \frac{\partial^2 f}{\partial\varphi^2} \right]$$
$$- \Gamma \cdot \left[ \frac{\partial[f\omega(\theta)\sin\theta]}{\sin\theta\,\partial\theta} + \frac{\partial[f\omega(\varphi)]}{\sin\theta\,\partial\varphi} \right] \qquad (3)$$

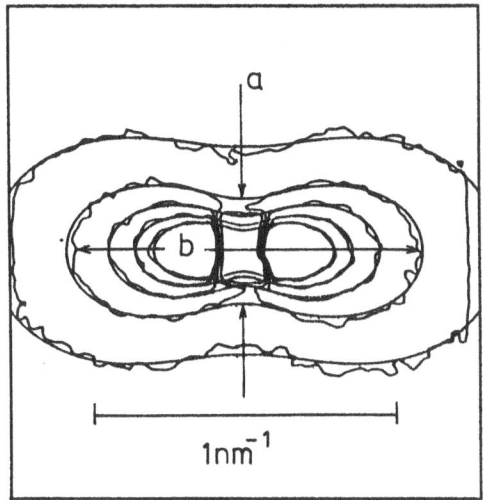

Fig. 10. Contour plot for the steady state anisotropic SANS scattering of CPS. The intensities of the lines are 2000, 4000, 6000, 8000 and 10000. The disturbance in the middle is due to the beam stop. The smooth curves are the result of a fit. $\Gamma = 20\ \mathrm{s}^{-1}$

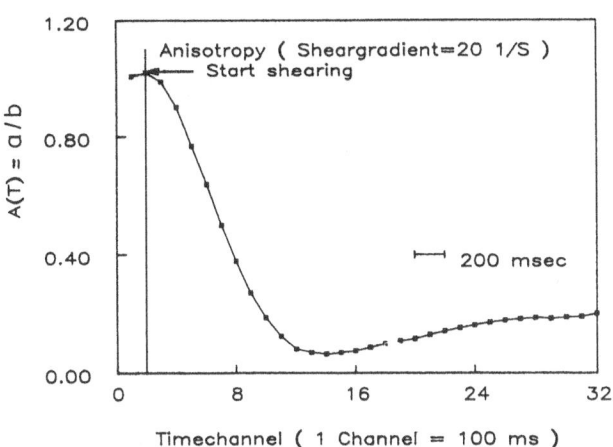

Fig. 11. The anisotropy $A = a/b$ (Fig. 10) as a function of time for CPS. The smooth line is only a guide to the eyes

$$\omega(\varphi) = -\sin^2\varphi\sin\theta$$
$$\omega(\theta) = 0.25\sin 2\varphi\sin 2\varphi\ .$$

$\theta$ is the angle between the rod axis and the $z$ axis of an orthogonal $x$, $y$, $z$ system. The velocity vector $\vec{v}$ is directed parallel to the $x$ axis and has the value $v = \Gamma \cdot y$.

$\varphi$ is the angle between the $x$ axis and the projection of the rod axis into the $xy$ plane. The function $f$ can be evaluated numerically for stationary ($\partial f/\partial t = 0$) and transient ($\partial f/\partial t \neq 0$) states. This equation is governed by two parameters, $\Gamma$ and $D$. $D$ is a rotational diffusion coefficient. More details about this experiment will be presented elsewhere [7].

Figure 8 shows, how the intensities of type I micelles increase and of type II micelles decrease during the $\Gamma = 0$ mode of operation. The continuous lines are exponential fits. Two relaxation times were found for each type of micelles. These relaxation times were $\tau_1^g = 1.74 \pm 0.08\ \mathrm{s}$, $\tau_1^d = 1.4 \pm 0.7\ \mathrm{s}$, $\tau_2^g = 15.3 \pm 2.8\ \mathrm{s}$, $\tau_2^d = 12.6 \pm 0.5\ \mathrm{s}$. The index $g$ stands for growth in intensity of type I micelles, the index $d$ for decay in intensity of type II micelles. It seems, that $\tau_1^g \approx \tau_1^d$ and $\tau_2^g \approx \tau_2^d$. The transient behaviour of the neutron scattering intensities can be correlated with macroscopic data, like the transient behaviour of the flow birefringence $\Delta n$ under the same conditions as used for the neutron measurements. $\Delta n$ as a function of time is shown in Fig. 9. A striking similarity of $\Delta n$ and of $I_2$, the intensity of type II micelles (see Fig. 7) is evident.

*Transient alignment of rods*

In 20 mM solution of Cetylpyridiniumsalicylate $C_{16}H_{33}(NC_5H_5)(C_6H_4OHCOO)$ and 20 mM of NaCl in $D_2O$ micellar rods are present. These rods have a radius of $2.15 \pm 0.05$ nm and are rather long [8].

At $t = 0$ a shear $\Gamma = 20\ \mathrm{s}^{-1}$ is applied and the transient behaviour of the anisotropic neutron scattering intensities is followed in time steps of 100 ms. After some time, which is around 4 s, a steady state situation is reached. The contour plot for the steady state intensity distribution is shown in Fig. 10. The smooth lines are due to a fit using Eq. (3) with $\partial f/\partial t = 0$. Only $D$, the rotation diffusion coefficient, was fitted. An excellent fit is obtained. For a simple description of the transient behaviour we plotted in Fig. 11 the anisotropy parameter $A = a/b$ (see Fig. 10) as a function of time. One observes an overshoot of $A$. Incidentially such an overshoot is expected to exist under certain conditions by solving Eq. (3).

In fact, a reasonable fit to the experimental curve of Fig. 11 can be obtained only by a rescaling of time. The reason for that is probably connected with the strong interaction of the micellar rods, which is not taken into account by theory and which is observed macroscopically by the existence of viscoelasticity. In this experiment the intensity of 1500 measurements were sampled.

*Acknowledgements*

The financial support from Bundesministerium für Forschung und Technologie (BMFT) under grant no. 03-

KA2BAY-7 is gratefully acknowledged. We thank W. Grießl, who was mainly involved in the production of a movie about time dependent SANS data and U. Ertl from ILL for his excellent assistance in electronic work.

## References

1. Folda T, Hoffmann H, Chanzy H, Smith P (1988) Nature 333:55−56
2. Angel M, Hoffmann H, Huber G, Rehage H (1988) Ber Bunsenges Phys Chemie 92:10−16
3. Baumann J, Kalus J, Neubauer G, Hoffmann H, Ibel K (1989) Ber Bunsenges Phys Chem 93:874−878
4. Peterlin A, Stuart A (1973) In: Euken A, Wolf KL (eds) Doppelbrechung, insbesondere künstliche Doppelbrechung, Hand- und Jahrbuch der chemischen Physik, Band 8, Abschnitt 1 B. Akademische Verlagsgesellschaft, Leipzig, pp 1−115
5. Peterlin A, Stuart A (1939) Z Phys 112:129
6. Kalus J, Hoffmann H, Chen SH, Lindner P (1989) Phys Chem 93:4267−4276
7. Jindal VK, Kalus J, Pilsl H, Hoffmann H, Lindner P to be published in Journal of Physical Chemistry 1990
8. Hoffmann H, Kalus J, Thurn H, Ibel K (1983) Bunsenges Phys Chemie 87:1120−1129

Authors' address:

Prof. Dr. J. Kalus
Experimentalphysik I
Universität Bayreuth
D-8580 Bayreuth, FRG

**Progress in Colloid & Polymer Science**

Progr Colloid Polym Sci 81:107 – 112 (1990)

# Drag-reducing surfactant solutions in laminar and turbulent flow investigated by small-angle neutron scattering and light scattering

P. Lindner[1]), H. W. Bewersdorff[2]), R. Heen[3], P. Sittart[3]), H. Thiel[3]), J. Langowski[4]) and R. Oberthür[5])

[1]) Institut Laue Langevin, Grenoble
[2]) ETH Zürich, Institut für Hydromechanik und Wasserwirtschaft, Zürich
[3]) Universität Dortmund, FB Chemietechnik, Dortmund
[4]) EMBL-Outstation, Grenoble
[5]) Institut für Festkörperforschung, Jülich

*Abstract:* Surfactant solutions containing rodlike micelles have been investigated by small angle neutron scattering and by light scattering at rest and under laminar and turbulent flow conditions. At rest and at room temperature long rods of average length 249 nm are found with static light and small angle neutron scattering. To interpret the orientation of these rods in a laminar shear gradient the assumption of an increased average rod length is necessary. In turbulent pipe flow the rod-like micelles are initially oriented in the bulk flow direction with more pronounced orientation in the near wall regions compared to the core regions of the pipe. Eventually a destruction of the initially rod-like micelles takes place leading to a rearrangement of the surfactant molecules into smaller aggregates part of which, in turn, build up loose but lasting (over several months at room temperature) superstructures.

*Key words:* Drag reduction; rodlike micelles; laminar flow; turbulent flow; small angle neutron scattering; light scattering

## Introduction

Drag reduction is the reduction of the friction factor in a turbulent flow at a constant flow rate. It can be achieved by addition of high molecular weight polymers, surfactants or fibres to a Newtonian fluid like water at lowest concentrations (Toms' effect [1]). Results published so far suggest that drag reduction with surfactant solutions can only occur when the solution contains *rodlike* micelles [2]. However, a satisfactory description of the interaction between turbulent eddies and the additive which leads to the drag reduction is still missing. In order to study the molecular basis of drag reduction under real flow conditions, the conformation of rodlike micelles is investigated by small angle neutron scattering (SANS) experiments with a surfactant solution subjected to laminar Couette flow and turbulent pipe flow (exhibiting drag reduction), as well as under standard conditions at rest with SANS and light scattering (LS).

## Surfactant system

The investigated drag reducing additive was an equimolar mixture of the cationic surfactant tetradecyl-trimethyl-ammoniumbromide ($C_{14}H_{29}$-$N(CH_3)_3$)Br with sodiumsalicylate ($C_6H_4OHCOONa$) dissolved in heavy water ($D_2O$). The total concentration of the solute was 2.4 mmol/l corresponding to about 1 g/kg (1000 ppm w/w). At a temperature of 20 °C and above the critical micellar concentration $c_{cmc} = 0.25$ g/kg [3] this system assembles into aggregates of very long rodlike micelles (called "$C_{14}$TAB/NaSal-h") with a large aspect ratio (ratio length to diameter). In order to allow for an equilibrium conformation of the micellar system the solution was heated up to $T \approx 60$ °C for 12 hours immediately after the preparation and was then cooled down to the measuring temperature of $T = 20$ °C two days before the experiment (fresh solution).

## Friction behaviour

Surfactant solutions containing rodlike micelles reveal a characteristic friction behaviour when the pressure drop of such a solution (being proportional to the friction factor $f$) is measured under laminar and under turbulent pipe flow

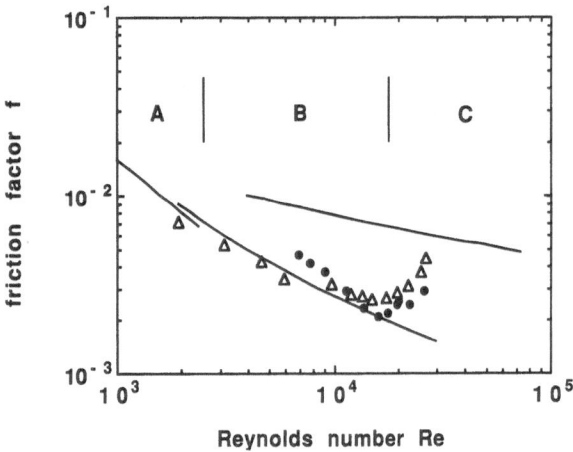

Fig. 1. Friction factor $f$ as a function of Reynolds number $Re$. Points: fresh solution, triangles: previously stressed solution (I), A: laminar flow region, B: turbulent, drag reduction flow, C: turbulent flow region (Prandtl-Karman law [17])

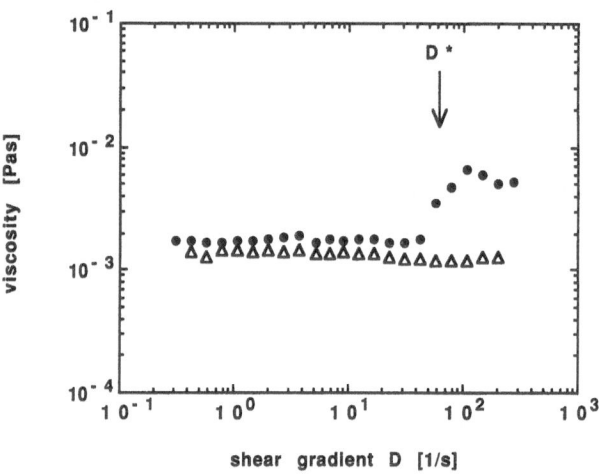

Fig. 2. Shear viscosity $\eta/Pas$ as a function of shear gradient $D/s^{-1}$. Points: fresh solution, triangles: previously stressed solution

conditions as a function of the flow velocity (being proportional to the Reynolds number $Re$). The friction factor $f$ of the surfactant solution is defined as $f = \Delta p\, d/(2\varrho w^2 l)$, where $\Delta p$ means the pressure drop over the pipe length $l$, $d$ the pipe diameter, $\varrho$ the density of the fluid, and $w$ its bulk velocity. The Reynolds number $Re$ which is based on the solvent viscosity $\eta$ is defined as $Re = w d\varrho/\eta$. The friction factor $f$ as a function of the Reynolds number $Re$ (Fig. 1) has been measured in a turbulent flow apparatus consisting of a gear pump, a surge tank for damping fluctuations produced by the pump, a steel pipe (diameter $d = 15$ mm, length $l = 2.5$ m) with an insertion made of quartz glass, serving as a window for the neutron scattering experiment (see below) and a heat exchanger for controlling the temperature [4]. The temperature was measured by a resistance thermometer, the pressure drop in the pipe by a pressure transducer, and the flow rate by an inductive flowmeter. Three characteristic regions can be distinguished in Fig. 1:

A. The friction behaviour of a Newtonian fluid in the laminar flow region (A) is described by $f = 16/Re$ (Hagen-Poisseuille law).

B. In the intermediate turbulent regime the friction factor $f$ of the solution is substantially lowered with respect to the friction factor of the pure solvent (drag reduction). Virk [5] has shown by an analysis of available literature data that the limit of the maximum attainable drag reduction is given by $1/\sqrt{f} = 19 \log(Re\sqrt{f}) - 32.4$ (lower drawn line in Fig. 1).

C. Exceeding a critical wall shear stress at higher Reynolds numbers $Re$, the drag reduction starts to break down and the friction factor of the solution approaches the friction behaviour of a turbulent Newtonian liquid (pure solvent). The turbulent flow region of a Newtonian fluid is described by the Prandtl Karman law [17] $1/\sqrt{f} = 4 \log(Re\sqrt{f}) - 0.4$ (upper drawn line in Fig. 1).

## Shear viscosity

Shear viscosity measurements of the surfactant solution in laminar Couette flow reveal a sudden increase of the viscosity above a critical shear rate $D^*$ (cf. Fig. 2, filled points).

The effect is immediately reversible. Above this threshold shear rate the solution shows viscoelastic behaviour. This feature has been found with different surfactant solutions containing rodlike micelles and has been attributed to a so called "shear induced state" (SIS [2], [6], [7], [8]). A simple shear alignment of long rod-like particles should lead to a decrease of the viscosity with increasing shear rate. Therefore shear induced new structures have to build up in the SIS. The shear viscosity of our solutions was measured in a Couette viscometer (gap width $d = 0.5$ mm and $d = 0.235$ mm). The results (shear viscosity $\eta$ as a function of shear gradient $D$) are shown in Fig. 2. The freshly prepared surfactant solution behaves like a Newtonian fluid at low shear gradients. For shear rates above $D^* \approx 60$ s$^{-1}$ the shear viscosity increases due to the formation of the shear induced state (filled points in Fig. 2). After subjecting the same system to turbulent flow for several hours (stressed solution), however, the ability to form a shear induced state has disappeared (triangles in Fig. 2).

## Small angle neutron scattering (SANS) measurements

The SANS experiments in laminar and turbulent flow were performed at the instrument D11 [9] of the Institut Laue-Langevin in Grenoble, France, with a neutron wavelength of $\lambda = 0.6$ nm, a wavelength distribution of $\Delta\lambda/\lambda = 9\%$ and sample-to-detector distances $L = 5$ m and 10 m, thus covering an effective range of the momentum transfer $Q = (4\pi/\lambda)\sin(\theta/2)$ of $0.08 \leqslant Q/\text{nm}^{-1} \leqslant 0.75$. Background correction of the solution spectra was done with the solvent and the empty cell measurement and the data were calibrated with a 1 mm standard water sample to allow for

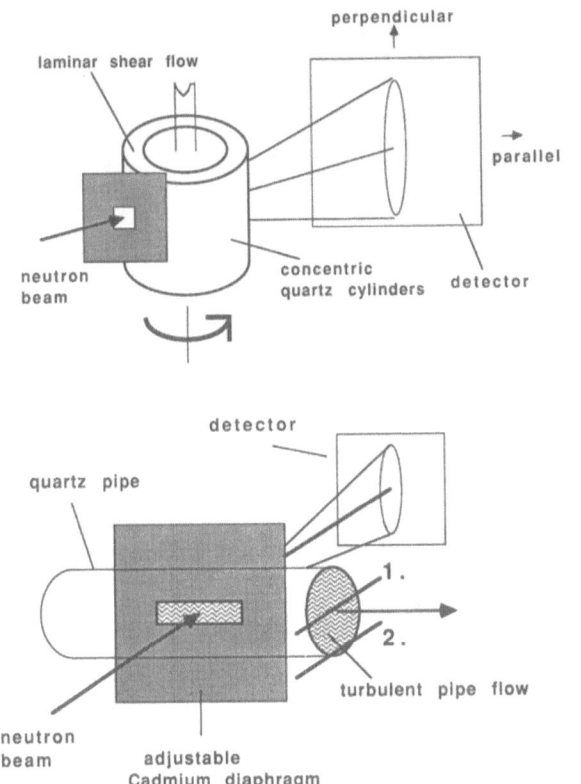

Fig. 3. Geometry of the SANS-experiment. a: laminar shear flow in the Couette type shear apparatus with the solution confined in the gap between both cylinders, b: pipe-flow geometry for the turbulent flow experiment (1: beam path centred to the pipe axis with centred diaphragm, 2: beam path only through near wall region with diaphragm in the lower position)

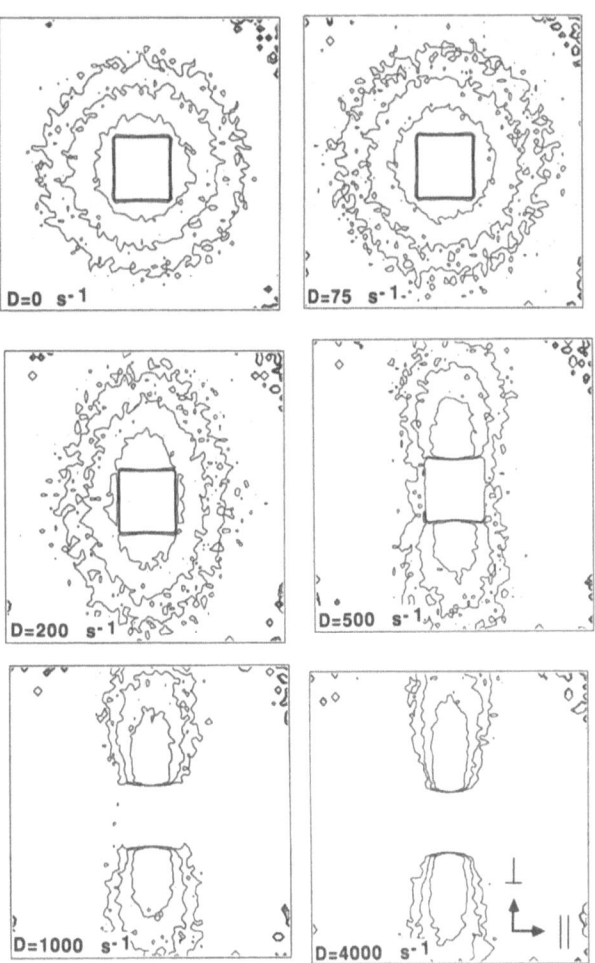

Fig. 4. Isointensity contour plots of the normalized 2 d multidetector data (SANS, $L = 5$ m, $\lambda = 0.6$ nm) for the surfactant solution (2.4 mmol/l $C_{14}TAB$/NaSal-h in $D_2O$ at $T = 20\,°C$) in *laminar shear flow* at various shear gradients $D/s^{-1}$

calculation of the differential scattering cross-section of the solute ($d\Sigma/d\Omega$) in absolute units (cm$^{-1}$) [10]. During the SANS experiments the temperature in the flow devices was kept at $(20.0 \pm 0.1)\,°C$.

The surfactant solution was measured under laminar shear flow conditions in the ILL Couette type shear apparatus (Fig. 3a, [11]) using concentric quartz cylinders with a gap width of $d = 0.5$ mm at different shear gradients in the range $75 \leqslant D/s^{-1} \leqslant 4000$.

The SANS experiment under turbulent, drag reduced flow conditions at Reynolds numbers in the range $850 \leqslant Re \leqslant 23220$ was performed with the same apparatus that had been used for the measurement of the friction behaviour (friction factor as a function of Reynolds number). The neutron beam path was perpendicular to the pipe axis and passed through the quartz section transparent to neutrons (Fig. 3b). The slit-like Cadmium diaphragm in front of the pipe is adjustable to different heights with respect to the pipe axis. In particular two positions of the diaphragm were used, (i) the centred position with the beam passing through the centre of the pipe and (ii) a lower position with the beam passing mainly through the near wall region of the pipe. In addition, SANS measurements of the fresh solution, the stressed solution and

the solvent $D_2O$ were done at rest using standard HELLMA quartz cells with a thickness of $d = 5$ mm at temperature $T = 20\,°C$ and sample-to-detector distances $L/m = 2.5, 5, 10$ and $30$ (neutron wavelength $\lambda = 0.6$ nm), hence covering an effective $Q$-range of $0.028 \leqslant Q/nm^{-1} \leqslant 1.5$.

## Light scattering measurements

Static light measurements of the fresh solution and stressed solution at rest were done at the EMBL-outstation, Grenoble with a Spectra Physics 2025 laser running at 488 nm with a power of $400 - 600$ mW, an AMTEC goniometer and a BI 2030-AT autocorrelator. Samples were measured in 1 cm cylindrical cells immersed in a water index matching bath. Data were collected from $30°$ to $135°$ scattering angle, corresponding to a range of scattering vectors $0.0074 \leqslant Q/nm^{-1} \leqslant 0.032$. The influence of dust on the

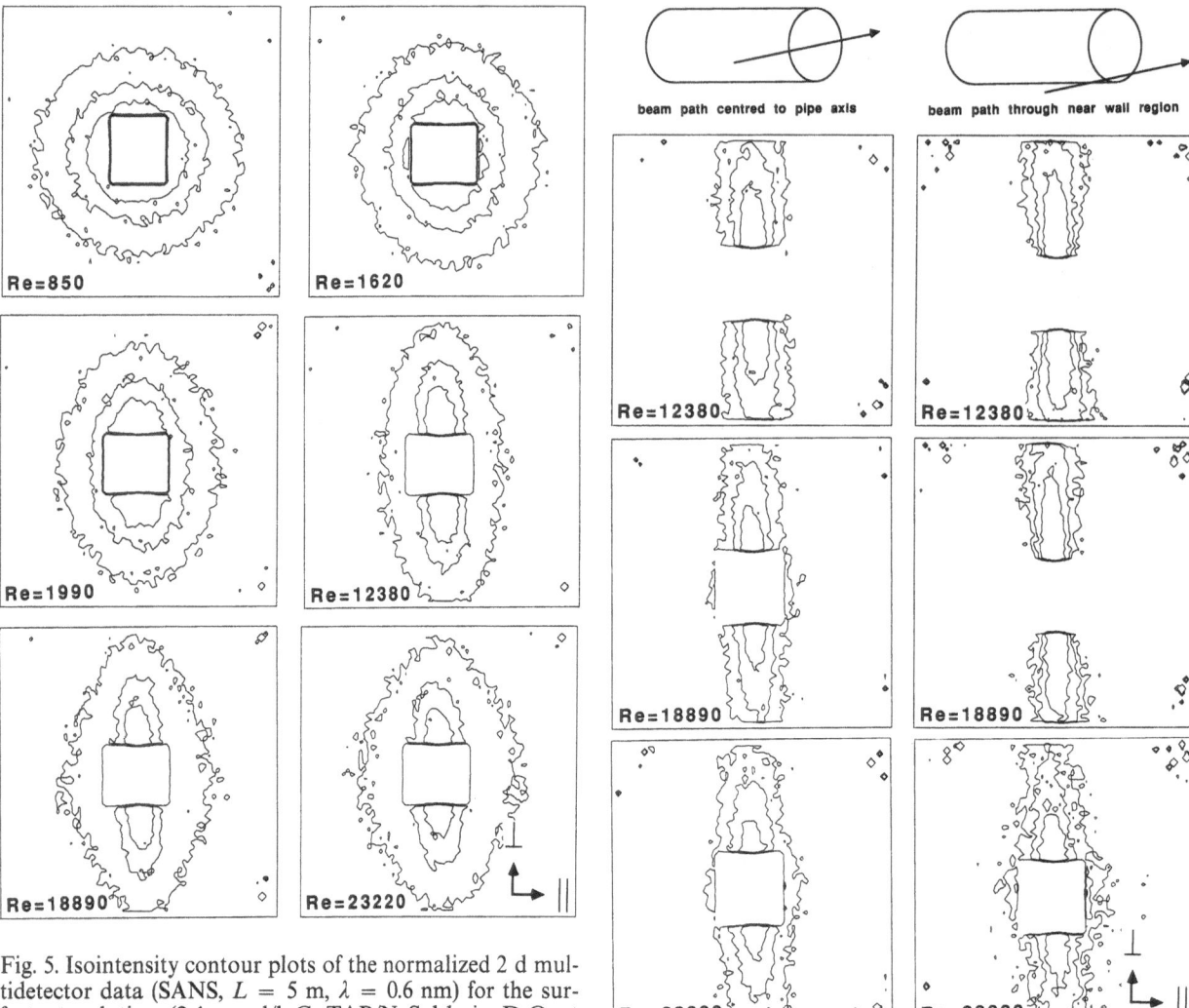

Fig. 5. Isointensity contour plots of the normalized 2 d multidetector data (SANS, $L = 5$ m, $\lambda = 0.6$ nm) for the surfactant solution (2.4 mmol/l $C_{14}TAB/NaSal$-h in $D_2O$ at $T = 20\,°C$) in *turbulent pipe flow* at various Reynolds numbers $Re$. Beam path is centred with respect to the pipe axis

data was minimized by filtration of the solutions through a 0.4 μm-filter and a dust elimination procedure in the correlator control program.

## Results and discussion

### SANS in laminar Couette flow

Figure 4 shows the results of SANS measurements of the surfactant solution in laminar shear flow, using the Couette type shear apparatus at various shear gradients $D$. The results are presented as isointensity-contour plots of the 2-dimensional multidetector data. With increasing shear gradient $D$ the anisotropy of the scattering pattern increases. This corresponds to an increasing alignment of the rodlike particles with

Fig. 6. Isointensity contour plots of the normalized 2 d multidetector data (SANS, $L = 10$ m, $\lambda = 0.6$ nm) for the surfactant solution (2.4 mmol/l $C_{14}TAB/NaSal$-h in $D_2O$ at $T = 20\,°C$) in *turbulent pipe flow* at the three highest measured Reynolds numbers $Re$. Left column: beam path centred with respect to pipe axis, right column: beam path through the near wall region

their long axis parallel to the direction of flow. However, as a comparison with calculation [13] shows [3] the obtained alignment cannot be explained by a rodlength of $L \approx 248$ nm (cf. Kalus et al. [14]) and suggests an increase of the average rod length by a factor $> 10$.

### SANS in turbulent pipe flow

Figure 5a – g shows the results of the SANS measurements, using the turbulent flow apparatus at various Reynolds numbers $Re$ with the Cadmium diaphragm adjusted to the centred position, with the beam passing mainly through the core region of the pipe. At first sight the resulting contour plots show a similar increasing anisotropy with increasing Reynolds number as in the Couette experiment with *laminar* shear flow. This means that even in *turbulent,* drag reduced pipe-flow the orientation of the rodlike micellar particles in flow direction increases with Reynolds number.

A more detailed inspection of the data obtained at higher Reynolds numbers $Re$, however, reveals some differences. Figure 6 shows the contour plots of the surfactant solution in turbulent flow at the three highest Reynolds numbers $Re = 12380$, $18890$ and $23220$ for the two experimental geometries where (i) the neutron beam mainly monitors the core region (left column) and (ii) the near wall region of the pipe (right column).

A more pronounced anisotropy is observed when the neutron beam passes mainly through the near wall region of the pipe. Moreover, in the breakdown region of the drag reduction the anisotropy of the scattering pattern clearly *decreases* with increasing Reynolds numbers for both beam paths (core and near wall region). Furthermore, the break down of drag reduction accompanied with a breakdown of the orientation of the rodlike particles is confirmed [15]. These results prove that even in turbulent (drag reduced) flow a difference in ordering of the rodlike particles exists between the near wall region and the core region of the pipe in agreement with the differences of the turbulent velocity profile between near wall and core region found with LDA measurements [16].

*Influence of different treatments in the pipe*

The data presented here were obtained during two different experimental campaigns at the ILL in February (I) and in August 1989 (II). In I the solution had been pumped for several hours under conditions where usually the break down of drag reduction occurs. The friction behaviour of the surfactant solution I measured afterwards is given by the triangles in Fig. 1. But the following SANS measurements in *turbulent* pipe flow starting at low Reynolds numbers, as well as in *laminar* Couette shear flow revealed no anisotropy of the scattering pattern. Although, the solution still showed drag reducing behaviour (Fig. 1). On the other hand, measurements of the shear viscosity of the stressed solution show an almost constant viscosity in the measuring range with no increase of the viscosity above the critical shear rate of $D^* \approx 60$

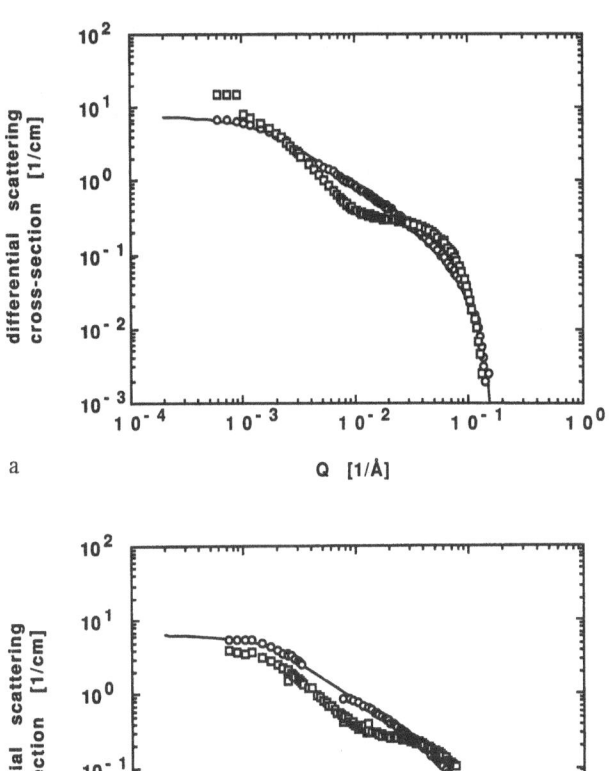

a

b

Fig. 7a–b. Differential scattering cross-section $d\Sigma/d\Omega/\text{cm}^{-1}$ as a function of the momentum transfer $Q = (4\pi/\lambda)\sin(\theta/2)$ from SANS and LS of the fresh and stressed surfactant-solution at rest (2.4 mmol/l $C_{14}TAB/NaSal$-h in $D_2O$ at $T = 20°C$). 7a: experiment I, fresh solution (open circles) and solution previously stressed before the turbulent flow-/SANS-experiment above the critical wall shear stress (open squares), drawn line: best fit for a homogeneous cylinder [12] with length $L = 267$ nm, radius of gyration $R_G = \sqrt{L^2/12} \approx 77$ nm and diameter $D = 4.18$ nm(aspect ratio $L/D = 64$); 7b: experiment II, fresh solution (open circles) and solution stressed below the critical wall shear stress (open squares), drawn line: best fit for a homogeneous cylinder [12] with length $L = 231$ nm, radius of gyration $R_G = \sqrt{L^2/12} \approx 67$ nm and diameter $D = 4.10$ nm (aspect ratio $L/D = 56$)

(see Fig. 2, open triangles). The anisotropic data under turbulent flow conditions (Figs. 5 and 6) were obtained with experiment II. In II the SANS measurements in turbulent flow were performed with the fresh solution up to a Reynolds number where complete break down of the drag reduction was *not* yet attained.

Both solutions that had been subjected to the turbulent flow in different ways (experiment I and II) were measured in separate SANS and LS experiments at rest in order to compare their scattering with the original solution. Figures 7a and b (open circles) show the normalized scattering cross-section as a function of scattering vector $Q$ resulting from small angle neutron scattering- (radial average over the 2-dimensional detector) and light scattering measurements of the freshly prepared surfactant solution at rest (2.4 mmol/l $C_{14}TAB/NaSal$-h in $D_2O$ at $T = 20\,°C$).

The drawn lines are the result of model calculations, using the Mittelbach-Porod formula [12] for a homogeneous, rigid cylinder. The best fit gives a length of $L = 267$ nm for the fresh solution of experiment I and $L = 231$ nm for II, corresponding to radii of gyration $R_G = \sqrt{L^2/12} \approx 77$ nm (I, Fig. 7a) and $R_G \approx 67$ nm (II, Fig. 7b). The light scattering data have been normalized to the absolute SANS scale by a combination of the radius of gyration of the whole particle from light scattering with the mass per unit length from SANS assuming monodisperse, homogeneous cylinders. For both stressed solutions (open squares) a remarkable change of the scattering curves is observed. The change is most pronounced for the solution which was stressed once above the breakdown of drag reduction (Fig. 7a). From the shape of the scattering curve it is evident that a simple cylindrical model does no longer fit the data. Moreover, a Zimm extrapolation of the scattering data at low $Q$ reveals a significant increase of the radius of gyration from $R_G \approx 77$ nm (I) and $R_G \approx 67$ nm (II) for the rodlike micelles in the fresh solution to $R_G \approx 112$ nm for the dissolved particles from experiment I, and $R_G \approx 78$ nm for experiment II. These structures, once

formed, do not change significantly over several months at room temperature as revealed by successive static scattering measurements.

## References

1. Toms BA (1948) Proc 1st Intern Congr on Rheology, Vol 2, p 135, North Holland, Amsterdam
2. Ohlendorf D, Interthal W, Hoffmann H (1986) Rheol Acta 25:468
3. Lindner P et al., (to be published)
4. Bewersdorff HW, Dohmann J, Langowski J, Lindner P, Maack A, Oberthür R, Thiel H (1989) Physica B156 & 157:508
5. Virk PS (1975) AIChEJ 21:625
6. Bewersdorff HW, Ohlendorf D (1988) Coll Polym Sci 266:941
7. Löbl M, Thurn H, Hoffmann H (1984) Ber Bunsenges Phys Chem 88:1102
8. Rehage H, Hoffmann H (1982) Rheol Acta 21:561
9. Ibel K (1976) J Appl Cryst 9:296
10. Ragnetti M, Oberthür RC (1986) Coll Polym Sci 264:32
11. Lindner P, Oberthür RC (1984) Revue Phys Appl 19:759
12. Mittelbach P, Porod G (1961) Acta Phys Austriaca 14:185
13. Hayter JB, Penfold J (1984) J Phys Chem 88:4589
14. Kalus J, Hoffmann H, Ibel K (1989) Coll Polym Sci 267:818
15. Bewersdorff HW, Frings B, Lindner P, Oberthür RC (1986) Rheol Acta 25:642
16. Bewersdorff HW (1989) In: Gyr A, Structure of turbulence and drag reduction, Springer Verlag, Berlin (in press)
17. Prandtl L (1933) Z VDI 77:105

Authors' address:

Dr. Peter Lindner
Institut Laue-Langevin
B.P. 156 X
F-38042 Grenoble cedex

**Progress in Colloid & Polymer Science**　　　　　Progr Colloid Polym Sci 81:113−119 (1990)

# The rheology of hard-sphere dispersions: the micro-structure as a function of shear rate

J. C. van der Werff and C. G. de Kruif

Van't Hoff laboratorium, Utrecht, The Netherlands

*Abstract:* Steady-shear viscosity measurements on colloidal hard-sphere dispersions were paralleled by light and neutron scattering experiments on sheared dispersions. In the viscosity experiments, we found concentrated dispersions to be shear-thinning but no shear-thickening was observed. The viscosity is determined by two dimensionless groups only: the volume fraction and the Peclet number. This underlines the hard-sphere nature of these dispersions. We designed a parallel − plate flow cell which we used in neutron scattering experiments with the D11-diffractometer (ILL, Grenoble). In these experiments we were able to probe a scattering plane containing the flow velocity and vorticity axis (planes of constant velocity) as well as the velocity − gradient plane (or shear plane) which contains the flow velocity and the gradient axis. These experimental results show an increasing deformation of the liquid-like equilibrium micro-structure with increasing shear rate. However, no evidence for strong ordering phenomena is found. In this paper we relate the observations made in the steady-shear viscosity measurements to those made in the scattering experiments.

*Key words:* Hard-spheres; shear viscosity; neutron scattering; micro-structure

## Introduction

The non-Newtonian behaviour of dense colloidal dispersions arises from interactions between the dispersed particles. Several types of interactions can be distinguished [1,2]. Firstly, the fluid-dynamical interactions. The movements of the colloidal particles generate velocity fields in the solvent and the particles interact via these velocity fields. The (relative) velocities of the particles stem from the shear flow as well as from the Brownian motion. Apart from its effect on the fluid dynamical interactions, Brownian motion in a sheared dispersion leads to another type of interaction. The spatial distribution of the particles is affected by the shear and the diffusion counteracts this entropically less favourable state. The third type of interactions that we can distinguish are the so-called potential interactions, for instance steric interactions or electrical double-layer interactions, which are usually present in stable dispersions. Furthermore, interactions due to Van der Waals − London forces may cause flocculation or gelation which is then accom-

panied by a dramatic change in the rheological properties.

The complexity of the observed rheological behaviour does often not allow for a straightforward explanation in terms of the interactions mentioned. The reason for this is that direct potential interaction, Brownian motion and fluid dynamical interactions are coupled via the flow dependent particle distribution in space. The theoretical problem can be simplified by neglecting one of these interactions alltogether, leading to unphysical systems. We have choosen to study the hard-sphere dispersion, a dispersion in which the particles do not interact via long-range direct interactions. The rheological behaviour can then be explained in terms of the competition of the disturbance of the particle distribution by the flow and the relaxation through Brownian motion.

The disturbed particle distribution can be studied by means of light and neutron scattering experiments. The particle distribution is represented by the radial distribution function $g(r)$. This function $g(r)$ gives the probability density for finding a particle at a distance

$r$ from a given test particle. In a dispersion under shear, $g(r)$ is a function of the direction and length of the position vector and the shear rate. In a scattering experiment, we observe the static structure factor $S(k)$, which is the Fourier transform of $g(r)$. A change in the (shear induced) micro-structure is reflected in a change in the structure factor $S(k)$.

Van Helden, Jansen and Vrij [3] synthesized a monodisperse colloidal dispersion of sterically stabilized silica particles in cyclohexane. It was found that the equilibrium properties of these dispersions could very well be described by the hard-sphere model. For instance, the osmotic compressibility of this system closely follows the compressibility of a hard-sphere dispersion [4]. The equilibrium structure factor, measured by neutron scattering, indicates a liquid-like order in these dispersions up to very high volume fractions. At all volume fractions, the measured structure factor can be described by the structure factor which follows from the Percus Yevick solution for the radial distribution function of hard-spheres [5]. Finally, the diffusive properties of the colloidal particles were measured and were found to be in agreement with those calculated for hard-spheres [6].

We measured the steady-shear and complex viscosity of these silica dispersion as a function of particle size, volume fraction, shear rate or frequency of the harmonically oscillating shear flow. We found that scaling laws which are applicable to hard-sphere dispersions were satisfied [7–9] and concluded that the non-equilibrium properties as well as the equilibrium properties of these dispersions can be described by the hard-sphere model.

These measurements provide a starting point for non-equilibrium structure factor measurements because at every volume fraction, the shear thinning curve is a unique function of the Peclet number. The rheological measurements were paralleled by light [10] and neutron [11–13] scattering experiments on dispersions under shear in which we observe the deformation of the dispersion micro-structure. Our aim is to relate changes in micro-structure to the shear rate dependent viscosity. In this contribution we will review the results of the steady shear viscosity measurements and the scattering experiments. The results of both studies will be compared and discussed here.

## Steady shear viscosity of hard-sphere dispersions

The steady shear viscosity of four dispersions of sterically stabilized silica particles in cyclohexane was measured as a function of shear rate and volume fraction [7, 8]. The dispersions differ in particle radius ($a = 28, 46, 76$ and $110$ nm). For the measurements we used a Deer rheometer with a Couette geometry as well as a parallel-plate geometry. Both set-ups are equipped with a vapour-lock to prevent evaporation of the volatile solvent. Details of sample preparation, measuring procedure and measuring cells are described elsewhere [7, 8].

Figure 1 shows that the relative viscosity in the two limiting cases $Pe = 0$ and $Pe = \infty$ is a function of volume fraction only and does not depend on particle size. The drawn lines are according to Quemada's semi-empirical expression [14]:

$$\eta_r = \left(1 - \frac{\phi}{\phi_m}\right)^{-2} \qquad (1)$$

where $\phi_m$ is the maximum packing volume fraction and $\eta_r$ is the relative viscosity e.g. the viscosity of the dispersion divided by the viscosity of the solvent. The functional form of this equation is identical to the Krieger-Dougherty [15] equation where the exponent is given by $k_m \phi_m$. Fitting our data to the Krieger-Dougherty equation, and thus to an extra adjustable parameter leads to $k_m \phi_m = 2$. Thus, the extra parameter is not needed. In Fig. 1 we used $\phi_m = 0.63$ in the low-shear limit and $\phi_m = 0.71$ in the high-shear limit. The low-shear value for $\phi_m$ suggests that in this limit, the dispersion is randomly close packed. This is in agreement with the idea that the Brownian motion determines the micro-structure in the low shear limit. The rather high value for $\phi_m$ in the high-shear limit suggests that in this case a dense packed, highly ordered structure is induced by the flow. For instance layers of particles in which the particles are hexagonally packed as was observed by Hoffman [16]. The layers are formed in the planes of constant velocity. It has been put forward that shear-thickening would result from the appearance of a new random state (for instance the breaking-up of the layers). In our dispersions, we have never observed shear-thickening, not even at Peclet numbers $Pe = 1200$, this being in contrast to dispersions of particles which interact via other than hard-sphere potentials [17]. The absence of shear-thickening indicates that the micro-structure of the dispersion becomes practically independent of shear rate at high enough Peclet numbers. In section III we will describe experiments in which we study the shear induced structure by means of small angle neutron scattering.

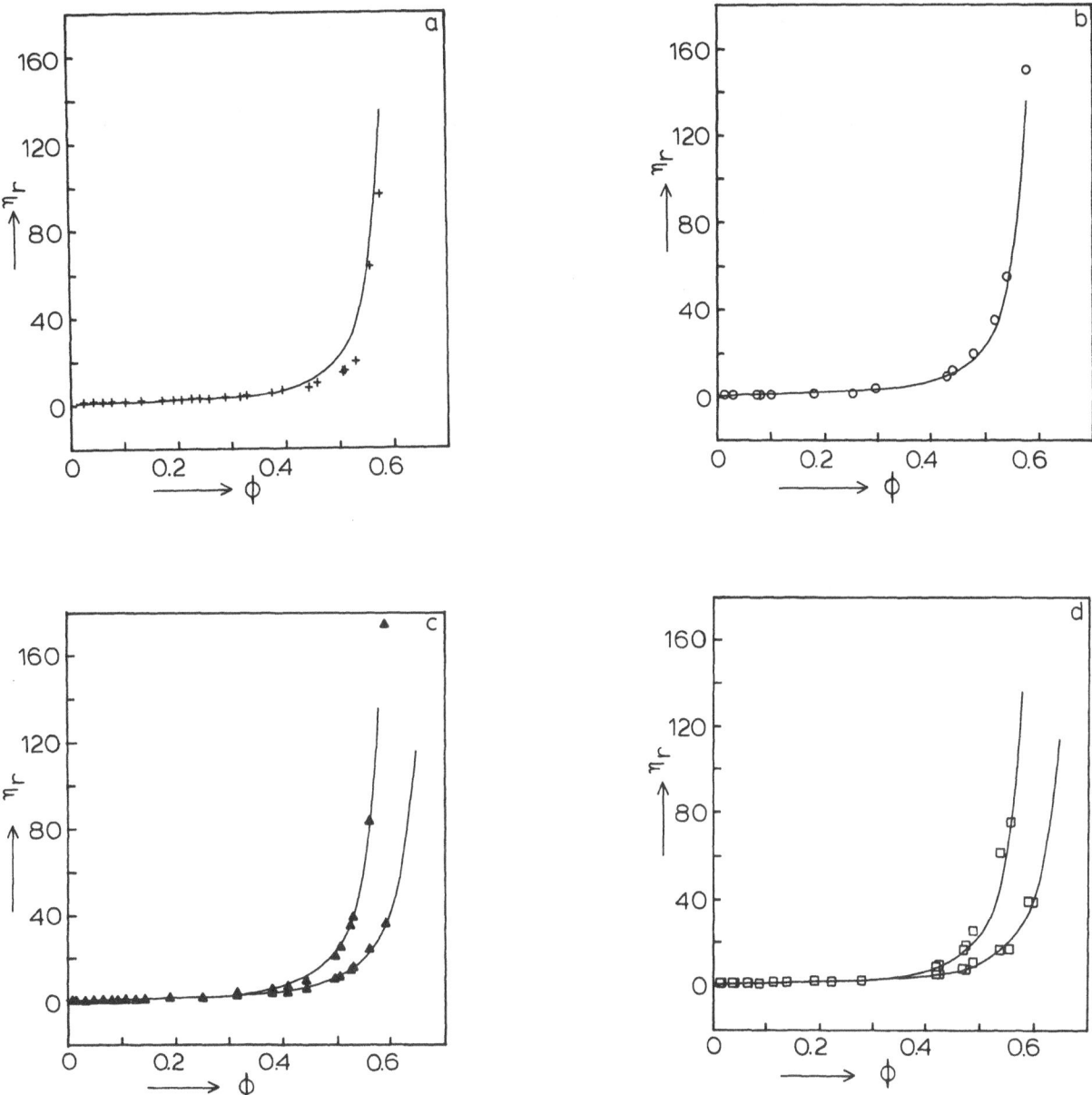

Fig. 1. Relative viscosity as a function of volume fraction $\phi$ in the limiting cases $Pe \to 0$ and $Pe \to \infty$. The drawn lines are according to Eq. (1) with $\phi_m = 0.63$ in the low-shear limit and $\phi_m = 0.71$ in the high shear limit. Particle radii: 28 nm (+); 46 nm (○); 76 nm (▲) and 110 nm (□)

The shear rate dependence of the viscosity was studied also. We found that the viscosity − Peclet number curves of four dispersions containing different sized particles are a function of volume fraction only. In Fig. 2 we show the characteristic Peclet numbers for dispersions of different sized particles as a function of volume fraction [9]. The characteristic Peclet number is defined as the Peclet number at which the viscosity has dropped half-way between the first and second Newtonian plateau: $\eta(Pe_c) = (\eta_r^0 + \eta_r^\infty)/2$. This characteristic Peclet number is seen to be independent of the particle size and varies with the volume fraction. $Pe_c$ shifts to lower Peclet numbers with increasing volume fraction. This is a result of the slow-down of the particle diffusion in more concentrated dispersions, the Brownian diffusion can less adequately counteract

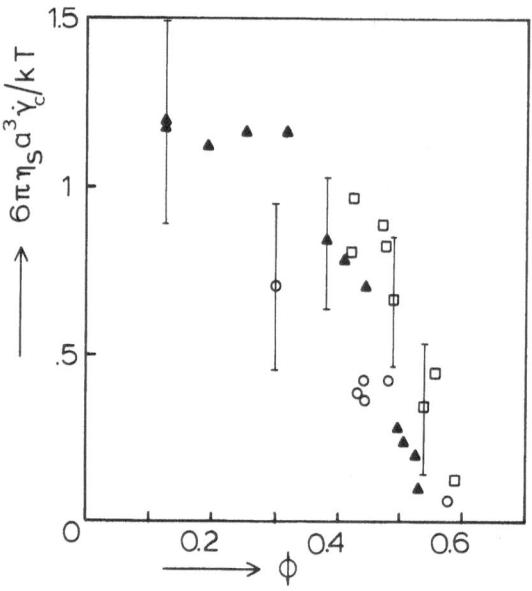

Fig. 2. The characteristic Peclet number as a function of volume fraction. Particle radii: 46 nm (○); 76 nm (▲) and 110 nm (□)

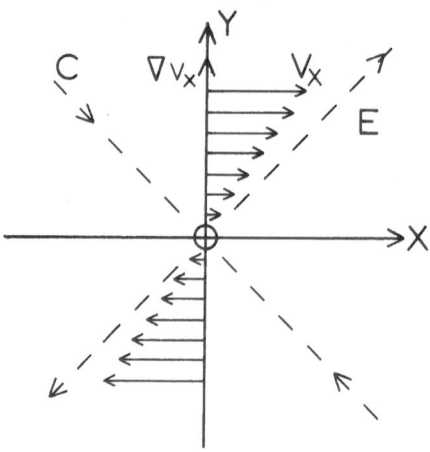

Fig. 3. Geometry of the simple shear flow field. $v_x$ is the velocity in the $x$-direction. The gradient of $v_x$ (indicated by $\nabla v_x$) is directed along the $y$-axis and the vorticity axis is along the $z$-axis. $E$ and $C$ mark the extensional and compressional components of the flow field. In the text we focus on the distortion in the velocity − vorticity plane ($xz$-plane) and the velocity − gradient plane ($xy$-plane)

the deformation of the particle distribution by the shear. The scaling of the relative viscosity with volume fraction and shear rate shows that these silica dispersions behave as hard-sphere dispersions indeed.

In the next section we will discuss scattering experiments that probe the static structure factor of dispersions subjected to steady shear flow. We will focus on the structure just after the first Newtonian plateau, well into the shear-thinning region, and in the second Newtonian plateau respectively.

### Determination of the structure factor by means of neutron scattering

In Fig. 3 we schematically sketched the simple-shear flow-field. The simple-shear flow can be seen as a superposition of a rotational and a pure straining component. The pure straining motion is responsible for the increase in viscosity when particles are brought into the fluid. Firstly, the undeformable spherical colloids cannot follow this component of the flow field and the flow is therefore locally disturbed. Secondly, this straining component changes the particle distribution. This can easily be seen if we define the axes of extension and compression of the pure straining motion as is done in Fig. 3. Along the compressional axis, we expect the particles to be pushed closer together, and along the extensional axis the particles are

driven further apart. This implies that we expect that the main peak of the structure factor $S(k)$ moves to higher $k$ along the compressional axis, and to lower $k$ along the extensional axis. So far, we discussed effects of the flow in the $xy$-plane to which we will refer in the following as velocity − gradient plane. The deformation of the structure factor in this plane seems to be most interesting but this plane is not easily probed with light or neutron scattering experiments.

In a neutron scattering experiment, the scattering vector $k$ is almost perpendicular to the incoming neutron beam. This is due to the fact that in these experiments, the scattering angles are very small. For instance, in the experiments we will discuss here, $\vartheta = 0.017$ rad. The vector length is, $k = |k| = \dfrac{4\pi}{\lambda}\sin(\vartheta/2)$, where $\lambda$ is the neutron wavelength and $\vartheta$ is the scattering angle.

In Fig. 4 we sketched the experimental geometry of our first neutron scattering experiment [11] in which we used a Couette-type flow-cell and the D11 diffractometer (ILL, Grenoble). In these experiments we probe the velocity − vorticity plane ($xz$-plane). The volume fraction of the dispersions studied was varied between 0.36 and 0.52 and the Peclet range studied was $Pe = 0.14 − Pe = 1.2$, so we studied structures that correspond to positions on the shear thinning curve starting in the first Newtonian plateau and ending beyond the characteristic Peclet number. It was

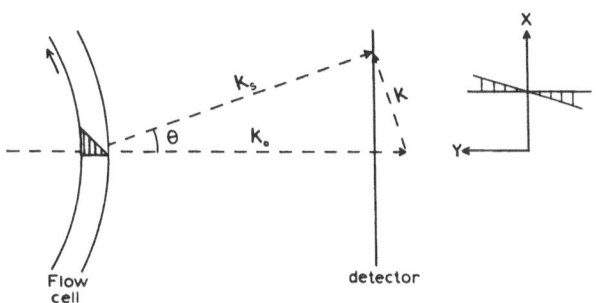

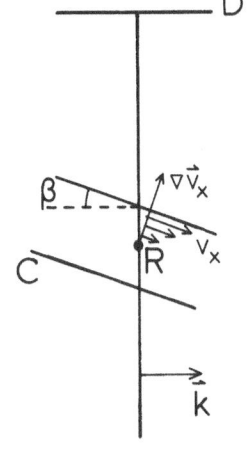

Fig. 4. Scattering geometry in neutron scattering experiments with the Couette-type flow-cell. **k** denotes a scattering vector. D is the two-dimensional detector, perpendicular to the neutron beam

Fig. 5. Top view on the scattering geometry in neutron scattering experiments with the parallel plate flow-cell. C is the vertically mounted shear cell, R is the rotation axis for the cell as a whole, D is the 2-dimensional detector and k is a scattering vector

found that the structure factor was shear dependent in the velocity direction as well as in the vorticity direction. This in contrast to theoretical calculations of the shear distorted structure factor which leave the vorticity direction unaffected [18−20]. The limited information obtained suggested that possibly string-like structures are formed. The particles appear to line up in strings and the strings lie parallel to the vorticity direction. No high intensity scattering spots in the velocity − vorticity plane were observed. Such bright spots or lines would immediately identify strong ordering effects. These measurements in the velocity − vorticity plane are relatively easy to do and of course necessary to map out the scattering space. It is however more interesting to study the velocity − gradient plane because theory [18−20] and simulations [21−23] give results for this plane only. With our Couette cell and the silica dispersions we cannot study this plane and we therefore designed a parallel-plate flow cell [12, 13].

The experimental set-up is shown in Fig. 5. The incoming neutron beam (D11 diffractometer, ILL, Grenoble) is horizontal and we therefore have to place the cell gap C in a vertical plane. The cell as a whole can rotate around a vertical axis R. This axis coincides with the vorticity axis of the flow. At a certain cell orientation, indicated by the angle $\beta$, the k-vectors probe a plane that contains the vorticity axis and intersects the velocity − gradient plane or shear plane. At different cell orientations $\beta$, we have different intersections with the velocity − gradient plane. From the runs at different angles $\beta$ we can construct the deformation of the structure factor in the velocity − gradient plane. The maximum angle $\beta$ is $\beta = 52°$, thus we can observe large parts of the velocity − gradient plane but not along the gradient axis (y-axis in Fig. 5).

Note that at $\beta = 0$ we probe the velocity − vorticity plane, as in the experiments with the Couette cell.

In Fig. 6 we show levels of the difference $S(k, Pe) − S(k)$ in the velocity − gradient plane (xy-plane). $S(k, Pe)$ denotes the non-equilibrium structure factor at a given Peclet number and $S(k)$ denotes the equilibrium structure factor. The figures are given for a dispersion of volume fraction $\phi = 0.48$ and the Peclet numbers range from 1 to 20. Increasing the Peclet number from $Pe = 1$ to $Pe = 10$ leads to a further deformation of the structure factor, whereas a further increase from $Pe = 10$ to $20$ does not lead to further deformation. Thus, the effect of the shear saturates. This is consistent with the observation of a second Newtonian plateau in the viscosity measurements. If we present the results as levels of the actual structure factor, then we do not observe "bright spots" which are characteristic for strong ordering, instead we still observe a deformed fluid-like structure. In Refs. [12] and [13] we compare our results to experiments by Clark and Ackerson [21, 22] theoretical predictions by Dhont [18], Ronis [19] and by Schwarzl and Hess [20] and to simulations by Hanley et al. [21]. These comparisons show that theoretical results of Ronis and Dhont give qualitatively the same picture if we present the results as contour plots of levels of the structure factor. However, if we make a more detailed comparison by presenting plots of the levels of $S(k, Pe) − S(k)$ or $S(k, Pe)/S(k)$, than some correspondence is found at very low Peclet numbers. A detailed comparison with simulation results cannot be made because only $S(k)$-pictures are available. The same can be said about the results of Clark and Ackerson. Our results for the deformation in the velocity − gradient plane indicate that the high shear structure of a hard-sphere dispersion is still deformed fluid-like.

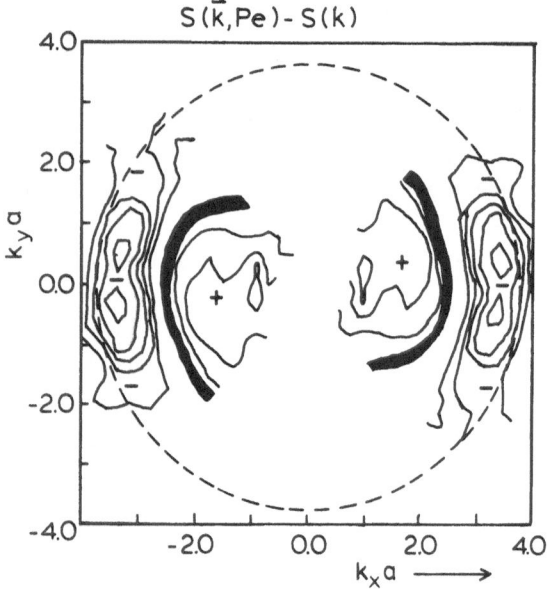

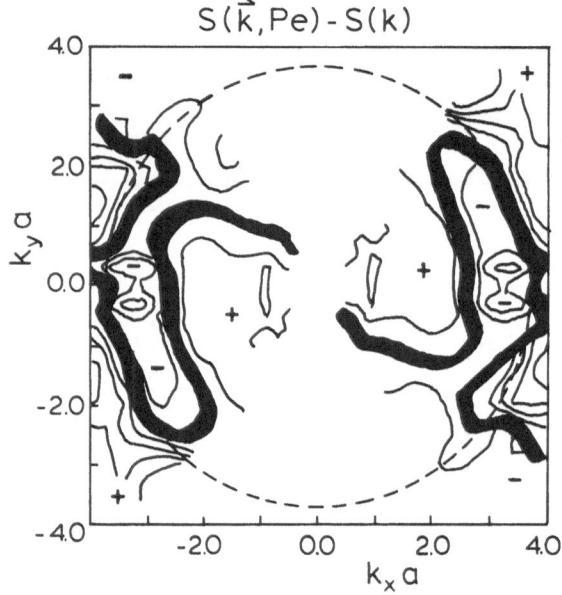

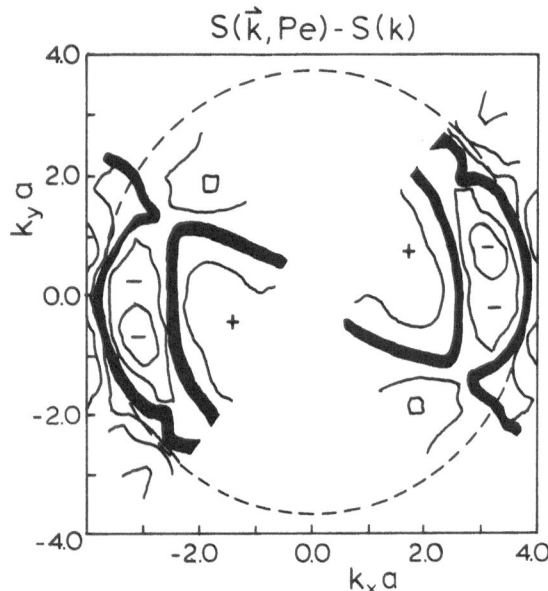

Fig. 6. Levels of $S(k,Pe) - S(k)$ for a dispersion with $\phi = 0.48$ and different Peclet numbers. The bold lines indicate the zero-level: it marks the positions where the numerical value of $S(k)$ is unaffected by the shear. The $(+)$ sign marks regions in which the value of $S(k)$ has increased, whereas the $(-)$ sign indicates that $S(k)$ has decreased in value.
a) $Pe = 1$
b) $Pe = 10$
c) $Pe = 20$

## Discussion and conclusions

We designed a new parallel-plate flow-cell [12] that we used in neutron scattering experiments. With this cell we can study many scattering planes that do have the vorticity direction in common. From the different intersections with the velocity − gradient plane (shear plane) we can construct the deformed structure factor in a large part of the velocity − gradient plane.

Hard-sphere dispersions which have liquid-like order in equilibrium have a deformed liquid-like struc-

ture in shear flow. Up to a Peclet number $Pe = 20$ and volume fractions between $\phi = 0.35 - \phi = 0.53$, no strong ordering is observed. At these Peclet numbers we are well into the second Newtonian plateau.

Viscosity measurements have shown that in the high shear limit, the viscosity seems to diverge at a volume fraction $\phi_m = 0.71$. This large value suggests that the particles would be orderly packed. This seems to be in contradiction to the observations made in the scattering experiments but we have not studied the gradient − vorticity plane yet. Perhaps evidence for strong ordering is observed in this plane. On the other hand, the viscosity experiments have furthermore shown that hard-sphere dispersions do not show shear-thickening. Shear-thickening is thought to be a result from a transition from an ordered state into another random state. The absence of shear-thickening can thus be seen as a confirmation of the conclusion that the hard-sphere dispersion do not strongly order at high shear.

Simulations of hard-sphere colloidal dispersions by Bossis and Brady [23, 24] did not show highly ordered shear induced structures at high Peclet numbers. Instead, they observe the formation of clusters. In the absence of Brownian motion, one (simulation) cell spanning cluster is formed. A small Brownian contribution breaks this cluster into smaller ones. The absence of an ordered shear induced state is in agreement with our results.

# References

1. Batchelor GK (1970) J Fluid Mech 41:545
2. Batchelor GK (1977) J Fluid Mech 83:97
3. Van Helden AK, Jansen JW, Vrij A (1981) J Colloid Interf Sci 81:354
4. Vrij A, Jansen JW, Dhont JKG, Pathmamanoharan C, Kops-Werkhoven MM, Fijnaut HM (1983) Faraday Discuss Chem Soc 76:19
5. De Kruif CG, Briels WJ, May RP, Vrij A (1988) Langmuir 4:668
6. Van Veluwen A, Lekkerkerker HNW, De Kruif CG, Vrij A (1987) Faraday Discuss Chem Soc 83:59
7. De Kruif CG, Van Iersel EMF, Vrij A, Russel WB (1987) J Chem Phys 83:4717
8. Van der Werff JC, De Kruif CG (1988) J Rheol 33:421
9. Van der Werff JC, De Kruif CG, Blom C, Mellema J (1989) Phys Rev A39:795
10. Ackerson BJ, Van der Werff JC, De Kruif CG (1988) Phys Rev A37:4919
11. Johnson SJ, De Kruif CG, May RP (1988) J Chem Phys 89:5909
12. De Kruif CG, Van der Werff JC, Johnson SJ, May RP (1990) Physics of Fluids A
13. Van der Werff JC, Ackerson BJ, May RP, De Kruif CG (1990) Physica A
14. Quemada D (1977) Rheol Acta 16:82
15. Krieger IM, Dougherty TJ (1959) Trans Soc Rheol 3:137
16. Hoffman RL (1972) Trans Soc Rheol 16:155
17. Barnes HA (1989) J Rheol 33:329
18. Dhont JKG (1989) J Fluid Mech 204:421
19. Ronis D (1984) Phys Rev A29:1453
20. Schwarzl JF, Hess S (1986) Phys Rev A33:4277
21. Hanley HJM, Rainwater JC, Clark NA, Ackerson BJ (1983) J Chem Phys 79:4448
22. Clark NA, Ackerson BJ (1980) Phys Rev Lett 44:1005
23. Bossis G, Brady JF (1984) J Chem Phys 80:5141
24. Bossis G, Brady JF (1989) J Chem Phys 91:1866

Authors' address:

C. G. de Kruif
University of Utrecht
Van't Hoff laboratory
Padualaan 8
3584 CH Utrecht, The Netherlands

**Progress in Colloid & Polymer Science**　　　　Progr Colloid Polym Sci 81:120 – 125 (1990)

# Field-induced structure in a colloidal suspension

Z. Mimouni, G. Bossis, C. Mathis, A. Meunier and C. Paparoditis

Laboratoire de Physique de la Matière Condensée, Université de Nice, Nice

*Abstract:* The formation of chains of polystyrene spheres in an aqueous suspension submitted to an electric field is studied by light scattering. We analyse the change in the diffraction pattern from an isotropic to a very anisotropic structure. The dynamics of the breakdown of the chains by Brownian motion, when the field is switched off, is studied by recording the decrease of the scattered intensity. The coefficient of diffusion perpendicular to the chain is obtained and compared to some theoretical predictions.

*Key words:* Light scattering; colloidal suspension; electric field; diffusion coefficient

## Introduction

Numerous observations on suspensions submitted to a high electric field have shown the formation of alignments of particles in the direction of the field. This effect is due to the difference in permittivity between the solid particles and the suspending fluid. It has been used for many purposes, for instance in order to fuse the biological cells [1] or to remove water from oil; these alignments are also responsible for the change of viscosity of a suspension by applying an electric field. This last effect was found by Winslow [2] in 1949 who demonstrated the possibility to monitor electronically the viscosity of a fluid. Many industrial applications related to hydraulic devices are possible, and interest in this field is rapidly growing [3]. The rheological behaviour of these fluids is explained on the basis of a Bingham model, with a yield stress related to the force which can break the structure. Except for some qualitative information obtained by direct observation with a microscope, very little things are known about the rigidity of the chains as a function of the field or about the time necessary to release the stress when the electric field is switched off. This paper is a first attempt to characterize the formation (or the deformation) of chains by the study of scattered light. In the first section we shall see how light scattering can help us to characterize the change from an isotropic structure to an oriented one. In the second section experimental results will be presented

and in the third section these results will be discussed with the help of some numerical results concerning ideal chain of spheres.

## Scattered light by a chain of spheres

In this section we shall assume that the suspension is diluted, so there is no multiple scattering. The scattered intensity is then given by:

$$I(\vec{k}) = C I_s(\vec{k}) \sum_{\alpha=1}^{N} e^{\vec{k} \cdot \vec{r}_\alpha} \sum_{\beta=1}^{N} e^{-i\vec{k} \cdot \vec{r}_\beta} \tag{1}$$

$\vec{k}$ is the difference between incident and scattered wave vectors:

$\vec{k} = \vec{k}_i - \vec{k}_d$; $\quad k = 2k_i \sin(\theta/2)$, where $\theta$ is the angle between $\vec{k}_i$ and $\vec{k}_d$

$I_s(\vec{k})$ is the intensity scattered by a single sphere; this is the usual form factor [4] if the sphere is small relatively to the wavelength, or a more complicated function given by Mie's theory for larger spheres [6].

The average over all the possible configurations will give:

$$\langle I(\vec{k}) \rangle = C I_s(k) N \left( 1 + \left\langle \sum_{\beta \neq \alpha}^{N} e^{i\vec{k} \cdot (\vec{r}_\alpha - \vec{r}_\beta)} \right\rangle \right). \tag{2}$$

If there is no position correlations between the particles, relation [2] becomes:

$$\langle I_0(\vec{k}) \rangle = C N I_s(k)(1 + O(\varphi)) \tag{3}$$

where $\varphi$ is the volume fraction of spheres. The scattered intensity is then isotropic and consists of a diffuse ring pattern. On the other hand, if all the spheres are perfectly aligned with a separation $d = 2a$, we get from (2):

$$I(\vec{k}) = C I_s(\vec{k}) \left[ \frac{\sin N\delta/2}{\sin \delta/2} \right]^2 \tag{4}$$

where $\delta = \vec{k} \cdot \vec{d}$.

For a phase shift $\delta = \pm 2p\pi$ we get the Bragg condition: $nd\sin\theta = \pm p\lambda$ where $n$ is the refractive index of the suspending medium, and the intensity reaches a maximum:

$$I(\vec{k}) = C I_s(\vec{k}) N^2. \tag{5}$$

Taking into account the refraction at the suspension-air interface, the locus of these maxima are a set of cones of angle $\pi - 2\theta_m$, where $\theta_m$ satisfies the relation:

$$d\sin\theta_m = \pm p\lambda.$$

When the field is switched off, the structure will be destroyed by Brownian motion, and the scattered pattern will progressively transform into the one characteristic of uncorrelated scatterers (cf. Eq. (3)). If we put:

$$\vec{r}_\alpha(t) = \vec{r}_\alpha(0) + \Delta\vec{r}_\alpha(t); \quad \vec{r}_\beta(t) = \vec{r}_\beta(0) + \Delta\vec{r}_\beta(t)$$

Eq. (2) will become:

$$\langle I(\vec{k}) \rangle(t) = \langle I_0(\vec{k}) \rangle \left( 1 + \left\langle \sum_{\beta \neq \alpha} \cos \vec{k} \right. \right.$$
$$\times \vec{r}_{\alpha\beta}(0) \cos \vec{k} \Delta\vec{r}_{\alpha\beta}(t)$$
$$\left. \left. - \sin \vec{k} \cdot \vec{r}_{\alpha\beta}(0) \sin \vec{k} \cdot \Delta\vec{r}_{\alpha\beta}(t) \right\rangle \right). \tag{6}$$

If we follow the change of intensity during a short enough time interval so that the relation $\vec{k} \cdot \Delta\vec{r}_{\alpha\beta}(t) < 1$ is satisfied, we can expand the cosine term:

$$\cos(\vec{k} \cdot \Delta\vec{r}_{\alpha\beta}(t)) = 1 - \frac{(\vec{k} \cdot \Delta\vec{r}_{\alpha\beta}(t))^2}{2} + \ldots \tag{7}$$

$$\sin(\vec{k} \cdot \Delta\vec{r}_{\alpha\beta}(t)) = \vec{k} \cdot \Delta\vec{r}_{\alpha\beta}(t) + \ldots . \tag{8}$$

On the other hand the average motion of the particles is governed by the Smoluchowski equation and for a given starting configuration we shall have, to first order in $\Delta t$, an average displacement given by:

$$(\Delta r^i_{\alpha\beta} \Delta r^j_{\alpha\beta})(\Delta t) = 2(D^{ij}_{\alpha\alpha} + D^{ij}_{\beta\beta} - 2D^{ij}_{\alpha\beta})\Delta t \tag{9}$$
$$= 2D^r_{\alpha\beta}\Delta t$$

$$\Delta r^i_{\alpha\beta}(\Delta t) = v^i_{\alpha\beta}(0)\Delta t \tag{10}$$

where $D_{\alpha\alpha}$ is the self diffusion and $D_{\alpha\beta}$ is the cross diffusion.

The velocity $\vec{v}_{\alpha\beta}$ in Eq. (10) is a convective velocity which (in the absence of interparticle forces) comes from the divergence of the diffusion function [5]. Inserting Eqs. (7)−(10) into Eq. (6) gives:

$$\langle I(\vec{k}) \rangle(\Delta t) = \langle I_0(\vec{k}) \rangle$$
$$\times \left( 1 + \left\langle \sum_{\beta \neq \alpha} \cos \vec{k} \cdot \vec{r}_{\alpha\beta}(0) \right\rangle \right)$$
$$- \langle I_0(\vec{k}) \rangle \left\langle \sum_{\beta \neq \alpha} \{ \cos \vec{k} \cdot \vec{r}_{\alpha\beta}(0) \right.$$
$$\times \vec{k} \cdot \boldsymbol{D}^r_{\alpha\beta}(0) \cdot \vec{k}]$$
$$\left. - \sin \vec{k} \cdot \vec{r}_{\alpha\beta}(0)[\vec{k} \cdot \vec{v}_{\alpha\beta}(0)] \} \right\rangle \Delta t. \tag{11}$$

This expression is valid for a short time internal $\Delta t$, and the relative diffusion tensor $\boldsymbol{D}_{\alpha\beta}$ or the relative convective velocity $\vec{v}_{\alpha\beta}$ are supposed to be constant during this time interval. In the following we also assume that $\boldsymbol{D}^r_{\alpha\beta} = 2\boldsymbol{D}$ is a constant whatever the distance $\vec{r}_{\alpha\beta}$. This approximation will be discussed in the last section. Furthermore for $\vec{k}$ perpendicular to the field direction, we can neglect the convective term (the convective velocity $\vec{v}_{\alpha\beta}$ is directed along $\vec{r}_{\alpha\beta}$ and $\vec{r}_{\alpha\beta}$ is almost perpendicular to $\vec{k}$).

It remains:

$$\langle I(\vec{k}) \rangle(\Delta t) = \langle I(\vec{k}) \rangle(0) - \langle I_0(\vec{k}) \rangle$$
$$\times \left\langle \sum_{\beta \neq \alpha} \cos \vec{k} \cdot \vec{r}_{\alpha\beta}(0) \right\rangle 2k^2 D_\perp \Delta t$$

or

$$\frac{\langle I(\vec{k})\rangle(\Delta t)}{\langle I(\vec{k})\rangle(0)} = 1 - 2k^2 D_\perp$$

$$\times \left(1 - \frac{\langle I_0(\vec{k})\rangle}{\langle I(\vec{k})\rangle(0)}\right)\Delta t. \quad (12)$$

So the decrease of the scattered intensity when the field is switched off allows to get the diffusion coefficient perpendicular to the chain by measuring the slope of the curve.

## Experimental

The sample is a suspension of monodisperse polystyrene sphere in water manufcatured by Rhône-Poulenc. Their diameter $d = 3$ µm has been verified by electronic microscopy. A microcell is formed by sandwiching two platinum wires between microscope slides. The preparation is hermetically sealed by means of a parafilm sheet fused between the two slides, thus trapping suspension and electrodes. In order to prevent a slight sedimentation, we match the density of polystyrene by mixing $D_2O$ with water. The volume concentration of the suspension is 0.2%. The electric field is applied between the platinum wires (whose separation is typically 2 mm) through a generator which is controlled by

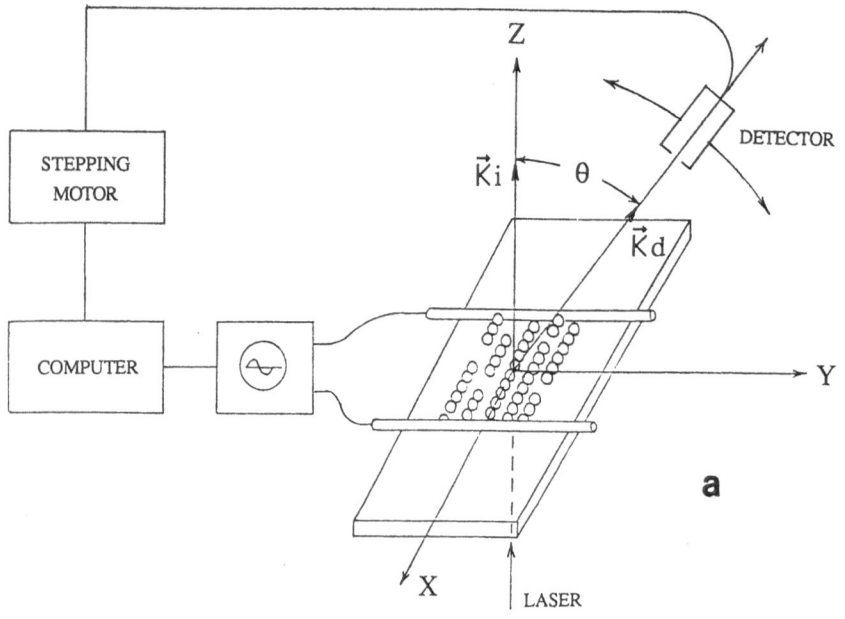

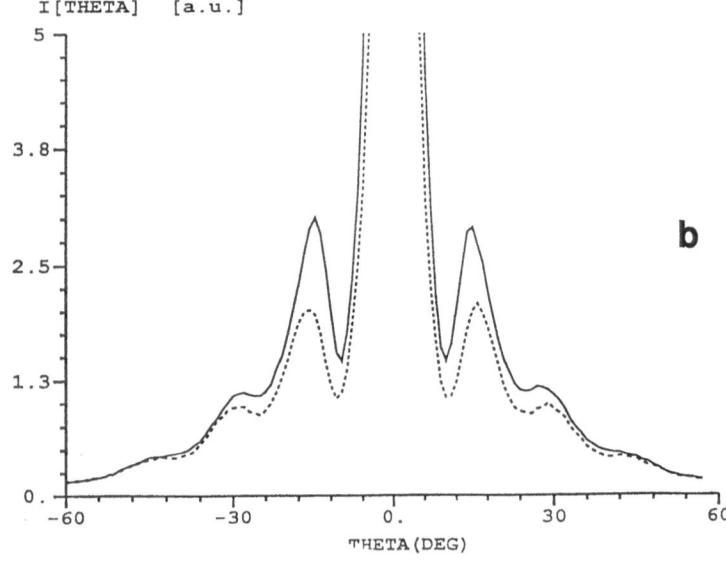

Fig. 1. a. Schematic view of the experiment. When the detector is in the plane YOZ, $k = k_i - k_d$ is perpendicular to the direction of the field; b. Scattered intensity in the plane YOZ (——) without field (– – –) with field

a microcomputer. The scattered intensity is recorded with a photodiode whose position is also monitored by the computer. In Fig. 1a we have sketched the experiment and in Fig. 1b the intensity in the YOZ plane is plotted as a function of the angle $\theta$, with and without field. The curve without field is well represented by Mie's theory with a first maxima for $\theta = 16°$. If the spheres were perfectly aligned, the ratio of the intensities would be equal to the average number of particles per chain (compare Eq. (3) to Eq. (5)) and the first maxima would be located at $\theta = 12°$. Actually in the central part the ratio of intensities is only 1.6 and the position of the first maximum is only slightly shifted towards the low angles. This indicates that for the maximum attainable field ($E = 300$ V/cm), the chains are not very rigid. This is evident from observations on a microscope. In Fig. 2a we can see a picture of a chain of spheres and get an idea of the defects relatively to an ideal chain of spheres. On the right, we have the scattered pattern which consists of a modulation of the

single sphere intensity $I_0(\vec{k})$ by the interference function whose maxima are located on different cones of angle $\pi - 2\theta_m$. The intersection of these cones with the plane of the screen gives the hyperbolas that we can see on the diffraction pattern. When the field is switched off the chains disintegrate as shown in Fig. 2b and the scattering pattern becomes isotropic and identical to that of isolated spheres. In Fig. 3a we have plotted the recording of the intensity as a function of time for an angle $\theta = 4°$. At a given time represented by the arrow, the field is turned off by the computer and we can follow the decrease of intensity with time. The initial slope is obtained by fitting the initial decrease by an exponential. The slope of the curve obtained for small angles is represented in Fig. 3b as a function of $k^2(1 - I_0/I)$. For these values of $k$ the expansion leading to Eq. (12) is justified. Furthermore the ratio of intensities without and with field is a constant equal to 0.6 and we get from Eq. (12): $D_\perp = (0.62 \pm 0.05)D_0$, where $D_0 = kT/6\pi\mu a$.

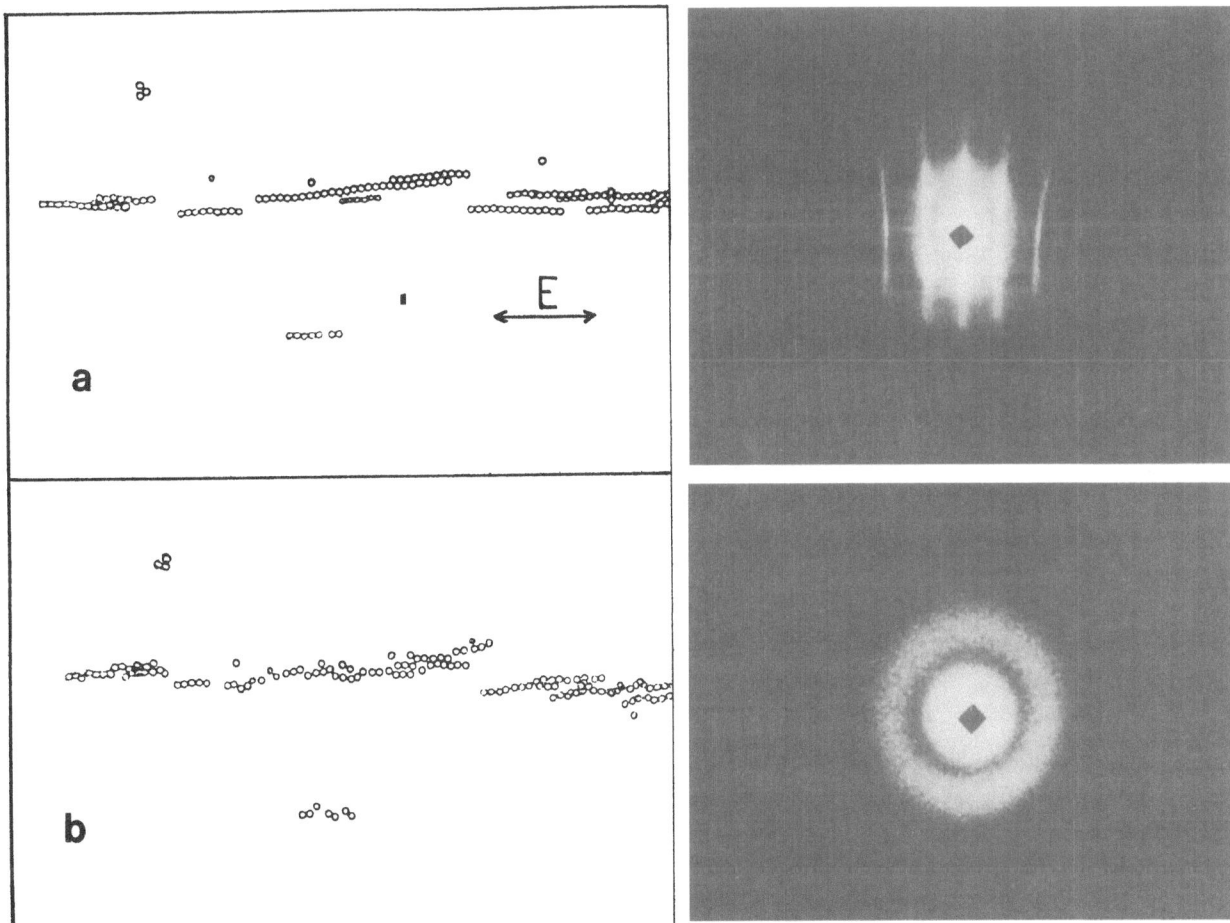

Fig. 2. a. Microscopy image of the chains of spheres for a field $E = 300$ V/cm and of the corresponding scattered intensity; b. Same picture a few tens of seconds after the field has been switched off

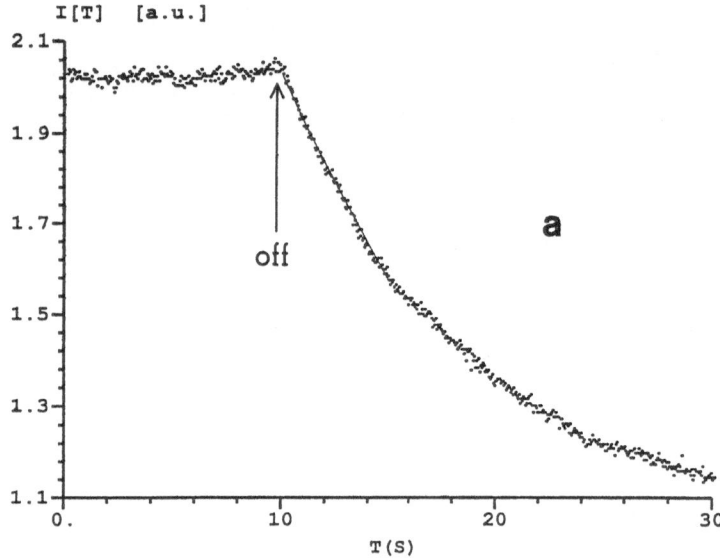

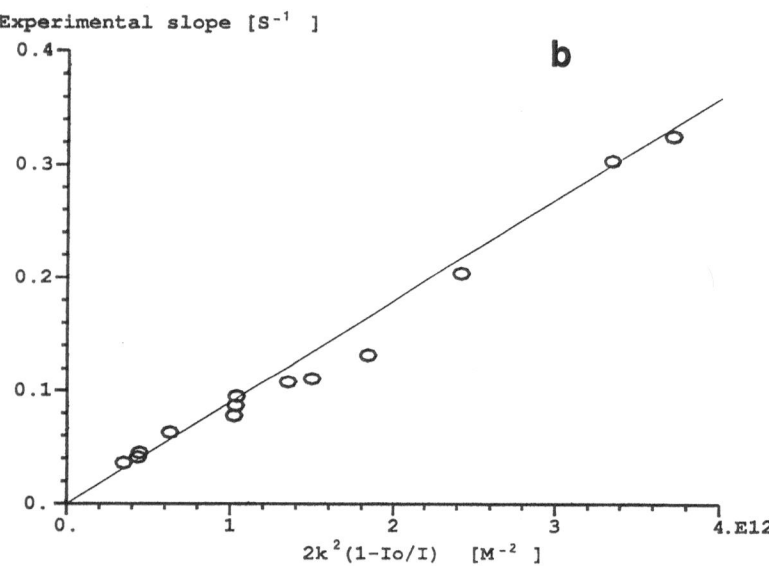

Fig. 3. a. Decrease of the intensity after switching off the field. The detector is in the YOZ plane with an angle $\theta = 4°$; b. Initial slope of the decreasing intensity as a function of $k^2(1 - I_0/I)$ of Eq. (12). A least square fit gives $D_\perp = (0.62 \pm 0.05)D_0$

## Discussion

In order to compare this value of $D_\perp$ with the theoretical predictions we can use the known results for two spheres where the relative diffusion tensor is given by [7]:

$$D_{\alpha\beta}^{ij}(\vec{r}) = 2\frac{kT}{6\pi\mu a}[G(r)e_ie_j + H(r)(\delta_{ij} - e_ie_j)]$$

where $\vec{e}$ is the unit vector joining the two centers of particle $\alpha$ and $\beta$. The functions $G(r)$ and $H(r)$ are known for any distance. At short distances ($\xi < 10^{-2}$) we have $G(r) = 2\xi$; $H(r) = 0.402 - 0.532/\ln\xi$ where $\xi = r - 2a$ is the separation between the spheres. For two spheres the perpendicular diffusion $D_\perp/D_0$ is just the function $H(r)$. This function increases with the separation of the spheres (its values is 0.55 for $\xi = 0.1$ and 0.66 for $\xi = 0.5$). The characteristic Brownian time for the spheres of diameter 3 μm in water is $T = a^2/Do = 15$ s. The initial slope is obtained on a time interval $\Delta t$ of about one second. During this time the Brownian displacement perpendicularly to the chain will be:

$$\langle r_\perp^2 \rangle / a^2 = 4(D_\perp/a^2)\Delta t;$$

so with $\Delta t \approx 1\,\text{s}$   we get:   $r_\perp \approx 0.4a$.

Consequently the experimental determination $D_\perp = (0.62 \pm 0.05)D_0$ seems to be close to the theoretical prediction. Nevertheless, in a chain, each sphere is surrounded by two other spheres and so the quantity $D_\perp$ should be different. We can calculate numerically this function with a many-body approach [8]. We find that the quantity $D_{\alpha\beta}$ is different depending on the proximity of the two particles $\alpha\beta$, since the cross diffusion $D_{\alpha\beta}$ decreases as $1/r_{\alpha\beta}$, so the decoupling approximation involved in the passage from Eq. (11) to Eq. (12) is not completely justified. Actually this decoupling approximation is justified in the limit of a perfectly aligned chain (in that case $\cos \vec{k} \cdot \vec{r}_{\alpha\beta} = 1$ whatever $r_{\alpha\beta}$) or in the limit of low correlation range such that only the two first neighbours of a particle give rise to coherent interference in average. In this last situation we have for any pair of spheres inside the chain:

$$D^\perp = D_{\alpha,\alpha}^\perp - D_{\alpha,\alpha+1}^\perp$$

and we find a theoretical value of 0.53 for a separation $(r - 2a) = 0.1$ and 0.66 for a separation $(r - 2a) = 0.5$. The experimental result $D_\perp/D_0 = 0.62 \pm 0.05$ agrees quite well with this hypothesis. On the other hand we can get an idea of the correlation range on the $Z$ axis by looking at the ratio of the intensities with and without field (cf. Fig. 1b). This ratio does not exceed a value of 2 whereas the theoretical ratio would be equal to the average number of particles in the chain which is at least one order of magnitude larger

(cf. Fig. 2a). It confirms our hypothesis of a next-neighbour correlation range.

The limit of a perfectly aligned chain would probably be attainable since the field we apply is rather low (300 V/cm; 2 MHz) and we could experimentally apply a field of 20 KV/cm but only at low frequencies. Unfortunately at low frequencies aqueous suspensions are electrolysed so we are looking for non aqueous suspensions with a good match in density and a large difference in permittivity between the particles and the suspending medium. In any event these properties are also required for a good electro rheological fluid. A better alignement of the spheres would also allow us to test the convective part of Eq. (13) by recording the time evolution with $\vec{k}$ parallel to the chain axis.

*Acknowledgements*

This work has been realized with the financial support of the CNRS (A.T.P.P.I.R.M.A.T.) and of the D.R.E.T. We are also grateful to Rhône-Poulence Recherches who kindly supplied us with the samples of polystyrene spheres.

# References

1. Schwan HP, Sher LD (1969) J Electrochem Soc 116:22C
2. Winslow WM (1949) J Appl Phys 20:1137
3. Block H, Kelly JP (1988) Phys D, Appl Phys 21:1661−1677
4. Berne BJ, Pecora RH (1976) "Dynamic Light Scattering", John Wiley, New York
5. Ermack DL, McCammon JA (1978) J Chem Phys 69:1352
6. Drake RM, Gordon JE (1985) Am J Phys 53:955
7. Batchelor GK, Green JT (1972) J Fluid Mech 56:375
8. Brady JF, Bossis G (1985) J Fluid Mech 20:111

Authors' address:

G. Bossis
Laboratoire de Physique de la Matière Condensée
Université de Nice
Parc Valrose
06034 Nice Cédex

**Progress in Colloid & Polymer Science**

Progr Colloid Polym Sci 81:126 (1990)

# Interactions between direct dyes in the dyeing of cellulosic substrates

H. Gerber

*Key words:* Adsorption; cellulose; direct dyes; substantivity; dyeing theory

One of the central problems of textile dyeing is matching a specific shade to a given pattern. As a rule, this is not possible with one Dyestuff alone. Dye mixtures or combinations consisting of 3—4 components are practically always used.

As experience shows, the sorption behaviour of a particular dyestuff is influenced by the other components of the combination that are always present. Thus, up to now it has not seemed possible to predict the equilibrium absorption of a component dye of a mixture on the basis of the absorption isotherms of the individual dyes.

To describe dyeing equilibria we use a Gouy-Chapman Model where the substrate is to be regarded as an absorbing surface. The partitioning of a component dye of the combination can be represented by the following expression:

$$\ln\frac{C_f}{C_b} = \ln K - \frac{zF\sigma_0}{RT\varepsilon_0\varepsilon}\frac{1}{\varkappa}$$

$$- \frac{zF^2}{RT\varepsilon_0\varepsilon Osp}\frac{1}{\varkappa}\sum z_i C_{fi}$$

where

| | |
|---|---|
| $C_f, C_{fi}$ | = Conc. of dyestuff absorbed |
| $C_b$ | = Conc. of Dyestuff in dyeliquor |
| $K$ | = Sorption constant |
| $\sigma_0$ | = Surface charge density |
| $Osp$ | = Specific surface |
| $z, z_i$ | = Charge numbers of Dyestuff anions |
| $1/\varkappa$ | = Debye-length |
| $R, T, \varepsilon_0, \varepsilon, F$ | = Gas constant, temperature, permittivity of vacuum, dielectric constant, Faraday constant |

The influence of electrolyte on the sorption behaviour is represented by the Debye-length $1/\varkappa$. Dye absorption is favoured by increasing ionic strength but increasingly hindered the more dyestuff is already absorbed. The indicated equation was used to

— determine affinities of direct dyes for different kinds of cellulosic substrates ($-RT\ln K$).
— evaluate interactions in dye combinations
— increase the reliability of recipe formulations
— determine surface charge $\sigma_0$ and specific surface $O_{sp}$ which is accessible to the dyestuff anions of a variety of cellulosic substrates.

Author's address:

Dr. H. Gerber
c/o Sandoz AG
CH-4002 Basel

**Progress in Colloid & Polymer Science**          Progr Colloid Polym Sci 81:127 – 130 (1990)

# Macrostructure of petroleum asphaltenes by small angle neutron scattering

J. C. Ravey[1]) and D. Espinat[2])

[1]) Laboratoire de Physico-chimie des Colloides, LESOC, Université Nancy, Vandoeuvre, France
[2]) Institut Français du Pétrole, Rueil Malmaison, France

*Abstract:* The size and shape of petroleum asphaltenes in solvents have been determined by small angle neutron scattering. Their most likely morphology is a membrane like particle. Their size decreases with temperature, this change being quite reversible. It is very solvent dependent: the more polar the solvent the smaller the aggregates. Similarly, the more polar the asphaltene the larger the particles.

*Key words:* Neutron scattering; asphaltene; solvent effect; temperature effect

## Introduction

The current definition of asphaltenes in based on the solution properties of petroleum residua or bitumen in various solvents; that is, asphaltenes are soluble in light aromatics and insoluble in light paraffins [1]. The various interests in the heavier petroleum industry have led to characterization studies, in order to understand the nature of the constituents in these systems [2 – 18]. The asphaltene molecules associate with each other in most of the solvents, even in "good solvents" like benzene, especially at higher concentrations and low temperature. Moreover, it has been suggested that they can also associate to other (lighter) fractions of the oils, namely the resins [19]. There is a clear indication for their very large polydispersity, and facts responsible for this variability have been discussed in recent papers, e.g. in a critical review by Speight et al. [20].

Although this "micellar" behavior of the asphaltenes has been known for a very long time, as yet detailed mechanisms of the intermolecular associations remain obscure. The most recent works tend to ascribe the mechanism of that aggregation to dipole-dipole interactions [21]. Indeed, the asphaltenes are the most polar fraction of the oils, in particular due to the presence of heteroatoms ($N, O, S$) which leads to charge imbalances on a local level; therefore the elementary molecular unit of asphaltenes ("monomer") may carry up to several dipoles. In fact, this elementary unit itself should be viewed as a polydis-perse set of bidimensional molecules, which are polycyclic, with some aliphatic chains, heteroatoms and a few basic/acidic groups. In pure solvents, many of the available data suggest a bidimensional structure [7 – 14]. A few small angle X-ray scattering investigations have revealed an average size of colloidal species in several oils, their radius of gyration being in the range of 3 – 7 nm [13 – 15], suggesting for the ultimate particles (no size variation) mean molecular weights of about $6 – 10 \times 10^3$ and radii of 2.5 – 3.0 nm. Using perdeuterated solvents makes small angle neutron scattering more powerful than SAXS by increasing the scattering contrast between medium and scatterer. Hence, relatively precise measurements can be performed even on rather dilute solutions [18]. In this paper we report data we have obtained from small angle neutron scattering for concentration about 1% for different types of asphaltenes and different solvents; the influence of temperature and additives (resins, nonionic surfactants) should also be considered, but will be reported elsewhere.

## Experimental

### Materials

Asphaltenes occurring in most crude oils, they are concentrated by vacuum distillation. In brief, the part of the residuum which is soluble in benzene, but insoluble in hot heptane constitutes the asphaltene fraction [1]. Its soluble part (but of low solubility in *n*-pentane) is the resin fraction.

Asphaltenes of various origins (Arabia, Venezuela ...) have been studied. They are named according their geographical source: Bati Raman, Boscan, Cerro Negro, Arab light, Safanya. The solvents were perdeuterated o-xylene, toluene, benzene, tetrahydrofuran, pyridine, nitrobenzene. Some of these systems were subjected to a fractionation by gel permeation chromatography [23]. According to the solvent used in the GPC process, the asphaltene fractions are noted $F_i$ (in THF), $f_i$ (in pyridine), or $\varphi_i$ (HVB asphaltene in pyridine). After the fractions were collected, they were brought to dry state, and then resuspended in the various solvents. Unfractionated samples are denoted global ($G$) samples.

*Small angle neutron scattering*

The measurements were performed at the Leon Brillouin Laboratory (L.L.B. − CEA CNRS, Saclay, France), on the PACE instrument. Scattered intensities were recorded in the $q$-range $0.005 - 0.3$ Å$^{-1}$, and evaluated on an absolute scale (cm$^{-1}$), after substraction of the incoherent noise, as usual. The scattering length of asphaltene is calculated from its atomic composition. Its density is 1.115. For modeling the mean scatterer morphology, we have tried to use simple homogeneous discrete particles:

$$I(q) = C K^2 M_W \{\langle A^2(q)\rangle + [S(q) - 1]\langle A(q)\rangle^2\}. \quad (1)$$

$A(q)$ is the amplitude scattered by the particle,
$S(q)$ is the structure factor of the fluid,
$K$ is the particle ($p$)/solvent ($s$) contrast parameter.
$K = (b/v)_p - (b/v)_s$, ($v$ is the molecular volume, $b$ the scattering length).
$C$ is the weight concentration. When $S(0) \to 1$ (no marked interparticle effect), the intensity at smaller scattering angle is proportional to the weight average molecular weight. If good fits can be obtained by having recourse to a mean representative particle only, then we have the simplest model with two parameters (*e.g.* two mean dimensions of a spheroid or a cylinder). If we want to introduce polydispersity or short range interparticle attractive potential, we have to add extra parameters.

We have also tried to use a fractal formalism for a floc of $N_C$ elementary units, using Texeira formula, where $N_C = 1 + \Gamma(D + 1)(\xi/r_0)^D$. But in that case, we have to introduce many parameters: the (two) dimensions of the elementary unit (from which the mean radius $r_0$, $\langle A^2\rangle$ and $\langle A\rangle^2$ can be calculated), $N_C$ and the fractal dimension $D$. Whatever the case, we calculate theoretical spectra, and perform a fitting over $\log(I(q))$, in order to get the same relative precision over the whole $q$-range ($\sim 5\%$).

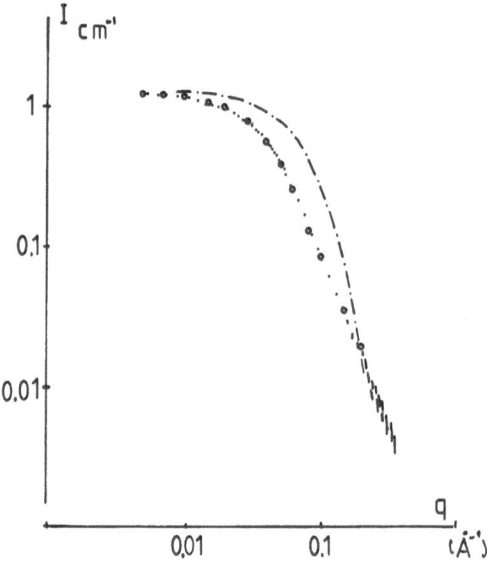

Fig. 1. Experimental spectrum for 2% Safanya asphaltene in pyridine at 50 °C, fitted to the spectrum (○) of a flat homogeneous spheroid ($5 \times 100$ Å, $M_w = 22000$), in a log log representation, and compared to the spectrum (— · —), of a spheroid of axial ratio 1/3 (same $M_w$)

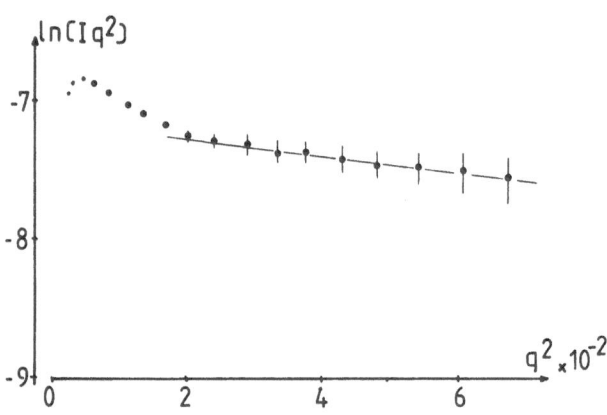

Fig. 2. Ln($I \cdot q^2$) vs $q^2$ representation for the same sample as in Fig. 1

**Results**

In Fig. 1, there is an example of experimental spectrum (in the log-log representation) fitted to the spectrum for a very flat spheroid (about $5 \times 100$ Å), hence quite different from the spectra for more globular spheroids. To further check this model, we have used the classical $\log(q^2 \cdot I(q))$ vs $q^2$ representation, from which a thickness of $7 \pm 3$ Å can be estimated (Fig. 2).

The spectra for relatively narrow fractions of Safanya asphaltene in THF solutions can also be fitted to the spectra of flat particles having nearly the same thickness (about $6 - 8$ Å). It can be seen that both F 3 and unfractionated samples can be represented by roughly the same mean particle. In other words, the macrostructure is such that, at fixed weight average molecular weight, the spectra are very insensitive to the polydispersity. We have checked this property in

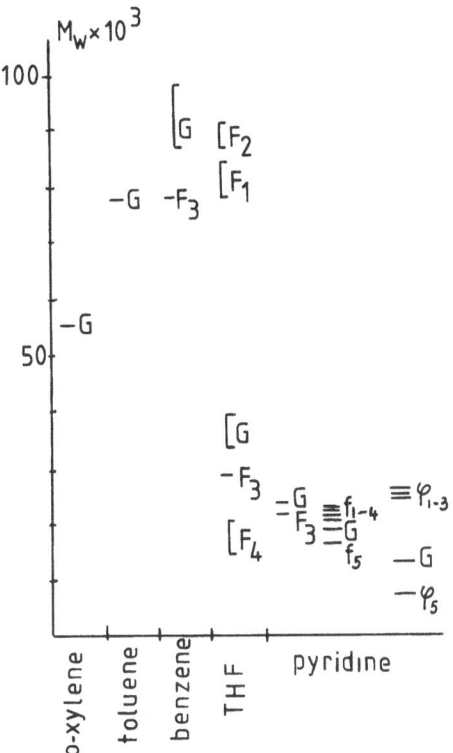

Fig. 3. $M_w$ values for fractionated ($F, f, \varphi$) and unfractionated ($G$) Safanya samples in various solvents

the case of the flat particle model. We made calculations by using different cumulative distribution functions similar to data from the work of Reerink obtained by ultracentrifugation [10] (polydispersity index $I = M_w/M_n \approx 3$). Computed spectra for nearly monodisperse ($I \sim 1.2$) and highly polydisperse ($I \approx 3.2$) ($M_w = 34\,000$) effectively appear very similar, but only if the thickness of all the particles is kept small (7 Å in these calculations). Using the fractal model would also explain, probably, such an invariability of the spectra. At any rate, the $M_w$ values from the fits are independent of the exact repartition of the dimensions of the aggregates, and practically all the data presented in this paper will also be model independent.

As far as $M_w$ is concerned, most of the data concerning the Safanya asphaltenes (for the concentration $C = 1\%$) are represented in the diagram of the Fig. 3. Different and stable fractions have been obtained by GPC fractionation (bars represent the repeatability of the measures over several months). Clearly, when THF is used as the solvent (eluent) in the GPC process [18], the resulting solutions in THF contain very different species (fractions $F_i$) except for $F_1$ and $F_2$, the heaviest fractions, although they are obtained at

marked different elution volumes. Hence, in THF, only heaviest fractions rapidly reorganize according to some equilibrium, leading to a decrease of their molecular weight. Now, when GPC is performed in pyridine, the different fractions are much less dissimilar, and they all contain much smaller aggregates (fractions $f_i$). Therefore, this polar solvent would bring conditions necessary to disrupt intermolecular associations, and $f_5$ could contain species not far from "monomers".

As far as HVB samples are concerned (fractions $\varphi_i$), in pyridine these systems appear more polydisperse than normal asphaltene; but this could result from a breaking of some of the largest monomers themselves (fraction $\varphi_5$). This could also be ascribed to the lower polarity of the HVB samples (resulting from the hydrotreatment).

It is also interesting to note that, for systems fractionated in THF and resuspended in various solvents, the $F_3$ fractions remain similar to the unfractionated sample ($G$) (i.e. they have about the same value of $M_w$) whatever the solvent. Therefore, all the above mentioned results are a clear indication of the micellar behavior of asphaltenes, the equilibrium between "units" and "micelles" depending on the polarity of both the solvent and the asphaltene.

### Effect of the polarity of the asphaltenes ($20°C, C = 1\%$)

From the chemical analysis of asphaltenes of various sources [23], the content (W/W) in heteroatoms can be evaluated. As a rule, the largest aggregates are obtained for the most polar asphaltenes, which are the richest in O, N, S atoms (the dielectric constant and the number of heteroatoms per 100 carbon atoms are roughly expressed by the same number; this can be shown by also taking into account the sulfur atoms in the data of Ref. [21]).

### Effect of the polarity of the solvent

$M_w$ values for global Safanya samples in various solvents are 100 000, 80 000, 65 000, 35 000, 20 000, for, respectively benzene, toluene, o-xylene, THF and pyridine. We have chosen to characterize the solvent by its dipole moment (in the liquid state). Clearly, the higher the dipole moment of the liquid the smaller the aggregates. Moreover, for smaller particles (highly dipolar solvents, HVB low polarity asphaltene, resins ...) attractive interaction effects between the aggregates become apparent on the spectra, as a steep increase of the intensity scattered at low $q$. In pyridine, the effect (at low temperature) becomes just noticeable; in nitrobenzene, which is known to be the most efficient to disrupt intermolecular associations, e.g. from vapor pressure osmometry experiments, the increase is so steep that it prevents any evaluation of the size of the aggregates. We have to note that the use of an "equivalent hard sticky spheres" model cannot account for a quantitative representation of the experimental structure factor $S(q)$ [24]: this would strengthen the highly anisometric shape model for which this approximation is certainly quite unsuitable. On the other hand, this low $q$ part of the spectra (e.g. in nitrobenzene) is markedly steeper than $q^{-2}$; this makes a model of a "fractal superstructure" quite unlikely.

## Conclusion

In conclusion, all our results appear coherent with the aggregation model proposed by Maruska et al. [2]: dipoles in the asphaltene units align head to tail, and then are effectively shielded as the asphaltene molecules stack-up pair wise, and form two dimensional aggregates. Indeed, when asphaltenes are more polar, we get larger aggregates (more numerous dipoles). But when the solvent is more polar, the dipoles of the units are less effective (or shielded by the solvent), and lead to smaller aggregates, with a correlative change of the interaggregate potential.

In summary, global asphaltene in solution are highly polydisperse systems of micellar aggregates made of a set of polydisperse units. Nevertheless, stable and relatively narrow fractions can generally be obtained. In the association process, the fundamental parameters are the polarity of both the solvent and the asphaltenes. For the shape of the particles, we could have some choice, for example between the membrane-like model (typical dimension $100-150$ Å $\times 5-10$ Å) or a fractal system with four parameters: $N_C$, the dimensions of the unit: $15-20$ Å $\times 2.5$ Å, and $D \approx 2-2.6$, depending on the systems. Indeed both these models can fit the data equally well. As a matter of fact, from other experimental data, in particular the very strong light scattering depolarization as measured by Gottis [17], and from the dipole-di-

pole mode of association, may be we could prefer the membrane like model.

*Acknowledgements*

The authors wish to thank Mme Buzier for her kind help during the experiments. They are indebted to LLB for allowing them to perform these studies on asphaltenes, and to Mrs Ducouret and Herzog for the preparation of the samples.

## References

1. Speight JG, Long RB, Trowbridge TD (1984) Fuel 63:616
2. Yen TF (1981) In: Bunger JW, Li NC (ed) "The Chemistry of Asphaltenes", Advances in Chemistry Series No 195, Am Chem Soc, Washington DC 39
3. Moschopedis SE, Fryer JF, Speight JG (1976) Fuel 55:227
4. Moore WJ (1972) Physical Chemistry, 4th ed, Prentice-Hall, Englewood Cliffs, NJ
5. Speight JG, Moschopedis SE (1977) Fuel 56:344
6. Albaugh EW, Talarico PC (1972) J Chromatogr 74:233
7. Neumann HJ (1965) Erdöl & Kohle 865
8. Ray BR, Witherspoon PA, Grim RE (1957) J Phys Chem 61:12
9. Eldib IA, Dunning HN, Bolen RJ (1960) J Chem Eng Data 5:550
10. Reerink H, Lijenzga J (1973) J Inst Petroleum 59:211
11. Altgelt KH, Harles OL (1975) Ind Eng Chem Prod Res Dev 14:240
12. Reerink H (1973) Ind Eng Chem Prod Res Dev 12:82
13. Dwiggins CW (1978) J Appl Cryst 11:615
14. Kim H, Long RB (1979) Ind Eng Chem Fundam 18:60
15. Tchoubar D, Herzog P, Espinat D (1988) Fuel 67:245
16. Gourlaouen C, PhD Thesis ENSPM (1984) Rueil Malmaison France
17. Gottis PG, Lalanne JR (1989) Fuel 68:804
18. Ravey JC, Ducouret G, Espinat D (1988) Fuel 67:1560
19. Pfeiffer JP (1950) In: "The properties of asphaltic bitumen", Elsevier, Amsterdam, Netherlands
20. Speight JG, Wernick DL, Gould KA (1985) Rev Institut François Petrole 40:52
21. Maruska HP, Rao BM (1987) Fuel Science Technology 5:119
22. Dickie JP, Yen TF (1967) Anal Chem 39:1847
23. Ducouret G, PhD Thesis (1987) University Paris VI, France

Authors' address:

J. C. Ravey
Laboratoire de Physicochmimie des Colloides
UA CNRS 406, Université Nancy I
BP 239, 54506 Vandoeuvre Cedex, France

**Progress in Colloid & Polymer Science**

Progr Colloid Polym Sci 81:131−135 (1990)

# Chemical reactions in microemulsions: Tracing the progress of a reaction from the influence of the reactants on the phase behaviour

R. Schomäcker

Max-Planck-Institut für biophysikalische Chemie, Göttingen

*Abstract:* Nucleophilic displacement reactions and ester hydrolysis have been performed in microemulsions stabilized by a nonionic amphiphile. The effect of the reactants added on the phase behaviour was studied as well as the effect of the progress of the reaction. The latter causes changes of the phase boundaries of the microemulsions with time. These changes were followed for tracing the progress of the reaction. The knowledge of the phase behaviour of mixtures of water, oil and nonionic amphiphile permits separating reactants and products.

*Key words:* Microemulsions; nonionic amphiphile; phase behaviour; phase transfer catalysis

## Introduction

Microemulsions appear to be excellent media for facilitating chemical reactions. They solubilize a large number of very different compounds, they possess a large internal interface, and they form spontaneously. As a consequence a variety of chemical reactions have been studied including ester hydrolysis [1, 2], the formation of macrocyclic lactones [3], nucleophilic displacement reactions [4], the Wacker reaction [5], photochemical reactions [6], acid-base equilibria [7, 8, 10], and a large number of enzyme catalyzed reactions [9−12].

The majority of the studies on chemical reactions in microemulsion media published so far deals mainly with the physical chemistry of the systems themselves. In a large number of publications, reactions were studied either as probes for claryfying the physical properties of the microemulsions, or for investigating the influence of an organized reaction medium on the kinetics of the reactions. Only a small number of studies describe the application of a microemulsion as a medium for performing chemical synthesis [4, 13−16].

The reason for this discrepancy is the following: For a kinetic study the reactants are added to the microemulsion as traces (in the range of $10^{-3}$ mole per liter or less), which do not cause a change of the phase behaviour. However, for performing a synthesis the concentrations of the reactants have to be much higher (in the range of one mole per liter). Increasing concentrations of additives cause increasing problems to control the phase behaviour of the microemulsion. This is especially true if high concentrations of electrolytes are added to a microemulsion stabilized by an ionic amphiphile. Still most studies are carried out with ionic amphiphiles like AOT or CTAB. An investigation of the effect of electrolytes on the phase behaviour of ionic and nonionic amphiphiles is described in references [17]−[19]. Since nonionic amphiphiles are much less sensitive to electrolytes than ionic amphiphiles the first group was chosen for this study.

First, the temperature dependent phase diagram of the microemulsion stabilized by a nonionic amphiphile is determined without any reactants. In a second step the reactants are added separately, for determining their influence on the phase behaviour of the microemulsion, without the reaction occuring. In a third step all reactants are added. During the reaction the concentrations of the reactants change, which chause a change of the phase boundaries with time. The progress of the reaction is traced following the changes of the temperatur interval limiting the one phase region.

The two reaction described are the nucleophilic displacement of bromide by iodide at 1-bromooctane and the hydrolysis of diethyl adipate and diethyl butylmalonate, respectively. The traces obtained from the phase behaviour are compared with traces obtained

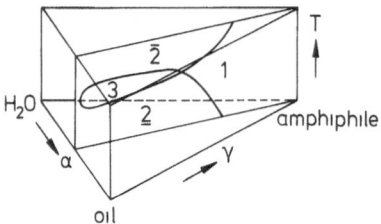

Fig. 1. Schematic phase prism of ternary mixtures of water, oil and nonionic amphiphile

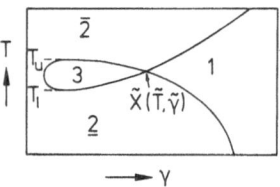

Fig. 2. Pseudobinary phase diagram of a ternary mixtures of water, oil and nonionic amphiphile at a constant ratio $a$

from GC-analysis and conductivity measurements, respectively.

## Experimental section

All substances used were of analytical grade. The water as doubly destilled. The amphiphile $C_8E_5$ was supplied by Bachem, Bubendorf (Switzerland) and used without further purification. The amphiphile $C_8E_6$ was synthesized by Dr. R. Laugwitz, University of Hannover. The phase diagrams were determined visually with the samples contained in closed test tubes thermostated in a water bath.

## Results

The phase diagram of a ternary mixture can be represented in a phase prism, with the Gibbs triangle as base and the temperature as vertical axis (Fig. 1). For simplifying the representation of the experimental results only one representative vertical section through the phase prism is plotted (Fig. 2). This section yields a pseudobinary phase diagramm. To determine this diagram the number of phases of mixtures of oil, water and amphiphile is observed as a function of temperature. The fraction of oil in the mixture of oil and water, $\alpha$, is kept constant, while the fraction of amphiphile, $\gamma$, is changed in the mixture of water, oil and amphiphile. This pseudobinary phase diagram is also used to describe the phase behaviour of the multi-component reaction mixtures as pseudo-ternary

mixtures of (water + salts) − (oil + nonpolar reactant) − nonionic amphiphile.

The phase behaviour of mixtures of water, oil and nonionic amphiphile follows a general pattern. At low temperatures these mixture separate into two phases ($\underline{2}$) with the amphiphile dissolved in the water phase. At high temperatures the amphiphile is dissolved in the oil phase ($\bar{2}$). Within a well defined intermediate temperature interval the mixtures separate into three coexisting liquid phases (3). In the amphiphile rich middle phase the mutual solubility between oil and water reaches a maximum. This phase is regarded as the microemulsion phase that is applied as solvent for the bimolecular reactions.

*Example 1: The nucleophilic displacement reaction*

When studying the reaction

$$C_8H_{17}Br + KJ \xrightarrow{H_2O/C_8E_5} C_8H_{17}J + KBr$$

in a ternary microemulsion, the mixture has six components. For examining the effect of the species involved on the phase behaviour of the microemulsion stabilized by $C_8E_5$, three- and four-component mixtures were studied with no reaction occuring. Figure 3a is the pseudobinary phase diagram of mixtures of $H_2O$, 1-bromooctane and $C_8E_5$ and $H_2O$, 1-iodooctane and $C_8E_5$, respectively. The mean temperature $\tilde{T}$ of the three-phase body is 36 °C for the mixtures containing 1-bromooctane, for the mixtures containing 1-iodooctane $\tilde{T}$ is 43 °C. The effect of the salts (E) involved is studied at $H_2O$-hexane-$C_8E_5$-mixtures (Fig. 3b). Is the water fraction of the mixture replaced by a one molar solution of potassium iodide $\tilde{T}$ increases from 55 °C to 62 °C. For a one molar solution of potassium bromide, $\tilde{T}$ decreases to 49 °C. Figure 3c represents the pseudobinary phase diagram of mixtures containing the products of the reaction, 1-iodooctane and potassium bromide. The ratio $\alpha$ of nonpolar to polar component is changed from $\alpha = 50$ wt% to $\alpha = 16$ wt% to give a 1:1 stochiometric ratio between the reactants. The two open cycles at $\gamma = 11$ wt% and 41.5 °C and 44 °C, respectively, represent the phase boundaries of a mixture of a one molar solution of potassium iodide and 1-bromooctane ($\alpha = 16$ wt%) stabilized by 11 wt% of $C_8E_5$.

This temperature interval shifts over a period of about 200 hours to 36 °C − 38 °C (Fig. 4a). This experiment is repeated with the more hydrophilic amphiphile $C_8E_6$ instead of $C_8E_5$. In this case the initial one phase temperature interval is 49.5 °C − 51.5 °C.

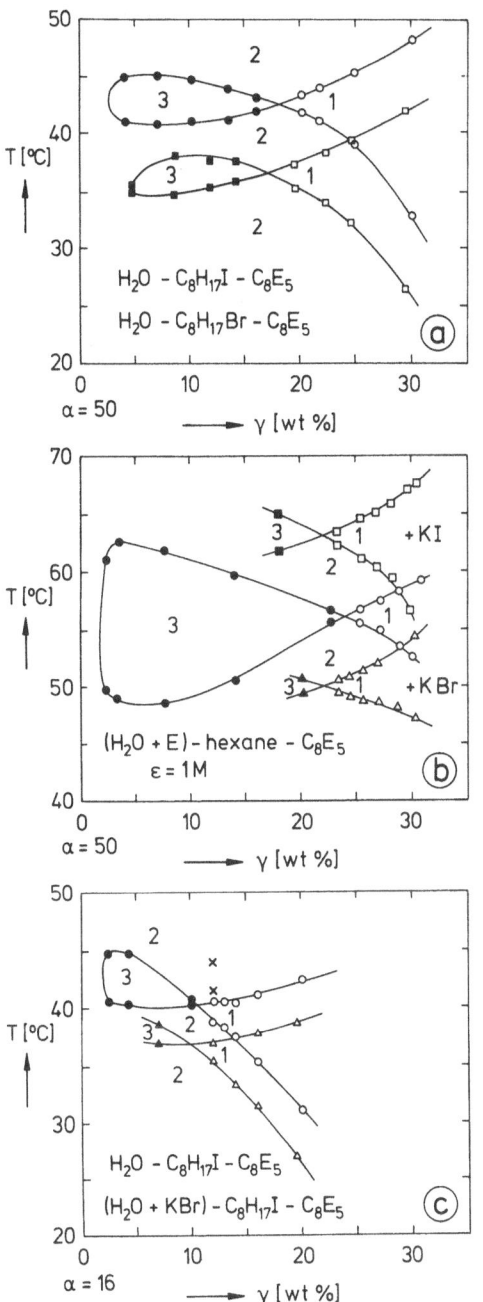

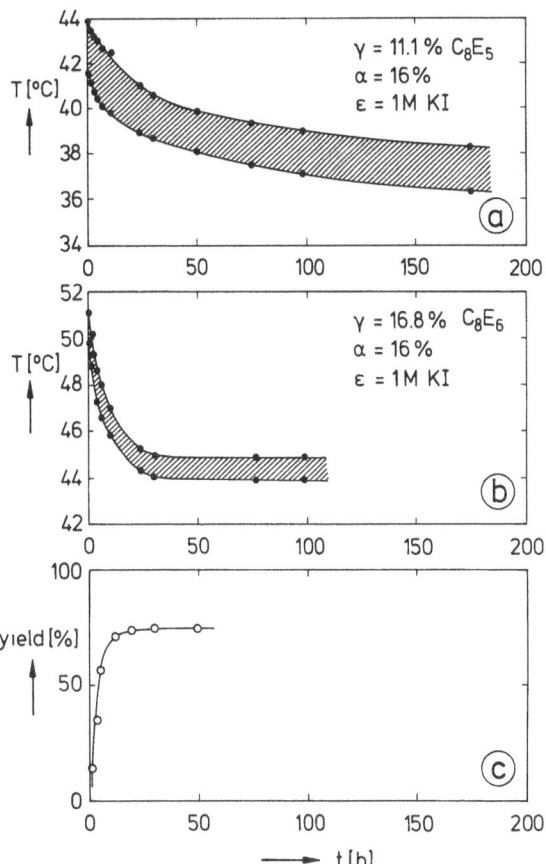

Fig. 4. a, b) Traces of the shift of the one phase interval with time. c) The yield of the reactions as a function of time

Fig. 3. Pseudobinary phase diagrams of ternary and quarternary mixtures of the components involved in reaction of 1-bromooctane with potassium iodide in a microemulsion stabilized by $C_8E_5$

Over a period of 50 hours this interval is observed to shift to 44 °C – 45.5 °C (Fig. 4b).

Is the initial temperature of 50 °C kept constant phase separation is observed after 2 hours. The rate of the reaction slows down until, finally, the reaction is stopped. The progress of the reaction performed in the microemulsion stabilized by $C_8E_6$ is also analyzed by gas chromatography.

The results are plotted in Fig. 4c. Both traces level off after about 30 hours. The yield does not exeed 75% because the reaction reaches an equilibrium. The equilibrium constant calculated is in good agreement with an equilibrium constant calculated from the rate constants of the reaction in protic polar solvents [20]. For isolating the product the temperature is decreased to 35 °C. The solution separates into two phases with the amphiphile mainly dissolved in the water rich phase. The organic layer is again extracted with water to remove all the amphiphile. The product is purified by a distillation. With ethanol as solvent the reaction is complete after about 10 hours. Reference [21] describes this reaction performed with phase transfer catalysis. Different crown ethers are testes as catalysts. The yields determined after 48 hours are between 7 and 63%.

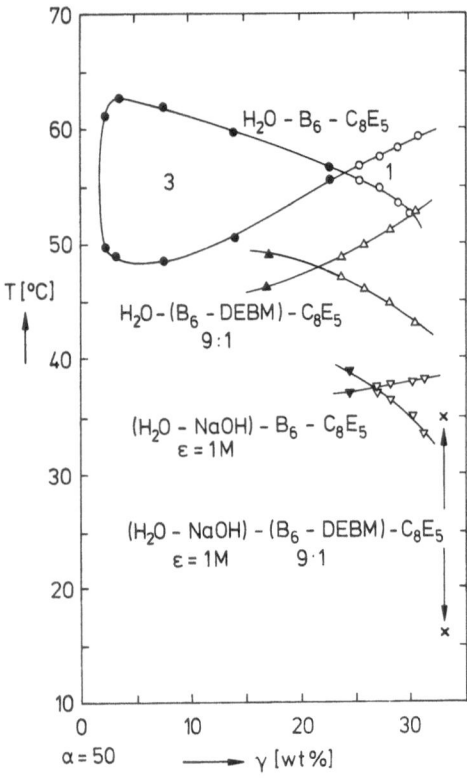

Fig. 5. Pseudobinary phase diagramms of ternary and quarternary mixtures of the components involved in the hydrolysis of diethylbutylmalonate (DEBM)

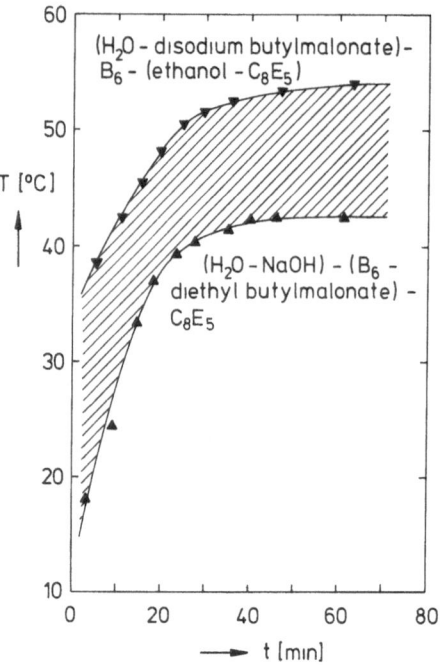

Fig. 6. Trace of the one phase interval of a microemulsion with the hydrolysis of DEBM in progress

The nucleophilic displacement reaction was repeated with other nucleophilic anions. 1-Octyl phenyl sulfide and 1-octyl cyanide were also prepared in a microemulsion medium from the reaction of 1-bromooctane with thiophene oxide and potassium cyanide, respectively.

*Example 2: The hydrolysis of esters*

The hydrolysis of diethyl butylmalonate and diethyl adipate was performed in microemulsions of water, hexane ($B_6$) and $C_8E_5$. These two compounds where chosen since diethyl butylmalonate is a typical representative of a intermediate product of a multi-step synthesis. The hydrolysis of diethyl adipate was studied to compare the experiment with a PTC experiment described by Starks in his fundamental paper on the PTC method [22]. Figure 5 shows the pseudobinary phase diagram of water, hexane and $C_8E_5$ mixtures at $\alpha = 50$ wt%. The addition of both reactants causes a decrease of $\tilde{T}$. For a microemulsion containing both sodium hydroxide and diethyl butylmalonate only the one phase interval is determined for a mixture with

$\alpha = 50$ wt% and $\gamma = 33$ wt%. This single phase interval is followed with time (Fig. 6). The mean temperature of this interval shifts from about 25 °C to 48 °C. This increase in $\tilde{T}$ is caused by the exchange of the lyotropic electrolyte sodium hydroxide against the hydrotropic electrolyte disodium butylmalonate [19].

In the experiment with diethyl adipate no strong $\tilde{T}$-shift is observed (Fig. 7a). At a temperature of 30 °C the mixture is homogeneous during the whole reaction time. At this temperature the reaction was followed also by conductivity measurements (Fig. 7b). The conductivity decreases due to the consumption of hydroxide during the reaction. Both tracers show that the reaction is complete after about 50 minutes. In a two-phase medium with a phase transfer catalyst added and under reflux the reaction time is about 30 minutes [22].

**Conclusion**

A variety of chemical reaction between ions and nonpolar molecules can be performed in microemulsion media. Microemulsions stabilized by nonionic amphiphiles allow the solubilisation of high concentrations of electrolytes and nonpolar compounds. The multi-component reaction mixture can be described

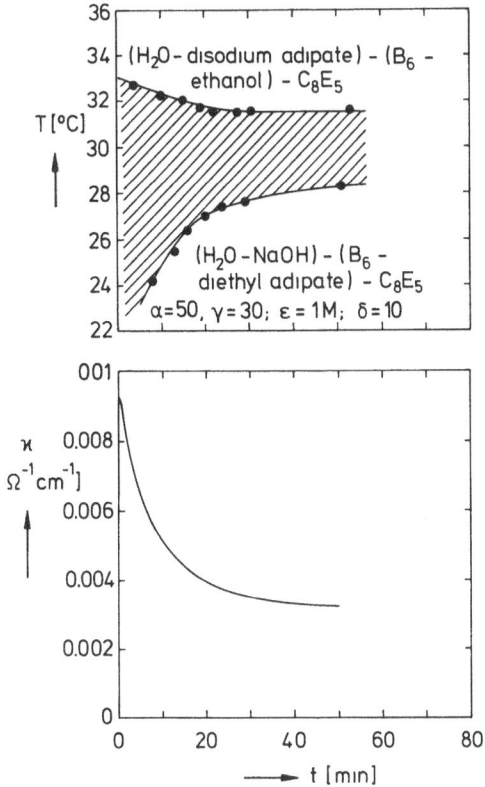

Fig. 7. a) Trace of the one phase interval of a microemulsion with the hydrolysis of diethyl adipate in progress. b) Trace of the conductivity at 30 °C

as a pseudo-ternary mixture of (water + electrolyte) − (oil + nonpolar reactant) and nonionic amphiphile. The stability of the microemulsion has to be controlled to maintain the progress of the reaction. The progress of the reaction can be traced from shifts of the phase boundaries. The reactions described show that the procedure has to be elaborated again for every reaction to be performed. Similar observations were made studying reactions like styrene polymerisation or the hydrolysis of triglycerides [23]. These results suggest applying microemulsions as substitute for toxic and hazardous polar organic solvents like chlorinated hydrocarbons, that are widely used in industrial scale organic synthesis.

*Acknowledgements*

This work was carried out in the laboratory of Prof. M. Kahlweit, to whom I am indebted for his support and helpful discussions.

**References**

1. Mackay RA (1981) Advances in Colloid Interface Science 15:131
2. O'Connor CJ, Lomax TD, Remage RE (1984) Advances in Colloid Interface Science 20:21
3. Jaeger DA, Ippoliti JT (1981) J Org Chem 46:4964
4. Martin CA, McCrann PM, Angelos GH, Jaeger DA (1982) Tetrahedron Letters 23:4651
5. Rico I, Couderc F, Perez E, Laval JP, Lattes A (1987) J Chem Soc, Chem Commun: 1205
6. Kiwi J, Grätzel M (1978) J Amer Chem Soc 100:6314
7. El Seoud OA (1989) Adv Colloid Interface Science 30:1
8. Knoche W, Schomäcker R (eds) (1989) Reactions in Compartmentalized Liquids, Springer Verlag, Berlin, Heidelberg
9. Luisi PL (1985) Angewandte Chemie 24:439
10. Luisi PL, Giomini M, Pileni MP, Robinson BH (1988) Biochim Biophys Acta 947:209
11. Fletcher PDI, Freedman RB, Robinson BH, Rees GD, Schomäcker R (1987) Biochim Biophys Acta 912:278
12. Schomäcker R, Robinson BH, Fletcher PDI (1988) J Chem Soc Faraday Trans 84:4203
13. Fendler JH, Fendler EJ (1975) Catalysis in Micellar and Macromolecular Systems, Academic Press, New York
14. Menger FM, Rhee JU, Rhee HK (1975) J Org Chem 40:3803
15. Jaeger DA, Ward MD, Martin CA (1984) Tetrahedron 40:2691
16. Gonzales A, Holt SL (1982) J Org Chem 47:3186
17. Kahlweit M, Strey R, Schomäcker R, Haase D (1989) Langmuir 5:305
18. Kahlweit M, Strey R, Firman P, Haase D, Jen J, Schomäcker R (1988) Langmuir 4:499
19. Firman P, Haase D, Jen J, Kahlweit M, Strey R (1985) Langmuir 1:718
20. Parker AJ (1969) Chemical Reviews 69:1
21. Anchisi C, Corda L, Fadda AM, Maccioni A, Podda G (1988) J Heterocyclic Chem 25:735
22. Starks CM (1971) J Amer Chem Soc 93:195
23. Walde P (1989) unpublished result

Author's address:

Dr. R. Schomäcker
MPI für biophys. Chemie, Abt. 040
Postfach 2841
3400 Göttingen

**Progress in Colloid & Polymer Science**            Progr Colloid Polym Sci 81:136 – 139 (1990)

# Modification of water-soluble polymers constituting multipurpose fire-fighting foams by new reactive fluorinated surfactants

S. Szönyi[1,2]), F. Szönyi[1]), I. Szönyi[2]) and Pr. A. Cambon[1,2])

[1]) Laboratoire de Chimie Organique du Fluor
[2]) Centre de Recherches Anti-Incendie de l'Université de Nice

*Abstract:* Modification of water-soluble polymer such as Xanthan gum is made by means of a synthetic reactive fluoralkylating agent: FLUOTAN B830. The resultant fluorinated macromolecule greatly increases the polar liquid repellency of the multipurpose foam compound while decreasing the polymer quantity and as a result the concentrate viscosity. Comparative tests are made with commercial and FLUOTAN B830-containing foams. Results concerning foam resitance on isopropylic alcohol and fire tests on acetone and heptane are given and discussed.

*Key words:* Interactions; multipurpose foam; reactive surfactant; water-soluble polymer

## Composition and function of multipurpose fire-fighting foams

*Aqueous foams* [1] used in the fire-fighting field are generally classified according to the nature of the foaming base used (Proteinic or Synthetic), according to the range of fuel to extinguish (Hydrocarbons or Polar liquids) and finally according to the possible value of their expansion rate (low, medium, high) [2].

Only multipurpose fire-fighting foams can extinguish all kinds of flammable liquids (hydrocarbons, polar solvents, other organic compounds ...) [3]. At present these foams are going to know a large development because of the increasing number of harzardous chemicals and because of the presence of oxygenated compounds (5 to 10%) such as Tertio Butylic Alcohols (TBA) or Methyl Tertio Butyl (MTBE) and Tertio Amyl Methyl Ethers (TAME) in the new gazolines without lead [4, 5].

The principal constituents used in this kind of foam are water-soluble polymers and highly fluorinated surfactants. Polymers have the property to slow the water drainage of the foam lamellae and to give a good resistance to the foam against the dehydrating action of hydrophilic flammable liquids such as alcohols, ethers, ketones, aldehydes, esters ... By insolubility then precipitation, polymers form on the hydrophilic flammable liquid surface a polymeric pellicle which extinguishes the fire.

Highly fluorinated surfactants give to the foam a better fluidity but always they protect the foam against the contamination by the hydrophobic flammable liquids. By rapid water drainage of the foam, highly fluorinated surfactants form on the hydrophobic flammable liquid surface a very thin aqueous film which extinguishes the fire.

## Modification of water-soluble polysaccharide-type polymers: Results and discussion

The polymers contained in the multipurpose foam compounds are polysaccharide-type biopolymers with very high molecular weights such as dextran, scleroglucan or xanthan gum [6]. In water solution, these polymers have a characteristic rheological behaviour of pseudoplastic fluids. Generally, they must be used in large proportion (1 to 2% in the foam concentrate) to give to the foam compounds a good extinguishing efficiency on polar liquid fire, but very often such a polymer concentration increases too much the foam compound viscosity; then there is a few suction problem with regard to the foam nozzles and a danger of

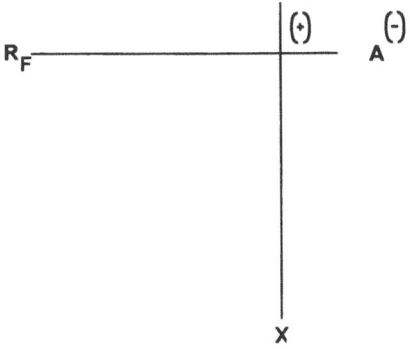

Fig. 1. FLUOTAN B830 agent structure

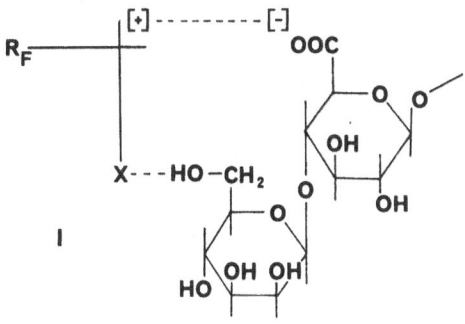

Fig. 2. Ionic interaction (I)

Table 1. FLUOTAN B830 characteristics

| | |
|---|---|
| Form | dark-brown liquid |
| Specific gravity at 20 °C | 1.13 ± 0.02 |
| pH | ~6 |
| Actives | ~30% |
| Solubility | water-soluble in all proportions |
| Surface tension (mN/m, 25 °C) | |
| 0.1% actives | 18 |
| 0.01% actives | 20 |
| Interfacial tension (mN/m, 25 °C) | |
| on cyclohexane | 4 ± 1 |

Fig. 3. Covalent interaction (II)

phase separation during the storage of foam concentrates.

In view of these inconvenients, we conclude that a structural transformation of polymers must be carried out [7—9]. Our experience in the synthesis of new highly fluorinated surfactants [10—14] allowed us to settle an original highly fluorinated surface active molecule possessing a functional group able to react with macromolecules' functional groups (Fig. 1). This new type of surfactant which we called FLUOTAN B830 is composed of a perfluoroalkylated chain, a polar group with positive charges, a negative counter-ion and a reactive functional group. Its physico-chemical characteristics are collected in Table 1.

At first we take an interest in the study of interactions between FLUOTAN B830 agent and xanthan gum. As soon as FLUOTAN B830 agent is added to Xanthan gum, an ionic interaction (I) gives rise between the carboxylic groups of xanthan gum sides chains and the perfluoroalkyl quaternary ammonium salt of FLUOTAN B830 agent (Fig. 2). Then there is a covalent interaction (II) between the hydroxyl groups of xanthan gum side chains and the reactive group X of FLUOTAN B830 agent (Fig. 3). This last interaction depends on the ageing, the mixing temperature or the base action.

Although there is grafting of hydrophobic perfluoroalkyl chains on xanthan gum, the new polymer obtained stays water-soluble. This solubility is explained by the branches degree of the polymer reducing the intermacromolecular junction zones and as a result improving its fluidity and its stability in the foam compound.

These results encourage us to treat proteins in hydrolysat form with FLUOTAN B830 agent according to the same process than previously:

— immediat formation of an ionic interaction (III) between the carboxylic groups of polypeptides and the perfluoroalkyl quaternary ammonium salt of FLUOTAN B830 agent (Fig. 4)
— progressive building of a covalent interaction (IV) between the characteristic groups of polypeptides (amine, thiol) and the reactive group X of FLUOTAN B830 agent (Fig. 5).

That last interaction depends as previously on the ageing, the mixing temperature and the base action.

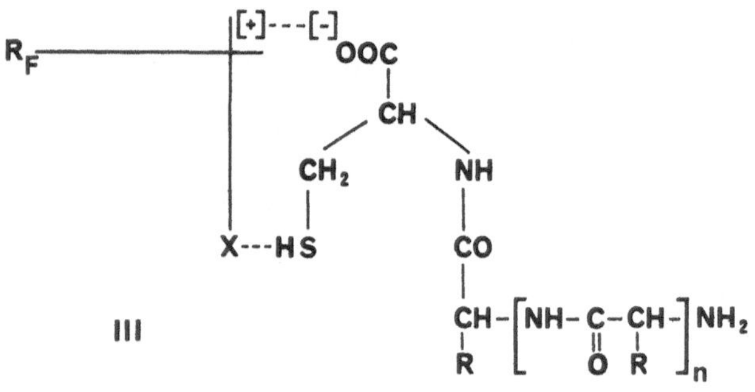

R_F —————— [+]---[-] OOC
                        |
                        CH
                   /         \
              CH_2           NH
               |              |
         X---HS              CO
                             |
                        CH-[NH-C-CH-]NH_2
                         |      ‖    |
                         R      O    R  n

III

Fig. 4. Ionic interaction (III)

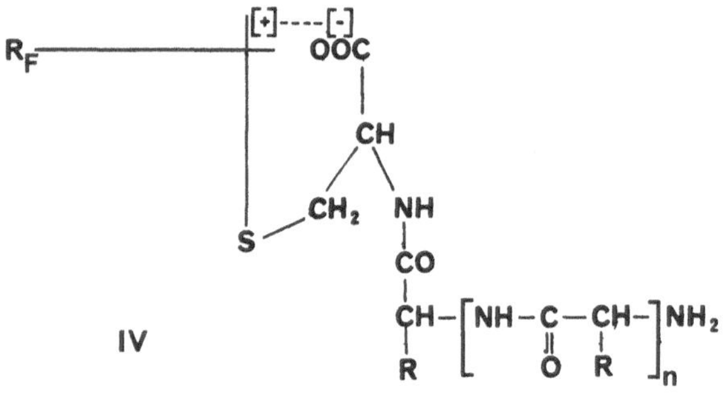

R_F —————— [+]----[-] OOC
                         |
                         CH
                    /         \
                 CH_2          NH
               /                |
              S                CO
                                |
                           CH-[NH-C-CH-]NH_2
                            |      ‖    |
                            R      O    R  n

IV

Fig. 5. Covalent interaction (IV)

Table 3. Fire tests experimental conditions

| Tests conditions on | Heptane | Acetone |
|---|---|---|
| Fire tray surface | 4.5 m$^2$ | 1.75 m$^2$ |
| Fuel quantity | 150 l | 125 l |
| Foam nozzle discharge | 11.4 l/min | 11.4 l/min |
| Application method | direct | indirect |
| Application rate | 2.5 l/min/m$^2$ | 6.5 l/min/m$^2$ |
| Preburn time | 1 min | 2 min |
| Dicharged time | 2 min 30 sec | 2 min 30 sec |
| Pause before burn-back | 2 min | 5 min |

crease considerably when the foam compound contains macromolecules treated by FLUOTAN B830 agent. At the same time, the macromolecules concentration used and as a result the foam compound viscosity, decrease. For example the AFFF pseudoplastic synthetic or proteinic foam compound including FLUOTAN B830 agent, gets on alcohol repellency value of 160 seconds or 600 seconds and the viscosity decreases from 7000 to 700 or 1700 cP.

Tables 3 and 4 collect the experimental conditions and the results of tests on acetone and heptane fires according to AFNOR standards: we notice for example when AFFF pseudoplastic synthetic foam treated by FLUOTAN B830 agent is used instead of a commercial AFFF pseudoplastic synthetic foam compound, the acetone fire is extinguished approximately twice faster at 15 °C (extinct. 100% 48 seconds, instead of 80 seconds); burn-back time on acetone is also improved (960 seconds instead of 520 seconds). Table 4 shows the limiting values of AFNOR standard and the results of heptane fire tests. In the case of multipurpose proteinic foam compound, the extinction is rapid and the foam produced is very resistant next to burn-back test.

Table 2. Alcohol repellency and viscosity values of several FLUOTAN B830 containing foams

| Multipurpose foam compound | Viscosity (cP) | Alcohol repellency (S) |
|---|---|---|
| Pseudoplastic synthetic | 8000 | 10 |
| Pseudoplastic AFFF synthetic | 7000 | 12 |
| Pseudoplastic AFFF synthetic + FLUOTAN B830 | 700 | 160 |
| Newtonian proteinic | 4 | <2 |
| Newtonian AFFF proteinic + FLUOTAN B830 | 6 | 50 |
| Pseudoplastic AFFF proteinic + FLUTOAN B830 | 1700 | 600 |

## Conclusion

The new FLUOTAN B830 agent synthesized have the power to make macromolecules of polypeptide or polysaccharide-type alcohol repellent. With this agent, it is possible to reduce considerably the polymer quantity in the foam compound concentrate, making it too much less viscous. Now, we work on other highly fluorinated reactive surface active molecules more suitable for the treatment of newtonian AFFF proteinic or synthetic foam compounds: usually, these foam compounds are known to produce only a foam

Finally, we had the idea to make a mixture composed of FLUOTAN B830 agent and two kinds of macromolecule: xanthan gum and polypeptide. We obtained the same type of interactions than previously.

We mesured the alcohol repellency (foam resitance time on isopropylic alcohol at 25 °C, expressed in seconds) of different mixtures performed above. We notice in Table 2 that the alcohol repellency values in-

Table 4. Fire tests results

| Multipurpose foam compound | [C] | Heptane | | | | Acetone | | |
|---|---|---|---|---|---|---|---|---|
| | | Extinct. | | Burn back | $T^0$ (air) °C | Extinct. 100% | Burn back | $T^0$ (air) °C |
| | | 99% | 100% | | | | | |
| Newtonian AFFF proteinic | 6% | 60 | 114 | 460 | 15 | 57 | 1440 | 15 |
| Pseudoplastic AFFF synthetic | 3% | 50 | 110 | 780 | 14 | – | – | – |
| | 6% | – | – | – | – | 48 | 960 | 15 |
| Pseudoplastic AFFF proteinic | 6% | 60 | 90 | 640 | 14 | 50 | 900 | 14 |
| Commercial pseudoplastic | 3% | 40 | 110 | 900 | 7 | – | – | – |
| AFFF synthetic | 3% | 64 | 176 | 720 | 22 | – | – | – |
| | 6% | – | – | – | – | 80 | 520 | 7 |
| Limiting value according to AFNOR standard | | | <270 | >240 (600) | 5−25 | <210 | >240 (480) | 10−20 |

for the extinction of hydrocarbons; until now they are unsuitable for the extinction of polar liquids.

## References

1. Bickerman JJ (1973) Foams: Theory and Industrial applications. Springer-Verlag Berlin
2. Gouezec R (1976) CTIF International Symposium (Berlin Ouest) pp 30−55
3. Szönyi S (1981) Revue Générale de Sécurité 5:35−38
4. De Gaudemaris G, Arlie JP, Guibet JC (1986) La Recherche 175:377−384
5. Brondel G (1989) La Recherche 211:836−844
6. Molyneux P (1983) In: Finch CA (ed) Chemistry and Technology of water-soluble polymers. Plenum Press, New York, pp 1−20
7. Szönyi S, Szönyi F, Szönyi I (1988) Fr. Demandes 8811345, 8812152, 8813612, 8813613
8. Szönyi S, Cambon A (1989) Fire Safety Journal (submitted)
9. Szönyi S, Szönyi F, Szönyi I (1989) Revue Générale de Sécurité (in press)
10. Szönyi S, Cambon A (1984) Fr. Demande FR 2,539623
11. Lampin JP, Cambon A, Szönyi F (1985) Eur Pat Appl EP 165,853
12. Cambon A, Szönyi F (1988) Fr. Demande FR 2,605,628
13. Szönyi S, Vandamme R, Cambon A (1985) J Fluorine Chem 30:37−57
14. Szönyi F, Cambon A (1987) J Fluorine Chem 36:195−209

Authors' address:

Dr. S. Szönyi
Laboratoire de Chimie Organique du Fluor −
Université de Nice
Parc Valrose 06034 Nice cedex (France)

**Progress in Colloid & Polymer Science**                    Progr Colloid Polym Sci 81:140−143 (1990)

# Adsorption of 5,8-diethyl-7-hydroxydodecan-6-oxime at toluene/ water interface and the rate of copper extraction

K. Prochaska and J. Szymanowski

Institute of Chemical Technology and Engineering, Poznań Technical University, Poznań, Poland

*Abstract:* The interfacial tension data were determined for 5,8-diethyl-7-hydroxydodecan-6-oxime at toluene/water interface and used to discuss the mechanism of copper extraction is very probable. The adsorbed hydroxyoxime molecule reacts in the first quick step with copper hydrated ion forming hydrophilic partly hydrated and charged intermediate 1:1 complex, which in the slow step reacts with hydroxyoxime molecule present in aqueous layers near the interface.

*Key words:* Interfacial tension; copper extraction; kinetics

## Introduction

5,8-Diethyl-7-hydroxydodecan-6-oxime (DEHDO) of the following structure is the active component of commercial extractant LIX 63.

$$
\begin{array}{c}
\text{HO} \\
\diagdown \\
\text{N} \quad \text{OH} \\
\parallel \quad | \\
\text{CH}_3(\text{CH}_2)_3\text{CH} - \text{C} - \text{CH} - \text{CH}(\text{CH}_2)_3\text{CH}_3 \\
| \qquad\qquad\qquad | \\
\text{CH}_3 \qquad\qquad \text{CH}_3
\end{array}
$$

It is used alone for germanium extraction on industrial scale [1, 2] and as the additive present in LIX 64N reagent is used in several high-scale industrial installations for copper extraction from acidic media [3, 4]. In this last case DEHDO acts as the accelarator increasing significantly the rate of copper extraction. Although several papers were published upon this problem the mechanism of metal extraction with LIX 63 and LIX 64N was not satisfactorily explained [4−6].

DEHDO can extract several metals from acidic solutions, including germanium, molibdenium, vanadium, palladium and uranium. From slightly acidic and ammonia solutions it can extract copper (at pH above 3), and nickel and cobalt (at pH above 5.5) as well as some other metals. It extracts also Au(III), Ag(I), Ti(VI), Zn(II), Ga(III), Cd(II), Sb(III), Hg(II), Bi(III) and Fe(III). DEHDO can also extract small amounts of copper from highly acidic solutions containing $20−200$ g dm$^{-3}$ H$_2$SO$_4$. Thus, its extraction properties are significantly different than those of aromatic hydroxyoximes used for copper extraction from acidic sulphate solutions.

In contradiction to aromatic hydroxyoximes dissociation of both the hydroxylic group and oximino group of DEHDO can occur in extraction systems, and the acidities of these two groups are comparable. As a result different types of complexes can be formed.

DEHDO has two hydrophilic groups ($-$OH and $=$NOH) in the middle of the molecule and a long hydrophobic hydrocarbon chain. Due to this long hydrocarbon chain the hydroxyoxime molecule is highly hydrophobic and DEHDO dissolves poorly in the aqueous phase. Its solubility in aqueous solutions is estimated as $4−6$ ppm [7, 8]. It also adsorbs at water/ hydrocarbon interfaces decreasing significantly the interfacial tension. Thus, it is highly probable that the metal complexation occurs at the interface as in the case of aromatic hydroxyoximes [4−6]. In such a case the interfacial population of DEHDO should be an important parameter influencing the rate of metal extraction.

The aim of this work is to determine the surface activity of 5,8-diethyl-7-hydroxydodecan-6-oxime at toluene/water interface and to use this data to discuss the kinetics and mechanism of copper extraction.

## Experimental

5,8-Diethyl-7-hydroxydodecan-6-oxime obtained according to the general method [9] was used.

The interfacial tension was measured by the drop volume method at 20°C by means of semiautomatic equipment. Drops of the aqueous phase were formed into the organic phase. The time of surface aging was 5–20 minutes. Redistilled water was used as a water phase and toluene as the organic one. Before use it was redistilled over sodium using a Vigreux distillation column. Mutually presaturated phases were used. Equal volumes (25 cm³) of both phases were automatically shaken at room temperature for 6 h and then left for 24 h for full separation. They were separated directly before measurements, disregarding layers near the interface of 1 cm depth each.

The following adsorption isotherms/equations were matched to the experimental data; the spline function [10], the logarithmic polynomial of the third order and the Szyszkowski, Frumkin and Temkin isotherms. After differentiating the term $d\gamma/dc$ was introduced to the Gibbs isotherm and the surface excess was computed. The predicted reaction orders in respect to hydroxyoxime concentration were computed according to the standard program MINEX using a IBM PC microcomputer [11, 12].

The hydrophile lipophile balance was estimated from our polarity data determined previously for aromatic hydroxyoximes by means of gas chromatography [13].

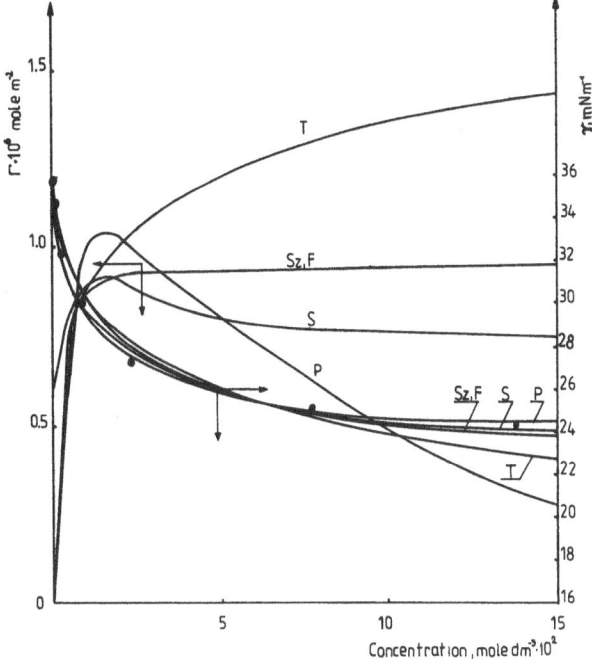

Fig. 1. Interfacial tension and surface excess isotherms (*Sz, F, T, S* and *P* denote the Szyszkowski, Frumkin and Temkin isotherm, the spline function and the polynomial, respectively)

## Results and discussion

The interfacial tension isotherm is given in Fig. 1. It can be quite well matched by various isotherms and functions considered, although deviations are significant and different for various models. The errors of model and the values of regression coefficients with the exception of the spline function are given in Table 1. The meanings of regression coefficients is the same as in our previous works [11, 14]. The errors of model are in the range of $10^{-8}-10^{-7}$.

DEHDO is very poorly soluble in water due to its high hydrophobicity. Its HLB coefficients estimated according to previously derived equations for various hydroxyoximes [13]: $HLB^G = 518.6/M$ and $HLB^D = 9.90-0.475\,n$, where $G$ and $D$ denotes the Griffin and Davies scale, respectively, and $M$ and $n$ the molecular mass of hydroxyoxime and the total number of carbon atoms in the hydroxyoxime molecule, respectively, are 2.3 both on the Griffin and Davies scale. This proves their high hydrophobicity and is in an agreement with their low solubility in water, estimated from the $HLB$ value as 4.3 ppm, according to the previously determined Eq. (15). This value is also in agreement with experimental values (4–6 ppm) determined by independent authors [7, 8]. As a result, the partition coefficient is high, $\log P$ is approximately 4, and the total

hydroxyoxime concentration is actually equal to its concentration in the organic phase and can be used to compute the surface excess isotherm.

The surface exces isotherms increase monotonously only as the Szyszkowski, Frumkin and Temkin isotherms are considered. The isotherms computed according to the Szyszkowski and Frumkin isotherms are quite the same as a result of weak interactions between adsorbed neutral molecules of hydroxyoximes.

However, the maximum is observed as the surface excess is computed according to the spline function (Gibbs isotherm) and the polynomial. In these two cases only the increasing parts of the isotherms up to maximum can be considered. We must assume that at this maximum the interface is already saturated and the surface excess does not further change. These maximum values of the surface excess computed according to the isotherms considered are somewhat different but they are of the same order (Fig. 1). Similar values were previously determined for aromatic hydroxyoximes [14, 16] and for the commercial LIX 63 ekstractant containing DEHDO as the active component.

Table 1. Regression coefficients ($a_i$) and errors of model ($e$) for matching of different isotherms to the interfacial tension isotherm (the meaning of parameters is the same as in previous work [14]

| Parameter | Szysz-kowski | Frumkin | Temkin | Poly-nomial | Spline |
|---|---|---|---|---|---|
| $a_1$ | $6.117 \cdot 10^{-2}$ | $6.117 \cdot 10^{-2}$ | $6.894 \cdot 10^{4}$ | $6.739 \cdot 10^{-5}$ | — |
| $a_2$ | $8.207 \cdot 10^{-4}$ | $8.207 \cdot 10^{-4}$ | $4.138$ | $9.383 \cdot 10^{-4}$ | — |
| $a_3$ | — | $1.264 \cdot 10^{-5}$ | — | $1.936 \cdot 10^{-3}$ | — |
| $a_4$ | — | — | — | $2.557 \cdot 10^{-2}$ | — |
| $e$ | $1.86 \cdot 10^{-7}$ | $1.85 \cdot 10^{-7}$ | $5.50 \cdot 10^{-7}$ | $1.51 \cdot 10^{-8}$ | $8.92 \cdot 10^{-8}$ |

The maximum observed on the surface excess isotherms computed according to the spline function and the polynomial can be interpreted as the result of extractant aggregation in the organic phase. The change in the free energy of the extractant association in organic solvents is lower than that of the adsorption at the water/hydrocarbon interface. It means that there is no association in the solution, and the activity coefficients are equal to unity for extractant concentrations smaller or equal to the concentration at which the maximum is observed. Then, assuming that only monomers are adsorbed at the interface the average degree of hydroxyoxime association ($m$) can be estimated from the modified Gibbs Eq. (17): $\Gamma = -m/RT d\gamma/d(\ln c)$. At the maximum $\Gamma^{sat} = -1/RT d\gamma/d(\ln c)$. Thus, $m = \Gamma^{sat}/\Gamma$ for $c > c_{max}$.

The average degree of association computed according to the spline function is 1.3, while it increases to 3 in the concentration region up to 0.1 mol $dm^{-3}$ as the polynomial is considered. The values estimated according to the spline function are in a good agreement with the independent experimental data ($m = 1.4-1.8$ at oxime concentration of $0.05-0.2$ mol $dm^{-3}$ [4]) determined by means of osmometric method for Lix 63 in benzene. Thus, the hydroxyoxime dimerization should be further considered, as in the case of aromatic hydroxyoximes in aliphatic hydrocarbons [4–6, 14–16]. However, as it was previously demonstrated, its effect upon final values of the predicted reaction orders is weak and can be neglected [11, 12, 14, 16].

DEHDO can form with copper both 1:1 and 2:1 complexes [18] according to the following equations:

$$2 H_2A_o = (H_2A)_{2o} \tag{1}$$

$$Cu_w^{2+} + H_2A = Cu(HA)_{2o} + 2 H_2^+, \tag{2}$$

$$Cu_w^{2+} + H_2A = CuA_o + 2 H_w^+, \tag{3}$$

where $H_2A$ stands for hydroxyoxime and subscripts $w$

and $o$ denote the water phase and the organic one, respectively.

In the first typical case assuming that the reaction proceeds at the interface of molecular dimensions then the following reaction scheme can be considered:

Scheme 1

$$Cu_w^{2+} + H_2A_{ad} = CuHA_{ad}^+ + H_w^+ \tag{4}$$

$$CuHA_{ad}^+ + H_2A_{int} = Cu(HA)_{2ad} + H_w^2 \tag{5}$$

$$Cu(HA)_{2ad} + 2 H_2A_o = Cu(HA)_{2o} + H_2A_{ad} + H_2A_{int} \tag{6}$$

Scheme 2

$$Cu_w^{2+} + H_2A_{ad} = CuHA_{ad}^+ + H_w^2 \tag{7}$$

$$CuHA_{ad}^+ + H_2A_{ad} = Cu(HA)_{2ad} + H_w^2 \tag{8}$$

$$Cu(HA)_{2ad} + 2 H_2A_o = Cu(HA)_{2o} + 2 H_2A_{ad} \tag{9}$$

where: subscripts $ad$ and $int$ denote the molecules in the interfacial monolayer and at the border of the sublayer from which the hydroxyoxime is transferred into the monolayer without diffusion.

As diffusion is neglected and the oxime interfacial concentration is assumed as equal to the surface excess then the following kinetic equations can be obtained: $r_{4,7} = k\Gamma$, $r_5 = k[HB]_o\Gamma$, $r_6 = k[HB]_o^3\Gamma$, $r_8 = k\Gamma^2$, and $r_9 = k[HB]_o^2\Gamma^2$, for reactions 4 and 7, 5, 6, 8 and 9 assumed as the limiting steps, respectively. Thus, the equations are the same as those considered previously for aromatic hydroxyoximes [11, 12, 14–16].

Exemplary reaction orders in respect to hydroxyoxime as predicted from the interfacial tension data in the same way as in our previous works [11, 12, 14–16] are given in characteristic Fig. 2. Depending upon the reaction assumed as the slowest one quite

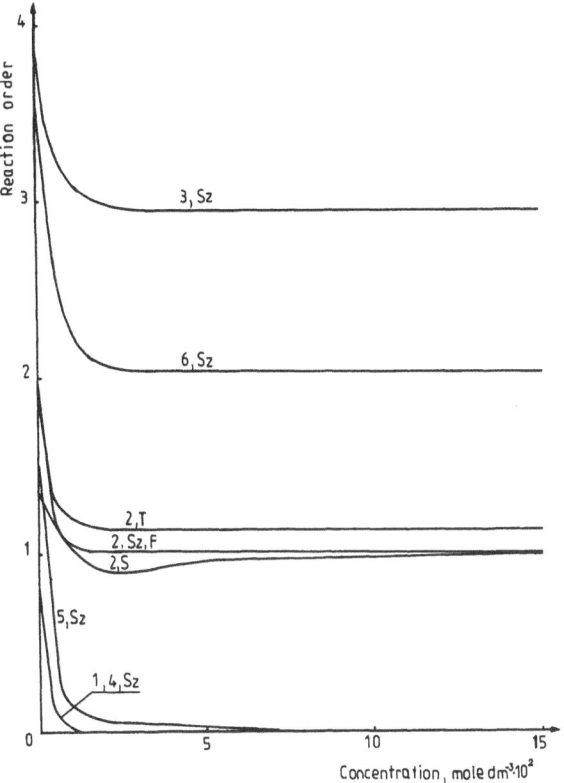

Fig. 2. Reaction orders in respect to hydroxyoxime as predicted from theinterfacial tension data (curve number denotes the reaction assumed as the slowest step)

different reaction orders are predicted. The differences between these reaction orders are quite important and, due to this, they can be used to verify the most probable mechanism scheme. It seems that scheme 1 with reaction 2 as the limiting step is the most probable. In this case the predicted reaction order is approximately 1 over almost the whole range of hydroxyoxime concentration. Only for very low hydroxyoxime concentrations the reaction order increases to about 1.5. Thus, quite a good agreement is observed between predicted and experimental reaction orders [4]. Moreover, the differences in the reaction orders computed according to the various adsorption isotherms are the smalest.

## Conclusions

5,8-Diethyl-7-hydroxydodecan-6-oxime probably reacts with copper according to the interfacial mechanism with the formation of the stable 2:1 complex as the slowest step. In the first quick step the adsorbed hydroxyoxime molecule reacts with copper ion, forming intermediate partly hydrated and charged complex which in the limiting step reacts with a hydroxyoxime molecule present in the aqueous film near the interface.

The spline function, the logarithmic polynomial, the Szyszkowski, Frumkin and Temkin isotherms match quite well the interfacial tension data and they can be used to predict the reaction order in respect of hydroxyoxime for copper extraction with 5,8-diethyl-7-hydroxydodecan-6-oxime.

*Acknowledgements*

This work was supported by Polish Research Program no. 03.08.

## References

1. De Schepper A (1976) Hydrometallurgy 1:291–298
2. De Schepper A, Coussement M, von Peteghen A (1983) Eur Pat 68 540
3. Fisher JFC, Notebaart CW (1983) In: Lo TC, Baird MHI, Hanson C (ed) Handbook of Solvent Extraction, John Wiley and Sons, New York, pp 649–665
4. Szymanowski J (1984) Wiad Chem 38:371–399
5. Flett DS (1977) Acc Chem Res 10:99–104
6. Danesi PR, Chiarizia R (1980) Critical Reviews in Analytical Chemistry 10:1–126
7. Foakes HJ, Preston JS, Whewell RJ (1978) Anal Chim Acta 97:349–356
8. Ashbrook AW (1972) Anal Chim Acta 58:115–121
9. Olszanowski A, Alejski K (1981) Przem Chem 60:537–538
10. Bogacki MB, Szymanowski J, Prochaska K (1988) Anal Chim Acta 206:215–221
11. Prochaska K, Alejski K, Szymanowski J (1989) Progr Colloid Polym Sci 79:327
12. Prochaska K, Alejski K, Szymanowski J (1989) Proc International Conference on Separation Science and Technology, Hamilton, Canada, paper no. S3–1f:181–188
13. Szymanowski J, Voelkel A, Rashid ZA (1987) J Chromatogr 402:55–64
14. Prochaska K, Szymanowski J (1989) Progr Colloid Polymer Sci 79:321
15. Szymanowski J, Cox M, Hirons GH (1984) J Chem Tech Biotechnol 34A:218–226
16. Szymanowski J, Prochaska K (1988) J Colloid Interface Sci 123:456–465
17. Popov AN (1987) In: Kazarinov VE (ed) The Interface Structure and Electrochemical Processes at the Boundary Between Two Immiscible Liquids, Springer, Berlin, pp 179–205
18. Preston JS (1975) J Inorg Nucl Chem 37:1235–1242

Authors' address:

Jan Szymanowski, Professor
Poznań Technical University
Pl. Skłodowskiej-Curie 2
60965 Poznań, Poland

**Progress in Colloid & Polymer Science**          Progr Colloid Polym Sci 81:144–150 (1990)

# Lipoaminoacid surfactants: Phase behavior of long chain Nα-Acyl arginine methyl esters

C. Solans, M. A. Pés, N. Azemar, and M. R. Infante

Instituto de Tecnología Química y Textil (C.S.I.C.), Departamento de Tecnología Química, Barcelona, Spain

*Abstract:* Long chain Nα-Acyl-L-arginine methyl esters are cationic lipo-aminoacid surfactants obtained in our laboratories by condensation of a fatty acid with the α-amino group of L-Arginine. As part of systematically studying the properties of these surfactants, their phase behavior in binary and multicomponent systems has been investigated as a function of chain length of the fatty acid residue. It has been found that Krafft temperature and critical micelle concentration (CMC) increase and decrease respectively with the increase in the alkyl chain length, as expected in a homologous series of surfactants. The Nα-Myristoyl-L-arginine methyl ester (MAM)/water phase diagram showed three liquid crystalline regions with hexagonal, cubic, and lamellar structures. Phase behavior of ternary Nα-Acyl arginine methyl ester/water/n-alkanol systems was found to conform to that of ionic surfactant systems. Microemulsion formation was investigated in quaternary Nα-Lauroyl-L-arginine methyl ester (LAM)/water/n-alkanol/hydrocarbon systems.

*Key words:* Surfactant; lipoaminoacid; micelle; liquid crystal; microemulsion; phase behavior

## Introduction

The aim or this paper is to present the results of a study on phase behavior of cationic aminoacid-based surfactants synthesized in our laboratories by condensation of a fatty acid with the α-amino group of L-arginine [1]. The synthetic method employed makes use of very mild experimental conditions to form the amide linkage, which leads to pure optical compounds in a high yield and in the absence of by-products.

Aminoacid-based surfactants have some distinctive structural features as shown by the general chemical formula of Nα-Acyl arginine derivatives, the surfactants object of this study:

$$R-CO-NH-CH-(CH_2)_3-NH-C^+Cl^-. \quad (1)$$

with COOMe and NH₂ groups:

$$R-CO-NH-\overset{\displaystyle COOMe}{\underset{}{CH}}-(CH_2)_3-NH-\overset{\displaystyle NH_2}{\underset{\displaystyle NH_2}{C^+}}Cl^-. \quad (1)$$

The amide bond, located between the hydrophilic (aminoacid residue) and hydrophobic (R) parts of the molecule, is expected to provide special properties to these surfactants because of its high hydrogen bonding ability. Another characteristic of this type of surfactants is the presence of an asymetric carbon atom in the molecule, making possible the formation of chiral aggregates.

Preliminary tests carried out with cationic Nα-Acyl-L-aminoacid derivatives to investigate properties such as antimicrobial activity, toxicity, and degree of irritating action, showed that they are active against a wide range of microorganisms [2, 3] and can be considered as nontoxic [4] and mild irritants [5]. These properties make them of special interest for their use in pharmaceutical, food, and cosmetic formulations. In there fields there is a strong need for surfactants with biocompatible and multifunctional properties [6].

The first studies on phase behavior of Nα-Lauroyl arginine methyl ester (LAM), reported in a previous paper [7], showed that it assembled in micelles and liquid crystalline phases similar to conventional surfactants. In order to learn the properties of lipoaminoacid surfactants in a systematic way, a study of their phase behavior as a function of the alkyl chain length ($R$) has been undertaken.

## Experimental

### Materials

N$^{\alpha}$-Lauroyl-L-arginine methyl ester (LAM) and N$^{\alpha}$-Myristoyl L-arginine methyl ester (MAM) were synthesized in our laboratories according to the method described in previous papers [1].

The following materials were used with no further purifications: n-pentanol (Fluka, 99.5%), n-heptanol (Fluka, 99.5%), n-decanol (Merck, 99.7%), hexadecane (Fluka, 99.5%), toluene (Merck, purest grade), squalane (Glyco Ibérica, commercial grade under the denomination Glyco SH-14), pyrene (Merck, 99%), and N-cetylpyridinium chloride (Merck, BP 1973). The water was distilled twice.

### Phase diagrams

The solubility areas were determined by tritating mixtures of the components with water under thermostatic conditions. The solubility limits were confirmed by long-time storage of samples at constant temperature. Presence of liquid crystalline phases was detected observing the samples against crossed polarizers, and assignement of their structures was achieved by examination under a polarizing microscope.

### Microscopy

An optical microscope Zeiss D-7082 equipped with a hot stage, polarized light, and automatic photographic camera was used.

### Fluorescence

Steady state spectrofluorescence by means of a Shimadzu RF 540 spectrofluorimeter was used to determine CMC and micelle aggregation quantity according to the "static quencher" method [8] using pyrene as a probe and N-cetylpyridinium chloride as a quencher.

### Surface tension

De Nouy ring method by means of a Lauda 7201 tensiometer was used to carry out surface tension measurements.

## Results and discussions

In this section, phase properties of binary N$^{\alpha}$-Myristoyl-L-arginine methyl ester (MAM)/water and ternary MAM/water/n-alkanol systems will be described and compared to those of N$^{\alpha}$-Lauroyl-L-arginine methylester (LAM) systems reported earlier [7]. This will be followed by a description of microemulsion formation in quaternary LAM/water/n-alkanol/hydrocarbon systems.

Qualitative information on phase behavior of binary MAM/water system was obtained by means of optical microscopy experiments according to the method described by Lawrence [9] widely used in phase behavior studies [10]. In this type of experiments, gradients of surfactant composition are produced by allowing water to diffuse into a solid surfactant placed on a microscope slide. The different phases, which develop into separate rings as a function of temperature, can be identified by examination through a microscope under polarized light. In this study, attention was restricted to the temperature interval between 10 °C and 90 °C.

At temperatures below 35 °C, only hydrated solid surfactant and water were observed. At 35 °C, an anisotropic band with a typical hexagonal liquid crystalline texture developed around solid MAM (Fig. 1a). At 43 °C, a narrow, isotropic and very viscous band, a cubic liquid crystalline phase, appeared between the hexagonal liquid crystalline phase and solid surfactant (Fig. 1b). As temperature increased, the band corresponding to the liquid crystalline phase expanded at the expense of the solid surfactant, which disappeared completely at 50 °C. With a further increase in temperature, the hexagonal liquid crystalline phase expanded, while that with cubic structure shrinked and vanished at approximately 65 °C. At 75 °C, a lamellar liquid crystalline phase (Fig. 1c) started developing at the expense of the hexagonal liquid crystalline phase. Above 85 °C, both mesomorphic phases had disappeared and only an isotropic liquid phase remained.

The temperatures of forming hexagonal and cubic liquid crystalline phases for the MAM/water system, 35 °C and 43 °C respectively, were higher than the corresponding temperatures for the LAM system, 28 °C and 38 °C respectively [7]. This could be attributed to the longer length of the alkyl chain of the former surfactant. Due to the more hydrophilic character of LAM as compared to MAM, formation of a lamellar liquid crystalline phase is not allowed in the binary LAM/water system. It is well known [11, 12] that the major factor determining the structure of the mesophases is the balance between the type of polar head and restrictions on alkyl chain conformational packing. The bulkier head-group of these lipoaminoacid surfactants can cause a considerable increase in the area per molecule at the interface. Consequently, formation of ordered structures of low surface area is only energetically favored with an increase in hydrophobicity.

a

solid crystal

E

b

solid crystal

I

E

c

D

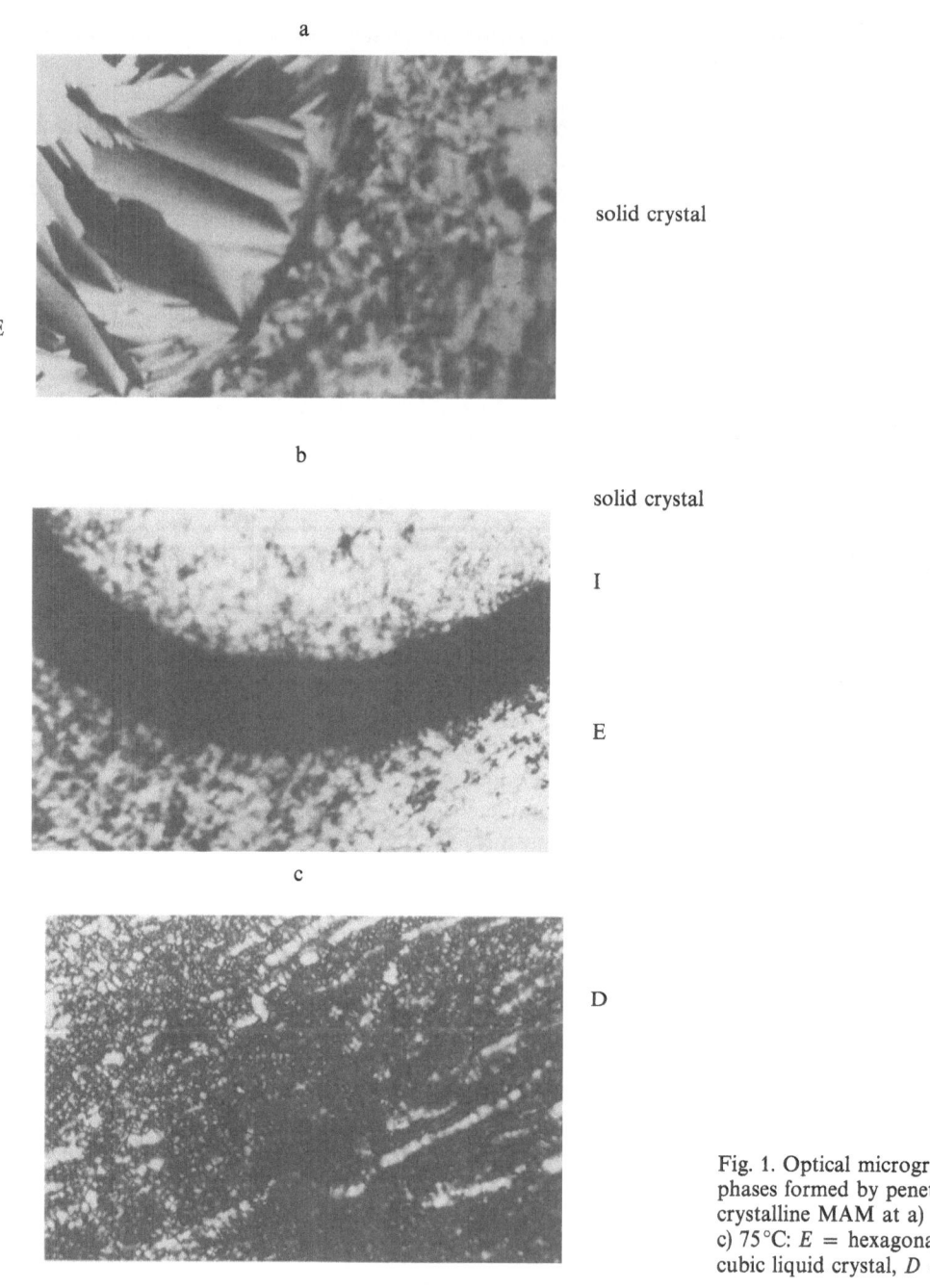

Fig. 1. Optical micrographies illustrating the phases formed by penetration of water into crystalline MAM at a) 35°C, b) 43°C, and c) 75°C: $E$ = hexagonal liquid crystal, $I$ = cubic liquid crystal, $D$ = lamellar liquid crystal

As expected in a homologous series of surfactants, an increase in the length of the alkyl chain resulted in an increase in the melting point of hydrated solid surfactant (Krafft temperature) and a decrease in critical micelle concentration (CMC). Figure 2 shows the solution behavior of MAM in water as a function of temperature. It can be observed that the pronounced increase in solubility, which corresponds to the Krafft temperature, is produced at 26.5°C. The corresponding value for LAM was 14.5°C [7].

In line with this trend, the CMC of MAM as determined by surface tension and fluorescence methods was found to be 1.1 mM and 2.0 mM respectively, while that of LAM, determined by surface tension measurements, was 6 mM [7]. Investigations to determine micelle properties, such as size and shape, are

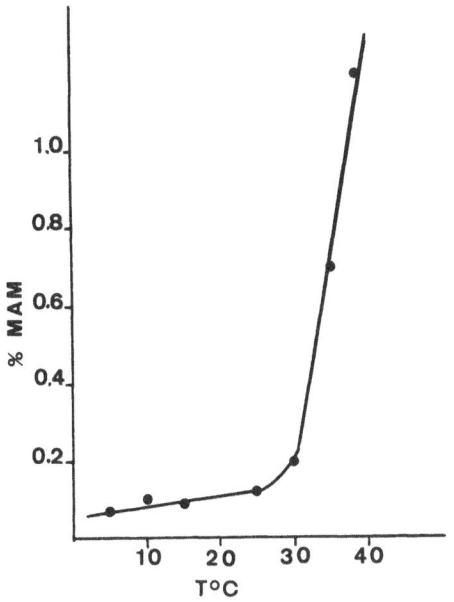

Fig. 2. Solubility curve for the MAM/water system as a function of temperature

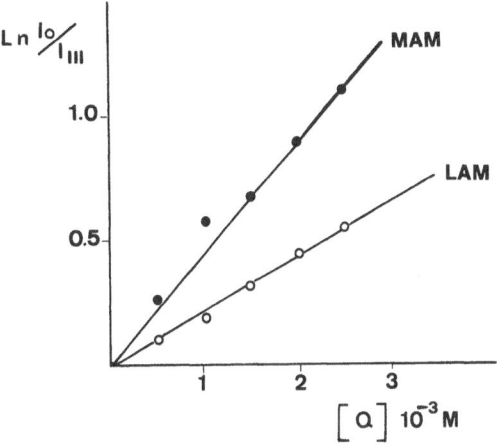

Fig. 3. Plots of the ratio of pyrene emission fluorescence intensity without ($Io$) and with ($I_{III}$) quencher versus quencher concentration for LAM and MAM at 25°C. LAM concentration: 9.45 mM; MAM concentration: 3.45 mM. Quencher: N-Cetylpiridinium chloride

still in progress. Therefore, complete data is not available yet. At lower surfactant concentrations, above but close to CMC, micelle aggregation quantity was determined by steady-state fluorescence using pyrene as a probe and cetylpyridinium chloride as a quencher [8]. At 25°C, the aggregation quantity of LAM, at a concentration of 9.45 mM, calculated from the slopes of the plots shown in Fig. 3 were 79 and 61, respectively. Microcope examinations and phase-diagram determinations of LAM and MAM showed that the favored structure − after exceeding the solubility limit of micelles in water − is that of a hexagonal liquid crystal, where cylinders of surfactant are arranged on a two-dimensional hexagonal lattice. Based on this information, it is reasonable to assume that at high surfactant concentrations, LAM and MAM assemble into cylindrical micelles close to the phase transition.

Phase diagrams for the ternary MAM/water/n-alkanol systems were determined as a function of alkanol chain length at 25°C and 40°C. Attention was focused on the monophasic regions in order to compare solubilization properties of lipoaminoacid surfactants to those of typical ionic surfactants. Three main regions corresponding to normal micellar solutions ($L_1$), inverse micellar solutions ($L_2$), and lamellar liquid crystalls ($D$) were obtained (Fig. 4). It can be observed that the phase regions were not significantly affected by a change of temperature from 25°C to 40°C.

The influence of the alkanol chain length was that anticipated for any ionic surfactant. The shorter chain homologue, n-pentanol (Fig. 4a), gave two large solution regions, $L_2$ and $L_1$, and a small lamellar liquid crystalline phase region, while the longer chain alkanol, n-decanol (Fig. 4b), stabilized the liquid crystalline phase giving rise to a reduction of $L_2$ and $L_1$ areas. The phase regions with n-heptanol (Fig. 4c) displayed an intermediate trend. Consequently, the maximum solubilization of water in $L_2$ phase decreased with the increase in the alkanol chain length, being 55%, 42%, and 15% for systems with n-pentanol, n-heptanol, and n-decanol, respectively.

The higher hydrophilic character of LAM was clearly revealed by the occurrence of a continuous solubility region from the pentanol corner to the water corner in the ternary LAM/water/n-pentanol system [7]. The replacement of n-pentanol with n-heptanol (Fig. 5) lead to a drastic reduction of the solubility regions and to an expansion of the lamellar liquid crystalline phase between two separated $L_1$ and $L_2$ phases.

It should be noted that the solubility of both lipoaminoacid surfactants in n-alkanols (Fig. 5) is considerably larger than commonly studied ionic surfactants.

Another aspect investigated was the capability of these aminoacid-based surfactants to form microemulsions. For this purpose, the solubility regions of quaternary lipoaminoacid surfactant/water/n-alkanol/hydrocarbon system were determined as a func-

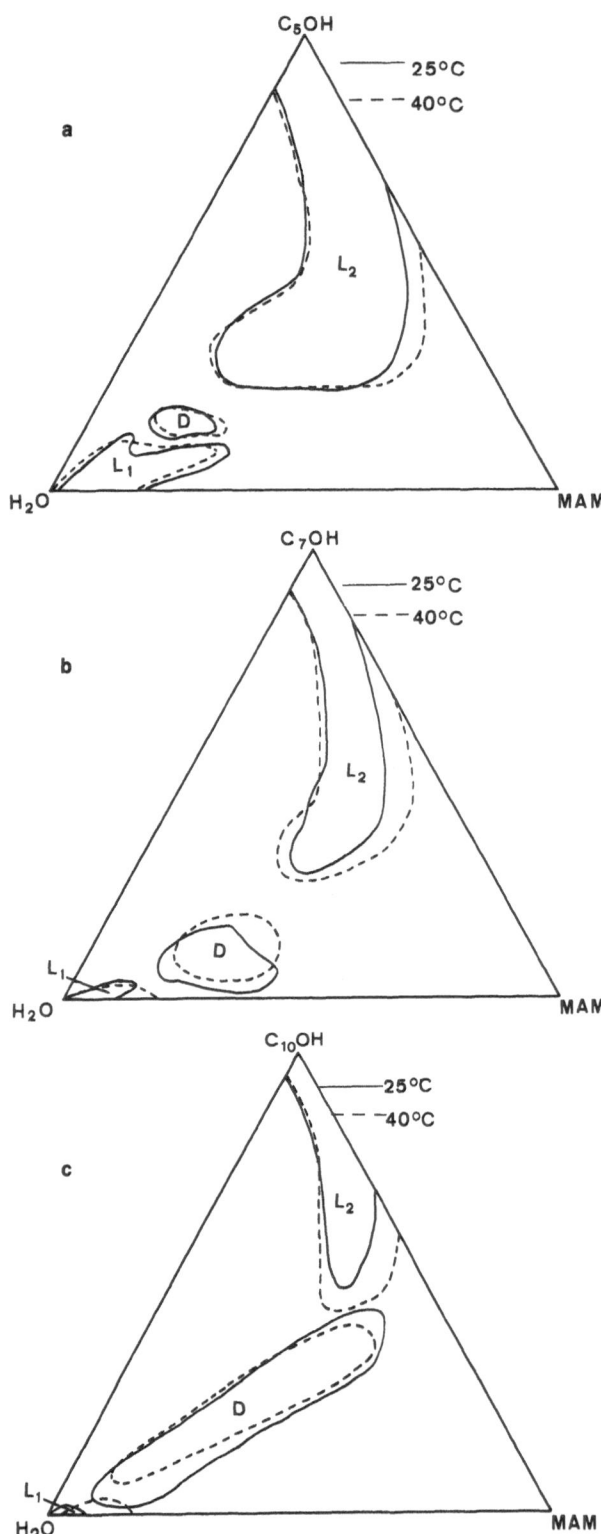

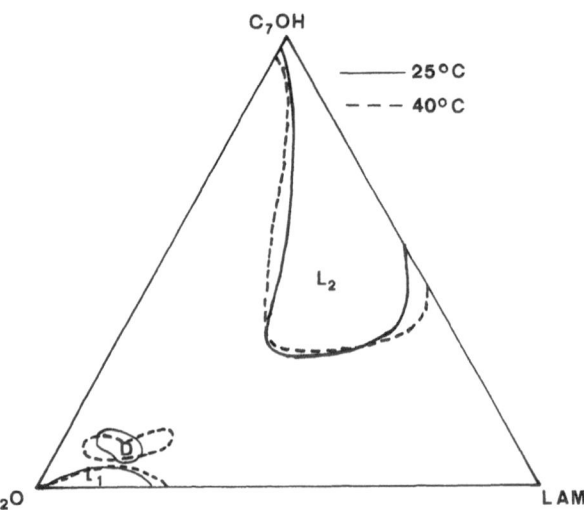

Fig. 5. Phase diagram of the ternary LAM/water/n-alkanol system at 25°C (——) and 40°C (— — —). $L_1$ = normal micellar solution, $L_2$ = inverse micellar solution, $D$ = lamellar liquid crystal, $C_7OH$ = n-heptanol

Fig. 4. Phase diagrams of the ternary MAM/water/n-alkanol systems at 25°C (——) and 40°C (— — —). $L_1$ = normal micellar solution, $L_2$ = inverse micellar solution, $D$ = lamellar liquid crystal, $C_5OH$ = n-pentanol, $C_7OH$ = n-heptanol, $C_{10}OH$ = n-decanol

tion of alkanol chain length, chemical nature of hydrocarbon, and temperature. In this paper, results concerning one surfactant (LAM), two n-alkanols (n-pentanol and n-heptanol), and three hydrocarbons (hexadecane, squalane, and toluene) are reported. Pseudoternary phase diagrams of the various systems with a constant surfactant/n-alkanol molar ratio equal to 0.25 are shown in Figs. 6—8, where the solubility regions obtained are displayed. In line with the behavior observed in the ternary lipoaminoacid surfactant/water/n-alkanol systems, the effect of temperature in the solubility regions of the quaternary systems is not significant. A general trend is observed: the large solubility regions emanate from the surfactant/alkanol apex and extend towards the hydrocarbon corner. This indicates that inverse-type microemulsions might be favored. Exceptions are systems with n-pentanol as alkanol and hexadecane and squalane as hydrocarbons, shown in Figs. 6a and 7a, respectively. Squalane (Fig. 7) is by no means optimal for the formation of large solubility regions in the systems considered in this study.

Considering the occurrence of liquid crystalline phases in the multiphasic areas of the pseudoternary phase diagrams might be indicative of the structuralization in the solution regions. When hexadecane or squalane was used as the oil, a lamellar liquid-crystalline phase was in equilibrium with a liquid phase only in the systems with n-heptanol (Figs. 6b, 7b). Therefore, the solubility regions displayed in Figs. 6a

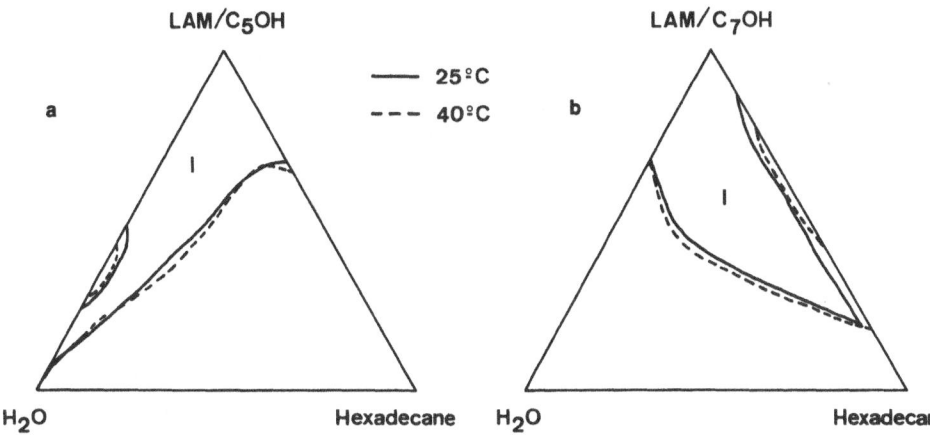

Fig. 6. Solubility regions in the pseudoternary phase diagrams of the systems LAM/water/n-alkanol/hexadecane with LAM/n-alkanol molar ratio equal to 0.25. $C_5OH$ = n-pentanol, $C_7OH$ = n-heptanol

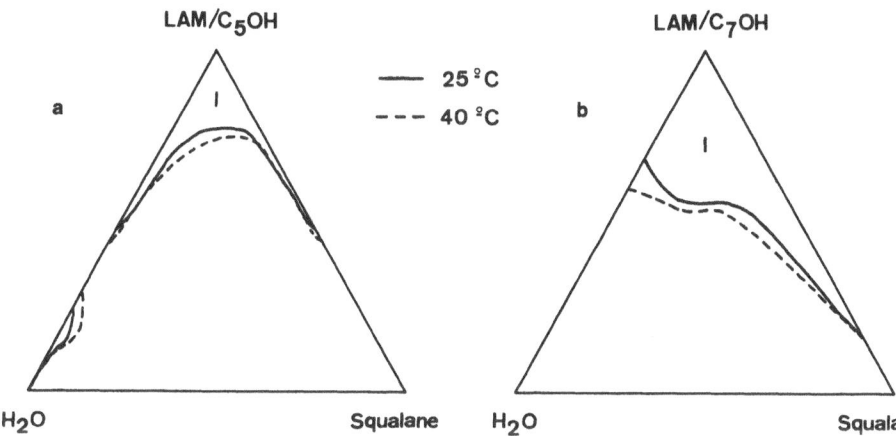

Fig. 7. Solubility regions in the pseudoternary phase diagrams of the systems LAM/water/n-alkanol/squalane with LAM/n-alkanol molar ratio equal to 0.25. $C_5OH$ = n-pentanol, $C_7OH$ = n-heptanol

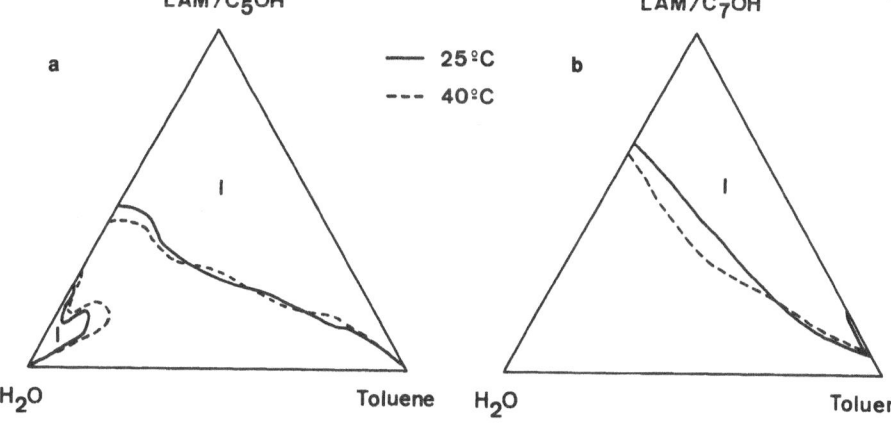

Fig. 8. Solubility regions in the pseudoternary phase diagrams of the systems LAM/water/n-alkanol/toluene with LAM/n-alkanol molar ratio equal to 0.25. $C_5OH$ = n-pentanol, $C_7$ = n-heptanol

and 7a, with n-pentanol, may not be due to the aggregation capacity of LAM, but to a co-solubility action. In contrast, using toluene as the oil, a liquid crystalline phase was present in the multiphasic regions of either n-pentanol or n-heptanol systems

(Fig. 8). This is an indication that both systems may possess a microstructure. Indeed, preliminary investigations [7] for structural characterization by multicomponent self-diffusion measurements of the LAM/water/n-pentanol/toluene system showed that by adding hydrocarbon to the ternary LAM/water/n-pentanol system, a gradual transition from bicontinuous structures to more ordered reverse structures was produced. As it has been reported [13, 14], this behavior is qualitatively analogous to that of common ionic surfactants.

## Conclusions

Properties of binary $N^\alpha$-Acyl arginine methyl ester/ water systems follow the trend expected in a homologous series of surfactants with the same polar group and different alkyl chain length. Phase behavior of ternary $N^\alpha$-Acyl arginine methyl ester/water/n-alkanol systems conform to that of ionic surfactant systems: normal and inverse micellar phases grow at the expense of the lamellar liquid crystalline phase with the decrease of the alkanol chain length. Addition of hydrocarbon to certain LAM/water/n-alkanol systems led to the formation of microemulsions.

*Acknowledgements*

We wish to thank Mrs. A. Vilchez and I. Carrera for their valuable assistance with the experiments. Financial support for this research project was provided by C.I.R.I.T. (with a grant to M. A. Pés) and C.I.C.Y.T. (Program PN-PGC No. PB 0445).

## References

1. Infante R, García Domínguez J, Erra P, Juliá R, Prats M (1984) Int J Cosm Sci 6:275
2. Infante MR, Molinero J, Erra P, Juliá MR, García Domínguez J (1985) Fette Seifen Anstrichmittel 8:309
3. Infante MR, Molinero J, Erra P, Juliá MR, García Domínguez J, Robert M (1986) Fette Seifen Anstrichmittel 88:108
4. Robert M, unpublished results
5. Vinardell MP, Molinero J, Parra JL, Infante MR, J of Cosm Sci, in press
6. Richtler HJ, Knaut J (1988) Proc 2nd World Surfactant Congress 1:33
7. Solans C, Infante R, Azemar N, Wärnheim T (1989) Progr Colloid Polym Sci 79:70
8. Zana R (1987) In: Zana R (ed) Surfactant Solutions. Marcel Dekker, New York, pp 241 – 294
9. Stevenson DG (1961) In: Durnham K (ed) Surface Activity and Detergency. McMillan, New York
10. Rendall K, Tiddy GJT, Trevethan MA (1983) J Chem Soc Faraday Trans I 79:C37
11. Mitchell DJ, Ninham BW (1981) J Chem Soc Faraday Trans II 77:601
12. Ekwall P (1975) In: Brown GH (ed) Advances in Liquid Crystals 1:1. Academic Press, New York pp 1 – 142
13. Lindman B (1984) In: Tadros TF (ed) Surfactants. Academic Press, London pp 83 – 109
14. Wärnheim T, Sjöblom E, Henriksson U, Stilbs P (1984) J Phys Chem 88:5420

Authors' address:

Dr. C. Solans
Instituto de Tecnología Química y Textil (CSIC)
C/Jorge Girona, 18 – 26
08034 Barcelona, Spain

**Progress in Colloid & Polymer Science**

Progr Colloid Polym Sci 81:151–155 (1990)

# Activation studies on human platelets using electrophoretic and quasi-elastic light scattering

O. Glatter and E. Spurej

Institut für Physikalische Chemie, Universität Graz, Graz, Austria

*Abstract:* Our interest in human platelets is based on the pathological activity of platelets during formation of thrombi in connection with atherogenesis. The shape change is the first step of platelet activation and aggregation. Thus starting with the appearance of blebs and pseudopods a change in surface charge density and hydrodynamic radius of the cells in physiological solution occurs. These parameters and their changes can be measured by quasi-elastic light scattering. We also found that platelets slightly activated by mechanical stress recover when incubated at 37 °C for about half an hour. Significant differences in the electrophoretic mobility as well as in the diffusion constant could be found when the sample was cooled down from 37 °C to 20 °C indicating shape change. These findings with laser light scattering correspond very well with electronmicroscopic results. Moreover, the temperature dependent activation of platelets is not occuring continuously but there seems to be a critical temperature range for activation of human blood platelets of healthy donors. To see activation caused by addition of an agent, we observed dose-dependent change of electrophoretic mobility with thrombin.

*Key words:* Human platelets; atherosclerosis; electrophoresis; light scattering; activation

## Introduction

A laser light scattering instrument for electrophoretic, diffusion and combined experiments was built and proved to be very potent and quick to investigate biological samples. Using this system we investigate human blood platelets under nearly physiological conditions.

Human blood platelets are discoid cells circulating with the blood stream. Besides other functions they play an important role during haemostasis by releasing clotting factors and forming haemostatic plugs. Unfortunately platelets act similarly during the pathological process of atherosclerosis that leads to various letal diseases [1].

The activation and aggregation of platelets takes place in several steps: appearance of blebs and pseudopods — socalled shape change, secretion of substances released from intracellular granules, and aggregation with crosslinking via fibrinogen bridges. It is well known that substances like collagen or thrombin cause platelet aggregation which can easily be followed in an aggregometer [2]. To investigate only platelet shape change it is necessary to use a much more precise method and instrumentation. Assuming that pseudopods increase the effective hydrodynamic volume and have special specific receptors influencing the surface charge of platelets, small differences in size and surface charge density have to be detected. For this application laser light scattering is an excellent method because it gives the electrophoretic mobility proportional to the surface charge density and the diffusion constant, which is a parameter for the size of the particles in solution. Thus we are able to distinguish different states of blebs and pseudopod formation in a short time, without long lasting preparations and under physiological conditions.

Being aware of the fact that the surface charge density could also be changed by adhesive substances it is a great advantage to be able to measure the hydrodynamic radius of the same platelet sample at the same

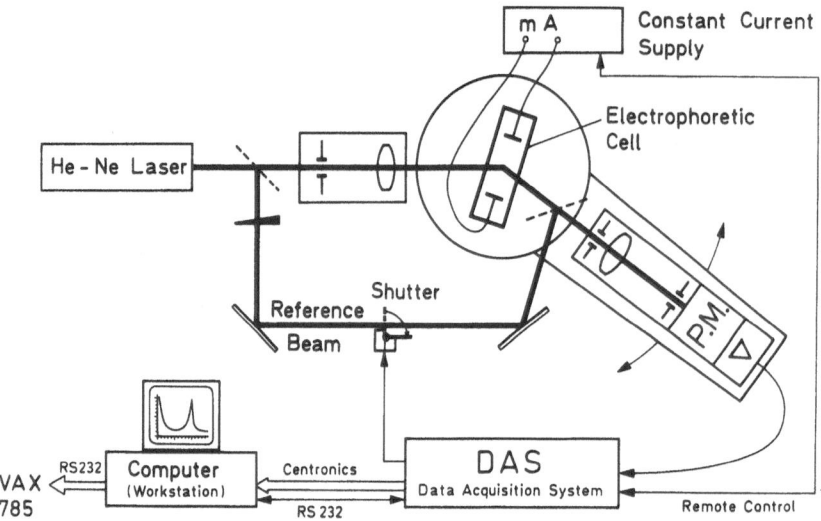

Fig. 1. The light scattering system for combined electrophoretic and diffusion experiments

time, avoiding manipulation with platelets and therefore the risk of unwanted activation. In this way the sensitivity of platelets to temperature, pH and mechanical stress is taken into account. Adhesive substances would influence the electrophoretic mobility very much but not the hydrodynamic radius.

There are general risk factors which can change platelet characteristics such as smoking, hypercholesterolemia [3] and stress. It is not yet fully understood in what way these risk factors influence platelet activation and aggregation, and therefore pharmacological regulation of the complex mechanism of atherosclerosis has been impossible.

In this context we want to follow platelet activation with laser light scattering and investigate the changes induced by different activating agents.

## Preparation

50 ml of fresh whole blood of young and healthy blood donors were anticoagulated with citrate (13.67 g $Na_3$ citrate, 0.74 g citric acid per 500 ml distilled water). After centrifugation at 100 g for 10 minutes the platelet rich plasma (PRP) was taken off and a pure fraction of platelets obtained by separation from plasma proteins on a column of Sepharose 2 B. Further centrifugation appeared to be unnecessary for our purposes and was therefore omitted. The suspension of platelets was immediately incubated at 37 °C to preserve the nonactivated and native properties of the platelets. In general it is necessary to maintain physiological conditions for the platelets during the

whole preparation procedure, that means to keep temperature close to 37 °C and pH at 7.4.

We also did comparative measurements with washed platelets according to the procedure of Radomsky and Moncada [4] with addition of $PGI_2$. Especially in light scattering results we could not find any differences between gelfiltered and washed platelets and so we prefer gelfiltration beeing the faster and easier method.

In order to study the influence of remaining leucocytes and/or erythrocytes in the sample we prepared leucocytes by centrifugation on a Hypaque — Ficoll gradient, and the erythrocytes by repeated washing procedures. Small amounts of these cells contaminating the platelet solution did not influence the experiment, because firstly they quickly sediment and secondly they have a very low scattering intensity at the optimum conditions for platelets.

## Methods

For investigation of biological samples laser light scattering is a convenient method. The sample volume is about 0.5 ml, only low concentrations are necessary ($9 \times 10^7$ platelets/ml solution), physiological conditions can be obtained and the measurement itself requires no chemical interaction with the particles.

Figure 1 shows the setup of the laser light scattering instrument. The light of a HeNe-laser is focused into the sample cell, which is thermostated. The scattered light is gathered by the detector optics into the photomultiplier. The voltage pulses are counted by an interactive data acquisition system [5], which also controls the experiment. Data are sent to a computer for further evaluation. Various parameters can be easily changed and adapted for a given sample.

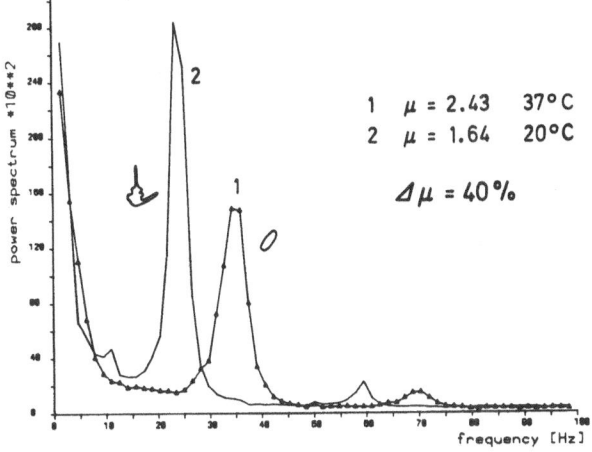

1  $\mu$ = 2.43    37°C
2  $\mu$ = 1.64    20°C

$\Delta \mu$ = 40%

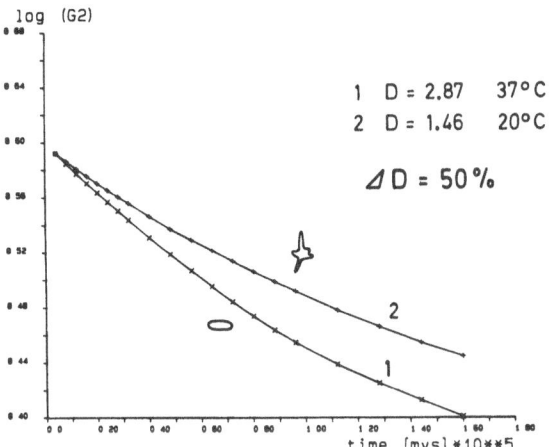

1  D = 2.87    37°C
2  D = 1.46    20°C

$\Delta$D = 50%

Fig. 2. Temperature dependence of the electrophoretic mobility and the diffusion coefficient of platelets in physiological solution by laser light scattering

We arranged our setup that way that it is possible to do electrophoretic and diffusion experiments quasi-simultaneously obtaining results within a few minutes.

The electrophoretic cell is a slightly modified version of the cell constructed by Smith & Ware [6]. The metal blocks are covered with gold to make the surface inert. The illuminated volume is located within a narrow gap (0.75 × 3 × 5 mm) to obtain a high field strength. The platinized platinum electrodes prevent formation of big gas bubbles. Because of the long distance between the gap and the electrodes products from electrode reactions can hardly influence the sample under investigation. The electrophoretic cell is connected do a constant current source which works under remote control of the data acquisition system. To minimize electroosmosis the surfaces of spacers as well as windows surrounding the gap are covered with a methylcellulose coating to obtain a neutral surface.

Diffusion measurements can also be carried out in the electrophoretic cell, which was proved by comparative measurements in a standard cuvette.

For evaluation of data it is optional to calculate power spectra with a gliding window FFT algorithm or correlation functions by software-correlator. ELS measurements give power spectra with peaks at the shift frequency due to the particles' different surface charge density. Homogeneous samples give single exponential correlation functions. For heterogeneous samples a logarithmic software-correlator is used to meet the different sample times. Performing an indirect Laplace transformation of the correlation function allows distinction between species of different size.

## Results

As already pointed out we were interested in changes of human blood platelets during shape change as the first step of activation [7].

The following conditions turned out to give the best results: scattering angle: 20°; temperature: 37°C, except for measurements of temperature dependent changes as described below. For ELS the applied electric field strength was 10−30 V/cm.

We found significant differences in the electrophoretic mobility, as well as the diffusion constant, by lowering the temperature of the sample from 37°C down to 20°C (Fig. 2). This finding corresponds with our electromicroscopic results (Fig. 3). The changes occur because of the formation of pseudopods during the activation process of platelets by temperature fall and therefore actually indicate the difference between activated and unactivated platelets. In this case the decrease in electrophoretic mobility was 40% and the decrease in diffusion coefficient was 50%.

We observed that there are quite significant differences between different blood donors, but because of the complexity of physiological processes and the numerous possibilities to influence platelets we are not yet able to assign abnormalities in platelet activation to special parameters. A temperature dependent activation as shown in Fig. 4 could not yet be found for smokers.

In that case the platelets had a spherical shape already at 37°C and did not change very much with decrease of temperature.

We also found that incubation of slightly activated platelets, caused by mechanical stress for at least half an hour at 37°C results in relaxation of platelets. Each filling process of the electrophoretic cell causes platelet activation and therefore we have to wait till the platelets are relaxed before studying changes with activating agents.

Furthermore, we found the decrease in the electrophoretic mobility of platelets after addition of thrombin was dependent on the concentration of the throm-

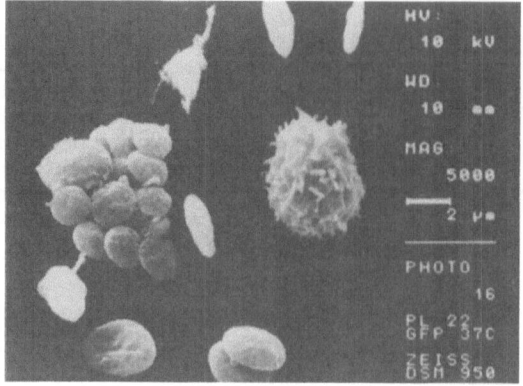

$\mu = 2.43$

$D = 2.87$

INTACT HUMAN BLOOD PLATELETS - DISCOID SHAPE

ACTIVATION

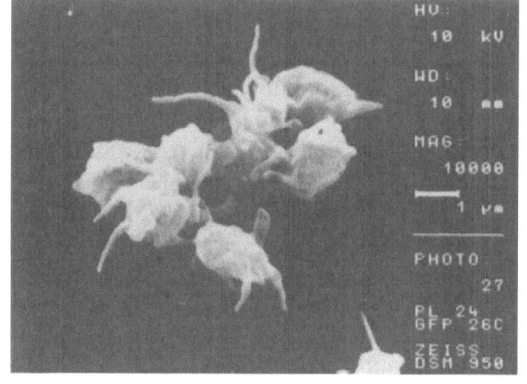

$\mu = 1.64$

$D = 1.46$

ACTIVATED HUMAN BLOOD PLATELETS - PSEUDOPODS

$\Delta\mu = 40\%$         $\Delta D = 50\%$

Fig. 3. Scanning electronmicrographs corresponding to Fig. 2 showing the temperature dependent activation of platelets

bin solution added (Fig. 5). This indicates that the rate of platelet activation can be controlled by the amount of activating agent. Being receptor-mediated the activation by thrombin is supposed to be dose-dependent.

In summary these results show that laser light scattering is a very powerful method to characterize the activation state of human platelets taking advantage of the speed and the possibility to work under physiological conditions.

In the future we want to use the temperature dependence of platelet activation to characterize the platelets of a special donor and start to investigate the influence of atherogenic agents.

Our future goals will be also to investigate other activating agents and the influence of several antagonists of the activation process. We also will try to find significant differences in the properties of human platelets from healthy and sick donors.

*Acknowledgements*

The authors wish to thank Claudia Polz for her expert preparation of the samples, Dr. Daniel Schneditz for his kind assistance and contribution of know how and Dr. Gisela Pfeiler for her excellent cooperation doing the scanning electronmicroscopy. We would also like to thank the Fonds zur Förderung der wissenschaftlichen Forschung for the financial support of this project, project number P6741C.

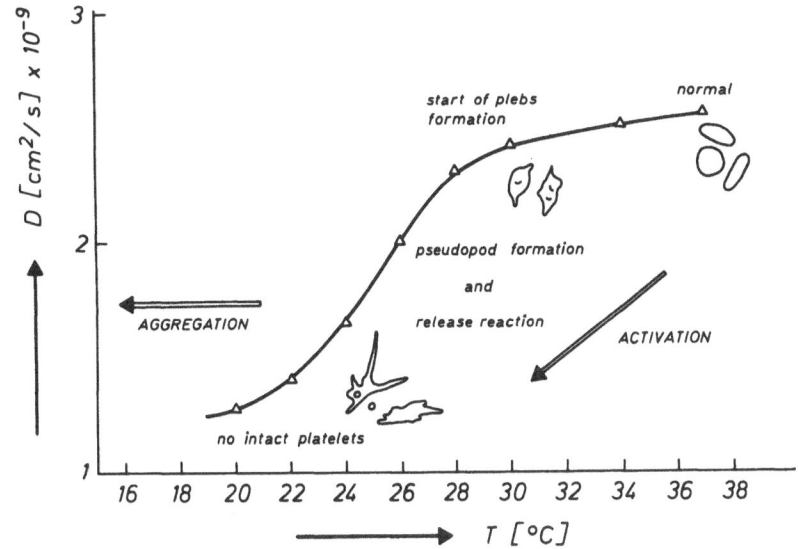

Fig. 4. Nonlinear course of temperature dependent activation of platelets shown as diffusion coefficient vs temperature

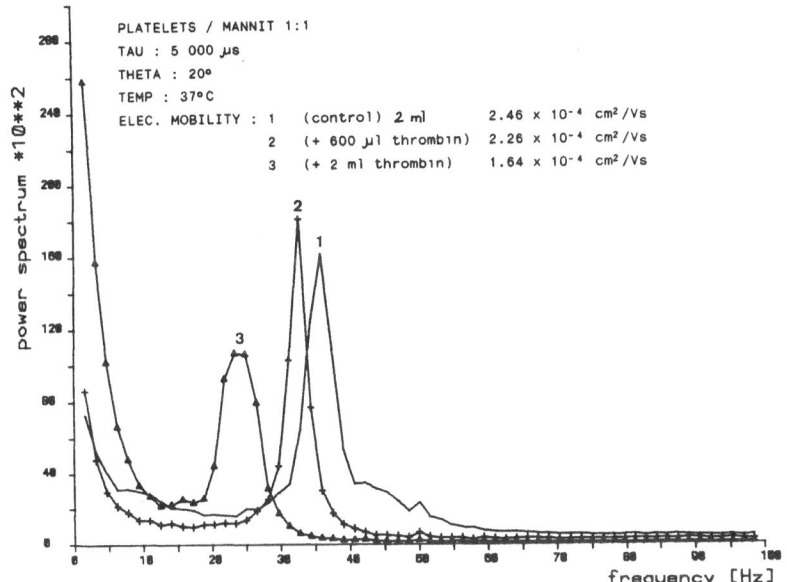

Fig. 5. Decrease in electrophoretic mobility with addition of an increasing dose of thrombin i.e. an activating agent

## References

1. Milton JG, Frojmovic MM (1987) Thrombosis Res 47:511−531
2. Latimer P, Born GVR, Michal F (1977) Arch Biochem Biophys 180:151−159
3. Nimpf J, Wurm H, Kostner GM, Kenner T (1986) Basic Res Cardiol 81:437−453
4. Vargas JR, Radomski M, Moncada S (1982) Prostaglandins 23:929−945
5. Fuchs HF, Jorde C, Glatter O (1989) Rev Sci Instrum 60 (5):854−857
6. Smith BA, Ware BR (1978) Contemporary Topics in Analytical and Clinical Chemistry 2:29−53
7. Steen VM, Holmsen H (1987) Eur J Haematol 38:383−399

Authors' address:

Otto Glatter
Institute of Physical Chemistry
University of Graz
Heinrichstraße 28
A-8010 Graz, Austria

**Progress in Colloid & Polymer Science**                    Progr Colloid Polym Sci 81:156 – 160 (1990)

# Quantitative aspects of labelling colloidal gold with proteins in immunocyto-chemistry

M. Horisberger

Nestlé Research Centre, Nestec Ltd., Lausanne, Switzerland

*Abstract:* Cytochemical markers prepared by labelling colloidal gold with a variety of bioactive molecules have gained wide acceptance both in transmission and scanning electron microscopy. However, details on the process and on the extent of adsorption of macromolecules to gold particles are scarce. The quantitative aspects of labelling colloidal gold with macromolecules were examined using proteins differing widely in size and shape such as protein A (single polypeptide chain), β-lactoglobulin (dimeric protein) and monoclonal IgE (four polypeptide chains). Adsorption isotherms were constructed at different pH's and the association constants evaluated by Scatchard plot analysis. At low protein concentration, practically all molecules were adsorbed and the isotherm was largely pH independent. Adsorption was maximal at the isoelectric point of the molecules (most compact form). At high coverage only β-lactoglobulin was irreversibly adsorbed. The proportion of labelled particles binding to immobilized ligands decreased sharply at low coverage. Particles labelled with IgE were unreactive towards the immobilized antigen, probably due to a restriction in the bending capacity of IgE. The results indicated that the empirical conditions generally used for preparing gold markers are valid in most cases.

*Key words:* Colloidal gold; electron microscopy; immunocytochemistry; IgE; β-lactoglobulin; protein A

## Introduction

Colloidal gold is a typical hydrophobic colloid of ionic nature whose particles carry a net negative charge in water caused by the presence of adsorbed anions. Since the work of Faraday in 1857 [1], the addition of a variety of polymers is well-known for stabilizing colloidal gold against coagulation by ions. These macromolecules do not appear to be adsorbed, especially at high coverage, over their entire length at the interface. It is thought that, at low concentration, the polymer coils attach themselves extensively in the form of partly flattened (random) coils. Further deposition causes more interpolymer contact and a reduction in surface contact per molecule. Without destroying their monolayer character, the coils can be further compressed until a plateau is reached [2].

Since 1971 [3], immunocytochemists have turned the polymer adsorption property of colloidal gold to their advantage. Indeed enzymes, lectins, antibodies, protein A, etc. ... can be adsorbed on to colloidal gold while retaining their specific binding properties [4]. These cytochemical markers have gained wide accep-

tance in transmission (TEM) and scanning electron microscopy (SEM). As cytochemical markers, gold particles are particularly interesting. Due to their opacity to electrons, the particles are easily detected by TEM (Fig. 1) [5]. As they are also good emitters of secondary electrons, they can be visualized by SEM (Fig. 2) [6] and the resolution of individual particles is improved on stereomicrographs [7]. This facilitates the interpretation of the results (Fig. 3) [8].

The adsorption of macromolecules on to a metal surface depends on a number of factors such as stability of the colloid itself, the concentration, shape, configuration and isoelectric point of the macromolecule, the ionic strength, pH and temperature of the suspending medium. Understanding these factors determines success or failure in the preparation and the utilization of colloidal gold markers. However, details on the process and on the extent of adsorption of macromolecules to gold particles are scarce. Only a few studies have appeared, as recently reviewed by Horisberger [9].

It should be noted that when gold particles are la-

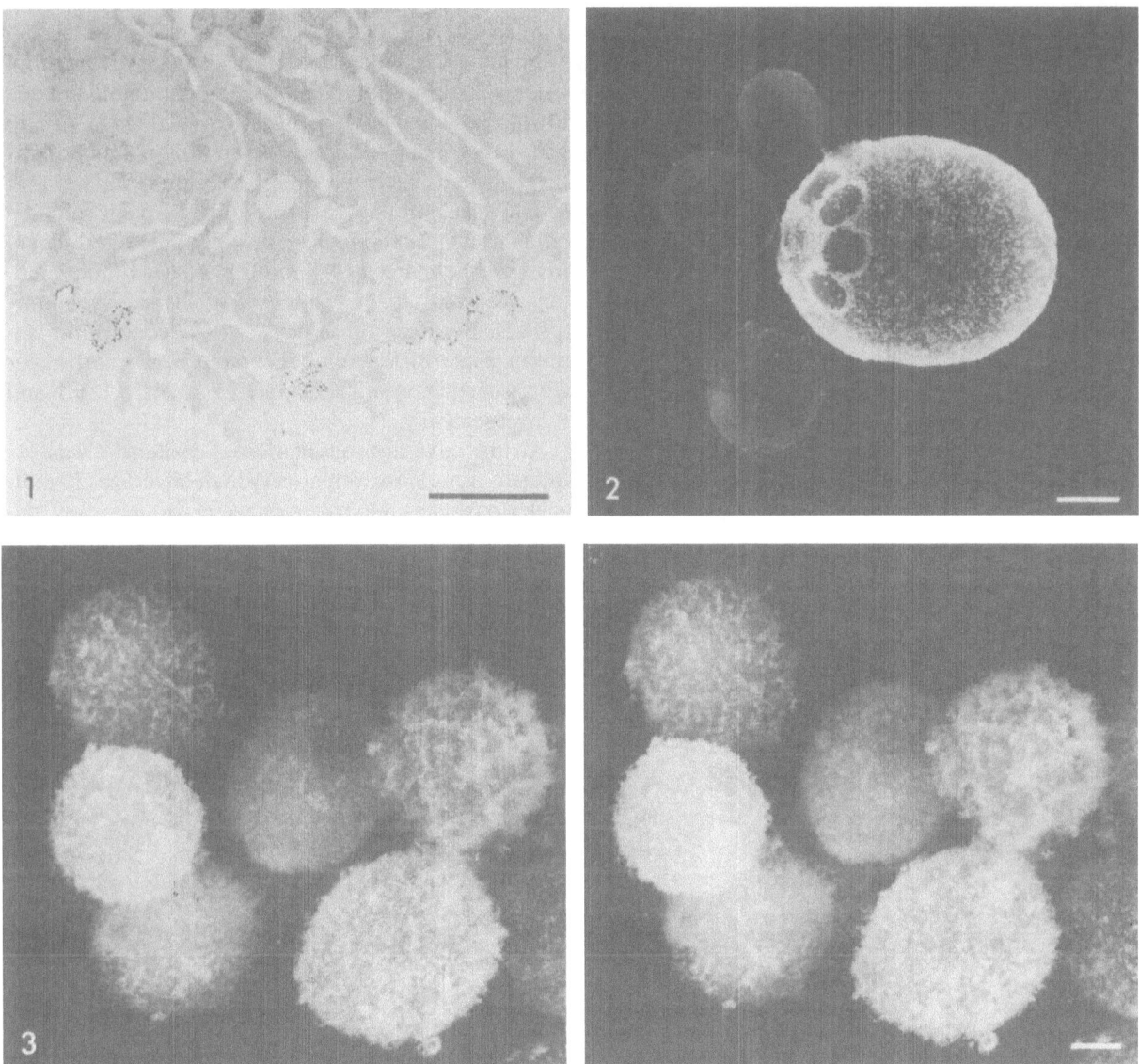

Fig. 1. Transmission electron microscopy. Distribution of *Fc* receptors for IgE on rat basophilic leukemia cells (RBL-1). Cells were sensitized with mouse IgE antibody monoclonal anti-bovine milk β-lactoglobulin and successively incubated at 4°C with anti-IgE antibodies and protein A–Au$_{12}$. The cells warmed for 30 min at 37°C developed surface clusters [5]. The subscript to the Au particles refers to their average diameter in nm. The bar represents 1 μm

Fig. 2. Scanning electron microscopy. Localization of anionic sites on *Candida albicans* blastoconidia. The yeast cells were marked with chitosan-Au$_{47}$, a cationic polymer of β(1→4)D-glucosamine. The mother cell is densely marked except for the bud scars. The emerging buds are almost free of marking. In yeast, the net negative surface charge is mainly associated with the phosphate groups of mannoproteins present on the cell surfaces. From reference 6. The bar represents 1 μm

Fig. 3. Scanning electron microscopy. Stereomicrograph of mouse mast cells marked with wheat germ lectin, a lectin specific for N-acetyl-D-glucosamine residues. A dense marking of WGL-Au$_{76}$ is observed only on some cells indicating that the surface distribution and/or accessibility of glycoproteins to the marker depends on cell maturity. From Ref. [7]. The bar represents 2 μm

belled with proteins, full stabilization against coagulation by electrolytes (such as those present in physiological buffers) is not always observed during storage especially with the large size markers used in SEM (>20 nm in diameter). Full stabilization can easily be achieved by the addition of polyethylene glycol, polyvinylpyrolidone or bovine serum albumin. These polymers are added only once the colloid has been

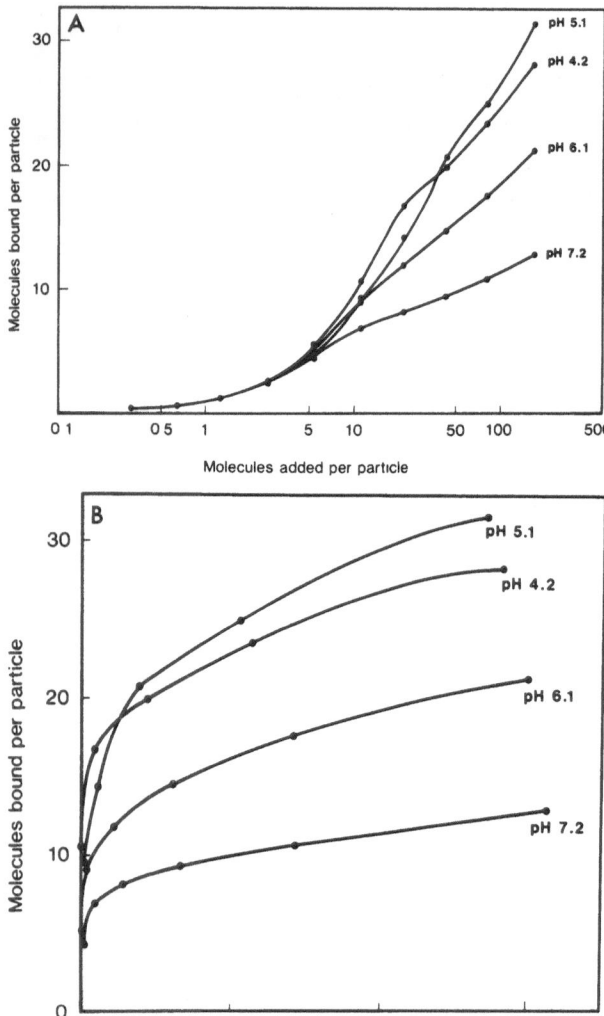

Fig. 4. Adsorption isotherms of [125]I-protein A adsorbed on Au$_{11.2}$ particles as a function of pH and molecules added per particle (A) or equilibrium concentration (B). Adapted from [12]

labelled with proteins. Full stabilization of the colloid is attributed to the fact that these macromolecules attach to the surface by relatively few segments extending in loops in the aqueous phase which keep the particles from coherence (steric stabilization). For a review, see [9].

In this communication, quantitative labelling of colloidal gold was examined using proteins differing widely in size and shape, i.e. protein A (single polypeptide chain, 42000 Da), β-lactoglobulin (dimeric protein, 36000 Da) and monoclonal IgE (four polypeptide chains, 203000 Da).

## Protein A

Protein A is widely used as a secondary reagent for the localization of antigen by the immunogold method [10]. This single polypeptide chain (42000 Da, pI 5.1.) has a very extended shape. Protein A is functionally bivalent and reacts with the Fc region of IgG [11].

The adsorption isotherms of protein A on colloidal gold (11.2 nm in diameter, Au$_{11.2}$) buffered at different pH's is shown in Fig. 4A and B [12]. Maximum binding occurred at the isolectric point of protein A (pI 5.1). By fitting the inverse of the adsorption isotherm (Fig. 4B), the number of molecules adsorbed per Au$_{11.2}$ particle was 38, 22 and 13 at pH 5.1, 6.1 and 7.2, respectively.

At low coverage, desorption of protein A was extremely slow. However, at very high coverage (25 molecules/particle), the particles were progressively depleted in protein A during storage, eventually leading to 100% depletion [12]. This indicates that at high coverage, protein A molecules are attached by short segments. However, being thus attached, the molecule can extend in solution and hence be available to combine with IgG (Fig. 5). A Scatchard plot analysis indicated that the association constant at low coverage ($1.5 \cdot 10^8$ M$^{-1}$ at pH 6.1) was reduced to $7.4 \cdot 10^6$ M$^{-1}$ at high coverage [12].

## β-Lactoglobulin

Goat β-lactoglobulin (36000 Da) is closely related to bovine; βL exists as a stable dimer dissociating below pH 3.5 [13].

Goat βL triated *in vivo* was adsorbed on Au$_{12.4}$ particles at pH 8 (Fig. 6) [14]. At the lowest concentration, practically all βL molecules were adsorbed. At high concentration, an equilibrium was established between free and adsorbed βL molecules.

Assuming that βL monomer is a sphere of 3.6 nm in diameter [15], one Au$_{12.4}$ particle would accomodate 20 βL molecules as a dimer in a compact monolayer (84% of the total surface occupied). However, under saturating conditions, βL molecules were adsorbed only to 64% of the theoretical maximum (12.9 ± 0.6 molecules/particle). This low value is probably due to the presence of a low concentration of electrolytes in the gold sol itself which inhibits the adsorption of macromolecules on to metallic surfaces. At both low and high coverage, βL molecules were irreversibly adsorbed [14].

The binding of colloidal gold labelled with βL to immobilized anti-βL antibodies is shown in Fig. 5.

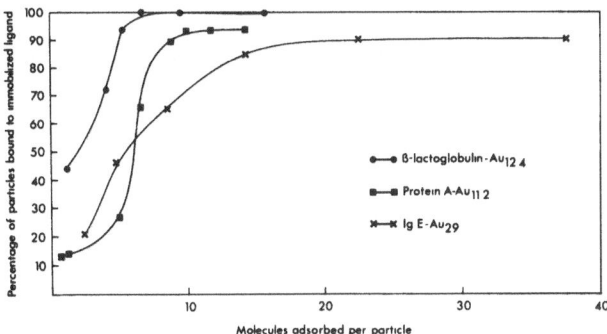

Fig. 5. Proportion of labelled gold particles binding to an immobilized ligand as a function of coverage density. Gold particles were labelled with an increasing amount of β-lactoglobulin molecules at pH 8, protein A at pH 6.1 and IgE at pH 8.8, and reacted respectively with anti-β-lactoglobulin IgG immobilized on protein A-Sepharose, IgG immobilized on Sepharose and anti-IgE immobilized on protein A-Sepharose. In all cases, the proportion of unreactive particles decreased as the number of protein molecules adsorbed per particle increased. Adapted from [5, 12 and 14]

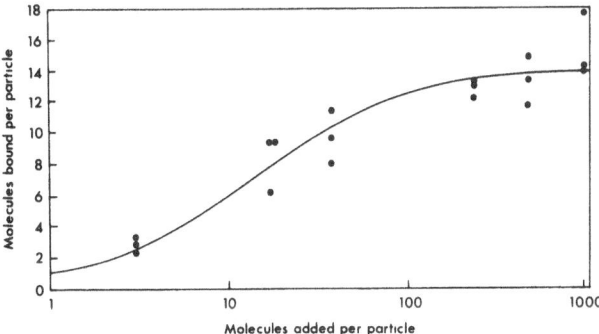

Fig. 6. Adsorption isotherm of goat $^3$H-β-lactoglobulin adsorbed at pH 8 on $Au_{124}$ particles (adapted from [14])

The lack of reactivity of particles at low coverage may be attributed to (1) absence of βL molecules on some particles, (2) inaccessibility of antigenic sites of adsorbed βL molecules, (3) steric hindrance due to polyethylene glycol (Carbowax 20 M) adsorbed to ensure full stabilization against coagulation by electrolytes and/or (4) necessity for stable binding of multiple interactions between the gold particle and the immobilized antibodies.

## Monoclonal immunoglobulin E

IgE has a molecular weight of 203000 Da [16] and is composed of four polypeptides: two identical heavy chains and two identical light chains rigidly linked by disulphide bonds in a *Y* or *T* shaped structure [17]. It is known that the binding of IgE molecules to their *Fc* receptor necessitates a change in the configuration of the antibody [18]. This phenomenon could thus preclude the use of labelled gold markers for receptor mapping.

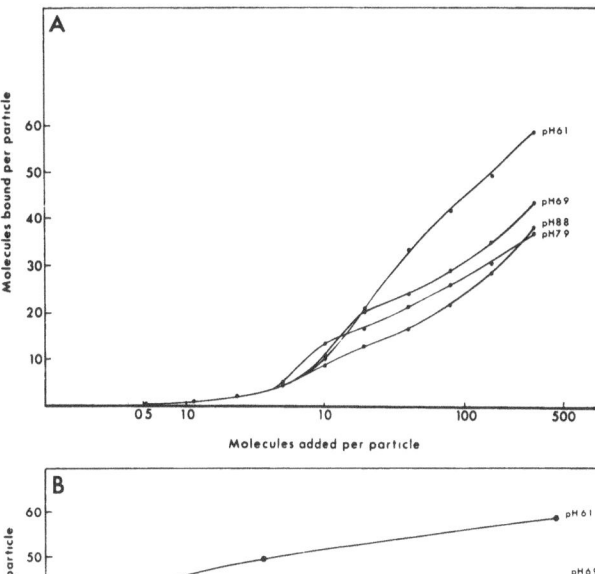

Fig. 7. Adsorption isotherms of monoclonal IgE adsorbed on $Au_{27.5}$ particles as a function of pH and molecules added per particle (A) or equilibrium concentration (B). Adapted from [5]

The adsorption isotherms of $^{125}$I-IgE antibody antibovine milk β-lactoglobulin adsorbed at different pH's on $Au_{27.5}$ particles is shown in Fig. 7A and B [5]. At low coverage, the isotherm was independant of pH (Fig. 7A). In the presence of a large excess of IgE, the highest coverage was obtained at pH 6.1 near the pI of IgE (5.2 – 5.8) (Fig. 7A and B).

The isotherms were examined at low and high surface coverage by Scatchard analysis (Fig. 8, Table 1). At low coverage, the binding constant (many contact points, side-on adsorption) is higher than at high coverage (fewer contact points) (Table 1). At low coverage (< 20 molecules adsorbed per particle), the release of radioactivity was less than 2% upon dilution which confirmed that IgE molecules are tightly adsorbed at low coverage. However, the proportion of reactive molecules towards immobilized anti-IgE antibodies (Fig. 5) decreased with a decreasing number of IgE molecules adsorbed.

It was observed that IgE-gold complexes did not bind to surface receptors of most cells [5]. This is not surprising since IgE must bend in order to bind to its

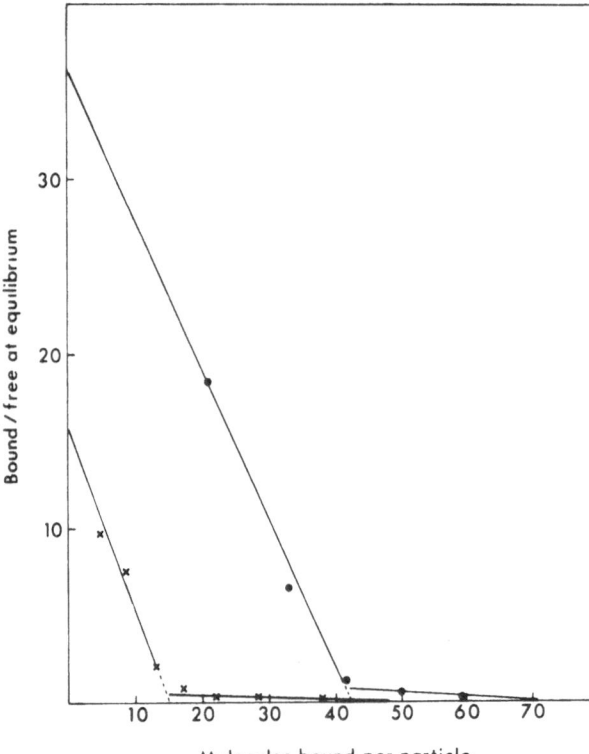

Fig. 8. Scatchard plots for IgE adsorbed on $Au_{275}$ particles at pH 6.1 (solid circle) and pH 8.8 (crosses). From Ref. [5]

Table 1. Number (N) of IgE molecules adsorbed per $Au_{275}$ particles at saturation, and binding constants (K) [5]

| pH of adsorption | N | K [$M^{-1}$] |
|---|---|---|
| 6.1 | 42 | $2.2 \cdot 10^9$ |
|  | 68 | $6.3 \cdot 10^7$ |
| 8.8 | 15 | $2.9 \cdot 10^9$ |
|  | 48 | $3.9 \cdot 10^7$ |

receptor through the *Fc* portion [18]. Such bending would be mostly prevented when IgE is adsorbed on a solid surface. We therefore used a three-step procedure to mark *Fc* receptors for IgE or RBL-1 cells (Fig. 1) [5].

## Conclusion

In all cases studied, the adsorption isotherms could be divided into two regions, i.e. of low and high coverage. At low protein concentration, practically all molecules were adsorbed and the isotherm was largely pH independant. At the higher concentration, an equilibrium was established between free and adsorbed molecules. Adsorption was maximal at the isoelectric point of the molecules where they assume the most compact form. In all cases reviewed, no evidence for the formation of a multilamellar shell was observed.

The markers were stable upon storage when prepared at low coverage. However, in the case of a single polypeptide chain such as protein A, as the protein concentration increased, the molecules became attached by short segment with a concomittant decrease in the affinity constant. Bioactivity of the macromolecules was maintained upon adsorption except for IgE where restricted bending of the molecules prevented its attachment to the *Fc* receptors. All these results indicate that the empirical conditions generally used for preparing gold markers [4, 10] are valid in most cases.

Acknowledgements

The author thanks Ms M. F. Clerc for the art work, Ms C. Mordasini for typing the manuscript and Dr. I. Horman for reviewing it.

## References

1. Faraday M (1857) Philos Trans R Soc 147:145–181
2. Eirich FR (1977) J Colloid Interface Sci 58:423–436
3. Faulk WP, Taylor GM (1971) Immunochemistry 8:1081–1083
4. Horisberger M (1981) Scanning Electron Microsc 2:9–31
5. Clerc MF, Granato DA, Horisberger M (1988) Histochemistry 89:343–349
6. Horisberger M, Clerc MF (1988) Eur J Cell Biol 46:444–452
7. Horisberger M (1988) Acta Histochemica, Suppl. Band 36:101–111
8. Horisberger M (1989) In: Hayat MA (ed) Colloidal gold: Principles, Methods and Applications, Academic Press, London, New York, pp 217–227
9. Horisberger M (1989) In: Verkleij AJ, Leunissen JLM (eds) Immuno-gold labeling in cell biology, CRC Press, Boca Raton, Florida, pp 49–60
10. Roth J (1982) In: Bullock GR, Petrusz P (eds) Techniques in immunocytochemistry, vol 1, Academic Press, London, New York, pp 107–132
11. Langone JJ (1982) Adv Immunol 32:157–252
12. Horisberger M, Clerc MF (1985) Histochemistry 82:219–223
13. Preaux G, Braunitzer G, Schrank B, Stangl A (1979) Hoppe Seyler's Z Physiol Chem 360:1595–1604
14. Horisberger M, Vauthey M (1984) Histochemistry 82:219–223
15. Swaisgood HE (1982) In: Fox PF (ed) Development in dairy chemistry: Proteins, vol 1, Applied Sciences Publishers, London, New York pp 1–59
16. Newman SA, Rossi G, Metzger H (1977) Proc Natl Acad Sci USA 74:869–872
17. Amsel LM, Poljak RJ (1979) Annu Rev Biochem 48:861–979
18. Holowka D, Conrad DH, Baird B (1985) Biochemistry 24:6260–6267

Author's address:
Dr. M. Horisberger
Nestec Ltd.
Nestlé Research Centre
P.O. Box 353
CH-1800 Vevey, Switzerland

**Progress in Colloid & Polymer Science**                                    Progr Colloid Polym Sci 81:161 – 168 (1990)

# Theory of scattering from colloidal aggregates

R. Klein[1]), D. A. Weitz[2]), M. Y. Lin[3]), H. M. Lindsay[4]), R. C. Ball[5]) and P. Meakin[6])

[1]) Fakultät für Physik, Universität Konstanz, Konstanz, West Germany
[2]) Exxon Research and Engineering Co., Annandale NJ, USA
[3]) Department of Physics, Princeton University, Princeton NJ, USA
[4]) Department of Physics, Emory University, Atlanta GA, USA
[5]) The Cavendish Laboratory, Cambridge, England
[6]) E. I. DuPont de Nemours Co, Experimental Station, Wilmington DE, USA

*Abstract:* Static and dynamic light scattering are major experimental tools to study colloidal aggregates. The theoretical methods for a proper analysis of such experiments are reviewed. It is shown how several interrelated features of the aggregation process determine the experimentally accessible quantities. These features are the structure of the clusters as characterized by their fractal dimension and their anisotropies and the shape of the cluster mass distribution. Using computer-generated clusters, obtained under the conditions of diffusion-limited and of reaction-limited cluster aggregation, and using results for the cluster mass distribution obtained from the Smoluchowski equation for irreversible growth, the static scattering intensity and the correlation function of quasi-elastic light scattering are calculated. The latter is shown to depend sensitively on rotational diffusion processes and on the cluster mass distribution. Finally, it is shown how the growth kinetics can be extracted from the angle dependence of the first cumulant.

*Key words:* Colloidal aggregation; fractals; kinetic growth processes; light scattering

## Introduction

Colloid aggregation is a prominent example of a random kinetic growth process, which results in highly disordered, complex structures [1, 2]. Considerable new insights into the problem of a quantitative description of the aggregation process have been obtained with the recognition that the clusters are statistical fractal objects, meaning that their mass scales with their size $R$ as $M \sim R^{d_f}$, where $d_f$ is the fractal dimension [3]. For three-dimensional clusters computer simulations and various experiments have shown that $d_f$ has typical values around 2, which reflects the rather open, tenuous structure of the aggregates.

Aggregation processes such as colloid aggregation are characterized by several interrelated properties: The first is the numerical value of the fractal dimension $d_f$ of the aggregates. The second is the cluster mass distribution $N(M; t_a)$, which gives the number of clus-

ters of mass $M$ at time $t_a$ after the aggregation has been initiated. A third important quantity is the time evolution of the radius of a suitably defined average cluster. It is important to realize that these three characteristics of the aggregation process are intimately related to each other. Furthermore, they can distinguish the two limiting growth processes of colloid aggregation. Traditionally these growth processes were called fast and slow aggregation; these two classes are now distinguished as diffusion-limited and reaction-limited cluster aggregation (DLCA and RLCA).

A colloid aggregation process begins with a monodisperse suspension of colloidal particles of radius $a$ at a low volume fraction. Initially, a sufficiently large repulsive potential barrier prevents the coagulation of two approaching colloidal particles. Aggregation can occur when the height of the barrier, $E_b$, is reduced. If $E_b \ll k_B T$, every collision will result in the particles sticking together, leading to very rapid aggregation. This process is limited only by the rate of diffusion-induced collisions between particles. The resulting clusters themselves continue to diffuse, collide with other particles and clusters, and aggregate further. The regime, where $E_b \ll k_B T$, is therefore called diffusion-limited colloid aggregation. If, by contrast, $E_b$ remains somewhat larger than $k_B T$, many collisions must occur before two particles can stick to one another. In thic case, the aggregation rate is limited by the probability $P$ to overcome the repulsive barrier, leading to much slower aggregation. This regime is therefore called reaction-limited cluster aggregation.

It is intuitively easy to understand that these two limiting aggregation processes will result in rather different behavior for the three main characteristics mentioned above. Since $P \approx 1$ for DLCA, a cluster of this type will preferentially grow at its tips. By contrast, if $P \ll 1$, as for RLCA, an incoming particle or small cluster will be able to explore the "fjords" of a cluster before sticking. As a result, the fractal dimension $d_f$ will be larger for RLCA than for DLCA. Furthermore, if $P \ll 1$, many contacts are required before two clusters stick together, resulting in a very slow aggregation rate for RLCA. It is also obvious that the cluster mass distributions will be quite different, since the sticking probability of two clusters is proportional to both $P$ and to the number of available bonding sites. It can therefore be expected that the growth kinetics in the RLCA case will leave a large number of small clusters behind, which is not to be expected for DLCA.

The major experimental methods to quantitatively investigate the characteristics of colloid aggregation processes have been both static and dynamic light scattering, since the clusters grow to sizes which are comparable or even larger than the wavelength of light. These experiments therefore probe both the structure and the dynamics on length scales which can be larger than the size of some of the clusters in the distribution and smaller than the size of other clusters. A proper analysis of scattering experiments therefore reveals the structure of the clusters, their cluster mass distribution and the growth kinetics, if such experiments can be done while the aggregation continues. The purpose of this paper is to review the theoretical methods which are required for such analysis.

## Static light scattering

A static light scattering experiment measures the angle dependence of the scattered intensity, $I(q)$, where $q = (4\pi n/\lambda)\sin(\theta/2)$. Here, $n$ is the refractive index of the solvent, $\lambda$ the wavelength of light in vacuum and $\theta$ denotes the scattering angle. If $N(M; t_a)$ denotes the number of clusters of mass $M$ in the scattering volume at aggregation time $t_a$, and if interactions between the clusters are negligible, the total scattered intensity is given by

$$I(q) = \sum_M N(M; t_a) I_M(q), \tag{1}$$

where $I_M(q)$ is the scattered intensity from clusters of mass $M$. The wavevector dependence of $I_M(q)$ reflects the structure of the clusters of mass $M$, when they are probed on the length scale $q^{-1}$. Choosing the radius of gyration $R_g$ as a measure of the overall size of these clusters, they appear to the light probe essentially as point particles, if $qR_g \ll 1$. In this case they scatter isotropically and the intensity is proportional to $M^2$, independent of $q$. Therefore we introduce the structure factor of clusters of mass $M$, $S_M(q)$, by $I_M(q) = M^2 S_M(q)$, so that

$$I(q) = \sum_M N(M; t_a) M^2 S_M(q) \tag{2}$$

and

$$S_M(q) \sim \int d^3 r \, e^{-i\vec{q} \cdot \vec{r}} g_M(r) \tag{3}$$

where $g_M(r)$ is the density correlation function. For a self-similar object of fractal dimension $d_f$, this function scales as $g(r) \sim r^{d_f-3}$. Since the colloidal clusters are of finite extent, self-similar scaling will only hold over a limited range of length scales. The lower limit of the scaling is given by the radius $a$ of a colloidal particle and is usually of no interest in light scattering. The upper limit occurs when $r$ reaches the radius of gyration, $R_g$, and the existence of this limit introduces a cutoff function $h(r/R_g)$ such that

$$g_M(r) \sim r^{d_f-3} h(r/R_g) \tag{4}$$

with $h(x) = 1$ for $x \ll 1$ and $h(x) \to 0$ for $x \gg 1$. The index $M$ on $g_M(r)$ emphasizes that the density correlation function for the clusters depends on their mass $M \sim R_g^{d_f}$.

From Eqs. (3) and (4) follows

$$S_M(q) \sim (qR_g)^{-d_f} \tilde{h}(qR_g) \tag{5}$$

where, as a consequence of the properties of $h(x)$, $\tilde{h}(y) \sim y^{d_f}$ for $y \ll 1$ and $\tilde{h}(y) \to$ const for $y \gg 1$. As a result, $S_M(q) \to 1$ for $qR_g \to 0$ and $S_M(q) \sim (qR_g)^{-d_f}$ for $qR_g \gg 1$. Since $S_M$ depends on $q$ only through the product $qR_g$, we will often write $S_M(q)$

$= S(qR_g)$. This emphasizes that $R_g$ is the only relevant length scale of a self-similar cluster of radial symmetry.

Since the scattered intensity $I(q)$, Eq. (2) is a "convolution" of $S_M(q)$ and the cluster mass distribution, there are, in general, always contributions to $I(q)$ arising from clusters for which $qR_g \approx 1$. Therefore, $I(q)$ will be influenced by the precise form of the cutoff functions in Eq. (4) or (5). Several models have been proposed for these functions [4–7]. To test these suggested forms we have calculated $S(qR_g)$ directly from computer-generated clusters obtained under both DLCA and RLCA conditons [8]. The simulations provide the positions $\vec{r}_k$ of the particles $k = 1, ..., M$ of a cluster of mass $M$, so that the structure factor can be calculated directly by

$$S(qR_g) = \frac{1}{M^2}\left\langle \sum_{k,l=1}^{M} \exp[i\vec{q}\cdot(\vec{r}_k - \vec{r}_l)]\right\rangle. \qquad (6)$$

The ensemble average is performed by averaging each cluster over many orientations and by averaging over several clusters of $M$ particles. We use 20 different clusters of each mass, with $M$ extending from 100 to 900. The results are parametrized by fitting to

$$S(x) = \left[1 + \sum_{s=1}^{n} c_s x^{2s}\right]^{-d_f/2n}; \quad x = qR_g. \qquad (7)$$

This form has the correct limiting behavior for $x \to 0$ and $x \gg 1$. The coefficient $c_1$ is determined from the requirement that (7) has to reduce to the Guinier structure factor for $x \ll 1$, which gives $c_1 = 2n/3d_f$. Good fits to the computed $S(x)$ are obtained for $n = 4$. For DLCA clusters with $d_f = 1.8$ we find $c_2 = 2.50$, $c_3 = -1.52$, $c_4 = 1.02$. For RLCA clusters ($d_f = 2.1$) these coefficients are 3.13, $-2.58$ and 0.95. The main differences compared to the analytically known forms of $S(qR_g)$ is that the latter do not adequately represent the fairly sharp crossover near $qR_g \approx 1$ from the $q$ independent behavior to the $(qR_g)^{-d_f}$ scaling form. It is interesting to note that the fits obtained here are in very good agreement with a stretched exponential form for $h(x)$ in Eq. (4) [9].

The other important quantity which determines the scattering intensity $I(q)$ is the cluster mass distribution $N(M;t_a)$. It has been studied by computer simulations [10–12] and by solving the Smoluchowski equation for irreversible aggregation [13],

$$\dot{N}_k = \frac{1}{2}\sum_{i+j=k} K(i,j)N_i N_j - N_k\sum_{j} K(k,j)N_j, \qquad (8)$$

where $N_k = N(M_k;t_a)$ and $K(i,j) = K(j,i)$ denotes the reaction rates for the aggregation of clusters of mass $M_i$ and $M_j$. Assuming a scaling form $N(M;t_a) = g(t_a)\psi(M/\bar{M}_n)$, where $\bar{M}_n$ is the $n$-th reduced moment of $N(M;t_a)$, $\bar{M}_n = M^{(n)}/M^{(n-1)}$, where $M^{(n)} = \sum_M N(M;t_a)M^n$, it follows from mass conservation that

$$N(M;t_a) = \frac{1}{\bar{M}_n^2(t_a)}\psi(M/\bar{M}_n(t_a)). \qquad (9)$$

The time dependence of the cluster mass distribution is entirely contained in the moments $\bar{M}_n(t_a)$, while the scaling function $\psi(x)$ is time independent.

The moments $\bar{M}_n(t_a)$ characterize the mass of an average cluster of the distribution at time $t_a$, where $n = 2$ is usually chosen. If the kernels are homogeneous of degree $\lambda$, $K(ai, aj) = a^\lambda K(i,j)$, it follows from (8) that $\bar{M}_2(t_a) = \exp(t_a/\tau)$ for $\lambda = 1$ and that $\bar{M}_2(t_a) = [1 + t_a/t_0]^z$ for $\lambda \neq 1$, where $z = (1 - \lambda)^{-1}$. Therefore, the case $\lambda = 1$, which has been shown [14] to correspond to RLCA, exhibits exponential growth of the characteristic cluster mass, whereas for $\lambda \neq 1$, the average cluster grows according to a power law.

Explicit expressions for the cluster mass distribution are known for certain classes of reaction kernels $K(i,j)$. For the DLCA case, the form $K(i,j) = 4\pi\sigma_{ij}(D_i + D_j)$ is appropriate [15], where $D_i$ denotes the translational diffusion coefficient of a cluster of mass $M_i$. It scales as $D_i \sim (R_H^{(i)})^{-1} \sim (R_g^{(i)})^{-1} \sim (M_i)^{-1/d_f}$. Here, $R_H^{(i)}$ is the hydrodynamic radius, which has been shown to scale like the radius of gyration [16]. The quantity $\sigma_{ij}$ is the effective collision radius of clusters of masses $M_i$ and $M_j$. Assuming $\sigma_{ij} = R_g^{(i)} + R_g^{(j)}$, the kernel becomes $K(i,j) \sim 2 + (M_i/M_j)^{1/d_f} + (M_j/M_i)^{1/d_f}$. Thus, $K(i,j)$ is homogeneous with $\lambda = 0$. Except for $M_i$ very different of $M_j$, this kernel can be approximated by a constant. In this case, the solution is [17]

$$N(M;t_a) = \frac{N_0}{\bar{M}_1^2(t_a)}\left(1 - \frac{1}{\bar{M}_1(t_a)}\right)^{M-1}, \qquad (10)$$

which represents a broad cluster mass distribution with a fairly sharp cutoff at large masses. $N_0$ in Eq. (10) is the total number of colloid particles. Since $\lambda = 0$, the average cluster mass grows linearly, $\bar{M}_1(t_a) = 1 + t_a/t_0$, where $t_0 = 3\eta V/(8k_B T N_0)$ with $\eta$ being the viscosity of the solvent and $N_0/V$ the inital particle concentration. Furthermore, defining by $\bar{R}_g = a(\bar{M}_1(t_a))^{1/d_f}$ the radius of the average cluster, $\bar{R}_g(t_a) \sim (t_a/t_0)^{1/d_f}$ is predicted for long times. Finally, the total number of clusters at time $t_a$, $N_t(t_a) = \sum_m N(M; t_a)$, decreases as $t_a^{-1}$ for $t_a > t_0$.

For the reaction-limited case the kernels are quite different. Ball et al. [14] used geometric arguments to determine the scaling of the kernel; they found that when the cluster masses are comparable, then $K(i, j) \sim M_i^\lambda$, but when $M_i \gg M_j$ the kernels behave as $K(i, j) \sim M_i M_j^{\lambda-1}$ with $\lambda = 1$ in both cases. Therefore, the scaling is the same as for the sum kernel $K(i, j) \sim M_i + M_j$, for which Eq. (8) has been solved analytically. Both kernels result in a power-law distribution with an exponential cutoff [13, 14],

$$N(M; t_a) \sim M^{-\tau} \exp(-M/2\bar{M}_2(t_a)) \tag{11}$$

with $\tau = 1.5$. Furthermore, exponential growth with time of the radius of the characteristic cluster is found.

These predictions, which are based on these solutions of the Smoluchowski equation, have been tested by computer simulations. Excellent overall agreement has been found, expect for the value of the exponent $\tau$ in Eq. (11). One study [11] suggested that $\tau \approx 1.5$, while another one [12] suggested that $\tau$ increases to a value closer to 2 for large $t_a$.

Experimentally, static light scattering has been studied for different colloids. In Fig. 1 we show results for gold colloids aggregated under DLCA and RLCA conditions, where measurements were performed at different aggregation times, $t_a$. The experimental results have been fitted to Eq. (2), where $S(qR_g)$ was taken from the computer-generated clusters and where the appropriate form for $N(M, t_a)$ was employed. The only unknown in each curve is then the moment $\bar{M}_2(t_a)$ at time $t_a$. The fitted values are given in the figure. At later stages in the aggregation the scattered light intensity is entirely dominated by clusters for which $qR_g > 1$ for experimentally available values of $q$, which results in the straight lines in the logarithmic plot. Their slopes determine the values of the fractal dimensions, which are found to be in very good agreement with those values of $d_f$ determined for the computer-generated clusters.

## Dynamic light scattering

Dynamic light scattering arises from the fluctuations of the scattered light caused by the diffusion of clusters, assuming the clusters to be rigid objects. When $qR_g < 1$, the fluctuations result from translational diffusion and for $qR_g > 1$ from both translational and rotational diffusion. Assuming single scat-

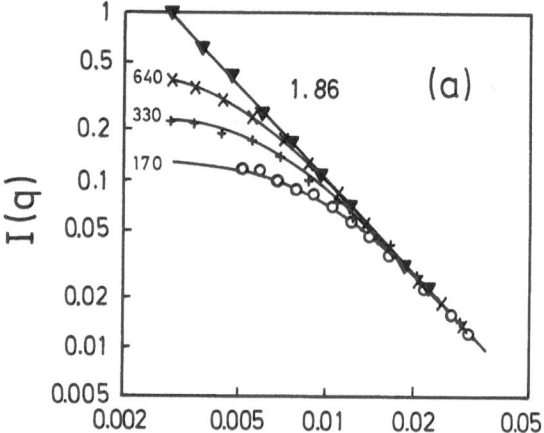

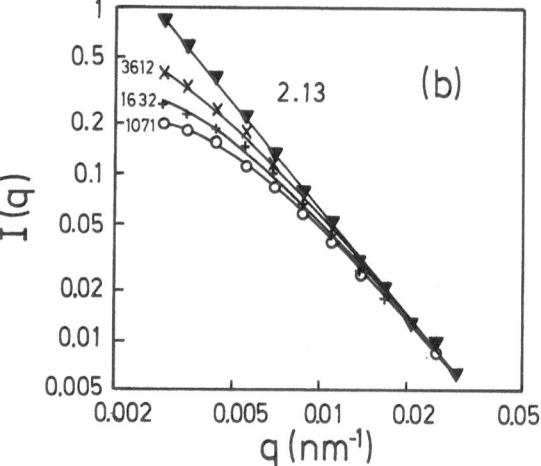

Fig. 1. The scattering intensity $I(q)$ obtained with colloidal gold aggregated by (a) DLCA and (b) RLCA. In each figure, different symbols represent data taken at different aggregation times $t_a$. The solid curves are the calculations described in the text. The values of $\bar{M}_2$ obtained in the fitting are labeled with each curve, except the upmost curves, which are linear, giving the fractal dimension $d_f = 1.86$ for DLCA and $d_f = 2.13$ for RLCA

tering from each colloidal particle, the scattered electric field is

$$E_s(q, t) = E_0 \sum_{\alpha=1}^{\mathcal{N}} \sum_{j=1}^{n_\alpha} p_j^\alpha \exp(i\vec{q} \cdot \vec{r}_j^\alpha(t)) \tag{12}$$

where $E_0$ is the incident field and $p_j^\alpha$ the scattering amplitude of particle $j$ in cluster $\alpha$, which has the position $\vec{r}_j^\alpha(t)$. Cluster $\alpha$ consists of $n_\alpha$ particles and $\mathcal{N}$ is the total number of clusters in the scattering volume. Decomposing $\vec{r}_j^\alpha(t) = \vec{R}_\alpha(t) + \vec{b}_j^\alpha(t)$ into the postion of the center of mass $\vec{R}_\alpha(t)$ of cluster $\alpha$ and the displacement $\vec{b}_j^\alpha$ of the particle from that center of mass, and assuming that translational and rotational diffusion processes can be decoupled, the normalized field autocorrelation function can be shown to be [18]

$$g_1(t) = \frac{1}{I(q)} \sum_M N(M) M^2 \\ \times e^{-q^2 D_M t} \sum_{l=0}^{\infty} S_l^M(qR_g) e^{-l(l+1)\theta_M t}. \quad (13)$$

Here, $D_M = \zeta/R_h$ is the translational diffusion coefficient and $\theta_M = 3\zeta/(4R_h^3)$ the rotational diffusion coefficient of a cluster of mass $M$, and $\zeta = k_B T/6\pi\eta$. The factors $S_l^M(qR_g)$ which appear in the second sum, are essentially the multipoles of the static structure factor, since $S_M(q) = S(qR_g) = \sum_l S_l^M(qR_g)$. They depend on the positions $\{\vec{b}_j^\alpha\}$ of particles in cluster $\alpha$ with respect to a cluster-fixed reference frame. These multipole terms reflect the anisotropies of the geometric structure of the clusters, expressed in Fourier space. Using computer-generated clusters, we have calculated $S_l^M(qR_g)$ for $0 \leqslant l \leqslant 7$ for clusters of 100 to 1000 particles [19]. For $qR_g \ll 1$, the $l = 0$ term dominates the scattering, implying that only translational diffusion contributes to the decay of $g_1(t)$. This result is expected, since at long wavelength the internal structure of the clusters is not resolved. For $l > 1$, the multipole terms rise as $(qR_g)^{2l}$, reach a main maximum when $qR_g \approx l$ and then decline. Therefore, rotational terms will contribute to the decay of $g_1(t)$ as $qR_g$ increases.

Using the $S_l^M(qR_g)$ from the computer-generated clusters and the cluster mass distribution Eq. (10) with $\bar{M}_1(t_a)$ determined from a fit to the static light scattering intensity, Eq. (13) was found [20] to provide an exellent representation of experimental data of the correlation function for gold colloids aggregated under DLCA conditions.

The multipole expansion terms, which appears as weighting factors of the rotational contributions in Eq. (13), reflect the anisotropies of the clusters. By scaling $S_l^M(qR_g)$ by $M^2$, it has been shown that $S_l^M(qR_g)/M^2$ is, for each value of $l$, independent of $M$. Thus, the anisotropies are found to be scale-invariant; they exist on all length scales [19].

The procedure used above is computationally intensive, making it difficult to apply to clusters larger than $M \approx 1000$ and $qR_g > 10$. To circumvent these practical limitations we approximate Eq. (13) by

$$g_1(t) = \frac{1}{I(q)} \sum_M N(M) M^2 S(qR_g) \exp(-q^2 D_{eff}^M t) \quad (14)$$

where

$$D_{eff}^M(q) = D_M f(qR_g) \\ = D_M \left[ 1 + \frac{3 \sum_l l(l+1) S_l^M(qR_g)}{4(\beta qR_g)^2 S(qR_g)} \right] \quad (15)$$

This form of $g_1(t)$ agrees with Eq. (13) up to the first cumulant. In Eq. (15) $\beta$ is the ratio of the hydrodynamic radius to the radius of gyration. At small angles, where $qR_g = qa\bar{M}^{1/d_f}$ is small, rotational diffusion plays almost no role, so that $D_{eff}^M(q) \approx D_M = \zeta/\beta R_g$. Using the results for $N(M)$ from static light scattering, the only unknown parameter in (14) is $\beta$. The analysis of experimental data, following this procedure, for different colloids aggregated under DLCA and RLCA conditions, has been performed [21], resulting in $\beta = 0.93$ for DLCA and $\beta = 1.0$ for RLCA. These values are in good agreement with numerical results for computer generated clusters [22, 23] and with theoretical predictions [16, 24, 25].

At larger values of $q\bar{R}_g$, the rotational contributions in $f(qR_g)$ become important, and the second term in the bracket in (15) contributes significantly.

The correlation function (14) is a sum of exponentials. Its deviation from a single exponential is due primarily to the polydispersity introduced by the cluster mass distribution, but the degree of this deviation is further influenced by the rotational contributions. Explicit calculations for DLCA [26] and RLCA [27] show that for $q\bar{R}_g \leqslant 1$ the difference between including rotations and neglecting rotations is small, whereas for $q\bar{R}_g \geqslant 2$ these differences become appreciable. These calculations, based on Eqs. (14) and (15), have been found to be in very good agreement with experiments for both small and large angle scattering.

Quasi-elastic light scattering can successfully be used to investigate the aggregation kinetics. This is difficult to achieve by static light scattering at later stages of the aggregation process, because of the limited $q$ range in the experiments. At late stages in the aggregation, scattering is dominated by large clusters with $qR_g \gg 1$, and $I(q) \sim (qR_g)^{-d_f}$ in the experimentally available $q$ range. Then, $\bar{M}$ and $\bar{R}_g$ can no longer be extracted from the data. Dynamic light scattering, however, offers the possibility to determine the hydrodynamic radius of the characteristic or average cluster, $\bar{R}_h$, at time $t_a$. It is important to realize that the two average radii are different moments of the cluster size distribution. A measurement of the first cumulant $\Gamma_1$ of the autocorrelation function yields an effective diffusion coefficient

$$\bar{D}_{eff}(q) = \frac{\Gamma_1}{q^2} = \frac{\sum_M N(M) M^2 S(qR_g) D_{eff}^M(q)}{\sum_M N(M) M^2 S(qR_g)}, \qquad (16)$$

from which an experimentally determined $\bar{R}_{eff} = \zeta / \bar{D}_{eff}$ follows. This quantity is, however, $q$ dependent. The true average hydrodynamic radius, whose dependence on aggregation time $t_a$ determines the growth kinetics, is defined by $\bar{R}_h = \bar{R}_{eff}(q = 0)$. However, since every quantity in Eq. (16) is known one can calculate $\bar{R}_h$ as a function of the measured $\bar{R}_{eff}$. These two quantities are identical only for a monodisperse sample of clusters without rotational diffusion effects, since the $q$ dependence of $\Gamma_1/q^2$ arises from both the cluster size distribution and rotational effects. Thus, calibration curves $\bar{R}_h = \bar{R}_h(\bar{R}_{eff}(q))$ are obtained. By determining $\bar{R}_{eff}(q)$ experimentally at different aggregation times $t_a$, and using the calibration curves, $\bar{R}_h(t_a)$ can be extracted. It is important to realize that the $t_a$-dependence of $\bar{R}_h$ will in general be different from the $t_a$-dependence of $\bar{R}_{eff}(q)$ for finite $q$. It is, however, $\bar{R}_h(t_a)$, from which the aggregation kinetics can be determined.

Power-law aggregation kinetics are characterized by the exponent $z$ in $\bar{M} \sim t_a^z$ at sufficiently long times. For the DLCA case it is convenient to consider the first reduced moment $\bar{M}_1(t_a)$. Then, $\bar{R}_g = a\bar{M}_1^{1/d_f} = a(M^{(1)}/M^{(0)})^{1/d_f} \sim t_a^{z/d_f}$. Dynamic light scattering and the use of the calibration curves leads to results for $\bar{R}_h(t_a)$. Therefore, to determine the exponent $z$, one has to relate $\bar{R}_h$ to $\bar{R}_g$. Since $\bar{R}_h = \zeta/\bar{D}_{eff}(q = 0)$, it follows from (16) that

$$\bar{R}_h = \beta a \frac{M^{(2)}}{M^{(2-1/d_f)}}. \qquad (17)$$

With the DLCA cluster mass distribution, Eq. (10), the required moments can be calculated with the result

$$\bar{R}_h = \frac{2\beta}{\Gamma(3 - 1/d_f)} \bar{R}_g \approx 1.44 \bar{R}_g$$

where $\Gamma(x)$ is the Gamma function. Therefore, $\bar{R}_h = 1.44 a(t_a/t_0)^{z/d_f}$ is predicted and the knowledge of $\bar{R}_h(t_a)$ from experiment can be used to determine the value of $z$. For three rather different colloids the values of $z$ thus obtained are all nearly unity [26], in agreement with the theoretical prediction, $z = 1$.

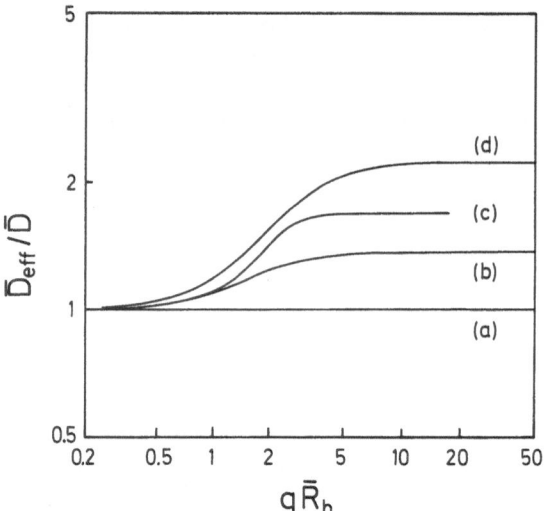

Fig. 2. Master curves for DLCA calculated for different conditions, illustrating the effects of polydispersity and of rotational diffusion

The above analysis determines the aggregation kinetics from the values of the first cumulants in the limit of $q \to 0$. There is, however, a considerable amount of additional information in the $q$ dependence of $\bar{D}_{eff}$, which is particularly relevant in the RLCA case, since it provides an experimental method to determine the value of the exponent $\tau$ in the cluster mass distribution, Eq. (11). The important observation is that $\bar{D}_{eff}(q)/\bar{D}$, where $\bar{D} = \bar{D}_{eff}(q = 0) = \zeta/\bar{R}_h$, is a function of $q\bar{R}_h$ only; this result reflects the fact that the cluster mass distribution is characterized by an average mass $\bar{M}$ and therefore by a single length scale $\bar{R}_h$. Therefore, all experimental results obtained from the first cumulant at any $q$ and at all $t_a$ should fall on the single master curve $\bar{D}_{eff}(q)/\bar{D}$ vs. $q\bar{R}_h$. To produce such master curve from experimental data for $\bar{D}_{eff}(q)$, one has to determine the value at $q = 0$, which is $\bar{D}$; furthermore, $q$ has to be scaled by $\bar{R}_h = \zeta/\bar{D}$.

It is instructive to discuss the shape of this master curve for several different conditions. In Fig. 2 we show results for the DLCA case, obtained from Eq. (16), where curve (a) applies for monodisperse spheres, so that $\bar{D}_{eff} = \bar{D}$, reflecting the fact that there is no $q$ dependence without polydispersity and rotational diffusion. Curve (b) takes into account the polydispersity as described by the DLCA cluster mass distribution, but still without rotations. As a result there is some $q$ dependence. Curve (c) is for a monodisperse cluster mass distribution, but for $D_{eff}^M(q)$ in Eq. (16) we use Eq. (15). The result is a somewhat greater $q$ dependence. Finally, in curve $(d)$ we take into account both $N(M)$ and rotations, which gives the strongest variation with $q\bar{R}_h$. This figure clearly shows the sensitivity to various features characterizing the aggregation.

For the RLCA case the shape of the master curve is more subtle due to the different form of the cluster mass distribution. This can be seen by calculating $\bar{D}_{eff}(q)$ analytically from (16) for a simplified case. With the assumptions $(i)\,D_{eff}^M = D_M$, which corresponds to the neglect of rotations, $(ii)\,N(M) = AM^{-\tau}$ for $M \leqslant M_c$ and $N(M) = 0$ for $M > M_c$, and $(iii)\,S(qR_g) = 1$ for $qR_g \leqslant 1$ and $S(qR_g) = (qR_g)^{-1}$ for $qR_g \geqslant 1$, Eq. (16) can be evaluated by converting the summations to integrals. The values of these integrals are, however, extremely sensitive to the precise values of $d_f$ and $\tau$, particularly when $d_f \approx 2$ and $\tau \approx 1.5$. The physical reason for this is the following: Although larger clusters scatter more strongly, this is almost exactly compensated by the larger number of smaller clusters resulting from the power-law cluster mass distribution, Eq. (11). Thus clusters of all masses contribute. The effective diffusion coefficient $\bar{D}_{eff}(q)$ under the above assumptions turns out to be $q$ independent at sufficiently small and at sufficiently large values of $q$; in the first case, all clusters are smaller than $q^{-1}$ and in the second case, $q > a^{-1}$, which is experimentally irrelevant. For $q$ values between these two limits, the $q$ dependence is complicated. Only if $M_c \to \infty$, the power law $\bar{D}_{eff} \sim q^\alpha$ with $\alpha = 1 - (2 - \tau)d_f$ is found for $\tau < 2$. This result would provide a simple means to determine $\tau$ experimentally. But a closer analysis shows that the simple power-law behavior cannot be observed if $\tau$ is either near to $2 - 1/d_f$, or near to 2, since the asymptotic behavior is only reached at unrealistically large values of $M_c$. Furthermore, since clusters with $qR_g \approx 1$ make a significant contribution to the scattering, rotational effects can not be neglected for real clusters.

To circumvent the assumptions made above, we have numerically evaluated Eq. (16), using Eq. (11) for $N(M)$, Eq. (7) for the structure factors, and the rotational effects as described by Eq. (15). In Fig. 3 we show the results, assuming $\tau = 1.5$, for three different values of the cutoff mass, $M_c = 10^4,\ 10^5,\ 10^6$, where $M_c = 2\bar{M}_2$. As expected, the initial $q$ independent part of $\bar{D}_{eff}(q)$, becomes smaller with increasing $M_c$. In the same figure, we have also performed the scaling of the three curves as described above.

The importance of such master curves is at least twofold. First, they extend the experimentally accessible $q$ range essentially to zero, thereby enabling the

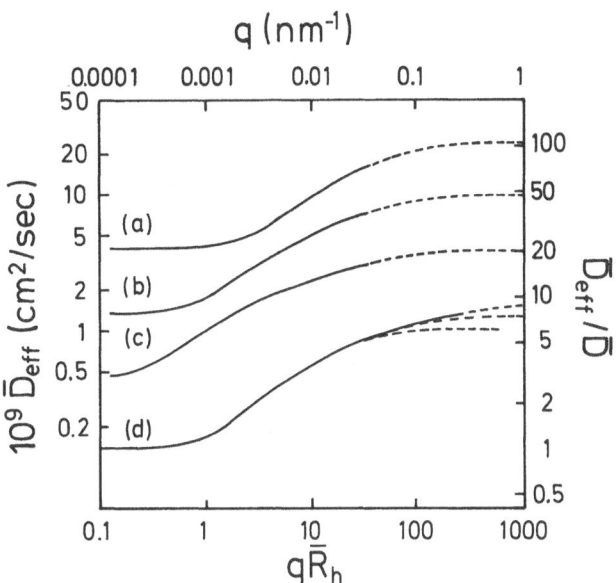

Fig. 3. Results for $\bar{D}_{eff}(q)$ for RLCA with $(a)\ M_c = 10^4$, $(b)\ M_c = 10^5$ and $(c)\ M_c = 10^6$. The scale for $\bar{D}_{eff}$ is on the left and the scale for $q$ is on the top. The experimentally accessible range of $q$ extends to about $0.03$ nm$^{-1}$, after which the calculated curves are represented by dashed lines. The three curves are scaled onto a master curve $(d)$ as $\bar{D}_{eff}/\bar{D}$ vs. $q\bar{R}_h$, whose scales are on the bottom and the right of the plot

determination of $\bar{R}_h(t_a)$, and hence the growth kinetics. Secondly, the shape of the master curve is very sensitive to various characteristics of the clusters. Some of these sensitive dependencies have been discussed already for the DLCA case in Fig. 2, where it became clear that the polydispersity described by $N(M)$ introduces some of the $q$ dependence of $\bar{D}_{eff}(q)/\bar{D}$. Similarly, for the RLCA case, the shape of the master curve is rather sensitive to the numerical value of the exponent $\tau$ in the cluster mass distribution. This is shown in Fig. 4 where the master curves are displayed for $\tau = 1.3,\ 1.5,$ and $1.7$ with and without rotations. This sensitivity of the master curve to the precise value of $\tau$ has been used to experimentally determine this exponent for three different colloids, and in all cases $\tau \approx 1.5$ has been found [21, 27] to accurately represent well the data obtained over a wide range of aggregation times and scattering angles.

## Conclusions

In this paper, theoretical methods are reviewed which are necessary for the quantitative analysis of colloidal aggregation processes studied by static and

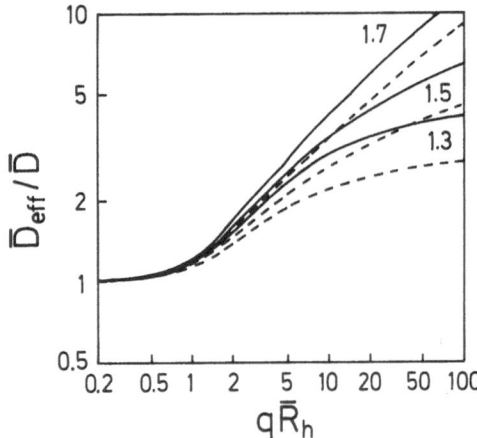

Fig. 4. Results for $\bar{D}_{eff}/\bar{D}$ for RLCA for several different values of $\tau$, each with (solid lines) and without (dashed lines) the effects of rotational diffusion

dynamic light scattering. It is shown that the combination of results of both types of scattering experiments allows the determination of several important features of the aggregation process, including the fractal dimension of the aggregates, their structure including their anisotropies, the cluster mass distribution, and the aggregation kinetics. Using detailed knowledge about the structure of the aggregates, which is obtained from computer simulations for the two limiting cases of DLCA and RLCA, it is demonstrated how sensitive (or insentive) the experimentally accessible properties are with respect to the characteristics of single clusters, of their mass distribution and of their growth. These methods lead to a self-consistent quantitative description of the DLCA and RLCA growth processes; they have been used [28] to analyse scattering experiments on various different colloids aggregated under both limiting conditions. This has provided strong evidence for the universality of both limiting regimes of colloid aggregation.

## References

1. Meakin P (1988) In: Lebowitz JL (ed) Phase Transitions, vol 12. Academic, New York
2. Vicsek T (1989) Fractal Growth Phenomena. World Scientific, Singapore
3. Weitz DA, Oliveria M (1984) Phys Rev Letters 52:1433
4. Freltoft T, Kjems JK, Sinha SK (1986) Phys Rev B 33:269
5. Dietler G, Aubert C, Cannell DS, Wiltzius P (1986) Phys Rev Letters 57:3117
6. Martin JE, Ackerson BJ (1985) Phys Rev A 31:1180
7. Hurd AJ, Flower WL (1988) J Colloid Inter Sci 122:178
8. Lin MY, Klein R, Lindsay HM, Weitz DA, Ball RC, Meakin P, to be published
9. Mountain RG, Mulholland GW (1988) Langmuir 4:1321
10. Meakin P, Vicsek T, Family F (1985) Phys Rev B 31:564
11. Brown WD, Ball RC (1985) J Phys A 18:L517
12. Meakin P, Family F (1987) Phys Rev A 36:5498
13. Van Dongen PGJ, Ernst MH (1985) Phys Rev Letters 54:1396
14. Ball RC, Weitz DA, Witten TA, Leyvraz (1985) Phys Rev Letters 58:274
15. Verwey EJW, Overbeek JTG (1948) Theory of the Stability of Lyophobic Colloids. Elsevier, Amsterdam
16. Hess W, Frisch HL, Klein R (1986) Z Phys B 64:65
17. Cohen RJ, Benedek GB (1982) J Phys Chem 86:3696
18. Berne BJ, Pecora R (1976) Dynamic Light Scattering. Wiley-Interscience, New York
19. Lindsay HM, Klein R, Weitz DA, Lin MY, Meakin P (1989) Phys Rev A 39:3112
20. Lindsay HM, Klein R, Weitz DA, Lin MY, Meakin P (1988) Phys Rev A 38:2614
21. Lin MY, Lindsay HM, Weitz DA, Ball RC, Klein R, Meakin P (1989) Proc R Soc Lond A 423:71
22. Chen ZY, Deutch JM, Meakin P (1984) J Chem Phys 80:2982
23. Chen ZY, Meakin P, Deutch JM (1987) Phys Rev Letters 59:2121
24. Wiltzius P (1987) Phys Rev Letters 58:710
25. Pusey PN, Rarity JG, Klein R, Weitz DA (1987) Phys Rev Letters 59:2122
26. Lin MY, Lindsay HM, Weitz DA, Klein R, Ball RC, Meakin P, to be published
27. Lin MY, Lindsay HM, Weitz DA, Ball RC, Klein R, Meakin P, to be published
28. Lin MY, Lindsay HM, Weitz DA, Ball RC, Klein R, Meakin P (1989) Nature 339:360

Authors' address:

Prof. Dr. R. Klein
Fakultät für Physik
Universität Konstanz
Postfach 5560
D-7750 Konstanz

**Progress in Colloid & Polymer Science**                    Progr Colloid Polym Sci 81:169 – 173 (1990)

# Dynamic scaling in colloid aggregation reaction limited process induced by electrolytes and polyelectrolytes

E. Pefferkorn[1]), J. Widmaier[1]), C. Graillat[2]) and R. Varoqui[1])

[1]) Institut Charles SADRON, Strasbourg, France
[2]) Laboratoire des Matérieux Organiques, Lyon, France

*Abstract:* We have studied the irreversible aggregation of colloidal polystyrene spheres using a particle counter technique. The mode of cluster growth is found to be modelled by a reaction limited aggregation process. Elementary grains disappear by collision with larger aggregates, which, in return, are only slowly formed. This complex process has one important characteristic: at each time, the concentration of small aggregates of $g$ elementary grains scales like $g^{-\tau}$, $\tau$ being a universal exponent in reaction limited processes.

*Key words:* Polystyrene latex; water soluble polymer; reaction limited aggregation; cluster size distribution; scaling exponents

## Introduction

Colloid aggregation is widely used in many industrial processes and polymers are gaining success in waste water treatment and mineral recovery [1]. Much however remains to be learned concerning the ultimate mechanism of flocculation processes. In fact, colloid instability may be induced by hard or soft surface modifications, depending on the nature and the concentration of the destabilizing agent. For polymers, their role as destabilizing agents is not properly known. On adsorption on colloid surfaces, they may operate by an effect of surface charge annihilation and/ or by inter-particle bridging [2 – 3]. In fact, neutralization of surface charges could be a major factor in the destabilization of colloids having opposite charge to the polymer, but the relative importance of the two effects needs to be established. In previous work, we studied the influence of excess electrolyte and optimum polymer concentration; in all cases charge screening or annihilation was of great importance. This demonstrates that electrical conditions play an essential role [4].

When a charged colloid is suspended in dilute electrolyte solution, the surface charge is not totally screened and the repulsive effect of residual electrostatic charges between colloid particles contribution to slow down the coagulation rate. In order to trigger polymer bridging, we chose a pH domain where the ionic surface charge of the latex is close to zero and the polymer relatively ionized. This scheme is novel and we hypothesize that the aggregation mechanism may show features of a chemically limited aggregation process, in the absence of hard interactions between polymer and colloid.

As in previous work, we attempted to compare electrolyte and polyelectrolyte induced aggregation [4 – 5]. The colloid characteristic of interest in our study was the temporal evolution of the cluster size distribution. The results are discussed in the light of characteristics of a reaction limited aggregation process (RLA) [6].

## Materials and methods

### Latex particles

Monosized latex particles were kindly provided by the Laboratoire des Matériaux Organiques of Lyon. They were polymerized under emulsifier-free conditions, using potassium persulfate as free radical initiator [7]. Two samples were used. Latex (a) had $SO_4^-$, $COO^-$ and OH surface groups. The point of charge (pzc) was found to be 3, so that at pH 3.5 the latex bore a net negative charge. Latex (b), obtained by hydrolysis of latex (a) at 90°C for 7 days, bore only carboxylic acid surface groups. The pzc was found to be 2.5. The density of the latex was 1.045 g/ml and the diameter of the elementary grain $D_1$ was 0.9 μm.

*Polymer-latex complex*

Poly(vinyl-4-pyridine) (PVP) was synthesized in ethylene glycol using azodi-iso butyronitrile as initiator and a fractionated sample of molecular weight $3.6 \ 10^5$ was used. The degree of protonation of PVP in aqueous solution was calculated on the basis of spectroscopic measurements at 253 nm and at pH 2.5, 73% of the pyridine groups are in the protonated form.

Taking into account the strong protonation of the PVP and the very small ionization of the latex at pH 2.5 and considering only the stoichiometry of charge-charge interactions, we note that the polymer segments are unable to interact with the latex surface near the pzc. However, we believe that on contact between polymer and latex, mutual interactions lead to increased surface ionization as observed, in the inverse direction, for the complex sulfonate-pyridine [8]. Our hypothesis is that the reaction between the polymer pyridine group and the latex carboxylic acid group limits the kinetics of the aggregation process.

*Flocculation experiments*

Aggregation. All experiments were performed at a constant concentration of 0.16 g latex/liter at 18 °C. The latex suspension was added to the electrolyte or polyelectrolyte solution and gentle tumbling was performed twice before perikinetic flocculation started. In order to eliminate obvious particle sedimentation during the slow electrolyte induced aggregation, these particular experiments were performed in a mixture of $(D_2O/H_2O)$ (57 g/47 g) having the same density as the latex.

Particle counting. The principle of the Coulter Counter is given in Refs. [1] and [9]. The aperture we used had a diameter of 50 μm and allowed determination of concentration of elementary grains in the second channel even in the presence of aggregates of larger sizes. Aggregates of two and more particles appear in the third and following channels. Using the fifteen channels available, one may count colloids having up to several thousand elementary particles. The relation (1) between the apparent cluster diameter $D_g$ and the number $g$ of elementary constituent particles of diameter $D_1$ holds:

$$g = (D_g/D_1)^3 .\qquad (1)$$

*Aggregate size analysis*

The crude histogram is then analyzed as indicated in [10]. One obtains the cluster size distribution $c_g(t)$, i.e., the concentration at time $t$ of aggregates constituted of $g$ elementary particles. To determine averaged characteristics of the aggregates, we calculated the different moments of the colloid size distribution [relation (2)], knowing that (i) the zeroth moment $N_0$ corresponds to the inverse of the number average size, (ii) the first moment $N_1$ is the initial particle concentration (corresponding to 1 g latex), and (iii) the second moment $N_2$ furnishes the weight average size. The colloid size polydispersity was obtained at each time by relation (3):

$$N_i(t) = \sum_g g^i c_g(t)\qquad (2)$$

$$Z_W/Z_N = N_0(t)N_2(t)/N_1^2 \sim t^p .\qquad (3)$$

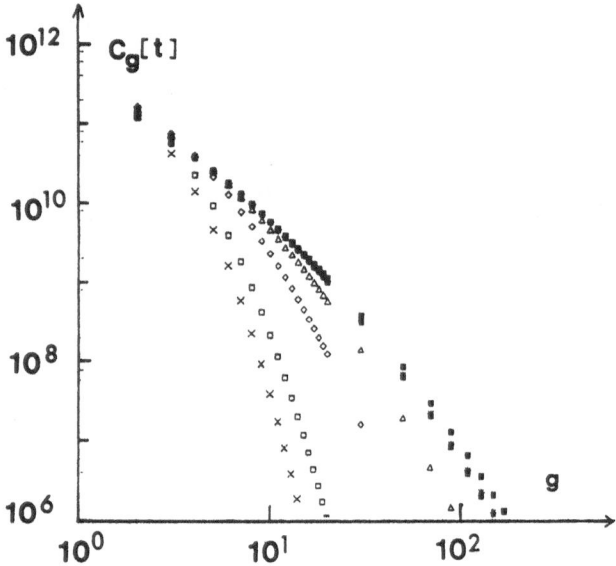

Fig. 1. Size distribution curve. Representation of the aggregate concentration $c_g(t)$ as a function of the number $g$ of constitutive particles for different time intervals $t_{min}$: ( × ), 350; (□), 990; (◇), 1225, (△), 2435; (⊞), 6635; (∗), 7840; (⊠), 11720: Latex $L(a)$/pH = 3.5/NaCl: 0.03 M, poor aggregation conditions

Relation (3) also defines the polydispersity factor $p$ which is used in the discussion.

**Results and discussion**

In Fig. 1 is represented the cluster size distribution for electrolyte induced aggregation of latex (a) suspended in 0.03 M NaCl solution at pH 3.5. The different curves correspond to increased flocculation periods. It is shown that (i) small aggregates are present at a maximum concentration during flocculation, and (ii) the concentration of the largest aggregates only slowly increases. Similar curves are obtained for polyelectrolyte induced aggregation of latex (b) at pH 2.5.

In Figs. 2 and 3 are represented on a log-log scale the variations with time $t$ of the zeroth and the second moments of the colloid size distribution for electrolyte induced aggregation of latex (a) at pH 3.5. After an initial time lapse, one observes the asymptotic scaling regime where $\log N_0$ and $\log N_2$ are linear functions of $\log t$. The different curves correspond to different electrolyte concentrations.

In Figs. 4 and 5 are represented on a log-log scale the variations with $t$ of $N_0$ and $N_2$ for polyelectrolyte induced aggregation of latex (b) at pH 2.5. As in the precedent case, linear variations with concentration of

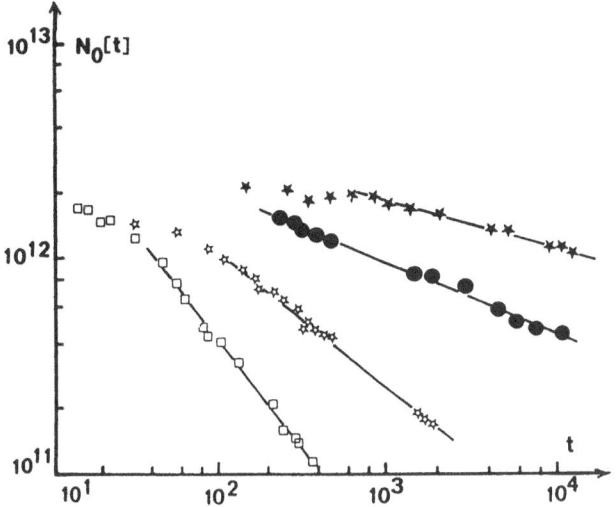

Fig. 2. Moment of order 0 of the cluster size distribution $N_0(t) = \sum_g c_g(t)$ as a function of the aggregation time $t$ (min), (log-log scale): Latex $L(a)/\mathrm{pH} = 3.5/\mathrm{NaCl}$: ($\bigstar$) 0.03 M; ($\bullet$) 0.05 M; ($\star$) 0.08 M; ($\square$) 0.15 M

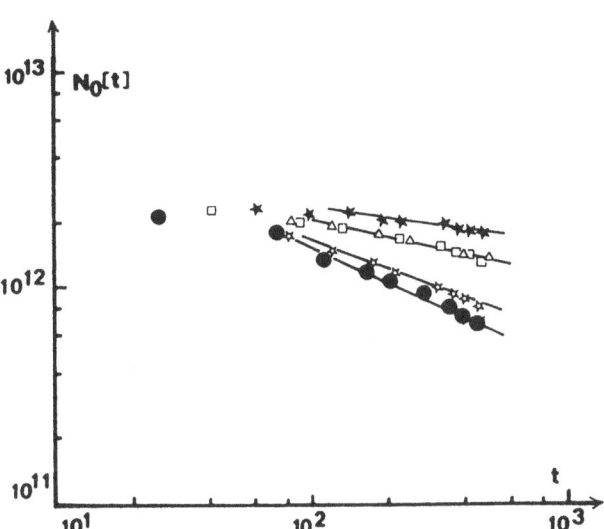

Fig. 4. Moment of order 0 of the cluster size distribution $N_0(t) = \sum_g c_g(t)$ as a function of the aggregation time $t$ (min), (log-log scale): Latex $L(b)/\mathrm{pH} = 2.5/\mathrm{P4VP}$: $C_p$ (g pol./$10^4$ g latex) = 0.7 ($\triangle$); 4.9 ($\star$); 7 ($\bullet$); 12 ($\square$); 14 ($\bigstar$)

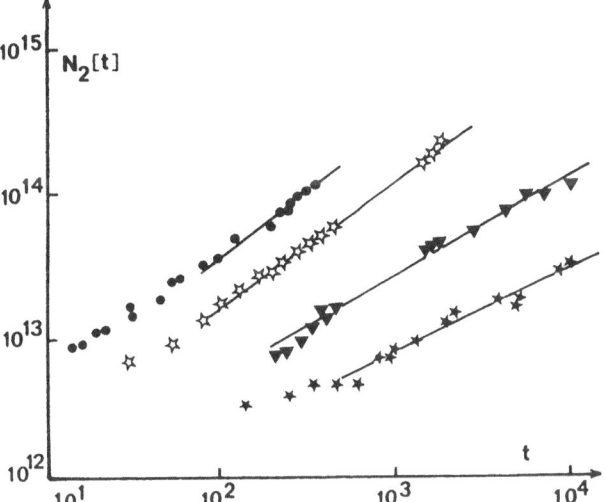

Fig. 3. Moment of order 2 of the cluster size distribution $N_2(t) = \sum_g g^2 c_g(t)$ as a function of the aggregation time $t$ (min), (log-log scale): Latex $L(a)/\mathrm{pH} = 3.5/\mathrm{NaCl}$: ($\bigstar$) 0.03 M; ($\blacktriangle$) 0.05 M; ($\star$) 0.08 M; ($\bullet$) 0.15 M

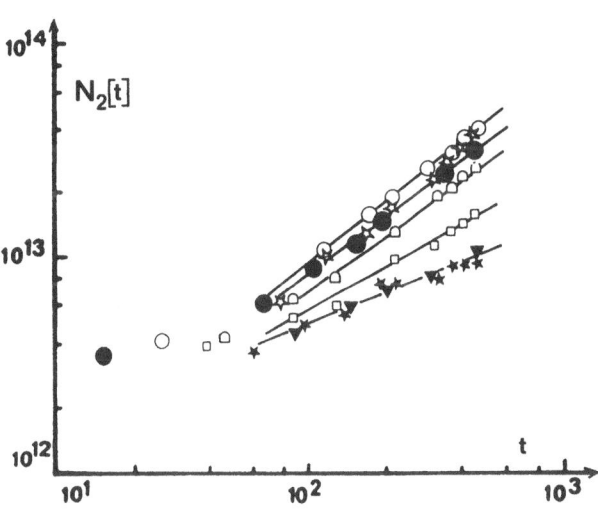

Fig. 5. Moment of order 2 of the cluster size distribution $N_2(t) = \sum_g g^2 c_g(t)$ as a function of the aggregation time $t$ (min), (log-log scale): Latex $L(b)/\mathrm{pH} = 2.5/\mathrm{P4VP}$: $C_p$ (g pol./$10^4$ g latex) = 0.7 ($\blacktriangledown$); 2.1 ($\square$); 4.9 ($\bullet$); 7 ($\bigcirc$); 9 ($\star$); 12 ($\square$); 14 ($\bigstar$)

the moments of the size distribution are determined in the asymptotic regime.

Our analysis is based on results established by computer simulation studies of aggregation processes. Our hypothesis, that slow aggregation has features of reaction limited aggregation (RLA) processes, is based on the continuously decreasing size distributions, shown in Fig. 1, which essentially are characteristic of a RLA process. From these studies, we take the classical relation for the reduced size distribution [6, 11 – 14]:

Table 1. Scaling exponents for electrolyte induced aggregation. Sulfate latex/pH 3.5/NaCl

| NaCl/ Molality | $z$ | $w$ | $p^{a)}$ | $\tau^{b)}$ | aggr. cond. |
|---|---|---|---|---|---|
| 0.03 | 0.655 | 0.226 | 0.429 | 1.65 | poor |
| 0.05 | 0.745 | 0.326 | 0.419 | 1.56 | poor |
| 0.08 | 0.909 | 0.655 | 0.254 | 1.28 | near opt. |
| 0.15 | 0.986 | 0.970 | 0.016 | – | opt. |

a) $t^p \sim N_2(t) N_0(t)$; b) $\tau = 2 - (w/z)$

Table 2. Scaling exponents for polyelectrolyte induced aggregation. Carboxylic latex/pH 2.5/P4VP

| Polymer conc. $C_p{}^{a)}$ | $z$ | $w$ | $p^{b)}$ | $\tau^{c)}$ | aggr. cond. |
|---|---|---|---|---|---|
| 0.7 | 0.551 | 0.269 | 0.282 | 1.51 | poor |
| 2.1 | 0.854 | 0.460 | 0.394 | 1.46 | poor |
| 4.9 | 0.924 | 0.443 | 0.481 | 1.52 | poor |
| 7.0 | 0.981 | 0.557 | 0.424 | 1.43 | poor |
| 9.0 | 1.007 | 0.460 | 0.547 | 1.54 | poor |
| 12.0 | 0.720 | 0.286 | 0.434 | 1.60 | poor |
| 14.0 | 0.443 | 0.139 | 0.304 | 1.69 | |

a) $C_p$ is expressed in g pol. P4VP/$10^4$ g latex;
b) $t^p \sim N_2(t) N_0(t)$; c) $\tau = 2 - (w/z)$

$$c_g(t) g^\tau t^w = f(g/t^z) \qquad (4)$$

where the exponents $z$ and $w$ are determined from the scaling laws for the variation of the second and zeroth moments of the size distribution respectively:

$$N_0(t) \sim t^{-w} \qquad (5)$$

and

$$N_2(t) \sim t^z \qquad (6)$$

with

$$\tau = 2 - (w/z). \qquad (7)$$

In Table 1 we present $w$, $z$ and $\tau$ values calculated for electrolyte induced aggregation. It is shown that $z$ and $w$ decrease with the electrolyte concentration. Aggregation of latex suspended in 0.15 M NaCl aqueous solution corresponds to diffusion limited aggregation, while in the lower concentrations 0.05 M and 0.03 M NaCl the process is characterized by a $\tau$ value of the order of 1.6.

In Table 2 we report $w$, $z$ and $\tau$ values determined from our experiments with carboxylic acid latices in

the presence of polyelectrolytes. For a given concentration of about $9.0 \ 10^{-4}$ g polymer/g latex, $w$ and $z$ pass through a maximum. For aggregation induced by polymer/particle bridging a unique $\tau$ value of the order of 1.55 is calculated (in agreement with the value derived from computer simulation under realistic conditions [15]).

A second result is relative to the temporal variation of the cluster size polydispersity under the poor aggregation conditions reported in Tables 1 and 2. In comparison to the value of $p$ close to zero which is found under optimum aggregation conditions, $p$ depends on the electrolyte or polyelectrolyte concentration. For $p$ equal zero a constant value of 2 is calculated for $Z_W/Z_N$ under optimum aggregation conditions [16]. Otherwise the size polydispersity depends on electrolyte or polyelectrolyte concentration. For electrolyte induced aggregation, the polydispersity factor $p$ increases monotonically with the dilution. For polyelectrolyte induced aggregation, $p$ becomes maximum at a concentration where the flocculation kinetics is relatively fast and leads to maximum size polydispersity ($p \sim 0.5$).

We now return to the observation of a unique $\tau$ value characterizing all aggregation processes performed under poor aggregation conditions. We use the relation (4) for the reduced size distribution and the scaling laws (5) and (6) to analyse these flocculation processes. The curve which merges all experimental results relative to the aggregate size distribution and its temporal evolution is represented in Fig. 6: all colloid size distributions corresponding to the asymptotic regime fit on this curve and one determines that, at each time, the concentration of aggregates of relatively small $g$, such as $g N_1/N_2(t) < 1$, is described by the following scaling law:

$$c_g(t) \sim g^{-\tau}, \qquad (8)$$

$\tau$ being of the order of 1.55.

This result was also found by computer simulation in the case of reaction limited aggregation with a collision efficiency smaller than 1 [17]. The implication for this work is that collisions between individual particles and small colloids do not systematically give rise to particle sticking, whereas each collision between large aggregates, despite their slow rate of formation, leads to cluster sticking. The real reason for the mechanism is not clearly established, but we believe that the electrical characteristics of charged colloids may be size dependent and hence may modify the aggregate reactivity [18].

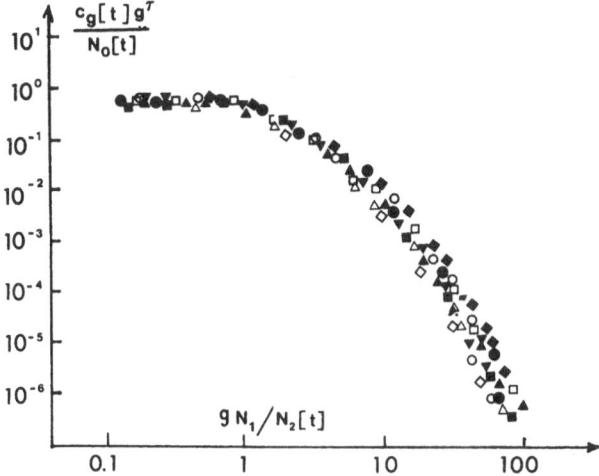

Fig. 6. Reduced size distribution curve corresponding to poor aggregation conditions in the asymptotic regime: for NaCl induced aggregation; (●) 0.03 M; (■) 0.05 M; for PVP induced aggregation, $C_p$ (g pol./$10^4$ g latex): (▼) 0.7; (△) 2.1; (▲) 4.9; (○) 7.0; (◇) 9.0; (□) 12.0; (◆) 14.0

## Conclusions

The presence of polymer adsorbed on a colloid surface modulates the collision efficiency for colloid sticking, but the aggregation mode itself does not depend on the degree of colloid surface coverage. We believe that the development of a reaction limited process, induced by soft colloid-polymer interactions, is observed if the aggregation mechanism results from interparticle polymer bridging. It is essentially different from the process resulting from colloid surface charge annihilation via polymer adsorption (hard colloid-polymer interactions) which induces fast aggregation for an optimum polymer concentration, the latter being similar to aggregation in the presence of excess electrolyte.

The other difference between fast and slow colloid aggregation is relative to the cluster size polydispersity. As small and large clusters are present at the same time under poor aggregation conditions, the process

leads to time and polymer concentration dependent size polydispersity shapes.

*Acknowledgements*

This work was performed under the auspices and with the financial support of the PIRSEM of the CNRS in the theme ARC "Flocculation".

## References

1. Ives KJ (1978) The Scientific Bases of Flocculation (Sijthoff and Noordhoff) Alphen aan den Rijn — The Netherlands
2. Napper DH (1983) Polymeric Stabilization of Colloid Dispersions. Colloid Science, Academic Press, New York
3. Vincent B (1982) Adv Colloid Interface Sci 4:193
4. Varoqui R, Pefferkorn E (1989) Progr Colloid Polym Sci 79:194
5. Varoqui R, Pefferkorn E (1988) Macromolecules 21:3096
6. Family F, Meakin P, Vicsek T (1985) J Chem Phys 83:4144
7. Goodwin JW, Ottewill RH, Pelton R, Vianella G, Yates DE (1978) Br Polym J 10:173
8. Varoqui R, Tran Q, Pefferkorn E (1979) Macromolecules 12:831
9. Walker PH, Hutka J (1971) Division of Soils, Technical paper $N^0$ 1, 3
10. Pefferkorn E, Pichot C, Varoqui R (1988) J Phys France 49:983
11. Meakin P, Vicsek T, Family F (1985) Phys Rev B 31:364
12. Jullien R, Botet R (1987) Aggregation and Fractal Aggregates World Scientific, Singapore
13. Kinetics of Aggregation and Gelation (1984) Family F, Landau DP (eds) North-Holland
14. On Growth and Form (1986) Stanley HE, Ostrowski N (eds) Martinus Nijhoff Dordrecht
15. Ball RC, Weitz DA, Witten TA, Leyvraz F (1987) Phys Rev Lett. 58:274
16. Ziff R in Ref. [13] p 191
17. Brown WD, Ball RC (1985) J Phys A 18:L517
18. Brochard-Wyart F, De Gennes PG (1988) CR Acad Sci II 307:1497

Authors' address:

E. Pefferkorn
Institut Charles SADRON
6. Rue Boussingault
F-67083 Strasbourg Cedex
France

**Progress in Colloid & Polymer Science**

Progr Colloid Polym Sci 81:174 (1990)

# Scattering properties of lyophylic colloidal silica particles in non-polar solvents. Effects of attraction forces

P. W. Rouw, C. G. de Kruif and A. Vrij

Van't Hoff Laboratorium, Universiteit Utrecht, Utrecht, The Netherlands

Specially prepared, small, lyophylized, silica spheres suspended in non-polar organic solvents are good model systems for the study of (concentrated) colloidal dispersions.

In the solvent cyclohexane the scattering properties can be described very well with a model in which the particles behave as a collection of hard spheres. This was found by us in many experiments of static-(SLS) [1] and dynamic light scattering (DLS) [2, 3] and also in experiments with small-angle-x-ray (SAXS) [4], neutron scattering (SANS) [5] and rheological properties [6]. In other, "poorer" solvents like benzene and toluene or dodecane, attractive forces become prominent when the temperature is chosen below a certain threshold [7–10].

In this presentation the effects of attractive forces between the particles on several scattering properties will be described. At small concentrations of silica dynamic light scattering is a sensitive method [8] to obtain accurate values of the diffusion coefficients as a function of temperature, from which the osmotic second virial coefficient can be deduced. An "adhesive hard sphere" with temperature dependent stickiness is introduced as a model for the pair potential. This potential is able to describe the temperature dependence of the diffusional and osmotic second virial coefficients. At higher concentrations, the light scattering of attracting particles is seriously troubled by effects of multiple scattering and this method therefore becomes unattractive. It is here that the small angle scattering methods have become powerful alternatives. Especially SANS is attractive because of its extra feature of contrast variation. SANS experiments of silica dispersions at 4 concentrations up to 30 volume percent and at 8 temperatures were performed [10]. For the interpretation of the measurements the same adhesive

hard sphere model was chosen for the pair potential. The structure factor was calculated with Baxter's theory [11] of the structure factor for this potential. This gives reasonable results although also an inconsistency in this interpretation scheme became apparent. The calculated stickiness parameter was found to be concentration dependent.

## References

1. van Helden AK, Vrij A (1980) J Colloid and Interface Sci 78:312
2. Kops-Werkhoven MM, Fijnaut HM (1981) J Chem Phys 74:1618
3. Vrij A, Jansen JW, Dhont JKG, Pathmamanoharan C, Kops-Werkhoven MM, Fijnaut HM (1983) Faraday Discuss Chem Soc 76:19
4. Moonen J, Vrij A (1988) Colloid and Polymer Sci 266:1140
5. de Kruif CG, Briels W, May RP, Vrij A (1988) Langmuir 4:668
6. de Kruif CG, van Iersel EMF, Vrij A, Russel WB (1985) J Chem Phys 83:4717
7. Jansen JW, de Kruif CG, Vrij A (1986) J Colloid Interface Sci 114:481, 492
8. Rouw PW, de Kruif CG (1988) J Chem Phys 88:7799
9. Rouw PW, Vrij A, de Kruif CG (1988) Progr Colloid Polymer Sci 76:1
10. de Kruif CG, Rouw PW, Briels WJ, Duits MHG, Vrij A, May RP (1989) Langmuir 5:422
11. Baxter RJ (1968) J Chem Phys 49:2770

Authors' address:

Prof. Dr. A. Vrij
Van't Hoff Laboratory
University of Utrecht
Padualaan 8
3584 CH Utrecht
The Netherlands

**Progress in Colloid & Polymer Science**          Progr Colloid Polym Sci 81:175−183 (1990)

# Ordering of latex particles during film formation*

M. Joanicot[1]), K. Wong[1]), J. Maquet[1]), Y. Chevalier[2]), C. Pichot[2]), C. Graillat[2]), P. Lindner[3]), L. Rios[4]) and B. Cabane[5])

[1]) Rhône Poulenc, Aubervilliers, France
[2]) CNRS-LMO, Vernaison, France
[3]) Institut Laue Langevin, Grenoble, France
[4]) Facultad de Quimica, UNAM, Mexico D.F.
[5]) CNRS-CEA, DLPC-SCM-URA331, CEN Saclay, Gif sur Yvette, France

*Abstract:* Small angle neutron scattering and electron microscopy are used to investigate the ordering of soft latex particles which have been dispersed in water and then brought into contact through evaporation of the solvent. We observe a succession of metastable states where the particles are kept separated by their hydrophilic surface layers: colloidal liquid (short range order only) to colloidal crystal (long range order) to foam (dense packing of polyhedral cells with latex interior and hydrophilic walls). At each stage the structure may collapse into a homogeneous latex phase through coalescence of the particles and expulsion of the hydrophilic material.

*Key words:* Film formation; neutron scattering; electron microscopy; ordering of particles; coalescence of particles

## Introduction

Films can be made by spreading a water based dispersion of soft particles onto a substrate, and then evaporating water until the particles come into contact and fuse together [1, 2]. This process may be viewed as a succession of 5 steps: (a) *ordering* of the particles as the dispersion is concentrated through evaporation; (b) *sticking* of the particles to each other; (c) *deformation* to fill all voids left by removal of water; (d) *coalescence,* i.e. fusion of the particles with expulsion of the surface active species which were used to stabilize the dispersion; (e) *interpenetration* of the cores of the particles, which is the process previously studied by Hahn et al. [3, 4].

Obviously the structure of the film will depend on how the particles are separated from each other, first in the dispersion (stabilization), and then in the film (contact). For instance, with hard particles (e.g. metal

oxides), the structure of the film is controlled by the quality of the dispersion [5]: when the particles repel each other strongly, they tend to order into a regular array which yields a dense, ordered film; when the particles do not repel they form fractal aggregates which yield a porous film.

In the opposite case of *particles which are soft and sticky,* the nature of the surface layers which stabilize the particles is equally important. We leave aside the trivial case of unstabilized particles which flocculate and coalescence into large lumps before the film is made. Instead we consider well stabilized particles with soft, hydrophobic, cores and hydrophilic surfaces, dispersed in water. In such dispersions coalescence may not occur until the particles are forced into direct contact by evaporation or drainage of water. Then, if coalescence does occur, it will proceed in a very concentrated medium where each particle is surrounded by 10 to 12 neighbors. Hence coalescence will be a collective phenomenon, unlike the type of coalescence that is observed in more dilute systems such as common emulsions.

---

*) This work used the neutron beams of ILL in Grenoble and LLB in Saclay.

## Materials and methods

### Latex particles

Our films are made from aqueous dispersions of polymer latex particles, which are widely used for coating paper, in paints, and in adhesives. The particles are spherical droplets with diameters in the 50–250 nm range; they are made of a hydrophobic polymer core surrounded by a hydrophilic layer. As mentioned in the introduction, we choose particles with a soft core: it is made of a statistical copolymer of styrene and butyl acrylate (PS-PBA), with equal amounts of styrene and butyl acrylate. The glass transition temperature of this copolymer is 15°C; hence at room temperature the core is indeed a viscous polymeric liquid.

The core of each latex particle is surrounded by a thin hydrophilic layer which generates repulsions between dispersed particles, thereby stabilizing the dispersion. For some of our particles this hydrophilic layer is made of zwitterionic surfactant molecules adsorbed on the core; the amount of surfactant is below a monolayer coverage of the core, and this generates weak repulsions between the particles. For the other dispersions this hydrophilic layer is made of acrylic acid (AA) sequences which are copolymerized with the core polymers; here the amount of AA groups at the surface amounts to a monolayer coverage of the particle. These AA groups have been neutralized at pH = 6.6, hence they are charged and generate strong repulsions between the particles.

The latex particles are made through a three-step emulsion polymerization designed to produce particles of homogeneous compositions and sizes. A typical receipe is as follows. Water 112.5 g; deuterated styrene 12.4 g; butyl acrylate 13.5 g; surfactant = N-N dimethyldodecylammonio propane sulfonate 0.26 g; buffer = $NaHCO_3$ 0.16 g; initiator = $S_2O_8K_2$ 0.16 g; temperature 70°C. A part of the monomer charge (20%) was initially batch polymerized to provide a primary seed latex (diameter = 55 nm), which was further diluted, swelled with 50% of the monomer mixture and then polymerized upon adding initiator and buffer. A third step allows one to polymerize the remainder of the monomer (30%), completing the conversion.

At the end of the reaction the particles are dispersed in an aqueous solution which also contains buffering salt and excess surfactant. These solutes screen the interaction between the particles, which results in poor ordering of the particles, especially so for the surfactant covered latex where the contact potential is low. These solutes can be removed through ion exchange with a mixed bed resin. In addition the solvent can be changed from $H_2O$ to $D_2O$ by exchange in an ultrafiltration cell.

### Film formation

Films are made by spreading a thin layer (0.1 to 0.3 mm) of a concentrated dispersion on a clean substrate, e.g. glass or quartz. In open air at 20°C such films take 1 to 2 hours to dry; the rate of water loss is controlled by the film thickness.

With dispersions of well stabilized particles there is an early stage where the dispersion just becomes more viscous as it is concentrated, then it crosses a gel point near $\phi$ = 0.5, and finally it dries uniformly by collapsing on the substrate. With poorly stabilized dispersions, islands of aggre-

gated particles can be seen floating on the surface of the liquid film, and the gel point is controlled by these aggregates.

The late stages of drying show significant differences according to the nature of the latex particles. The PAA covered latex gives films which remain uniform in concentration and transparent throughout the drying process; when fully dry the films show a number of cracks which indicate a crystalline structure. The surfactant covered latex gives films which show significant turbidity in the late stages, and also macroscopic separation into dry regions surrounding wet areas; during the course of drying these wet areas shrink in size and increase in thickness; this suggests that water is expelled from the drying film into the wet areas. At the end the film is smooth and has a slight residual turbidity, indicating an amorphous structure.

### Contrasts for small angle neutron scattering

We want to study the transition from a dispersion, where the latex particles are dispersed in water, to a film, where water and hydrophilic species are dispersed in a polymeric matrix. This requires the use of techniques which provide a good contrast between hydrophilic and hydrophobic species. In neutron scattering this contrast is provided by isotopic labeling: we use either protonated latex in $D_2O$, or deuterated latex in $H_2O$. The latter method is preferable for the following reason. In the late stages of drying, when the water content of the latex films is small, there is a two-way exchange of water between the film and ambient humidity; consequently the $D_2O$ in the film is progressively replaced by $H_2O$ from the atmosphere, and contrast is lost even though there is still water in the film. To some extent this can be avoided by working in an atmosphere which is saturated with $D_2O$; however it is simpler to use deuterated latex in $H_2O$, and this provides the additional advantage of reducing the incoherent scattering from protonated material.

With surfactant covered latex it is possible to use a third type of contrast, where the core polymer is deuterated, the solvent is deuterated as well, but the surfactant is protonated. In this way it is possible to follow the segregation of hydrophilic material even when all water has been removed from the film.

### Small angle neutron scattering experiments

Any scattering experiment on drying films must meet a certain number of constraints. Firstly, the film must be of uniform thickness, and this thickness must be small enough that the film can dry at a reasonable rate without forming a dry layer ("crust") on its outer surface. Secondly, the loss of water must be monitored continuously; this can be done either by weighing the sample or by calculating the integral of its scattered intensity; during data acquisition it is necessary to stop the evaporation altogether. Thirdly, there must be enough matter in the beam to scatter a significant fraction of incident neutrons.

These constraints are met by the following setup. The latex dispersion is spread on a quartz plate as a film of thickness 0.1 to 0.3 mm; the weight of this film is 0.3 to 1 g. The plate is placed in a plastic box with a cover which can be closed tightly. Water loss through evaporation is monitored by weighing continuously the box with the plate and film in it: with the cover off, a film of latex volume fraction $\phi$ = 0.5

Table 1. Dispersions of latex particles in water

| Name | Core | Surface | Solvent | Diameter (Å) | | Contrast |
|------|------|---------|---------|------|------|----------|
| | | | | EM | QELS | |
| JFD5 | PBA/PS | AA sequences | $D_2O$ | 1200 | 1190 | HHD |
| 118NL | PBA/PS | $C_{12}N^+C_3SO_3^-$ 0.8 mol/100 Å$^2$ | $H_2O$, salt $5 \times 10^{-2}$ M, $C_{12}N^+C_3SO_3^-$ | 1030 | 1180 | DHH |
| 118L | PBA/PS | $C_{12}N^+C_3SO_3^-$ 0.2 mol/100 Å$^2$ | $H_2O$ | 1030 | 1180 | DHH |
| R32NL | PBA/PS | $C_{12}N^+C_3SO_3^-$ | $H_2O$, salt, $C_{12}N^+C_3SO_3^-$ | | 470 | DHH |
| R32L | PBA/PS | $C_{12}N^+C_3SO_3^-$ | $H_2O$ | | 470 | DHH |
| R32L + C12 | PBA/PS | $C_{12}N^+C_3SO_3^-$ | $H_2O$, $C_{12}N^+C_3SO_3^-$ | | 470 | DHH |

dries completely in about one hour; with the cover on, the film can remain at constant $\phi$ for many hours.

The box is placed in the neutron beam with the quartz plate inclined at an angle $\omega = 10°$ from the beam, unless a specific orientation is required. Smaller angles are not acceptable because of a specular reflection of the neutrons on the film; larger angles are not practical because the effective thickness crossed by the neutrons is less, and also because the sample may flow more easily to the lower end of the plate.

Preliminary experiments were performed on the instrument PACE of LLB. With this instrument we used neutrons of 11 and 20 Å wavelengths and a sample to detector distance of 4.75 m to achieve the appropriate range of $Q$ values; the collimation of incident neutron is defined by a 1 cm circular diaphragm 5 m before the sample and another one right before the sample. The scattered neutrons are then collected according to scattering angle by an array of 30 circular detectors; thus interference patterns are available only in a radially averaged form. Final experiments were performed on the instrument D11 of ILL [6]. There we used neutrons of wavelength 6 Å $\pm$ 9%, the collimation of which is defined by a diaphragm of 3 $\times$ 5 cm after the velocity selector 40 m before the sample, and another one of 1 $\times$ 1 cm right before the sample. The neutrons are collected on the 2-d detector (64 $\times$ 64 cells), usually positioned at 35.7 m from the sample. After the 2-d patterns have been observed the data are radially averaged and normalized for the efficiency of detector cells; the normalization spectrum is a 2 mm thick $H_2O$ sample measured at a sample to detector distance of 2 m. The resulting scattering curves are used to measure peak positions and widths, asymptotic laws for scattering at $Q \rightarrow 0$ and at high $Q$, and integrated intensities.

*Scattering curves for dispersions of latex particles*

Scattering experiments are basically interference experiments where rays scattered at a given angle from all scattering centers in the sample are collected in one point of the detector [7, 8]. The scattered amplitudes which result from such interferences depend on the content of each scattering center (expressed as a density of scattering length) while the phase differences are determined by the distance between scattering centers and by the magnitude of the scattering vector $Q$.

For a given value of $Q$, pairs of scattering centers closer than $1/Q$ scatter rays which are in phase, and can therefore be lumped together; hence the interference pattern is determined by the relative contents of small volumes of dimensions $1/Q$ and separated by distances larger than $1/Q$. If all these volumes have the same content (homogeneous material) their contributions will cancel out in the interference process, and the material will be transparent; for this reason we get no small angle scattering from a homogeneous polymer film, or from a bulk solvent. However it is possible, through isotopic substitution, to achieve very different densities of scattering lengths between two hydrogenated materials, e.g. latex vs. water. Then small angle scattering will

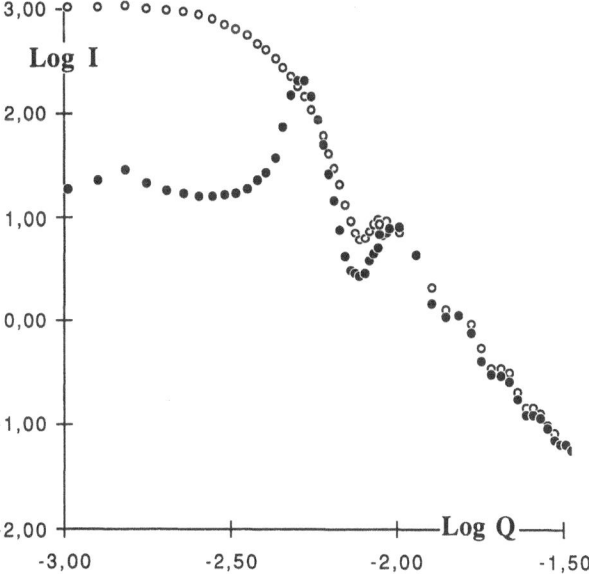

Fig. 1. Scattering curves for dispersions of PAA-covered latex in $D_2O$. Circles: dilute ($\phi = 0.01$); Dots: concentrated ($\phi = 0.57$). The intensities are radial averages of values measured in the 2-d scattering patterns; the $Q$ values correspond to distances between 0.04 Å ($\text{Log}\,Q = -1.5$) up to 6000 Å ($\text{Log}\,Q = -3$). The intensity scales of each spectrum have been shifted to the same intensity per particle at high $Q$

be produced, reflecting the distribution of deuterated material in a protonated matrix or vice versa.

It is instructive to see how the resulting interference patterns measure pair correlations within an array of latex particles. Figure 1 shows typical scattering curves for dispersions of PAA-covered latex in $D_2O$. The smallest $Q$ value in our experiments is $10^{-3}$ Å$^{-1}$; for such $Q$ values significant phase differences are obtained only at distances beyond 1000 Å. At such distances we probe large scale fluctuations of the number concentration of particles. In dilute dispersions these fluctuations are simply proportional to the particle concentration; the resulting intensity measures the product of mass times concentration. In concentrated dispersions the fluctuations are suppressed by electrostatic repulsions between particles; hence the values of intensity/concentration are depressed to an extent which measures the strength of repulsions.

At higher $Q$ values significant phase differences are obtained for neighboring particles. Hence the interferences will be constructive at $Q$ values corresponding to nearest neighbor spacings, and the intensity scattered by concentrated dispersions will rise to a peak near the corresponding Bragg distance. For well ordered systems this oscillation can be so strong that a second and third diffraction order are visible (Fig. 1). However various types of disorder may destroy this oscillation: for instance screening the interparticle repulsions with added salt will allow strong fluctuation in the values of distances between neighboring particles; then the depression and peak caused by regular ordering will vanish. The oscillation may also be blurred by a dispersion in particle sizes, which will result in a dispersion in their distances.

Finally for $Q$ values larger than $(2\pi/\text{diameter})$ we observe the distribution of distances within a particle. Because the particles are 3 dimensional, compact objects rather than flat or linear ones, the intensity decay in this range follows Porod's law $I(Q) = AQ^{-4}$, and the coefficient $A$ yields the surface area of the particles (Fig. 2). On top of this decay there is a damped oscillation, the period of which matches the particle diameter $2b$ (peaks near $2\pi/b$, $4\pi/b$ etc.). According to this period we calculate the core diameter of the latex particles, which is nearly equal to the overall diameter obtained from electron microscopy (1150 vs. 1190 Å).

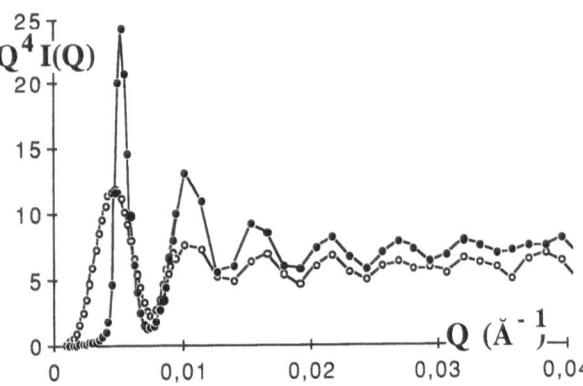

Fig. 2. Intensities scattered from dispersions of PAA-covered latex in $D_2O$. Dots: dilute ($\phi = 0.01$); circles: concentrated ($\phi = 0.57$). This plot reveals the asymptotic limit (Porod's law) and the shape oscillations whose period is $4\pi/\text{diameter}$. Up to 6 oscillations can be recognized: this shows that the particles are extremely monodisperse

## Observation of dry films through electron microscopy

For electron microscopy the samples must be thin, dry, and stained. They are prepared according to the following procedure. First a small piece of dry film is cut with a razor blade and glued on a stand. Then it is cooled to $-35\,^{\circ}C$ and sliced to 0.1 µ thickness with an ultramicrotome apparatus, finally this slice is placed on a grid and stained by contact with a drop of uranyl acetate; excess salt is eliminated by contact with drops of pure water. The uranyl acetate stains the hydrophilic material in the film, in particular the acrylic acid groups which form the hydrophilic membranes of the original latex particles.

## Ordering in concentrated dispersions

Concentrated dispersions of monodisperse particles have a bluish tint, indicating that light rays of short wavelengths are diffracted at observable angles. In neutron scattering the corresponding diffraction peaks are at very small angles: a dispersion poured into a scattering cell will produce a Debye-Scherer ring at 0.006 radian from the beam.

However most of our experiments were performed on thin films, and these yield a much more interesting information: indeed some of the scattering patterns

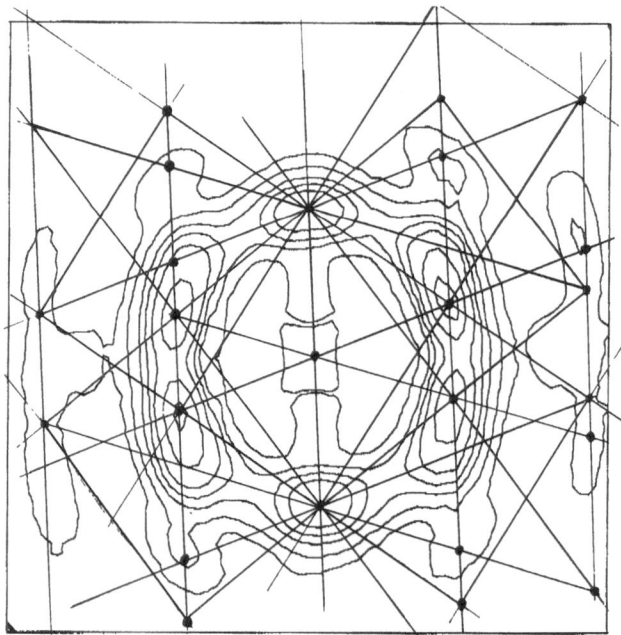

Fig. 3. Contour plot of the scattering pattern for a film at volume fraction $\phi = 0.2$, spread on a quartz plate oriented at 11° from the beam. The heights of successive contours are 1E3, 2E3, 5E3, 1E4, 5E4, 1E5, 2E5. The diffraction spots on the meridian of the figure are produced by diffraction on horizontal 111 planes aligned parallel to the plates; the vertical stripes on either side correspond to diffraction on other 111 and 200 planes, as indicated

consist of spots instead of rings (Fig. 3). The observation of diffraction spots implies that *the dispersion has long range order as in colloidal crystals* [9−11]. We find this long range order for dispersions of monodisperse particles which repel each other strongly, e.g. all dispersions of PAA covered latex (JFD5) regardless of solvent, and the large surfactant covered latex in pure water (118L).

For films which are spread on a plate nearly parallel to the beam, the interference pattern is always oriented as in Fig. 3: there are two spots on the vertical axis (meridian of the figure), and two vertical stripes on either side of this axis, with a large number of spots on either stripe. This orientation of the pattern must be caused by the quartz plates, i.e. some reticular planes of the colloidal crystal are aligned parallel to the quartz plates. This alignment must persist over distances at least as large as the sample thickness, otherwise a powder pattern from many disoriented crystallites would be observed.

Most colloidal crystals have a face centered cubic (fcc) structure: indeed this is the most efficient way of packing spheres; the less dense, body centered cubic (bcc) structure is only observed with low volume fractions and very long range repulsions [9−11]. Thus we try to assign the observed diffraction patterns according to a fcc structure of cell size $a$. In this structure the most dense planes are the 111 planes where the particles are arranged on a triangular lattice at a distance $d = a/\sqrt{2}$; the distance between consecutive 111 planes is $d_{111} = a/\sqrt{3}$ (Fig. 4).

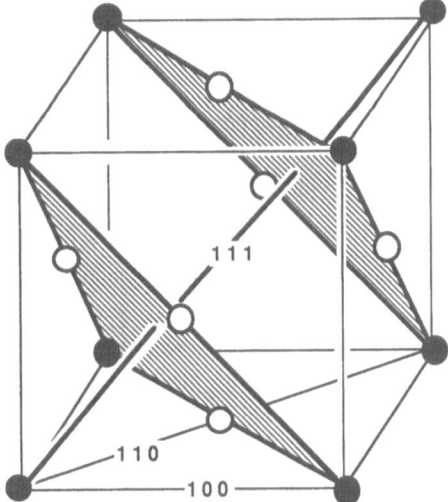

Fig. 4. The face centered cubic structure for colloidal crystals. The shaped planes are the 111 planes, which are normal to the 111 direction of the crystal. Because these planes are the most dense planes in the structure, they are aligned along the quartz substrate of the film

Because the 111 planes are the most dense planes in the structure, we assume that the colloidal crystal is built with such planes along the quartz substrate. In the typical experimental setup the quartz plates are nearly parallel to the neutron beam, hence the beam is in a direction of the 111 plane. In this plane, all crystallographic directions have indices 110; we now consider the diffraction by a crystallite which is oriented with a 110 direction along the beam (Fig. 5).

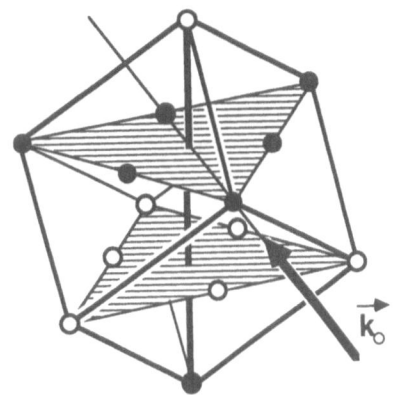

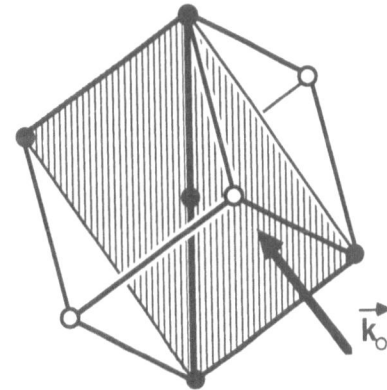

Fig. 5. Order in colloidal crystals of latex particles. Top: real space, face centered cubic structure, with particles at the vertices and at the face centers of a cube. The shaded triangle corresponds to a 111 plane, which is aligned along the quartz plates. The plate is parallel to the neutron beam, and the crystal is oriented with a 110 direction along the beam
Bottom: reciprocal space, body centered cubic symmetry, with diffraction spots at vertices and at the center of a cube. With the beam along a 110 direction, the scattering vectors are in a 110 plane, which is shaded. Hence the scattering pattern is rectangular, with one diagonal along the normal to the orienting plates (111 direction)

The reciprocal space of a fcc structure has a bcc unit cell of size $1/a$. Diffraction is observed when the scattering vectors $Q$ of the experiment fall on a lattice vector of this structure. For small angle scattering the scattering vectors are all in a plane normal to the beam; hence the diffraction pattern is a section of the reciprocal structure by a plane normal to the neutron beam. This section is shown in Fig. 5; it is a centered rectangular lattice with one diagonal of the rectangle along the 111 direction; the distance to the first spots is $= \sqrt{3}/a$.

Of course the crystallites can have any orientation as long as their 111 direction is normal to the quartz plate; the interference pattern will select those crystallites whose reciprocal structure has lattice vectors in the plane of scattering vectors; these correspond to two possible orientations, one as shown in Fig. 5, and another one flipped by $\pi$ around the normal to the plate, which is the diagonal of the rectangle. In this way we recover a set of 6 first order spots, all with indices 111. The neighboring cells of reciprocal space will also contribute diffraction spots close to these first order spots, in particular the 200 spots which are the centers of the neighboring rectangles. The vertical stripes of Fig. 3 are made of a succession of such spots, with indices 111, 200, and 220.

A good way to confirm the assignment of diffraction spots is to rotate the sample, and observe the resulting changes in the diffraction pattern. There are three main types of rotations, which we examine in turn.

— Firstly we may keep the quartz substrate along the beam, and rotate it around the beam; this simply produces a rotation of the whole scattering pattern aroud the beam, as expected.

— Secondly we may rotate the plate around its normal, which is almost normal to the beam. This produces a rotation of the structure around its 111 direction, and a similar rotation of the reciprocal space. If the sample contains a uniform distribution of crystallites with all orientations around this axis, this rotation will leave the scattering pattern unchanged. In fact there may be some large crystallites in the beam, and these will produce stronger scattering for one of the first order rectangles in Fig. 3; this effect will change with the rotation, as the scattering plane selects the diffraction spots from different crystallites.

— Thirdly we may change the angle between the beam and the plates. With the beam at 30° from the plates, the 111 spots which are on the meridian, normal to the quartz plates move out of the diffraction plane; the vertical stripes on either side begin to turn into crescents with 4 spots on them. At an angle of 45° there are still 4 spots, on top of a circular ring. At an angle of 60° there are 6 spots, unevenly spaced. At an angle of 90° (plate normal to the beam) the 6 spots form a regular hexagon, and higher order diffraction

spots can be identified which complete the triangular lattice (Fig. 6). This is consistent with the assumption that the 111 planes of the structure are aligned along the quartz plates; indeed the reciprocal lattice of a triangular lattice is also a triangular lattice. However the observation of diffraction spots instead of a ring in this configuration is surprising, as it implies that there is a preferential direction in the plane of the substrate. This may be because the structure is aligned by shear as it is spread [12, 13]; alternatively there may be a single crystallite which dominates the scattering; hence the crystallite size may be comparable with the beam size.

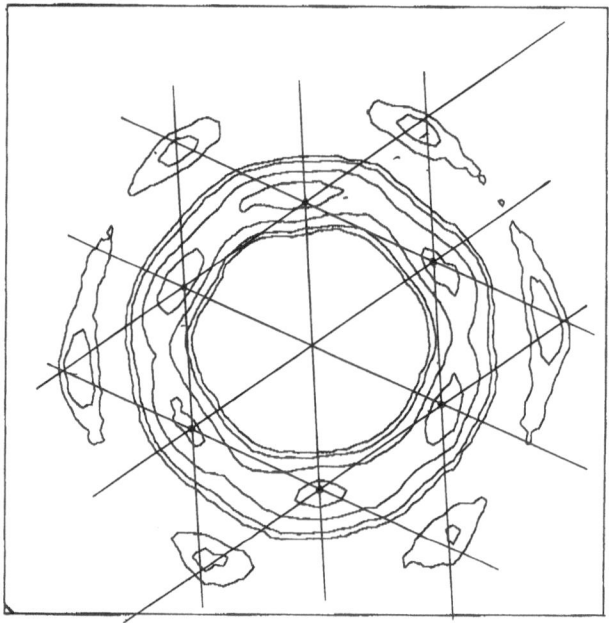

Fig. 6. Contour plot of the scattering pattern for the same film as shown in Fig. 5, but oriented normal to the beam. The heights of successive contours are 250, 400, 1000, 2000. The diffraction pattern reflects the reciprocal structure from the triangular array of particles in a 111 plane which is aligned along the quartz substrate

Finally we must briefly discuss the case of films which do not show macroscopic alignment. As mentioned above, diffraction patterns of the type shown in Figs. 3 and 6 are observed only if the latex particles are monodisperse and if they are in a solvent of low ionic strength. There are 3 instances where the diffraction spots are replaced by Debye-Scherer rings:

(i) When the latex volume fraction is too low ($< 20\%$) the dispersion does not have long range order.

(ii) When salt is present in the solvent, the electrostatic forces between latex particles are screened, and long range order is lost; this is the case for all unwashed latex dispersions, e.g. R32NL and 118NL. It is noteworthy that long range forces are necessary even at very high volume fractions.

(iii) When the particles are polydisperse in size the ordering becomes disorganized as well; this is the case for the small, R32 latex dispersions.

## Deformation of particles

The ordered structures presented above are produced by dispersions of latex particles where the volume fraction of latex is 20 to 50%, and the rest of the volume is occupied by water. Now this water is removed through evaporation, and we observe the resulting changes in the ordering and in the individual shapes of particles. Here the qualitative observation is that the scattering patterns retain the same appearance until the diffraction spots disappear. A more quantitative point of view is provided by a measure of the intensity and distance (position) of the first diffraction order as functions of the volume fraction of latex. These are presented in Fig. 7, where the measured interparticle distances are compared with those expected for a uniform fcc packing of the particles.

For some of the films the measured distances agree quite well with those predicted for a uniform fcc packing of the particles in the film; deviations are caused

by a non uniform drying of the film, where the perifery tends to dry before the center. For others, after an initial decrease there is a plateau where the weight of the film diminishes while the diffraction spots remain at the same position, and then the whole diffraction pattern collapses abruptly. It will be argued that the latter behavior is caused by coalescence of the particles in the film. Here we discuss the former behavior, where all distances keep shrinking regularly according to the overall volume fraction of latex in the film: this is the behavior observed with all acrylic acid covered particles.

The problem with this continuous shrinkage is that it proceeds beyond the point where the particles come into direct contact: for a fcc packing of the particles this occurs at $\phi = 74\%$. Further shrinkage beyond this point is possible only if the particles deform to fill the volumes formerly occupied by water; in this process the particles must acquire flat faces, and in the end transform into polyhedra. This transformation is best described by partitioning the space into polyhedral cells according to the symmetry of the lattice, and then filling those cells with droplets of increasing volume, as explained by Lissant [14]. For the fcc lattice the individual cells are rhombic dodecahedra (RDH); each one of the 12 faces is a rhombus, and the angle between faces is 120°; there are 6 of the vertices where 4 edges meet at an angle of 71°, and 8 vertices where 3 edges meet at an angle of 109° (Fig. 8). This cell can be derived from the fcc structure shown in Figs. 4 and 5 by intersecting each interparticle distance with a plane

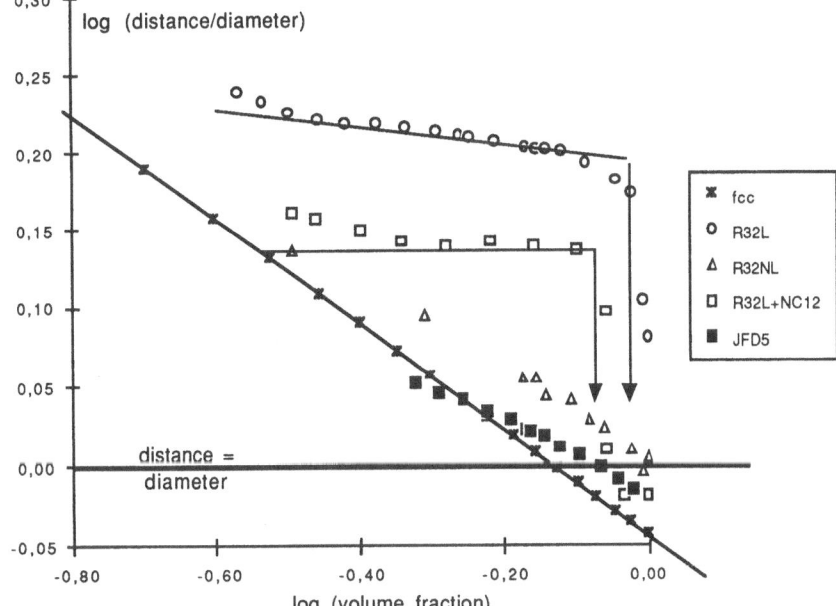

Fig. 7. Bragg distances for latex particles in drying films. The distance is calculated from the position of the diffraction peaks, assuming a fcc packing of the particles: $d = 2\pi/Q \times \sqrt{3/2}$. The volume fraction of latex is calculated from the measured weight of the film

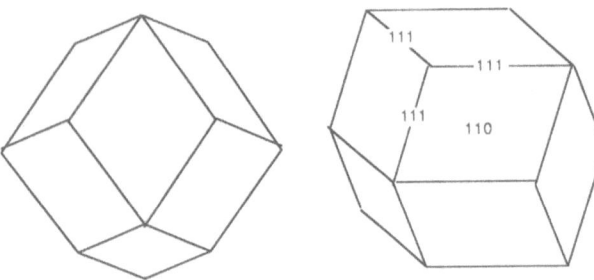

Fig. 8. A rhombic dodecahedron (RDH) is obtained when soft spheres are packed according to a fcc structure, and compressed until they fill all the available volume

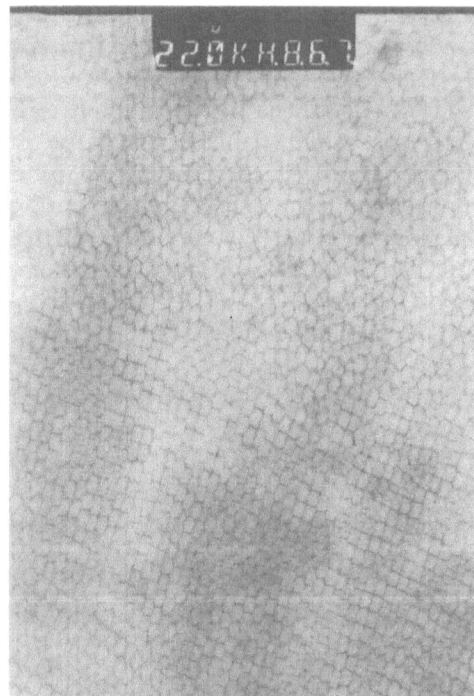

Fig. 9. Electron microscopy: a film of JFD5 latex has been dried, sliced to 0.1 μ thickness and stained on the hydrophilic membranes which separate the particles. The cells correspond to the original latex particles

oil/soap/water systems [15]. This foam structure may be observed through electron microscopy if the hydrophilic membranes are labeled with heavy atom salts. Figure 9 shows the electron microscopy of a film sliced at low temperature and stained in this way.

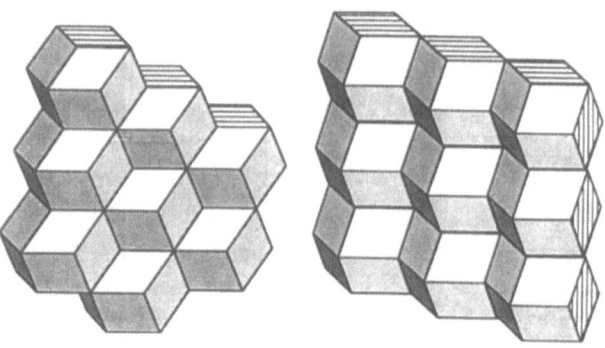

Fig. 10. Latex films which have lost most of their water have a foam structure made of RDH cells. The interior of each cell is the original latex core of the particle, and the cell walls contain the hydrophilic material, e.g. polyacrylic acid sequences or emulsifier molecules, plus some hydration water. The array on the left hand side is normal to a threefold, 111 axis; that on the right is normal to a fourfold, 100 axis

This picture shows the regular packing of rhomboedral cells, with a variety of orientations with respect to the electron beam. In some regions the cells are viewed along their 111 axis, showing a hexagonal array of cells with three additional edges joining the center of each hexagon; in other regions with the same orientation these top edges are missing, and only distorted hexagons are observed. In other regions the cells are viewed along a 100 axis, which is the axis joining two vertices with four edges each across the center of the cell; then the projection of the rhombic faces is a square, and indeed rectangular patterns are observed in the lower right hand corner of Fig. 9. Two models of these arrays are presented in Fig. 10.

normal to it; it is then found that the resulting faces are parallel to the 110 family of planes, while the edges are parallel to 111 type directions.

At the beginning of the deformation the latex particles touch the faces of the cells in their middle, where their surface is flattened to a small disk; at the end they fill the RDH cells completely, and the hydrophilic material is confined to the faces of the polyhedra. At this stage the structure may be described as a foam, similar in structure to the biliquid foams observed in

## References

1. Vanderhoff JW, Bradford EB, Carrington WK (1973) J Polym Sci Symp 41:155–174
2. Kast H (1985) Makromol Chem Suppl 10/11:447–461
3. Hahn K, Ley G, Schuller H, Oberthür R (1986) Colloid Polymer Sci 264:1092–1096
4. Hahn K, Ley G, Oberthür R (1988) Colloid Polymer Sci 266:631–639
5. Nelson RL, Ramsay JFD, Woodhead JL, Cairns JA, Crossley JAA (1981) Thin Solid Films 81:329–327
6. Ibel K (1976) J Appl Cryst 9:296

6. Ibel K (1976) J Appl Cryst 9:296
7. Jacrot B (1976) Rep Progr Mod Phys 39:911
8. Guinier A, Fournet G (1955) Small Angle Scattering of X rays Wiley
9. Pieranski P (1983) Contemp Phys 24:25
10. Goodwin JW, Ottewill RH, Parentich A (1980) J Phys Chem 84:1580—1586
11. Monovoukas Y, Gast AP (1989) J Colloid Interface Sci 128:533
12. Lindner P private communication
13. Ashdown S, Marković I, Ottewill RH, Lindner P, Oberthür RC, Rennie AC (1990) Langmuir 6:303—307
14. Lissant KJ (1966) Colloid Interface Sci 22:462—468
15. Ebert G, Platz G, Rehage M (1988) Ber Bunsenges Phys Chem 92:1158—1164

Authors' address:

M. Joanicot
Rhône Poulenc
52 rue de la Haie-Coq
93308 Aubervilliers
France

**Progress in Colloid & Polymer Science**                    Progr Colloid Polym Sci 81:184–188 (1990)

# Molercular dynamics simulations of colloids: Supercooled Yukawa systems

N. Pistoor[1]) and K. Kremer[2])

[1]) Institut für Physik, Johannes-Gutenberg-Universität, Mainz, FRG
[2]) Institut für Festkörperforschung, Forschungszentrum Jülich, Jülich, FRG

*Abstract:* We performed molecular dynamics simulations on one and two component Yukawa systems. Cooling the system down and inspecting pair distribution functions (pdf) and bond correlation functions (bcf) we found the one component system to crystallize into a bcc-like lattice rather than an fcc lattice which is the stable phase of the simulated system at low temperatures. Upon cooling the two component system freezes into a glassy state without exhibiting crystalline structure in pdf or bcf. We define particle excess functions which show that spacial fluctuations in the number density of particles of the different components decay quite slowly. Therefore we believe that a well defined state of the two component system can only be reached in simulations long enough to allow those fluctuations to decay.

*Key words:* Molecular dynamics; Yukawa potential; pair distribution function; bond correlation function

## The model

The effective interaction between two charged macroions in a dilute aqueous solution can be well described by a Yukawa-Potential [1]

$$U(r) = U_0 \frac{a}{r} \exp(-\lambda r/a) \qquad (1)$$

where $a = n^{-1/3}$ is a typical interparticle distance and $n = N/V$ is the particle number density. The prefactor $U_0$ is proportional to the effective charges of the interacting particles. The energy scale is set by $U_a = \langle U(a) \rangle = \langle U_0 \rangle \exp(-\lambda)$ where the average is performed over all pairs of particles in the system.

The state of a one component system (i.e. all particles carry the same effective charge) can be specified by two dimensionless parameters, the screening parameter $\lambda$ and the dimensionless temperature $kT/U_a$. The phase diagram (Fig. 1) which is known from previous simulations [2, 4] as well as from light scattering experiments on latex spheres [5] shows three different phases: fluid, bcc crystal and fcc crystal.

In a two component system one expects suppression of crystallization in favour of a frozen but disordered ("glassy") state.

## The simulation

We performed molecular dynamics (MD) simulations for two different systems with parameters as displayed in Tab. 1 by integrating the equations of motion of the macroions

$$m\frac{d^2 r_i}{dt^2} = -\frac{\partial}{\partial r_i}\sum_{j \neq i} U(|r_i - r_j|) - \Gamma\frac{\partial r_i}{\partial t} + W_i \qquad (2)$$

for $N$ particles in a cubic box of Volume $V = Na^3$ with periodic boundary conditions. The temperature was controlled by weakly coupling the particles to a heat bath by means of some friction constant $\Gamma$ and a stochastic force $W_i$ [6].

A typical timescale of the problem is given by the Einstein frequency $\Omega$ which is derived from a harmonic approximation to the mean effective potential

$$U_{eff}(r) = n\int g(r')U(|r - r'|)dr'. \qquad (3)$$

By performing a Taylor expansion around $r = 0$ which gives

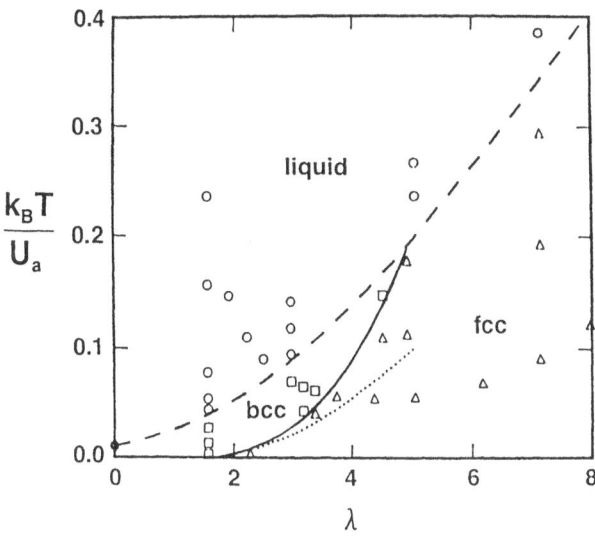

Fig. 1. The phase diagram of one component Yukawa systems. Open circles indicate points where the liquid phase is stable. Points where the fcc or bcc phase are stable are indicated by triangles or squares, respectively. The dashed line separating liquid and solid phases indicates the value of $kT/U_a$ where the rms displacement reached $0.19\alpha$. The dotted line is the harmonic result for the bcc to fcc phase transition, and the solid line interpolates the simulation results for this transition (from [4])

Table 1. Simulation parameters

|  | One component system | Two component system |  |
|---|---|---|---|
| Number of particles | $N = 512$ | $N_A = 170$ | $N_B = 342$ |
| Screening parameter | $\lambda = 6.0$ | $\lambda = 3.4$ |  |
| Effective charges |  | $q_A/q_B = 0.7$ |  |
| Temperature | $kT/U_a = 0.4...0.1$ | $kT/U_a = 0.2...0.025$ |  |

$$U_{eff}(r) = \langle U \rangle + \frac{1}{2}m\Omega^2 r^2 + \ldots \qquad (4)$$

we find

$$\langle U \rangle = 4\pi n \int_0^\infty g(r) U(r) r^2 \, dr$$

and

$$\frac{1}{2}m\Omega^2 a^2 = \frac{1}{6}\lambda^2 \langle U \rangle. \qquad (5)$$

The period $\tau_E = 2\pi/\Omega$ of one oscillation with that frequency can be used to compare the timescale of different systems.

## Calculated quantities

From the configurations produced by the simulations we calculated various quantities, among them pair distribution functions, particle excess functions and bond correlation functions [7].

The pair distribution functions $g_{\nu\mu}(r)$ give the probability of finding another particle of type $\mu$ a distance $r$ away from any given particle of type $\nu$.

$$g_{\nu\mu}(r) = \frac{1}{4\pi} \int \frac{1}{nN} \left\langle \sum_{n=1}^{N_\nu} \sum_{m=1}^{N_\mu} \delta(r - r_n + r_m) \right\rangle d\Omega. \qquad (6)$$

The particle excess functions $\Delta N_{\nu\mu}(r)$ give the difference in the number $N_{\nu\mu}(r)$ of particles of type $\mu$ contained in a sphere of radius $r$ around any given particle of type $\nu$ and the average number $n_\mu V(r)$ contained in any arbitrary volume $V(r) = \frac{4}{3}\pi r^3$ of the same size.

$$\Delta N_{\nu\mu}(r) = N_{\nu\mu}(r) - n_\mu V(r). \qquad (7)$$

The bond correlation functions probe correlations in the relative orientations of distance vectors ("bonds") connecting each particle with its 14 nearest neighbors. (This number was found by integrating the pair distribution functions up to the first minimum.) The definition is [7].

$$G_l(r) = \langle P_l(\cos\theta) \rangle_r / G_0(r) \quad l = 2, 4, \ldots, 10. \qquad (8)$$

Here $\theta$ is the relative tilting angle of two bonds and $P_l$ are the Legendre polynomials. The average is performed over all pairs of bonds at distance $r$ from each other, normalized by the bond pair distribution function $G_0(r)$.

## Some results

### One component system

Figure 2 shows the pair distribution function $g(r)$ at different stages of the simulation. The first one be-

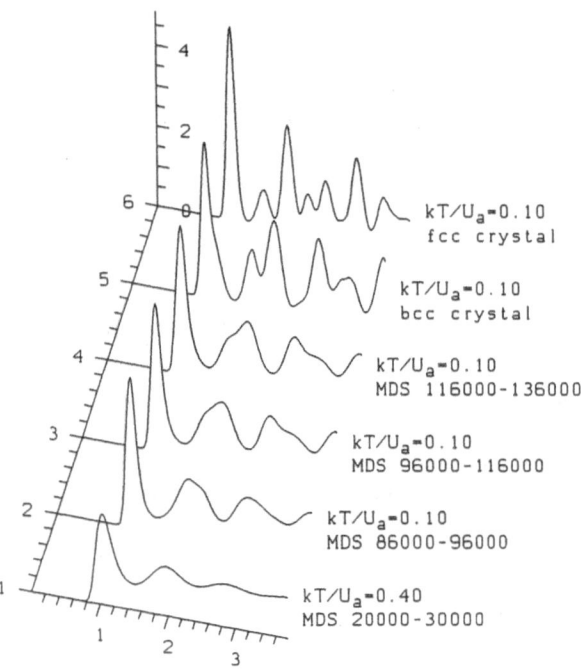

Fig. 2. Pair distribution functions $g(r)$ vs. $r/a$ of the one component Yukawa system at different stages of the simulation as indicated in the figure. The upper two curves are for a bcc and fcc crystal at $kT/U_a = 0.1$ (1000 MDS $\approx$ 7.8 $\tau_E$)

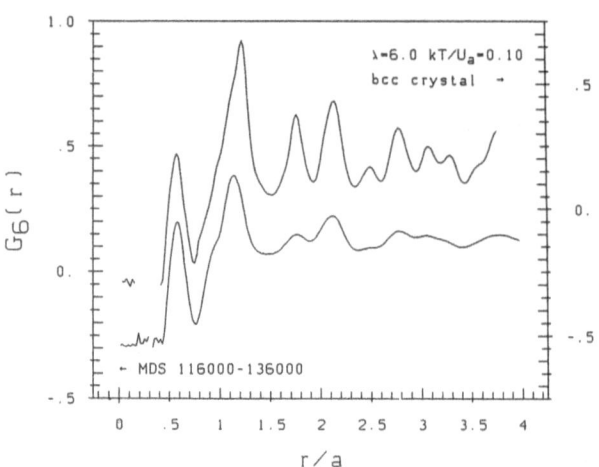

Fig. 3. Bond correlation functions $G_6(r)$ vs. $r/a$ for the one component system (lower curve, right scale) and a bcc crystal (upper curve, left scale) at $kT/U_a = 0.1$

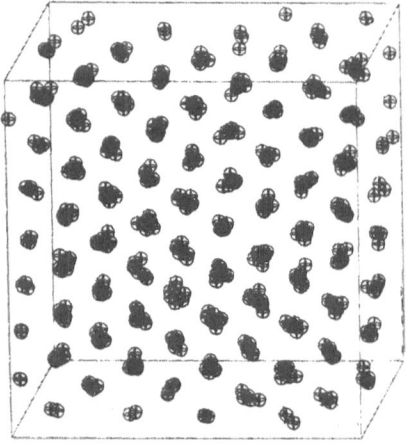

Fig. 4. Sample configuration of the one component system at $kT/U_a = 0.1$. The figure shows the projection of a three-dimensional cube along a direction which clearly exhibits the crystalline structure of the system

longs to the fluid state at temperature $kT/U_a = 0.4$ and reflects the local shell structure which rapidly decays after some interparticle spacings. The second one was recorded shortly after cooling the system down from $kT/U_a = 0.2$ to $kT/U_a = 0.1$. The second peak shows a small shoulder on the right which has previously been considered a signature of some glassy state (e.g. [9]). As one can see from the next two curves in Fig. 2 this is not the case: the maximum in the second peak is now on the right and also the third peak begins to exhibit some additional structure. Comparing this function to the pair distribution functions of bcc and fcc crystals at the same temperature (last two curves in Fig. 2) we find similarities to the bcc structure.

This is also supported by comparing the bond correlation functions (Fig. 3): the peak positions coincide with those of a bcc crystal, and $G_6(r)$ has not decayed to zero at $r/a = 4.0$ and will probably not do so for larger distances in larger systems. So there is long ranged orientational order in the system, which can be visualized by direct inspection of some sample configuration (Fig. 4). The crystalline structure of the system is evident.

What is interesting about these findings is the fact that the stable phase for this system at $\lambda = 6.0$ and $kT/U_a = 0.1$ should be the fcc crystal rather than the bcc crystal as can be seen from the phase diagram (Fig. 1). The system prefers the bcc structure instead, may be because its short range order is more similar to that of the fluid: integrating $g(r)$ up to the first minimum gives 14 nearest neighbors for the bcc crystal (first and second shell whose peaks are strongly overlapping at $kT/U_a = 0.1$) and almost 14 in the fluid in contrast to twelve for the fcc crystal. There alre also

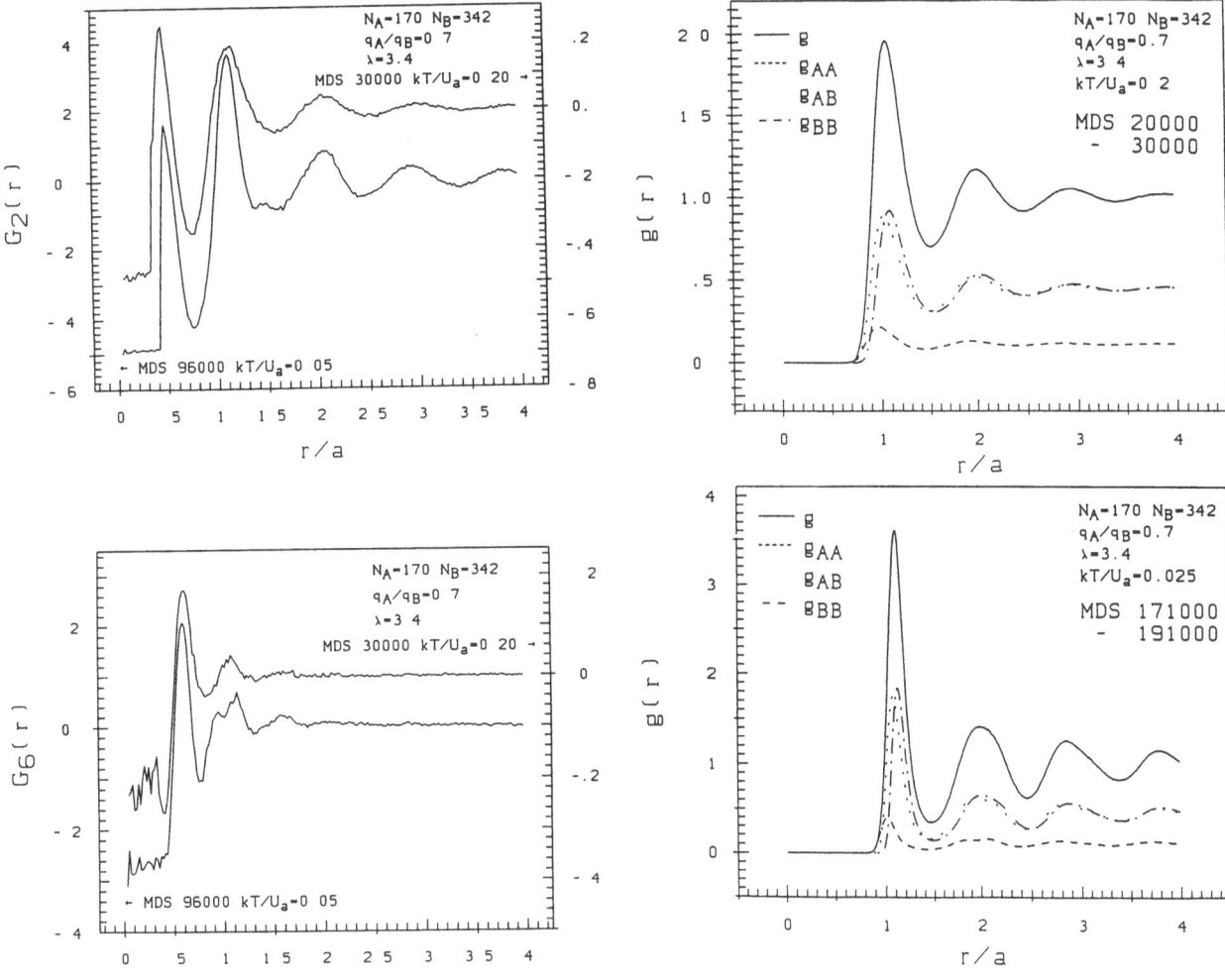

Fig. 5. Partial pair distribution functions $g_{v\mu}(r)$ vs. $r/a$ for the two component system (a) in the fluid phase at $kT/U_a = 0.2$ and (b) in a frozen state at $kT/U_a = 0.025$

Fig. 6. Bond correlation functions $G_l(r)$ vs. $r/a$ of the two component system in the fluid at $kT/U_a = 0.2$ (upper curve, right scale) and in a frozen state (lower curve, left scale) for (a) $l = 2$ and (b) $l = 6$

other theoretical considerations starting from a Landau expansion of the free energy which predict the first nucleated phase after rapid cooling to be bcc, irrespective of the stable phase [8].

*Two component system*

For the two component system partial pair distribution functions were calculated. Figure 5 shows some results for $kT/U_a = 0.2$ and $kT/U_a = 0.025$. The peaks are more pronounced at the lower temperature, but there are no structural changes visible comparable to those in the one component system. Although the applicability of the Yukawa potential to two component colloidal solutions is not as well established as

in the one component case, our results seem to agree with recent experimental findings [10].

The bond correlation functions (Fig. 6) do not show any particular long range orientational order: they all decay to zero for $r/a > 2$, only $G_2(r)$ exhibits oscillations that reflect the shell structure that is also visible in the pair distribution functions.

A somewhat more interesting feature of the two component system can be deduced from the particle excess functions. Figure 7a shows these functions for $kT/U_a = 0.2$, approximately 150 $\tau_E$ after starting the simulation. The function $\Delta N(r)$ which does not distinguish between the two types of particles shows decaying oscillations around zero, as one would expect if the number density of the particles is only modulated over a length scale comparable to $a$. In contrast to

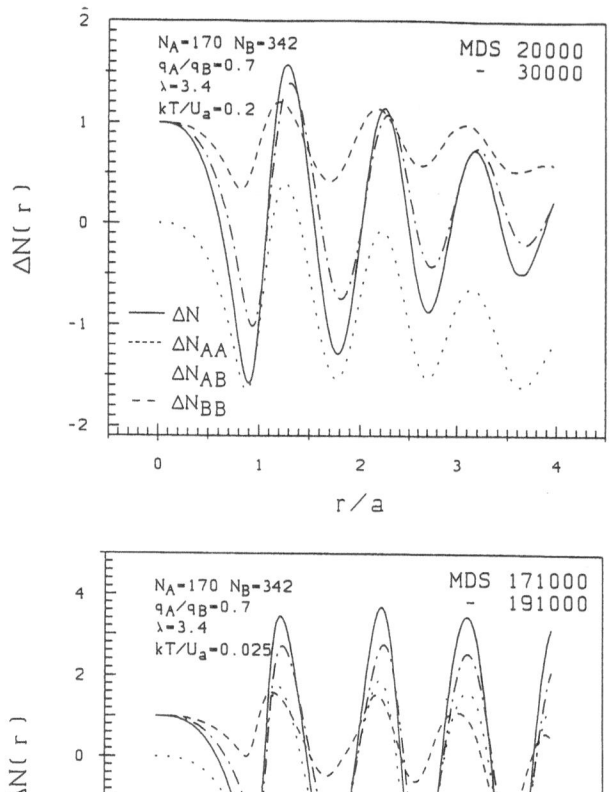

Fig. 7. Particle excess functions $\Delta N_{\nu\mu}(r)$ vs. $r/a$ (a) at $kT/U_a$ = 0.2 (during the first stage of the simulation) and (b) at $kT/U_a = 0.025$ (later stage of the simulation)

clude that in our system there is no tendency for phase separation or clustering of particles of the same type, but spacial fluctuations of the number density of particles of different types decay quite slowly because the particles have to move over several interparticle spacings. This explains why $\Delta N(r)$ reaches its equilibrium shape much faster than the partial $\Delta N_{\nu\mu}(r)$, because there only local adjustments of the particle positions are necessary.

Since the structure and other features of the glassy state might depend on whether fluctuations of the kind discussed above have decayed or not at the time when the particles cease to diffuse and the system freezes, much longer simulations are necessary in order to reach a well defined state of the two component system.

*Acknowledgements*

This research was supported in part by Deutsche Forschungsgemeinschaft, Sonderforschungsbereich 262 and by IBM Deutschland GmbH, Studienprojekt No. F 132.

This article contains material which will be part of a PhD thesis.

**References**

1. Alexander S, Chaikin PM, Grant P, Morales GJ, Pincus P, Hone D (1984) J Chem Phys. 80:5776
2. Kremer K, Robbins MO, Grest GS (1986) Phys Rev Lett 57:2694
3. Kremer K, Grest GS, Robbins MO (1987) J Phys A 20:L181
4. Robbins MO, Kremer K, Grest GS (1988) J Chem Phys 88:3286
5. Monovoukas Y, Gast AP (1989) J Coll Int Sci 128:533
6. Schneider T, Stoll E (1978) Phys Rev B 17:1302
7. Steinhardt PJ, Nelson DR, Ronchetti M (1983) Phys Rev B 28:784
8. Alexander S, McTague J (1978) Phys Rev Lett 41:702
9. Wendt HR, Abraham F (1978) Phys Rev Lett 41:1244
10. Versmold H, private communication

Author's address:

N. Pistoor
Institut für Physik
Johannes-Gutenberg-Universität
Postfach 3980
D-6500 Mainz
FRG

this, $\Delta N_{AA}(r)$ is always positive up to $r/a = 4$, which means that the density of particles of type $A$ in the neighborhood of another particle of type $A$ is larger than the overall density of the particles in the system. At the later stages of the simulation this asymmetry in the particle distribution has decreased considerably (Fig. 7b). After about 500 $\tau_E$, now at $kT/U_a = 0.025$, all four functions strongly overlap. We therefore con-

**Progress in Colloid & Polymer Science**                    Progr Colloid Polym Sci 81:189 – 197 (1990)

# Structural studies of phospholipid cubic phases

J. M. Seddon, J. L. Hogan[1]), N. A. Warrender and E. Pebay-Peyroula[2])

Chemistry Department, Southampton, UK
[1]) Department of Chemistry, Ohio State University, Columbus, Ohio, USA
[2]) Institut Laue-Langevin, Grenoble, France

*Abstract:* In our previous work, we have shown that a general feature of saturated phosphatidylethanolamines is that they exhibit lamellar − inverse hexagonal ($H_{II}$) phase transitions with increasing temperature, at all water contents. We have found that the effect of decreasing the lipid chainlength to $C_{12}$(didodecyl) is to cause the appearance of three intermediate bicontinuous cubic phases in different regions of the binary phase diagram. We have identified these phases as being of cubic aspects 4, 8 and 12, corresponding to probable spacegroups Pn3m ($Q^{224}$), Im3m ($Q^{229}$) and Ia3d ($Q^{230}$). Furthermore, we find that for the $C_{16}$(dihexadecyl) lipid, cubic phases are also induced in qualitatively the same regions of the phase diagram by methylation of the polar headgroup, which increases its hydrophilicity. We have investigated whether the chirality of the lipid influences the chirality of the cubic phase formed, and have found that, at least for these phospholipids, it does not. We have also found that cubic phases are formed by hydrated phosphatidylcholine/fatty acid 1/2 (mol/mol) stoichiometric mixtures, when the chainlength is reduced, and we report preliminary results obtained by X-ray and neutron diffraction on these systems.

*Key words:* Cubic phases; phospholipids; lyotropic liquid crystals; X-ray diffraction; periodic minimal surfaces

## Introduction

It is now quite well accepted (see [1] for a recent review) that the general form of an amphiphile/water binary phase diagram, in the liquid-crystalline region, is as shown schematically in Fig. 1. In this hypothetical case, the transitions between the phases are taken to be driven predominantly by changes in the water content, and any temperature dependences are neglected. The variable can either be the water content itself (e.g. for soaps), but more generally, for pseudobinary systems, represents the effective hydrophilicity of the interfacial region where the polar headgroups reside. This is determined primarily by the headgroup chemical structure, but may also be strongly dependent (for ternary or multi-component systems) on the presence of solutes, either polar, non-polar or amphiphilic, in the system [2]. Furthermore, for lipid mixtures, the effective hydrophilicity may deviate strongly from the average value of the various components, due to headgroup-headgroup interactions

which can compete with headgroup-water binding. Although for real systems Fig. 1 is a gross over-simplification (see Fig. 3 below), it is useful in that it focusses attention on the natural locations of the various phases within the generalised phase diagram. The factor that determines this location is the preferred interfacial mean curvature within each phase. In general, this curvature increases systematically with increasing hydration (or hydrophilicity), because the value of the optimal area per surfactant headgroup normally increases with increasing water content. The mean curvature increases from strongly negative for inverse structures (water-in-oil) such as the inverse hexagonal $H_{II}$ phase (note that the sign of the mean curvature is arbitrary), through zero for the lamellar $L_\alpha$ phase, to strongly positive for the "normal" (oil-in-water) structures such as the hexagonal $H_I$ phase.

However, there may be certain regions, labelled a, b, c and d in Fig. 1, where more complicated "intermediate" phases may be observed. In one case, the binary system sodium dodecyl sulphate (SDS)/water,

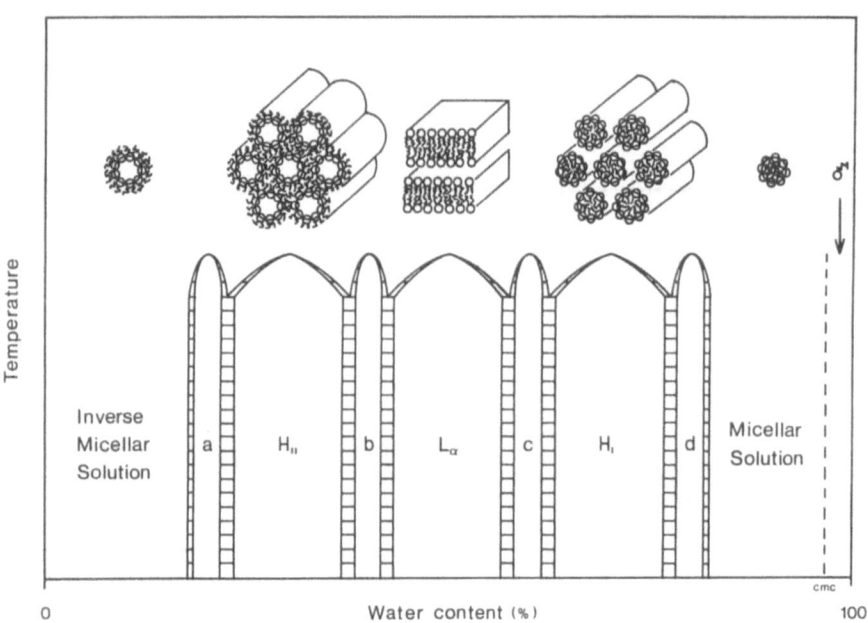

Fig. 1. Hypothetical lyotropic binary phase diagram. Taken from [1]

four such phases have been detected between the $H_I$ and $L_\alpha$ phases, i.e. in region c of Fig. 1 [3]. In general, many of the intermediate phases are cubic, and our present knowledge of their structures is largely due to the pioneering X-ray diffraction work of Luzzati and co-workers [4, 5]. The cubic phases in regions b and c of Fig. 1 are all bicontinuous, with inverse and normal topologies respectively [5, 6]. The structures of the cubic phases observed in region d are still a matter of considerable controversy, and it is not yet completely established whether they are based on packings of anisotropic micelles [6], or on a combination of micelles with a continuous monolayer [5]. Very little is known yet about the possible cubic phases in region a.

It is now becoming clear that cubic phases are far more common than was previously realized (although it had been assumed for many years (correctly) that the "viscous isotropic" phases seen in many binary and ternary phase diagrams are indeed cubic) [7, 8]. They have recently been detected by neutron diffraction in certain ternary surfactant/oil/water (microemulsion gel) systems, and in these cases their structure may be based on a surfactant monolayer, rather than a bilayer [9–11]. It was suggested by Scriven that bicontinuous microemulsions may be based upon minimal surfaces [12]. These are surfaces where the mean curvature $H$ is everywhere zero, and the Gaussian curvature $K$ is negative (or zero at certain special points). Such surfaces can be extended to fill space as

periodic minimal surfaces, some of which have cubic symmetry, which divide space into two congruent sub-volumes. This idea was developed to explain the structures of the various known bicontinuous cubic phases [13–17]. These authors have discussed how symmetry considerations lead directly to minimal surfaces, when a fluid bilayer separates two equal volumes, with no pressure difference between the sub-volumes. For inverse phases, the minimal surface defines the mid-surface of the hydrocarbon chain region.

The inverse bicontinuous cubic phases appear to arise from a frustration between the optimal mean curvature of the interface, and the optimal packing of the hydrocarbon chains [16, 18–21]. By adopting a negative interfacial Gaussian curvature with a dividing minimal surface at the centre of the hydrocarbon chain region, this frustration is reduced, because it allows for a lower lateral area per molecule at both polar interfaces than in the chain region (Fig. 2). There is some scope here for confusion in that the lateral area per lipid molecule is *maximal* at the minimal surface, and *decreases* on moving along the normal in either direction (i.e. towards either of the polar headgroup regions). Although it doesn't give perfect relief from frustration, since it is not possible to fill space with such a lattice, whilst simultaneously maintaining both a uniform average molecular length (i.e. bilayer thickness) and a uniform mean curvature of the layers, within certain regions of the phase diagram such structures should become favoured [22].

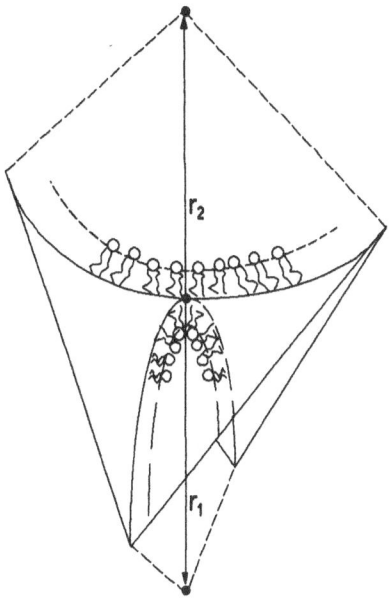

Fig. 2. Relief of bilayer packing frustration by formation of a surface of negative Gaussian curvature (saddle surface)

The significance of the mean and Gaussian curvature elastic energies for the stability of cubic phases was demonstrated by Helfrich [24], and has been extended by a recent statistical mechanical treatment [25]. At present, there is no reliable estimate for the Gaussian curvature elastic modulus of any real system, and little information is available as to the role of factors such as chainlength, headgroup structure, solutes etc. on modulating the cubic phase structure and stability; this is one of our aims in studying phospholipid systems.

Along with the phosphatidylcholines (PC), phosphatidylethanolamines (PE) are normally the other principal class of phosholipid in animal cell membranes, and it is thus of great importance to biology to characterise the lyotropic phase behaviour of this class of lipid, in order to understand its possible role in biological functions involving membranes. It is well known that they are much more weakly hydrophilic than the phosphatidylcholines, and we have previously shown for the saturated systems that this leads to a low hydration of the $L_\beta$ and $L_\alpha$ phases [26]. In addition, we found that, at high temperatures, they form the inverse hexagonal $H_{II}$ phase *at all water contents*. We now report on the formation of cubic phases by these phospholipid systems. This is an area of great current interest in membrane biology, since it is now becoming clear that cubic phases almost certainly play a crucial role in a range of processes such as membrane

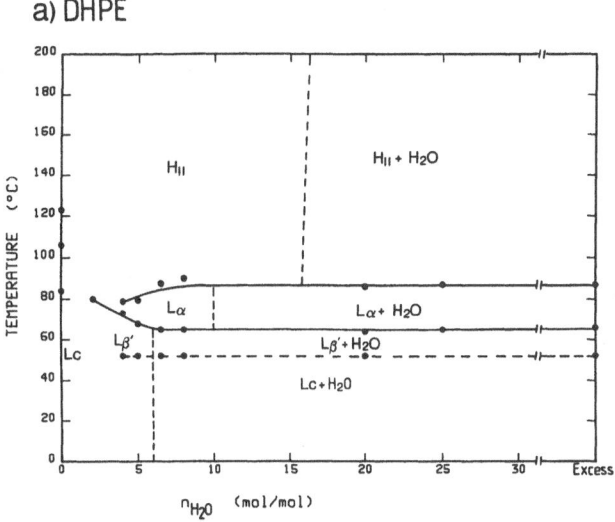

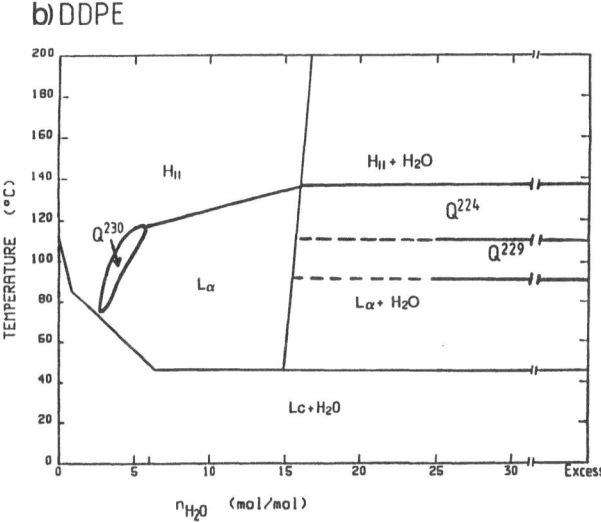

Fig. 3. Schematic binary phase diagrams of (a) DHPE and (b) DDPE in water

fusion, fat digestion, and the ultrastructural organization of cells [5, 6].

## Results and discussion

### Phosphatidylethanolamines and their methylated derivatives

Schematic binary phase diagrams for the $C_{16}$ and $C_{12}$ dialkyl (ether) compounds dihexadecyl phosphatidylethanolamine (DHPE) and didodecyl phosphatidylethanolamine (DDPE) in water are shown in Fig. 3. Our results for DHPE are in agreement with

DDPE / H₂O

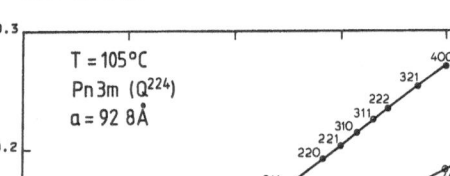

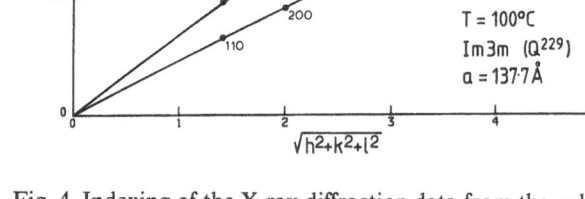

Fig. 4. Indexing of the X-ray diffraction data from the cubic phases Im3m (Q²²⁹) and Pn3m (Q²²⁴). The open circles show reflections which, although allowed by the space group, were not observed. The error in the measured $Q(hkl)$ values is less than the size of the plotted symbols

Table 1. Cubic phases observed in DDPE

| $T$ (°C) | Water volume fraction | Spacegroup type and number | Lattice parameter (Å) | Underlying periodic minimal surface |
|---|---|---|---|---|
| 91 | 12 | Ia3d (Q²³⁰) | 88 | Schoen's gyroid |
| 100 | ≥ 50 | Im3m (Q²²⁹) | 138 | Schwarz's P-surface |
| 105 | ≥ 50 | Pn3m (Q²²⁴) | 93 | Schwarz's F-surface |

a partial phase diagram previously reported for this lipid [27]. The phase diagram for DDPE extends that previously published [26]. It is seen that the effect of reducing the chainlength from $C_{16}$ to $C_{12}$ is to induce the appearance of three cubic phases, intermediate between the $L_\alpha$ and $H_{II}$ phases. At low hydration an optically isotropic phase occurs, which we tentatively identify as Ia3d (Q²³⁰), although the reflections with higher indices than (211) and (220) are extremely weak. In the excess water region, the phase sequence $L_\alpha$-Im3m (Q²²⁹)-Pn3m (Q²²⁴)-$H_{II}$ is observed upon heating. The assignment of the Pn3m cubic phase was previously reported by us [26]. The limiting hydrations of the two latter cubic phases are not yet accurately known, but are in excess of approximately 50 wt%. Severe difficulties were found in fully resolving this region of the phase diagram, probably due to metastable effects with the cubic phases. The indexing of the two highly-hydrated inverse cubic phases is shown in Fig. 4. The linearity of the plots (through the origin) gives a measure of the reliability of the indexing. The lattice parameter is given by $2\pi$ times the reciprocal gradient of the plots. The indexing as cubic aspects 8 ($T = 100\,°C$) and 4 ($T = 105\,°C$) is quite convincing. We assume that the correct space group is the most symmetrical one within a given cubic aspect, which leads to the assignments as Im3m and Pn3m respectively.

For the $C_{16}$ lipid, we have found that the effect on the binary phase diagram of headgroup methylation (which increases its hydrophilicity) is qualitatively sim-

ilar to that of reducing the chainlength, causing bicontinuous cubic phases to appear between the $L_\alpha$ and $H_{II}$ phases (J. L. Hogan and J. M. Seddon, manuscript in preparation). Structure determinations for the Pn3m and Im3m cubic phases of DDPE have been carried out using the pattern recognition approach of Luzzati and co-workers (J. M. Seddon, P. Mariani, H. Delacroix and V. Luzzati, manuscript in preparation). The results are consistent with the representations of the structures of the well-established inverse bicontinuous cubic phases, shown in Fig. 5. For each of these phases the minimal surface defines the mid-surface of the hydrocarbon chain region. Note that the representation of Ia3d in terms of networks of water/lipid rods is in no way inconsistent with the presence of the gyroid minimal surface (not shown). Rather, as the water content decreases, these rods shrink towards the edges of the skeletal graphs of the gyroid surface (the lines threading the two enantiomorphic sub-volumes which are separated by the minimal surface). It is clear that neither the interfacial mean curvature, nor the average chain length of the lipid molecules can be uniform in this case. On the other hand, for the Pn3m structure, where the water content is in the region of 50 wt%, a description where the lipid/water interfaces are nearly parallel to the F-minimal surface would be more appropriate than that shown here. However, such a representation makes the underlying symmetry elements of this phase less apparent.

Rather few examples of the Pn3m (Q²²⁴) and Im3m (Q²²⁹) cubic phases have so far been reported, and all of them, apart from the Im3m (Q²²⁹) cubic phase reported in SDS/water [3], are inverse. Pn3m (Q²²⁴) was first identified in a phosphatidylethanolamine extract from blow fly larvae, hydrated at low pH (see Ref. [5]). It has subsequently been observed in monoglycerides [13, 28], and in a polar lipid extract from the extreme thermoacidophilic archaebacterium *Sulfolobus solfataricus* [29]. Apart from DDPE, discussed above, a Pn3m (Q²²⁴) cubic phase, with a lattice parameter of $a = 122$ Å, has also recently been reported to be

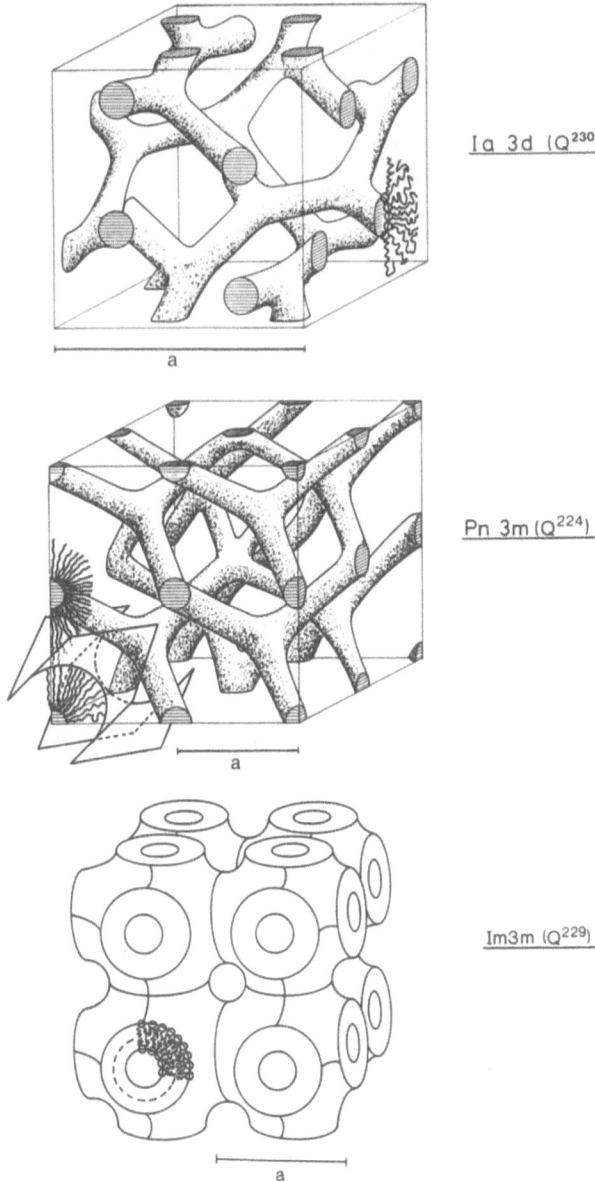

Ia 3d (Q²³⁰)

Pn 3m (Q²²⁴)

Im3m (Q²²⁹)

Fig. 5. Structures of the inverse bicontinuous cubic phases observed in hydrated DDPE

adopted by fully-hydrated dioleoyl phosphatidyle-thanolamine (DOPE), following repeated temperature cycling between −5 and 15°C [30]. For the mono-methylated derivative of DOPE, upon cooling from the $H_{II}$ phase to room temperature and then incubat-ing for very long times (months), the sample trans-formed to two coexisting cubic phases, of probable space groups Pn3m (Q²²⁴), with $a = 136$ Å, and Im3m (Q²²⁹), with $a = 175$ Å, respectively [20]. The only other example of the Im3m (Q²²⁹) cubic phase is that of the monoglycerides [28]. A striking feature of the

Pn3m (Q²²⁴) and Im3m (Q²²⁹) inverse cubic phases is that they can be stable in the presence of very large amounts of water. This is a crucial point when con-sidering the possible biological relevance of cubic phases, since biological systems normally function in a large excess of aqueous solution.

*Effect of molecular chirality*

A question which seems to us to be of great interest is that of whether molecular chirality of the lipid can lead to the formation of chiral cubic phases. For the cholesteric blue phases I and II of chiral thermotropic liquid crystals, which are peculiar cubic phases with lattice parameters in the optical region, the observed crystallographic space groups, I4₁32 (No. 214) and P4₂32 (No. 208) respectively, are chiral [31].

The first chiral *lyotropic* liquid-crystalline cubic phase, of space group P4₃32 (Q²¹²), has recently been identified in a ternary lipid/protein/water system [5]. This phase is derived from the cubic phase Ia3d (Q²³⁰) by replacing one of the networks of lipid/water chan-nels by discrete inverse micelles within which the pro-tein molecules reside. It appears that the chirality of the protein imposes the chirality of the cubic phase. However, phospholipids themselves have a chiral cen-tre in the glycerol group, which is located close to the polar/non-polar interface. Indeed, monolayers of the optically active isomer L-dipalmitoylphosphatidyl-choline were observed to form chiral solid domains, whereas the racemic mixture did not [32].

A question we have considered is whether the non-chiral cubic phase Ia3d (Q²³⁰), containing left- and right-handed interfacial networks, which we found at low water content for racemic DDPE, could also be formed by the chiral (sn-) form of the lipid. This crys-tallographic spacegroup is highly characteristic, hav-ing a large number of forbidden reflections. To try to answer this question, we studied the system dimethyl-DHPE at a water content of approximately 6.6 wt% (2.7 moles water per mole lipid), which we have found also to form the cubic phase Ia3d (Q²³⁰) between 6 and 25 wt% water at high temperatures (in the range 100−150°C). This system was more suitable than DDPE, since a larger number of Bragg reflections could be detected. The fact that the water content was very low meant that any chiral interactions should be maximal (water causes a lateral expansion of the in-terface, which will weaken any direct headgroup-head-group interactions). The indexing of the diffraction data from the cubic phase formed by the racemic mix-ture and by the optical isomer are shown in Fig. 6. Both forms index as Ia3d (Q²³⁰), with a lattice param-

## a) rac-2M-DHPE/2·6 H₂O

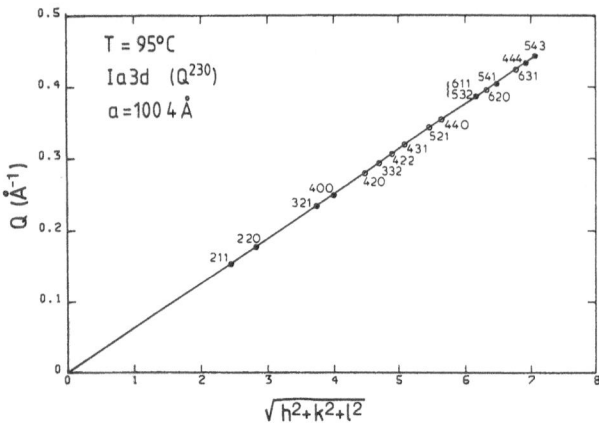

## b) sn-2M-DHPE/2·8 H₂O

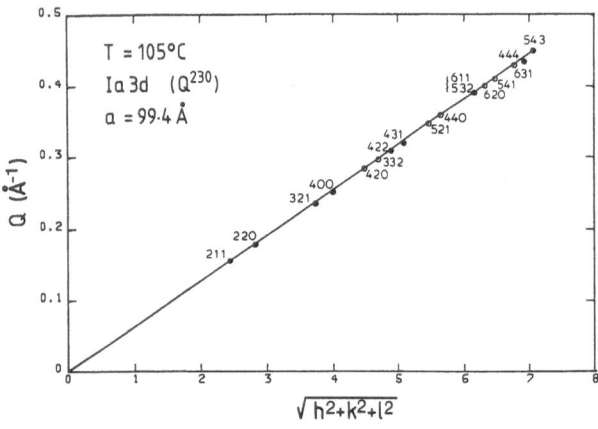

Fig. 6. Indexing of the cubic phase of both the (a) *rac*-, and (b) *sn*-forms of dimethyl-dihexadecylphosphatidylethanolamine as Ia3d (Q²³⁰). The water content was 6.3 and 6.8 wt%, and the temperature 95 and 105 °C, respectively

eter of $100 \pm 1$ Å, demonstrating that the molecular chirality has no effect on the structure of the cubic phase. It is of course possible that in the future, lipid systems with a larger molecular chirality will be be found, which will induce the adoption of chiral cubic phases.

*Phosphatidylcholine/fatty acid 1/2 (mol/mol) stoichiometric mixtures*

It has been found that fully hydrated stoichiometric 1/2 mol/mol mixtures of phosphatidylcholine (PC)

with fatty acids (FA) of the same chain length undergo melting transitions directly from the $L_\beta$ gel phase to the inverse hexagonal $H_{II}$ phase [33, 34; J. M. Seddon, G. Cevc and D. Marsh, manuscript in preparation]. We have found that, as in the case of the purely binary PE/water systems, the effect of decreasing the chainlength is to induce the formation of bicontinuous inverse cubic phases. For the $C_{12}$ mixtures, the X-ray data at 60 °C was consistent with the formation of the cubic phase Pn3m (Q²²⁴) [J. M. Seddon (1987) ILL Annual Report, p. 487]. Using this system, we have attempted to test the hypothesis that the structure of this cubic phase is indeed based upon Schwarz's F-surface. To do this we performed neutron diffraction experiments on samples where the lipid methyl end groups were either protonated, or deuteriated. Since the methyl end groups should be localized close to the minimal surface (with, for example, a Gaussian distribution along the normal to the surface), the transform of the difference structure factors should then yield an approximation to the minimal surface.

In order to see how well the underlying structure of the F-minimal surface could in principle be discerned in such a neutron experiment, assuming, as in the X-ray case, that it would be possible to measure out to the (321) reflection (i.e., a resolution of approximately one quarter of the lattice parameter), we have calculated the low-resolution density profiles (programs courtesy of H. Delacroix, P. Mariani and V. Luzzati) for the F-minimal surface, using the structure factors calculated by MacKay [17]. The results are shown in Fig. 7. Owing to computational limitations, the weakest structure factor, F(311), was omitted from the calculation. The underlying form of the minimal surface can be clearly discerned. The (110) section clearly shows the snaking tunnels, and the (001) sections show the plateaus at $z = 1/4$ and $z = 3/4$ (not shown), which are characteristic of the F-surface.

The neutron data measured at the ILL, Grenoble, on instrument D17, from 1/2 mixtures of dilauroyl phosphatidylcholine (DLPC), both with fully protonated lauric acid (LA), and with methyl end group deuteriated lauric acid (d₃-LA), in 50 wt% D₂O at 60 °C are shown in Fig. 8a and 8b, respectively. The observed peaks appear to index as the 110, 200 and 220 reflections of a cubic lattice, with a lattice parameter of $138 \pm 2$ Å, but disappointingly, only three peaks were clearly resolved. This result conflicts both with our previous X-ray measurements, which were consistent with a Pn3m (Q²²⁴) cubic phase, with a lattice parameter of $111 \pm 1$ Å at this temperature, and also with our previous preliminary neutron measurements

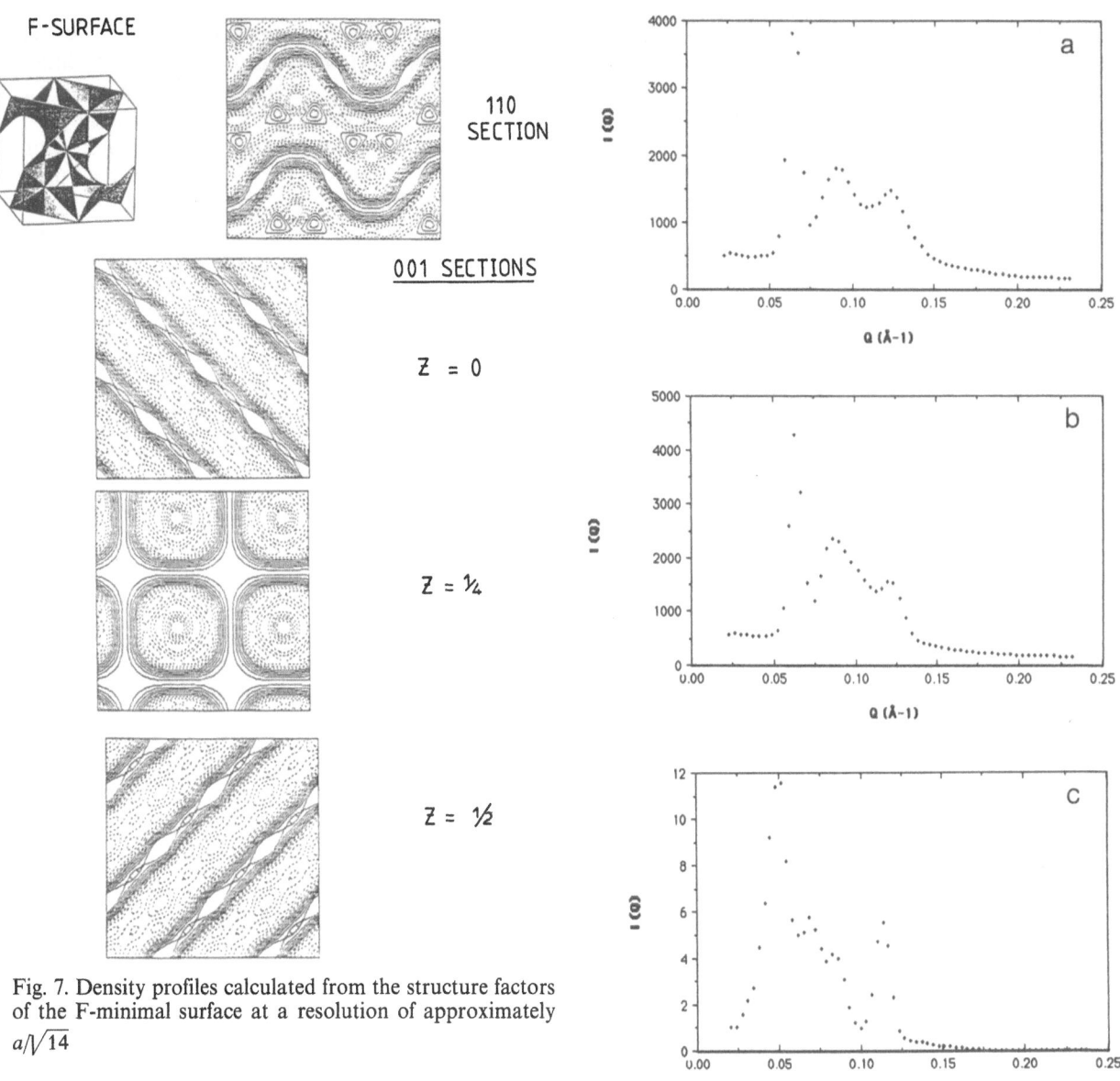

Fig. 7. Density profiles calculated from the structure factors of the F-minimal surface at a resolution of approximately $a/\sqrt{14}$

Fig. 8. Neutron diffraction patterns in 50 wt% $D_2O$ at 60 °C from 1/2 mol/mol mixtures of (a) DLPC/LA; (b) DLPC/$d_3$-LA; (c) DMPC/MA. The experiments were carried out at the ILL, Grenoble, on instruments (a,b) D17, and (c) D16. The intensity scale is in arbitrary units

at the ILL on instrument D 16, which gave peaks consistent with Pn3m ($Q^{224}$), with a lattice parameter of 108 ± 1 Å at 54 °C. The discrepancy in lattice parameter is in the wrong direction to be attributed to the steep inverse temperature dependence of this phase, and in any event a (111) reflection is not seen in the D 17 data. The data of Fig. 8a and b are actually more consistent with an Im3m ($Q^{229}$) cubic phase, and this is supported by the larger size of the lattice parameter (cf. data in Table 1). We can only conclude that these $C_{12}$ lipid mixtures probably exhibit *two* bicontinuous cubic phases, very close together in water content and/ or temperature, but more work will be required to resolve this point. Such difficulties in obtaining the

true equilibrium cubic phase in fully hydrated lipid systems has also been noted in monoglycerides [28] and unsaturated phosphatidyl-ethanolamines [30, 20], and thus seems to be inherent to these inverse bicontinuous structures. In any event, at present we cannot attempt to analyze our neutron data in terms of minimal surfaces.

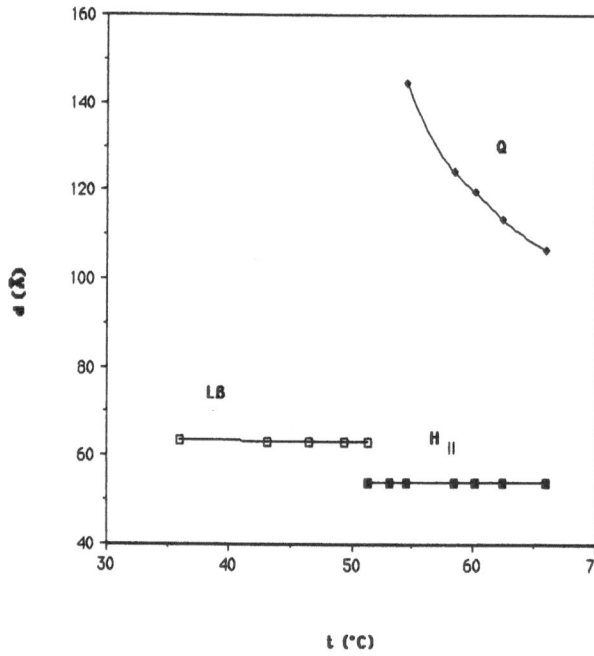

Fig. 9. Temperature dependence of the neutron diffraction d-spacings of 1/2 mol/mol mixture of DMPC/MA in 50 wt% $D_2O$. For the cubic phase, the spacings plotted are those tentatively assigned as the (110) Bragg peak

For 1/2 mixtures of the $C_{14}$ lipids dimyristoyl phosphatidylcholine (DMPC)/myristic acid (MA) in 50 wt% $D_2O$, the D16 neutron data at 60 °C presented in Fig. 8c show four Bragg reflections. The peak at $Q = 0.116$ Å$^{-1}$ corresponds to a d-spacing of $54 \pm 1$ Å, and is independent of temperature, as shown in Fig. 9. This peak is due to an inverse hexagonal $H_{II}$ phase, as confirmed by X-ray diffraction (data not shown), where higher order reflections were also detected. (Below 51 °C, two orders are obtained from an $L_\beta$ gel phase, with a d-spacing of $63 \pm 1$ Å). The other three lines index as the 110, 200 and 211 reflections of a cubic lattice, with a lattice parameter of $169 \pm 4$ Å. These could possibly be the first three allowed reflections of the cubic phase Im3m ($Q^{229}$). As seen from Fig. 9, the lattice parameter ($a = \sqrt{2} \cdot d_{110}$) of the cubic phase decreases sharply from 205 Å to 151 Å in the temperature range 54.5 to 65.9 °C. Thus the volume per unit cell decreases from $8.6 \times 10^6$ Å$^3$ to $3.4 \times 10^6$ Å$^3$ in this short temperature interval. This can only be due to an extremely steep decrease in the limiting hydration of the cubic lattice with increasing temperature. For these $C_{14}$ mixtures, we had previously only observed additional X-ray diffraction lines to those from the $H_{II}$ phase when the stoichiometry was reduced to slightly below 1/2. However, an isotropic phase was indeed recently detected by NMR for fully-hydrated $C_{14}$ 1/2 mixtures, although X-ray data indicated that it was a Pn3m ($Q^{224}$) cubic phase [35], which seems to disagree with our neutron data.

It thus appears that the behaviour of these shorter chainlength phosphatidylcholine/fatty acid mixtures is unusually complex in the excess water region, with coexisting $H_{II}$ and cubic phases occurring, and further experimental work is needed to clarify it.

## Conclusions

The formation of bicontinuous cubic phases in these phospholipid systems clearly arises from an interplay between the packing of the chains, and the hydration/hydrophilicity of the headgroups. The results reflect the way in which the lipid chemical structure influences the balance between the preferred mean and Gaussian interfacial curvatures, and the packing constraints, within these systems. One possible explanation for the appearance of intermediate bicontinuous cubic phases upon reducing the chainlength is that the difficulties of filling the interstitial hydrophobic regions of the $H_{II}$ phase become progressively more severe as the chains become shorter. A point is then reached where the system prefers to adopt phases in which the chainlength can remain more uniform, even though at the price of expressing a lower value of mean (inverse) interfacial curvature. However, another possible explanation is that the bicontinuous structure is stabilized directly, by the system developing a large value of Gaussian curvature elastic modulus relative to its mean curvature elastic modulus. A third factor which should be mentioned is that the magnitudes of the curvature elastic moduli decrease strongly with decreasing chainlength, and this may well affect the relative stability of phases. More extensive sets of data on such lipid systems are required before these different possible contributions can be reliably assessed. However, it appears from the results on the phosphatidylcholine/fatty acid mixtures that the induction of inverse bicontinuous cubic phases upon reducing the chainlength may be a general feature of those lipid systems which have a tendency to form the inverse hexagonal $H_{II}$ phase.

*Acknowledgements*

We would like to thank the staff of the Institut Laue-Langevin, in particular Dr. L. Braganza, for help in carrying out the neutron diffraction experiments. This work was supported by grant SERC GR/C/95428 to JMS.

# References

1. Seddon JM (1990) Biochim Biophys Acta 1031:1
2. Cevc G (1989) J Phys France 50:1117
3. Kekicheff P, Cabane B (1987) J Phys France 48:1571
4. Luzzati V (1968) In: Chapman D (ed) Biological Membranes. Academic Press, New York, p 71
5. Mariani P, Luzzati V, Delacroix H (1988) J Mol Biol 204:165
6. Lindblom G, Rilfors L (1989) Biochim Biophys Acta 988:221
7. Ekwall P (1975) Adv Liq Cryst 1:1
8. Tiddy GJT (1985) In: Eicke HF (ed) Modern Trends of Colloid Science in Chemistry and Biology. Birkhäuser Verlag, Basel, p 148
9. Anderson DM (1986) Ph. D. thesis, University of Minnesota
10. Tabony J (1986) Nature 319:400
11. Rädler JO, Radiman S, De Vallera A, Toprakcioglu C (1989) Physica B156 and 157:398
12. Scriven LE (1976) Nature 263:123
13. Longley W, McIntosh TJ (1983) Nature 303:612
14. Larsson K, Fontell F, Krog N (1980) Chem Phys Lipids 27:321
15. Hyde ST, Andersson S, Ericsson B, Larsson K (1984) Z Kristallogr 168:213
16. Charvolin J (1985) J Phys France 46:(C3-)173
17. MacKay AL (1985) Nature 314:604
18. Sadoc JF, Charvolin J (1986) J Phys France 47:683
19. Charvolin J, Sadoc JF (1987) J Phys France 48:1559
20. Gruner SM, Tate MW, Kirk GL, So PTC, Turner DC, Keane DT, Tilcock CPS, Cullis PR (1988) Biochemistry 27:2853
21. Gruner SM (1989) J Phys Chem 93:7562
22. Charvolin J, Sadoc JF (1988) J Phys Chem 92:5787
23. Anderson DM, Gruner SM, Leibler S (1988) Proc Natl Acad Sci USA 85:5364
24. Helfrich W (1981) In: Balian R (ed) Physics of Defects. North-Holland, Amsterdam, p 715
25. Huse DA, Leibler S (1988) J Phys France 49:605
26. Seddon JM, Cevc G, Kaye RD, Marsh D (1984) Biochemistry 23:2634
27. Caffrey M (1985) Biochemistry 24:4826
28. Caffrey M (1987) Biochemistry 26:6349
29. Gulik A, Luzzati V, De Rosa M, Gambacorta A (1985) J Mol Biol 182:131
30. Shyamsunder E, Gruner SM, Tate MW, Turner DC, So PTC, Tilcock CPS (1988) Biochemistry 27:2332
31. Jerome B, Pieranski P, Godec V, Haran G, Germain C (1988) J Phys France 49:837
32. Moy VT, Keller DJ, Gaub HE, McConnell HM (1986) J Phys Chem 90:3198
33. Marsh D, Seddon JM (1982) Biochim Biophys Acta 690:117
34. Koynova RD, Tenchov B, Quinn PJ, Laggner P (1988) Chem Phys Lipids 48:205
35. Ryba NJP, Heimburg T, Marsh D (1989) Biophys J 55:248a

Authors' address:

J. M. Seddon
Department of Chemistry
The University
Southampton SO9 5NH, UK

**Progress in Colloid & Polymer Science**                    Progr Colloid Polym Sci 81:198−202 (1990)

# The use of the specular reflection of neutrons to study surfaces and interfaces

J. Penfold

Neutron Science Division, Rutherford Appleton Laboratory, Chilton, Didcot

*Abstract:* The specular reflection of neutrons is now becoming an established technique for the study of surfaces and interfaces. A discussion of the experimental requirements is presented, and particular emphasis on the advantages for the study of problems in surface chemistry is included. A description of the reflectometer, CRISP, on the pulsed neutron source ISIS is given. Recent experimental results on the adsorption of mixed surfactants at the air-water interface are described.

*Key words:* Specular reflection; neutrons; interface surfactant

## Introduction

Most of the common optical phenomena, including refraction, reflection and interference, have been observed with slow neutrons [1]. It has been shown that the intensity of reflected and transmitted neutrons follow the same laws as electromagnetic radiation with the electric vector perpendicular to the plane of incidence [2]. Hence the refractive index at the boundary between two media is defined in the usual way as,

$$n = k_1/k_0 \tag{1}$$

where $k_1$, $k_0$ are the neutron wave vectors inside and outside the media.

The refractive index is commonly written as,

$$n = 1 - \lambda^2 A + i\lambda C \tag{2}$$

where $A = Nb/2\pi$, $C = N\sigma_a/4\pi$, $N$ is the atomic number density, $b$ is the bond coherent scattering length, $\sigma_a$ is the adsorption cross-section, and $\lambda$ is the neutron wavelength.

The refractive index for most media is less than unity; $(1 - n)$ being typically of the order of $10^{-6}$, and as a result total external reflection is observed at very small glancing angles. Since the first measurements of total reflection by Fermi and coworkers [3], it has been extensively used in neutron polarisers [4] and neutron guides [5]. In recent years, however, attention has focused on the application of specular reflection

of neutrons to study surface and interfacial problems. It was shown by Thomas and coworkers [6] that neutron reflection experiments give information about the neutron refractive index profile normal to the surface, and that a judicious use of hydrogen/deuterium contrast can provide unique information for a range of problems in surface chemistry. Due to the magnetic dipole interaction, magnetic materials exhibit a neutron spin dependent refractive index (this is the basis of neutron spin polarisers using critical reflection) and Felcher [7] has shown that the specular reflection of spin polarised neutrons is a particularly sensitive probe of surface magnetism.

The advent of dedicated spectrometers [8, 9] has been accompanied by a rapid expansion in the scientific application of the technique to surface chemistry, surface magnetism and solid films.

## Experimental

The essence of a neutron reflection experiment is to measure the specular reflection over a wide range of wave vector transfers ($Q = 4\pi \sin\theta/\lambda$, $\theta$ is the glancing angle of incidence) perpendicular to the reflecting surface. The wide $Q$ range can be achieved either by using a monochromatic beam and scanning a larger number of angles, or by using the broad band neutron time-of-flight (TOF) method to determine $\lambda$ at fixed $\theta$. As the critical glancing angles are small, narrow well collimated beams are required. To date, the majority of the reactor based measurements have been made using monochromatic long wavelength neutrons and a $\theta - 2\theta$ angular scan. However, on a pulsed source, such as ISIS, the natural way to make the measurement is the white beam TOF method. The fixed sample geometry ensures a constant sample illumination, and the $Q$ resolution (dominated by the $\Delta\theta$ contribution) is essentially constant over the wide $Q$ range available.

A schematic diagram of the CRISP reflectometer, on the ISIS pulsed neutron source, is shown in Fig. 1. Although the reflectometer has been described in detail elsewhere [8]: the important features are described here. It views the 20 K hydrogen moderator giving an effective wavelength range 0.5 to 13 Å. The beam is inclined at 1.5° to the horizontal (specifically for liquid surfaces, and a horizontal slit geometry is used giving typical beam dimensions 40 mm width and between 0.25 and 6 mm height; the beam size and divergence is variable and defined by two cadmium apertures (S1, S2). A single disc copper (C) defines the wavelength band ($\Delta\lambda$) and provides some frame overlap suppression. Additional suppression is provided by a series of frame overlap mirrors ($F$) which are set to reflect out of the main beam neutrons of wavelengths greater than 13 Å.

The detector ($D$) (a single well shielded He$^3$ detector, or a one dimensional multidetector with a positional resolution of <1 mm) is located some 1.75 mm from the sample position.

The experimental arrangement is extremely flexible and solid films can be studied over a range of angles from 0.25° to 3°; liquid surfaces can be studied at angles less than 1.5° by the insertion of a supermirror. The $Q$ range available is $\sim 4 \times 10^{-3}$ to 0.65 Å$^{-1}$, and the limiting reflectivity is $\leqslant 10^{-6}$. A further demonstration of the flexibility of the spectrometer is that it can, at ease, be converted to a polarised neutron reflectometer [11]. The incident beam is spin polarised by a cobalt-titanium supermirror, and the spin direction controlled by a static guide field and a Drabkin spin flipper; to give good polarising efficiency ($\geqslant 0.99$) and flipping efficiency ($\geqslant 80\%$) over the wavelength range $2-13$ Å.

*Model fitting*

To date, neutron reflection data has been analysed by model fitting, using standard optical equations (for example, Fresnel's law), and methods developed for multilayer optics [12].

In particular, the matrix method of Born and Wolf [13] provides a convenient framework to calculate reflectivities for systems that can be treated as a series of discreet layers. However, for multilayer systems where the interfaces are non-ideal (diffuse or roughened) then this approach rapidly becomes numerically unwieldy. A suitable alternative is the method of Abeles [14] which in conventional optics defines a characteristic matrix per layer in terms of Fresnel coefficients, and phase factors from the relationship between the electric vectors in successive layers such that the characteristic matrix per layer is,

$$C_M = \begin{bmatrix} e^{\imath\beta}_{m-1} & r_m e^{\imath\beta}_{m-1} \\ r_m e^{-\imath\beta}_{m-1} & e^{-\imath\beta}_{m-1} \end{bmatrix} \quad (3)$$

where, $r_m$ is the Fresnel reflection coefficient at the $m-1$, $m$th interface such that

$$r_m = \frac{p_{m-1} - p_m}{p_{m-1} + p_m} \quad (4)$$

and $\beta_{m-1} = (2\pi/\lambda)n_{m-1}\sin\theta_{m-1}$.

Following Cowley and Ryan [15] it is now possible to include a roughened or diffuse interface at each boundary (without dividing it into a series of discret layers) by introducing a Gaussian roughness factor of the form described by Nevot and Crocé [16] such that

$$r_m = \frac{p_{m-1} - p_m}{p_{m-1} + p_m}\exp - 0.5(q_{m-1}q_m\langle\sigma\rangle^2) \quad (5)$$

where $\langle\sigma\rangle$ is the root mean square roughness, $q_{m-1} = 2k\sin\theta_{m-1}$ and $q_m = 2k\sin\theta_m$.

For $N$ layers the matrix elements $M_{11}, M_{12}$ of the resultant matrix for the system gives the reflectivity

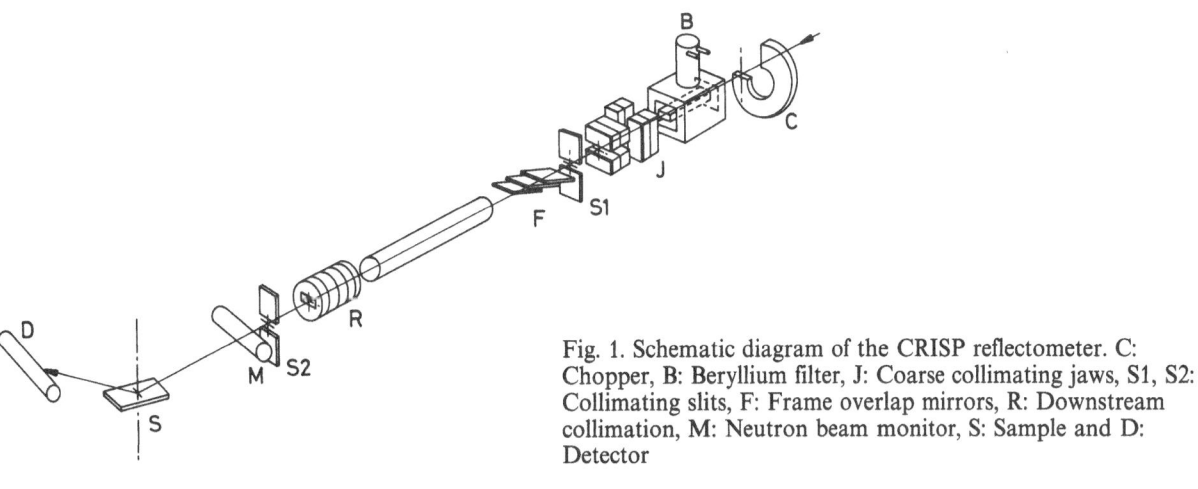

Fig. 1. Schematic diagram of the CRISP reflectometer. C: Chopper, B: Beryllium filter, J: Coarse collimating jaws, S1, S2: Collimating slits, F: Frame overlap mirrors, R: Downstream collimation, M: Neutron beam monitor, S: Sample and D: Detector

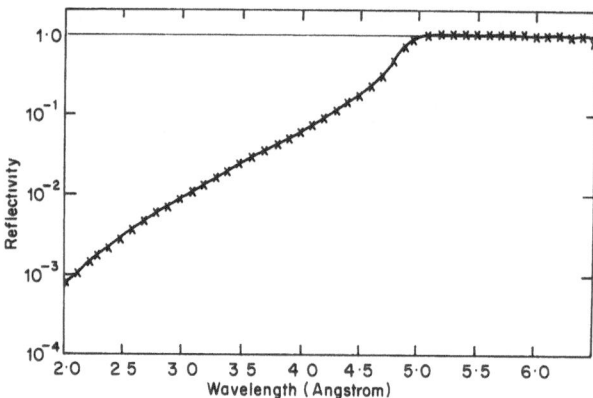

Fig. 2. Reflectivity versus wavelength for $\theta = 0.3°$ for $\lambda/10$ optical flat (i) + data points, (ii) solid line is a least squares fit with Nb = $0.36 \times 10^{-5}$, $\Delta\theta = 5.01\%$ and $\sigma = 32.12$ Å

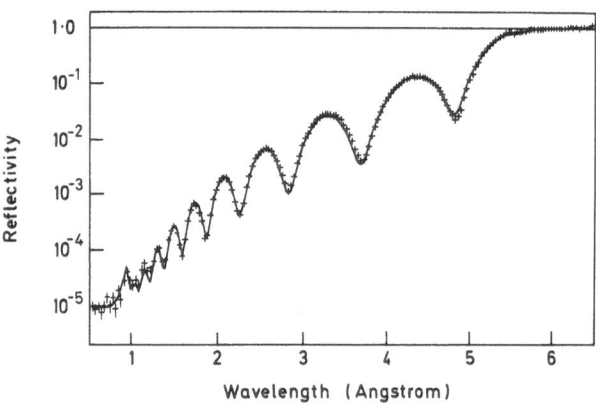

Fig. 3. Reflectivity versus wavelength for $\theta - 0.5$ for a 10 bilayer deuterated Cadmium docosanoate Langmuir-Blodgett film deposited onto a silicon wafer (i) + data points, (ii) solid line is a least squares fit for a single uniform film of $d = 967.6$ Å, Nb = $0.767 \times 10^{-5}$ Å$^{-2}$, $\sigma_1 = 14.77$ Å, $\sigma_2 = 14.3$ Å, $\Delta\theta = 4.58\%$ and level background of $2 \times 10^{-5}$

$$R = \frac{M^{21} M^{21*}}{M_{11} M_{11}*}. \qquad (6)$$

This method provides a convenient and closed form for calculating exactly reflectivity profiles for real systems [12].

An illustration of the application of these methods to some standard samples measured on the CRISP reflectometer are shown below.

Figure 2 shows the reflectivity versus neutron wavelength for a $\lambda/10$ optical flat. The solid line is a non-linear least squares fit assuming a surface roughness of the fit described above (the detailed parameters associated with the Fresnel fit are included in the figure caption).

Figure 3 shows the application to a 10 bilayer deuterated cadmium docosanote Langmuir-Blodgett film deposited onto a silicon wafer. The solid line is a non-linear least squares fit assuming a total layer thickness of 967.6 Å

(28.4 Å per layer), with a mean scattering density of $0.77 \times 10^{-5}$ Å$^{-2}$ (further details of the fitted model parameters are included in the figure caption).

### Scientific results

An extensive scientific programme [10] has emerged very quickly on He CRISP reflectometer, in surface chemistry, surface magnetism and solid films. In surface chemistry much of the initial interest has been in the adsorption of surfactants, polymers and fatty acids and the air-liquid interface, and more recently has been successfully extended to the liquid-solid interface, and liquid-liquid interfaces.

Since neutron reflection is related to the refractive index profile normal to the surface, and since the refractive index profile is directly related to the composition profile, it is clear that the specular reflection technique can be used to obtain both the amount adsorbed and structural information about the layer. It is particularly important for the study of adsorption at the air solution interface due to the possibility of using isotopic substitution. As a majority of chemical systems contain some protons, and protons and deuterons has opposite sign, H/D substitution is used extensively.

In the reflection experiment it is possible to choose the hydrogen/deuterium ratio for many solvents such that there is no specular reflection. In the case of surfactants, the solution will still be null reflecting even if the solute is fully deuterated. Any reflection will then of course result entirely from the surface adsorption of the surfactant, and so the technique will be specific to the adsorption of the solute. It is also possible to eliminate the reflection from the solute and determine the surface profile of the solvent. Isotopic substitution can also be used to highlight particular parts of a solute molecule by selective deuteration.

These features were used to some advantage in an early study on the adsorption of the surfactant decyltrimethyl ammonium bromide (DTAB) at the air-solution interface [17]; where it was possible to determine not only the surface excess but the detailed surface structure. It has also successfully been applied to the system tetramethylammonium dodecyl sulphate (TMDS) in water [18]. By selective deuteration of the tetramethyl ammonium counterion it was possible to obtain additional information about the degree of counterion binding, and the extent of the diffuse counterion layer.

Classically, the principle method to study surfactant adsorption is to combine measurements of the surface

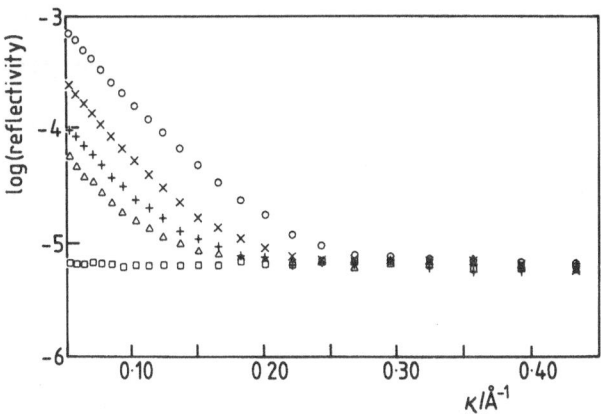

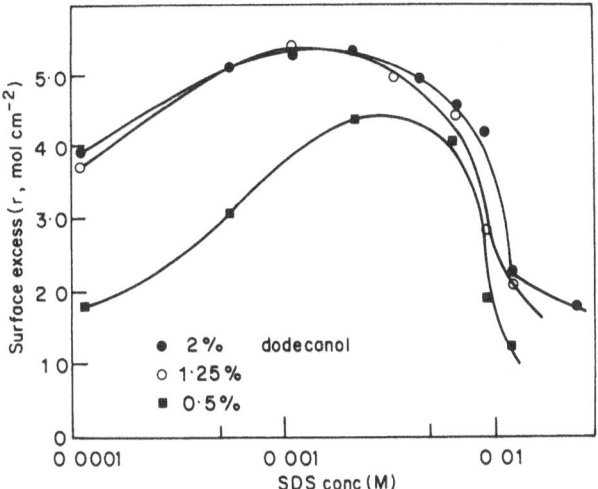

Fig. 4. Reflectivity profiles of deuterated dodecanol/sodium dodecyl sulphate (SDS)/null reflecting water mixtures (background not subtracted). Circles show the profile of a spread monolayer of deuterated dodecanol on null reflecting water, squares show the reflectivity of 0.009 M SDS in water. The remaining three profiles are for 1.25% dodecanol in SDS and a varying SDS concentration of 0.00675 M ($\times$), 0.009 M (+) and 0.012 M ($\triangle$). The cmc is 0.008 M

Fig. 5. Adsorption isotherm for dodecanol in dodecanol/sds mixtures derived from reflectivity measurements

tension and of the activity of one of the components using the Gibbs equation. Neutron reflection determines not only the surface excess but can provide additional information on the surface structure. This is particularly true for concentrations of surfactant above the critical micellar concentration (CMC) when the Gibbs equation loses its effectiveness and for mixed systems.

The nature of adsorption from mixtures of surfactants has been relatively little studied because the Gibbs equation becomes rather cumbersome to apply. A classical problem in surface tension measurements is the minimum in the surface tension often observed in the sodium dodecyl sulphate (SDS) water system, which arises from the presence of small amounts of the impurity dodecanol.

Dodecanol is almost insoluble in water and on its own forms a close-packed monolayer on the surface. Thus a very small amount in an SDS solution below the cmc will preferentially adsorb at the surface. Above the cmc the SDS solubilizes the dodecanol and removes it from the surface. Since the surface free energy of a monolayer of SDS is greater than one containing a proportion of dodecanol, the surface tension will then rise at the cmc. In principle the Gibbs equation may be used to determine independently the surface excesses of SDS and dodecanol, but the determination of the activity of the latter is not easy. Figure 4 shows how easy it is to measure the dodecanol concentration by reflectivity. The figure shows the reflectivity of SDS

(protonated) in null reflecting water and, as would be expected, since both surfactant and water are contrast matched to air, there is no reflectivity. On the other hand the spread monolayer of pure deuterated dodecanol gives a strong reflected signal. The addition of a trace amount of deuterated dodecanol in the system leads to a significant increase in the reflectivity because the dodecanol displaces SDS from the monolayer. The actual amount has been determined over a wide range of conditions, leading to a set of isotherms of dodecanol under conditions of constant SDS concentration [19] (see Fig. 5). The preliminary analysis of recent results with either the dodecanol or the SDS deuterated and with protonated and deuterated subphases indicate that the detailed surface structure of both components can be determined.

*Acknowledgements*

R. C. Ward, C. Shackleton, J. Herdman, W. G. Williams and R. Felici have contributed to the development of CRISP. R. M. Richardson provided the Langmuir-Blodgett film.

The mixed surfactant measurements have been carried out in collaboration with R. K. Thomas, E. Simister, E. M. Lee and A. Rennie.

## References

1. Klein AG, Werner GA (1982) Rep Prog Phys 46:259
2. Lechner J (1987) Theory of Reflection. Martinus Nijholf, Dortrecht
3. Fermi E, Zinn W (1946) Phys Rev 70:103
4. Hayter JB, Penfold J, Williams WG (1978) J Phys E 11:454

5. Maier-Leibnitz H, Springer T (1963) Reactor Sci & Tech, J of Nucl Energy, Parts A & B 17:217
6. Hayter JB, Highfield RR, Pulman BJ, Thomas RK, McMullen AI (1981) J Chem Soc Faraday, Trans 1 97:1437
7. Felcher GP (1981) Phys Rev B 24:1995
8. Penfold J, Ward RC, Williams WG (1987) J Phys E 20:1411
9. Felcher GP, Hilleke RD, Crawford RK, Haumann J, Kleb R, Ostroski G (1987) Rev Sci Inst 58:609
10. Penfold J, Thomas RK, J Cond Matt Phys 2 (1990) 1369
11. Felici R, Penfold J, Ward RC, Williams WG (1988) Appl Phys A45:169
12. Penfold J (1988) RAL-88-088
13. Born M, Wolf E (1970) Principles of Optics, Pergamon Press, Oxford
14. Heavens OS (1955) Optical Properties of Thin Films, Butterworths, London
15. Cowley RA, Ryan TW (1987) J Phys D: Appl Phys 20:61
16. Névok L, Croie P (1980) Phys Appl 15:761
17. Lee EM, Thomas RK, Penfold J, Ward RC (1989) J Phys Chem 93:381
18. Penfold J, Thomas RK, Lee EM (1989) Mol Phys 68:33
19. Thomas RK, Rennie A, Penfold J, Lee EM, Simistei E to be published

Author's address:

J. Penfold
Neutron Science Division
Rutherford Appleton Laboratory
Chilton Didcot
Oxon OX11−0QX

**Progress in Colloid & Polymer Science**    Progr Colloid Polym Sci 81:203−208 (1990)

# The specular reflection of neutrons from interfacial systems

E. M. Lee[1]), R. K. Thomas[1]) and A. R. Rennie[2])

[1]) Physical Chemistry Laboratory, University of Oxford, Oxford, UK
[2]) Institut Laue Langevin, Grenoble, France

*Abstract:* The application of new technique of specular neutron reflection to the study of the solution/air and solid/solution interfaces is reviewed. Examples are given of the determination of the structures of adsorbed layers of tetramethylammonium dodecyl sulphate (TMDS) and decyl trimethylammonium bromide (DTAB) at the air/liquid interface. In both of these systems the layer consists of an outer alkyl chain region, which is not close-packed and which contains no water, and a head group region containing the head groups, counterions, water and a small fraction of alkyl chains. The non-ionic surfactant hexaoxyethylene glycol monododecyl ether ($C_{12}E_6$) is shown to form a bilayer at the quartz/aqueous solution interface at concentrations close to the c.m.c. However, this bilayer is found to be defective and might be better described as a layer of flattened micelles. The structure persists down to coverages where less than half the geometric surface is covered with bilayer.

*Key words:* Reflection; neutrons; adsorption; surfactant; interface

## Neutron specular reflection

The variation in the reflectivity of an interface with glancing angle depends on the neutron refractive index profile normal to the interface. The neutron refractive index is simply related to composition by

$$\eta = 1 - (\lambda^2/2\pi)\varrho_s \qquad (1)$$

where $\varrho_s$ is the scattering length density given by

$$\varrho_s = \sum n_i b_i \qquad (2)$$

where $n_i$ is the number density of the $i$th nucleus and $b_i$ its scattering length. A measurement of the neutron reflectivity profile can therefore lead to information about interfaces [1, 2].

The reflectivity profile may be calculated exactly for any model structure using the optical matrix method for light polarized perpendicular to the reflection plane [3]. A clearer understanding of the phenomenon may be obtained by using the kinematic approximation [4] in which the reflectivity is given as a function of momentum transfer $\varkappa$ by the equivalent expressions

$$R(\varkappa) = \frac{16\pi^2}{\varkappa^2}|\varrho_s(\varkappa)|^2 \qquad (3)$$

or

$$R(\varkappa) = \frac{16\pi^2}{\varkappa^2}|\varrho_s'(\varkappa)|^2 \qquad (4)$$

where $\varkappa = 4\pi\sin\theta/\lambda$, $\theta$ being the glancing angle, $\varrho_s(\varkappa)$ is the one dimensional Fourier transform of $\varrho_s(z)$, and $\varrho_s'(\varkappa)$ is the Fourier transform of the derivative of $\varrho_s(z)$. It follows from [4] that the reflectivity of a simple interface with a step $\Delta\varrho_s$ between the two media will vary as $1/\varkappa^4$. A layer adsorbed at this interface will give rise to a reflectivity profile which is the product of the Fourier transform of the scattering length density gradient within the layer and the $1/\varkappa^4$ decay. For a uniform layer there will be a series of superimposed fringes, whose spacing is related inversely to the thickness of the layer and whose amplitude depends on the composition and density of the layer. If the layer consists of more than one component, for example both solvent and solute, it may not be possible to disentangle the density and composition.

Because the difference in neutron refractive index across any interface is small the reflectivity is also low except at very small glancing angles. In this region total external reflection may further enhance the signal. Since there is always some other source of scattering, for example incoherent scattering from proton containing systems, small angle scattering, or quasielastic scattering, the range of momentum transfer over which the reflectivity can be measured is restricted to below the value where the background scattering swamps the rapidly diminishing reflected signal. This will vary from sample to sample but at present $\varkappa_{max}$ is typically about 0.35 Å$^{-1}$, somewhat smaller than in the closely related X-ray reflection experiment.

Conceptually the neutron reflection experiment is simple. Since $R$ has to be known as a function of $\varkappa$ it can either be measured as a function of angle at a fixed wavelength or as a function of wavelength at a fixed angle. The latter has the advantage for liquids that the sample need not be moved. This is the method employed on the CRISP reflectometer at ISIS at the Rutherford-Appleton Laboratory [5]. The former method has features which may offer advantages for particular systems. For example, the wavelength can be increased beyond the Bragg cut-off to give extra transparency to a crystal in an investigation of the solid/liquid interface. The variation of angle at a fixed wavelength is the main method used at the Institut Laue-Langevin, Grenoble, on the D17 spectrometer [6].

## Contrast variation

There are two special features of the neutron reflection experiment, which distinguish it from the X-ray technique. The first is that condensed matter is generally much more transparent to neutrons than to X-rays and it is therefore possible to do neutron reflection experiments at solid/liquid and liquid/liquid interfaces. The technique is likely to be particularly effective at the solid/liquid interface because many crystalline materials become almost totally transparent to neutrons when the neutron wavelength is too long for Bragg diffraction to occur, typically about 6–7 Å. The neutron beam may then be passed through the crystal and reflected from the interface between crystal and solution. The second distinguishing feature of neutrons is the possibility of changing the reflectivity from a given chemical system by isotopic substituion. This technique of "contrast variation" has previously been used successfully in small angle scattering. In reflectivity experiments contrast variation can be used to make the technique genuinely specific to the surface layer.

Contrast variation is most widely applicable for proton containing systems, although it should also be effective for a number of other isotopes of which examples are chlorine, lithium and nickel. The scattering length densities of $D_2O$ and $H_2O$ are of opposite sign and it is therefore possible to make up an $H_2O/D_2O$ mixture whose refractive index is exactly matched to that of air. That there is no specular reflection at all from this mixture is illustrated in Fig. 1. Figure 1a shows the specular reflection from the $D_2O$/air interface as observed on a square multidetector. The specular beam occurs near the centre of the lower part of the detector and has an integrated intensity of the order of 10$^7$ counts. The signal from the null reflecting water (n.r.w.) is shown in Fig. 1b. At the position of the specular peak the level of signal is of the order of 100 counts, which is indistinguishable from the background incoherent scattering. For a dilute solution of a deuterated surface active species in n.r.w. there will be no reflection unless there is preferential adsorption of the solute at the air/solution interface, in which case the reflected signal depends only on the adsorbed layer. Under these circumstances the technique is surface specific.

## Surfactants at the air liquid interface

In the case of adsorption at the surface of null reflecting water the reflectivity profile gives both the structure of the layer and the absolute amount adsorbed at the interface, irrespective of the density profile of the solvent normal to the surface. The effect of change of amount adsorbed is illustrated in Fig. 2 where are shown the reflectivity profiles of alkyl chain deuterated tetramethylammonium dodecyl sulphate (TMDS) at different concentrations in n.r.w. [7]. The critical micelle concentration (c.m.c.) of TMDS is 0.057 M and the concentrations shown in Fig. 2 range from about 1/10th to just above this value. The incoherent scattering background has been subtracted so that the reflectivities plotted in the figure are entirely from the adsorbed layer of TMDS. It should be noted that the vertical scale is logarithmic and that the reflectivity changes by a factor of about two over the concentration range used. It is straightforward to derive the thickness $\tau$ of the layer and the area per molecule $A$ from this data. At the c.m.c. $A$ is 60 $\pm$ 2 Å$^2$ and at the lowest concentration shown in the figure, 0.00047 M, it increases to 108 $\pm$ 5 Å$^2$. Other isotopic data, to be discussed below, indicate that the model of a single uniform layer is not an adequate description of this interface. However, in this range of

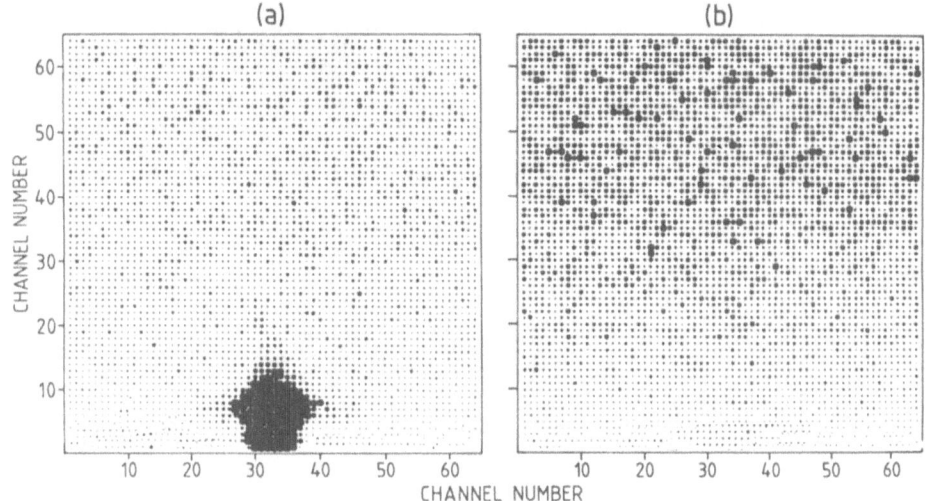

Fig. 1. Neutron scattering patterns registered on a multidetector for (a) $D_2O$ and (b) null reflecting water. The dots in increasing order of size represent intensities in 100 count intervals from 0 to 800. The completely filled points represent intensities greater than 800. The specular reflection peak from $D_2O$ reaches a maximum of 244000 counts. The momentum transfer is such that reflection from the $D_2O$ is close to 100%

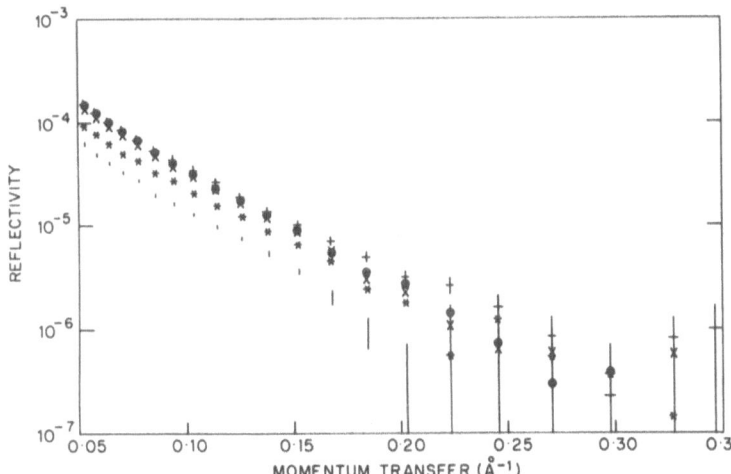

Fig. 2. Reflectivity profiles of hTMdDS in null reflecting water at concentration of 0.000475 (|), 0.00095 (*), 0.00285 (×), 0.0057 (○), and 0.012 M (+). The flat incoherent scattering background has been subtracted

momentum transfer the slope of a plot of $\log R$ against $\varkappa$ depends approximately upon $\tau$ and it can therefore be concluded that $\tau$ does not change significantly over this range of concentration. This is in contrast to results obtained for decyl trimethylammonium bromide (DTAB), to which we now turn.

A refinement of the contrast variation method is to eliminate the contribution not only of the solvent but of part of the adsorbed layer itself. This is illustrated by the reflectivity of different isotopic species of DTAB in solution in null reflecting water, shown in Fig. 3 [8]. The three methyl groups of the head group of DTAB or the alkyl chain may be deuterated. The scattering length densities of these groups when protonated are sufficiently close to that of n.r.w. that they make no significant contribution to the reflectivity.

Thus the whole layer may be studied in isolation if the reflectivity of a solution of fully deuterated DTAB (dDdTAB) in n.r.w. is measured. The individual parts of the layer may be studied by using dDhTAB or hDdTAB. The reflectivity profiles of these three species are completely different, there being changes in both the level of reflectivity and its shape. The greater slope of the fully deuterated species in comparison with the chain deuterated species shows that the whole molecule thickness is greater than just the alkyl chain, proving unequivocally that the molecule is oriented at the interface. The power of the method lies in the quantitative fit of a single structure to all three sets of data. This leads to the structure shown in Fig. 4a where two separate regions of the layer can be distinguished, the alkyl chain and head group regions. Un-

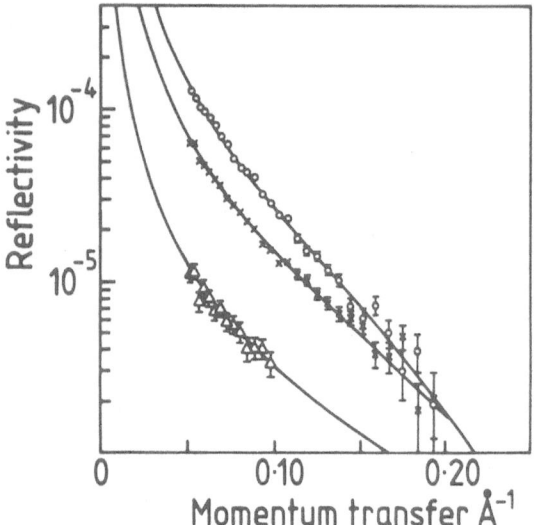

Fig. 3. Observed and calculated specular reflectivity profiles of isotopic species of 0.05 M DTAB in null reflecting water. In order of decreasing reflectivity the isotopic species are dDdTAB, dDhTAB, and hDhTAB. The continuous lines are calculated for the model of the interface shown in Fig. 4

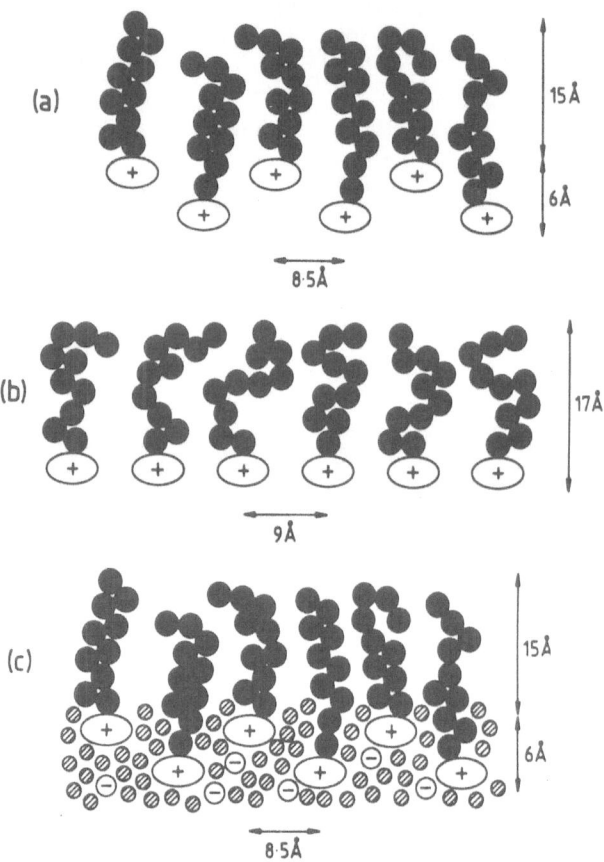

Fig. 4. Distribution of water and DTAB molecules at the air/solution interface. (a,b) Configuration of the surfactant molecules within the adsorbed layer at (a) 0.05 M and (b) 0.01 M or less. (c) Full structure of the layer at 0.05 M

like TMDS, where the thickness of the layer does not vary with surface concentration, the DTAB layer thins on lowering the concentration and the resulting structure is shown in Fig. 4b.

So far it has been shown that contrast variation may be used to study the surfactant in "isolation". At least as interesting is the role of the solvent at the interface. This can be studied by matching the contrast of the surfactant to that of air whilst enhancing the contrast of the solvent. Then only the density profile of the solvent at the interface is obtained. For dilute surfactant solutions this is approximately achieved by using the protonated surfactant in $D_2O$. It is reasonable to assume that the overall density of the interfacial region cannot exceed that of the bulk components by more than a percent or two. Then the two simplest possible models for the behaviour of the water at the interface are either that it occupies the residual volume not occupied by the surfactant and counterion in the head group region or that it occupies the residual volume in both parts of the surfactant layer. The residual volume is known from the measurements in null reflecting water and the reflectivity profiles for the two possible modes of adsorption of the water at the interface are calculated to be quite different. Experimental results on both TMDS and DTAB and on several other surfactants, including the non-ionic hexadecyl hexaoxyethylene mono ether

$(C_{16}E_6)$ [9] show that the water only occupies the head group region. The alkyl chain region therefore contains a significant proportion of air.

## Adsorbed layers at the solid liquid interface

Contrast variation can also be used to great effect in the study of adsorbed layers at the solid/liquid interface [10, 11]. For example, the scattering length density of amorphous quartz is $3.5 \times 10^{-6}$ Å$^{-2}$ compared with $6.35 \times 10^{-6}$ and $-0.58 \times 10^{-6}$ Å$^{-2}$ for $D_2O$ and $H_2O$ respectively. Just as the isotopic composition of water can be adjusted to match that of air, so it can be matched to quartz. The important difference is that a protonated surfactant is almost matched to air and hence is nearly invisible at the air/liquid interface but contrasts quite strongly with quartz. Thus, at the quartz/water interface it is not necessary,

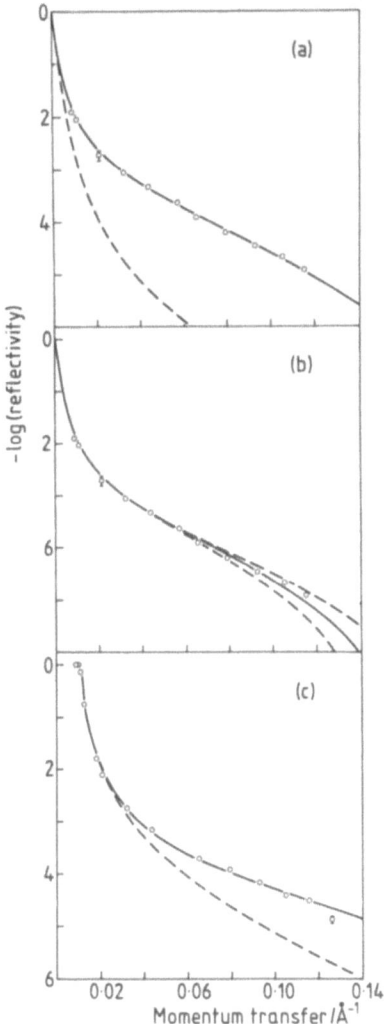

Fig. 5. Calculated and observed reflectivity profiles of $C_{12}E_6$ adsorbed at the quartz/water interface. The concentration of $C_{12}E_6$ was $8.7 \times 10^{-5}$ in (a,b) null reflecting water and (c) $D_2O$. The reflectivity of the clean surface is shown as a dashed line in (a) and (c). The continuous lines are those calculated for the structural model shown in Fig. 6. The additional dashed lines in (b) show the effect of varying the layer thickness by $\pm 10\%$

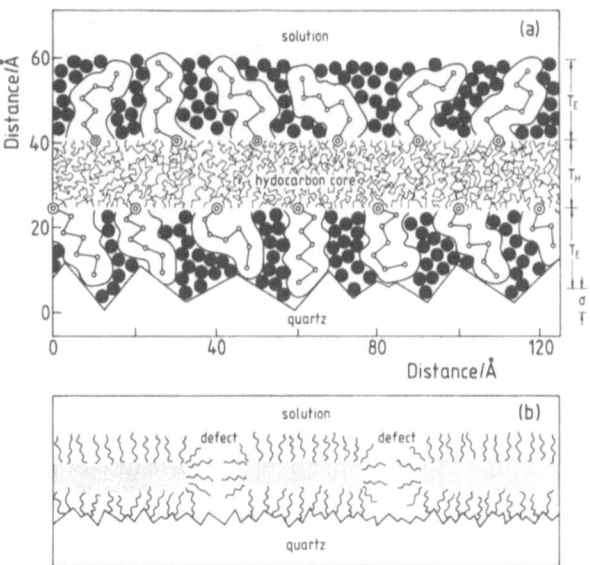

Fig. 6. Schematic diagram of the structure of a layer of $C_{12}E_6$ adsorbed at the quartz/water interface close to the c.m.c. (a) is drawn approximately to scale and also shows the main structural parameters. (b) illustrates the "defects"

although it may be useful, to use expensive deuterated surfactants. Adequate variations in contrast may be achieved by varying the isotopic composition of the solvent.

The ever present experimental problem with the solid/liquid interface, especially for a single surface, is that of cleanliness. Neutron reflection from the quartz/water interface, especially when the quartz and water are contrast matched, is sensitive to only a few percent contamination of the surface by a hydrogen contain-

ing species. The method is therefore more directly sensitive than any other technique to contamination by such species and this is a considerable asset. It is not particularly limited to hydrogen containing contaminants. Other contamination could be investigated by a wider range of solvent contrasts.

Lee et al. [10] have investigated the structure of a layer of hexaoxyethylene glycol monododecyl ether ($C_{12}E_6$) at the amorphous quartz/water interface using a polished ($\lambda/10$) face with an area of about $70 \times 30$ mm$^2$ actually in the neutron beam. Comparison between calculated and observed reflectivity profiles showed that the surface was clean before addition of the surfactant. The surface is expected to be saturated at the c.m.c. ($8.7 \times 10^{-5}$ M) and the observed reflectivity profiles indeed show large changes in reflectivity (Fig. 5a and c). In the case of the solvent being water approximately contrast matched to quartz (c.m.q.) the reflectivity increased by more than two orders of magnitude over that of the clean surface. Once again, the structure cannot be accurately characterised by a single reflectivity profile but the combination of the two sets of isotopic data leads unambiguously to the conclusion that the surfactant adsorbs as a bilayer. The model used to fit the data contained four adjustable parameters, the thickness of the ethylene oxide layers (taken to be identical), the thickness of the hydrocarbon region, the area per molecule in the bilayer, and the coverage. The thickness of the hydrocarbon core was found to be $17 \pm 3$ Å, close to the value for the fully extended chain, that of the ethylene oxide groups to be $16 \pm 3$ Å, about 90% of the length of the fully extended chain, and the area per molecule to be $23 \pm 3$ Å$^2$. The average hydration number of the ethylene oxide groups is $2.0 \pm 0.5$. The maximum coverage obtained was 0.75 of the surface. The area of surface not covered by the bilayer appeared to be an integral part of the structure suggesting that it results from defects in the bilayer structure. This could be interpreted either as a layer of flattened micelles or a defective bilayer. Unfortunately, the technique is not directly sensitive to the lateral structure and cannot give any more precise information. The structure of $C_{12}E_6$ is shown in Fig. 6.

## References

1. Hayter JB, Highfield RR, Pullman BJ, Thomas RK, McMullen AI, Penfold J (1981) J Chem Soc, Far Trans 1, 77:1437
2. Bradley JE, Lee EM, Thomas RK, Willatt AJ, Gregory DP, Penfold J, Ward RC, Waschkowski W (1988) Langmuir 4:821
3. Lekner J (1987) Theory of Reflection, Martinus Nijhoff, Dordrecht
4. Crowley TI (1984) D Phil Thesis, Oxford
5. Penfold J, Ward RC, Williams WG (1988) Appl Phys A 45:169
6. Rennie AR, Crawford RJ, Lee EM, Thomas RK, Crowley TL, Roberts S, Qureshi MS, Richards RW (1989) Macromolecules 22:3466
7. Penfold J, Lee EM, Thomas RK (1989) Mol Phys 68:33
8. Lee EM, Thomas RK, Penfold J, Ward RC (1988) J Phys Chem 93:381
9. Lee EM, Pearce Y, Simister EA, Thomas RK, Penfold J (unpublished)
10. Lee EM, Thomas RK, Cummins PG, Staples EJ, Penfold J, Rennie AR (1989) Chem Phys Lett 162:196
11. Lee EM, Thomas RK, Simister EA, Penfold J, Rennie AR, Langmuir in press

Authors' address:

Dr. R. K. Thomas
Physical Chemistry Laboratory
South Parks Road
Oxford, OX1 3QZ, England

**Progress in Colloid & Polymer Science**                 Progr Colloid Polym Sci 81:209−214 (1990)

# Film bending elasticity in microemulsions made with nonionic surfactants

L. T. Lee[1]), D. Langevin[1]), J. Meunier[1]), K. Wong[2]) and B. Cabane[2])

[1]) Laboratoire de Physique Statistique de l'ENS, France
[2]) DPC-SCM, CEA-CEN Saclay, Gif-sur-Yvette, France

*Abstract:* The interfaces between water and alkanes in the presence of a monolayer of nonionic surfactant (polyethylene glycol alkyl ether) have been studied as a function of temperature for surfactants of different chain lengths. At low temperatures and moderate surfactant concentration, the aqueous phase is an oil in water microemulsion; at high temperatures, the oil phase is a water in oil microemulsion. In our study, the middle of the studied temperature range corresponds to the inversion temperature where a middle phase microemulsion coexists with both oil and water. − The interfacial tensions, which are very low, have been measured with surface light scattering and the interfacial film roughness with ellipsometry. From these data, the bending elastic constant of the surfactant monolayer is determined. The calculated persistence length of the monolayer is compared to the characteristic size in the middle phase microemulsion measured with neutron scattering.

*Key words:* Non ionic surfactants; bicontinuous microemulsions; interfacial tension; persistence length; film bending elasticity

## Introduction

Microemulsions are mixtures of oil, water and surfactant molecules. Although the corresponding phase diagrams do not exhibit features qualitatively different from those of ternary mixtures of small molecules, when the surfactant chain length and concentration are sufficiently large, the formation of specific liquid crystalline phases is observed. These phases result from molecular segregation of oil and water, the oil and water microdomains being separated by surfactant monolayers. This type of segregation is expected to remain present in microemulsion phases formed at lower surfactant concentrations.

The present study was initiated with nonionic surfactants where considerable information on phase diagrams were available [1−3]. Our aim is to show that the surfactant film properties, in particular the bending elasticity, are essential in determining the microemulsion bulk structure. The data presented in this paper correspond to microemulsions where the film spontaneous curvature is about zero, and where it is expected that the size of the microstructural elements is an exponential function of the bending elastic modulus as in the case of the film persistence length [4, 5].

## Background

### Structural models

All microemulsions are characterized by well-defined sizes. Details of the various models describing microemulsion structures have been discussed elsewhere [6]. In general, for droplet microemulsions:

$$R = \frac{3\phi_o}{C_s \Sigma} \quad \text{or} \quad R = \frac{3\phi_w}{C_s \Sigma} \tag{1}$$

for oil-in-water or water-in-oil droplets respectively. $R$ is the radius of the microemulsion droplet, $\phi_o$ and $\phi_w$ are the volume fractions of oil or water in the microemulsions, $C_s$ is the number of surfactant molecules per unit volume and $\Sigma$ the parking area of the surfactant molecule.

For middle phase bicontinuous microemulsions where both oil and water microdomains are present and separated by surfactant films, the characteristic size, $\xi$, of the oil or water microdomain is proportional to both the volume fractions of oil and water as follows [7]:

$$\xi = \frac{6\,\phi_o\,\phi_w}{C_s\,\Sigma}. \qquad (2)$$

Furthermore, for these bicontinuous phases, there is a critical length scale which serves as a yardstick for determining whether the film surface is flat or wrinkled. This is the persistence length, $\xi_k$. Thus, at observation scales smaller than $\xi_k$, the surface is flat, and at scales larger than $\xi_k$, the surface appears randomly undulated. In the cubic space-filling model of de Gennes and Taupin [4], $\xi_k = \xi$ while in other models, $\xi_k \propto \xi$ [5]. The persistence length is related to the bending elastic modulus, $K$, of the film by [4]:

$$\xi_k = a\exp(2\pi K/kT) \qquad (3)$$

where $a$ is a molecular length. From this relationship, one can see the role of film rigidity in the long-range order of the microemulsion structure. If $K$ is small ($< kT$), $\xi_k$ is of the order of 100 Å and structural inversion from oil-in-water to water-in-oil droplet takes place via a bicontinuous phase. On the other hand, if $K$ is large ($\gg kT$), then $\xi_k$ is macroscopic and structural inversion takes place via a lamellar phase.

## Experimental

### Systems

The microemulsions studied here are ternary systems consisting of water, alkane and nonionic surfactant. Three surfactants of increasing chain lengths, $C_8E_3$ (from Bachem), $C_{10}E_4$ and $C_{12}E_5$ (from Nikko Chemicals) have been chosen in order to vary the bending elasticity properties of the surfactant film. The corresponding alkanes are decane, octane and hexane respectively. A different alkane is chosen in each case so that the inversion temperatures for all three systems fall in the same region, between $20-30\,^\circ\text{C}$.

The initial water-alkane weight ratio is 1:1 and the total surfactant concentrations are 4, 2 and 1 weight percent for $C_8E_3$, $C_{10}E_4$ and $C_{12}E_5$. For each system, the components are weighed in a tube and mixed gently. The mixture is then bubbled with $N_2$ and left to equilibrate in a temperature bath. After phase separation of the microemulsion, the water and oil phases are transferred carefully into the respective surface tension or ellipsometry cell. It is to be pointed out that in two phase droplet systems, a depletion layer in the vicinity of the interfacial monolayer due to steric repulsion has been observed in other systems by ellipsometry [8]. For the present systems studied, this effect is not expected to play an important role due to the low surfactant concentrations

in the case of $C_{10}E_4$ and $C_{12}E_5$, and for $C_8E_3$, most of the measurements are taken in the three-phase region. In three-phase systems, a drop of the surfactant rich middle phase is added to the water-oil interface to serve as a reservoir to ensure complete monolayer coverage of the interface by the surfactant film.

## Techniques

### Surface light scattering

Interfacial tensions are measured by surface light scattering which is a non-perturbing technique suitable for very low interfacial tensions. Details of this technique applied to the study of interfaces have been given elsewhere [9]. Briefly, in this technique, one measures the half-width $\Delta\nu$ of the frequency spectrum of the thermal fluctuations, where:

$$\Delta\nu = \frac{\gamma q + (\varrho_1 - \varrho_2)g/q}{4\pi(\eta_1 + \eta_2)}; \qquad (4)$$

$\gamma$ is the interfacial tension, $q$ the scattering wave vector, $\varrho_1$ and $\varrho_2$ are the bulk densities and $\eta_1$ and $\eta_2$ the bulk viscosities which have been measured using the Ubbelohde viscometer. In this study, the reported $\gamma$ is an average value of several measurements carried out at different $q$ values ranging from 400 to 2000 cm$^{-1}$.

### Ellipsometry

The bending elasticity or rigidity properties of the surfactant films are studied by ellipsometry. Details of experimental set-up and interpretation of data can be found elsewhere [8, 10, 11]. For the present systems, the measured signal, the ellipticity, is given by:

$$\bar{\varrho} = \frac{\pi}{\lambda}\frac{\sqrt{\varepsilon_1 + \varepsilon_2}}{\varepsilon_1 - \varepsilon_2}\eta \qquad (5)$$

where $\lambda$ is the wavelength of the light, $\varepsilon_1$ and $\varepsilon_2$ are the dielectric constants of the bulk phases, and $\eta$ is a measure of the interfacial properties where:

$$\eta = \eta^L + \eta^R; \qquad (6)$$

$\eta^L$ is an interfacial thickness term and $\eta^R$ a surface roughness term. The interfacial thickness term has been calculated by Drude for isotropic monolayers [12]. For optically anisotropic layers, supplementary terms have to be included [13].

The roughness term has the form [8]:

$$\eta^R = -\frac{3}{4}\frac{kT}{\pi\gamma}\frac{(\varepsilon_1 - \varepsilon_2)^2}{\varepsilon_1 + \varepsilon_2}q_e. \qquad (7)$$

$q_e$ is an upper cut-off value for capillary waves resulting from the curvature energy of the interfacial film and is given by:

$$q_e = \frac{\pi}{2}\sqrt{\frac{\gamma}{K}}. \qquad (8)$$

By combining Eqs. (7) and (8) and plotting $\eta$ versus $\frac{(\varepsilon_1 - \varepsilon_2)^2}{(\varepsilon_1 + \varepsilon_2)} \cdot \frac{1}{\sqrt{\gamma}}$, the slope and intercept yield information on $K$ and film thickness respectively. In the present investigation, ellipsometric measurements are carried out as a function of temperature and the $\eta$ values obtained plotted against the corresponding $\gamma$ measured by surface light scattering.

### Small angle neutron scattering

The structures of the microemulsions are studied by small angle neutron scattering (SANS) using the Orphée neutron facilities at the Centre d'Energie Atomique, Saclay. The wavelength of the neutrons is about 7 Å with corresponding $q$ range of 0.0063 to 0.13 Å$^{-1}$.

In this paper, only data for the middle phase bicontinuous microemulsion will be discussed. Characteristic distances are deduced from the positions of the scattering peaks by:

$$Q_{\text{max}} = \frac{2\pi}{d} \tag{9}$$

where $Q_{\text{max}}$ is the scattering peak position and $d$ the Bragg distance. In the above, we have introduced the characteristic size, $\xi$, for the oil and water microdomains. Since the scattering peak is a correlation peak, $d$ is the distance between two oil or two water domains such that $d = 2\xi$ [14, 15].

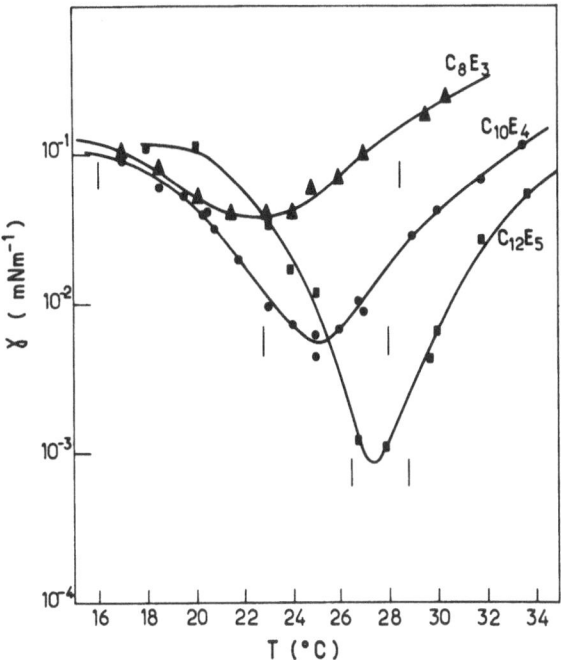

Fig. 1. Interfacial tension versus temperature for water-oil-nonionic surfactant systems

## Results and discussions

### Interfacial tension

The interfacial tensions of water and alkanes in the presence of the various surfactants are given in Fig. 1 as a function of temperature. In all cases, with increase in temperature, the interfacial tension goes through a minimum as the microemulsion undergoes structural inversion from an oil-in-water to a water-in-oil microemulsion. Such an inversion arises from a change in spontaneous curvature of the surfactant film; for nonionic surfactants, this is effected by an increase in temperature causing dehydration of the polar head, and for ionic surfactants, by an increase in salinity causing charge screening effect. The observed minimum surface tension then occurs at the optimal or inversion temperature, $T^*$, where the spontaneous curvature of the surfactant film approaches zero. In this context, since the surface tension corresponds to a large extent to the energy required to unfold the surfactant film covering the interfaces between the microdomains [4], it is clear as to why the interfacial tension is minimum when the film spontaneous curvature tends to zero. The microemulsion in this case is bicontinuous with both water and oil microdomains. In Fig. 1, the regions where bicontinuous phase occur are enclosed within vertical bars.

### Film bending elasticity

The ellipsometric signals, $\eta$, plotted as a function of $\frac{(\varepsilon_2 - \varepsilon_2)^2}{(\varepsilon_1 + \varepsilon_2)} \cdot \frac{1}{\sqrt{\gamma}}$ are given in Fig. 2. It can be seen from this figure that the points obtained for each surfactant can be fitted to a straight line, indicating that within the range of temperature studied, the bending elasticity properties of the films are not dependent on the temperature. From the slopes of these plots, the bending elasticity constants, $K$'s, are deduced to be about $0.31 \, kT$ and $0.51 \, kT$ for films containing $C_8E_3$ and $C_{10}E_4$ surfactants respectively. Preliminary measurements of the $C_{12}E_5$ system indicate that the bending elasticity constant is the highest of the three cases. Thus, as the surfactant chain length is increased, the rigidity of the film is also increased.

As mentioned earlier, from the intercepts of the above plots, the film thickness can be deduced. But in view of the low accuracy of the values of these intercepts, such deductions have not been made. It is to be pointed out also that the accuracy of the film rigidity, deduced from the surface roughness term, measured using this technique improves as the interfacial tension decreases. This is because under these conditions, the surface roughness term dominates due to large thermal fluctuations.

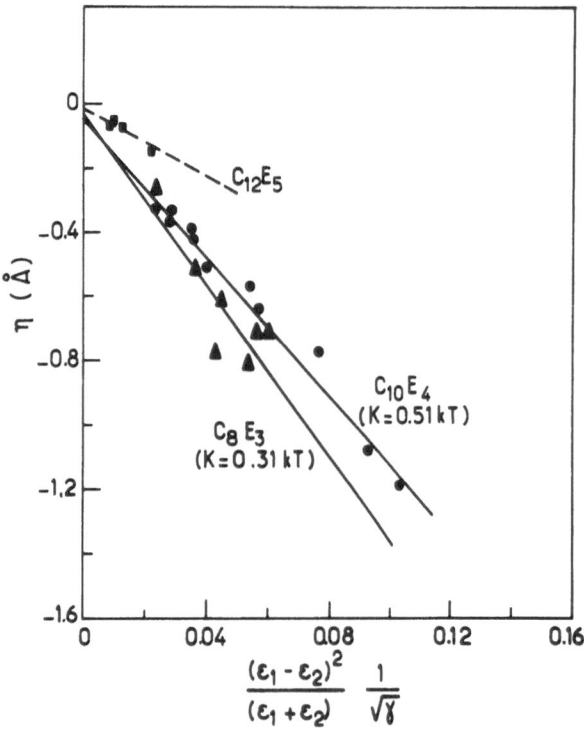

Fig. 2. Ellipsometric results and bending elasticity constants

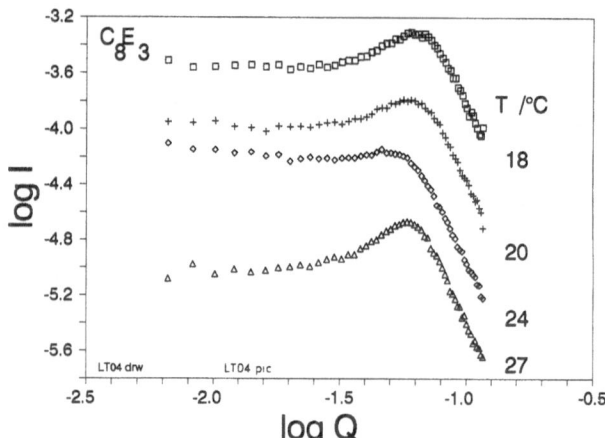

Fig. 3. Neutron scattering curves for $C_8E_3$ microemulsions at different temperatures

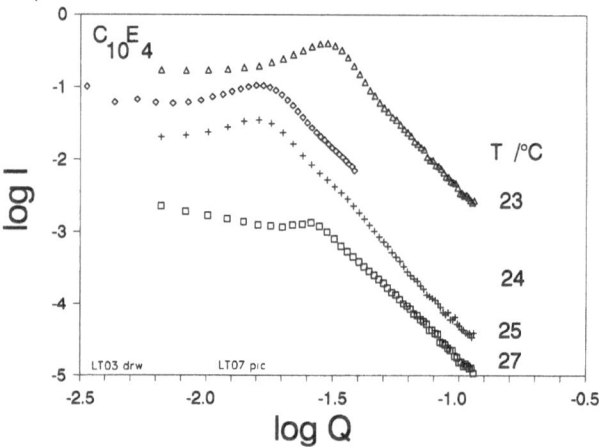

Fig. 4. Neutron scattering curves for $C_{10}E_4$ microemulsions at different temperatures

*Renormalization of K*

The bending elasticity constants discussed above have been obtained without taking into account the length scale or $q$-dependence of $K$. It has been shown, though, that $K$ is a function of the length scale of measurement [16, 17]. This concept is comprehensible considering the fact that $K$ depends on the degree of surface roughness and that this roughness depends on the scale of observation. A renormalization of both $K$ and the interfacial tension, $\gamma$, is required here. Using a coupled mode theory for thermal fluctuations, Meunier [10] has shown that $K(q)$ decreases with increasing scale of observation at large $q$, as obtained by Peliti and Leibler [17], and is constant at small $q$.

Equation (3) above of de Gennes and Taupin relating the bending elasticity constant to the persistence length is applicable for $K$ independent of the scale length. Taking into account the $q$-dependence of $K$, Meunier [10] derived a corresponding relationship for the persistence length as follows:

$$\xi_k = a(2.117)\exp[3.254\,K(q)/kT]. \qquad (10)$$

A molecular length $a$ has to be introduced but the value is not known. A choice of the value of $a$ is usually arbitrary, ranging from 5 Å [18, 19] to 10 Å [4, 20]. In this study, a value of 7.5 Å has been chosen. The physical significance of this value is that it approximates the diameter of the surfactant molecule, taking the parking area to be 43 Å$^2$ [21]. Thus, from the measured values of $0.31\,kT$ ($C_8E_3$) and $0.51\,kT$ ($C_{10}E_4$), the renormalized results are $0.35\,kT$ and $0.76\,kT$ respectively at the molecular scale of 7.5 Å.

*Structure*

Measurements of the characteristic sizes, $\xi$'s, of the microemulsions have been carried out using small angle neutron scattering. Figures 3 and 4 show the scattering spectra (Log$I$ versus Log$Q$) of the middle phase microemulsions of $C_8E_3$ and $C_{10}E_4$ respectively taken at various temperatures. The spectra in these figures have been displaced vertically in order to separate out the curves for easier distinction. It can be seen from the presence of definite scattering peaks that all these microemulsions possess well-defined characteristic domain sizes. In addition, while the peak positions for the $C_8E_3$ system seem to be almost independent of temperature, those for $C_{10}E_4$ show considerable change with temperature. It will be shown later that this is correlated to the respective temperature dependence of the interfacial properties.

From these measurements, the characteristic sizes of the microemulsions evaluated from the scattering peaks are compared with the persistence length calculated from the bending elastic modulus (Eq. (10)) and with the characteristic size calculated from Eq. (2). Comparisons are made at the optimal temperature where the interfacial tension is minimum and the spontaneous curvature approaches zero. Table 1 shows the measured and renormalized values of $K$, experimental values of $\xi$, and calculated values of $\xi_k$ and $\xi$ using Eqs. (10) and (2) respectively. $\xi_k$ is determined using $a = 7.5$ Å as discussed above, and $\xi$ calculated using $\Sigma = 43$ Å$^2$. $C_s$ has been estimated taking into account the amount of surfactant dissolved in the bulk oil phase using the phase diagrams in Refs. [1] and [2].

The results from Table 1 show that for both of the microemulsions, the experimentally determined sizes agree quite well with calculated ones, and that the persistence length, $\xi_k$, is close to $\xi$. Furthermore, the size of the microemulsion increases with film rigidity as the surfactant chain length is increased from $C_8E_3$ to $C_{10}E_4$.

The bending elastic moduli and sizes of nonionic bicontinuous microemulsions shown above are of the same order of magnitude as those obtained for single chain ionic surfactants combined with alcohol [19].

Figure 5 shows the experimentally measured values of $\xi$ (from Figs. 3 and 4) as a function of temperature.

It can be seen that the size goes through a maximum as a function of temperature, with the maximum occurring in the region of minimum interfacial tension and almost zero spontaneous curvature. The continuous solid lines in this figure are calculated from the dimensional relationship $\xi^2 \sim \dfrac{kT}{\gamma}$ using a prefactor 0.4 and corresponding interfacial tension values shown in Fig. 1. The good agreement between experimental and calculated results indicate that these nonionic microemulsion systems obey simple theories relating interfacial tension with structure. This has been observed also in other systems [19], but the value of the prefactor is not universal and depends on the system.

A consequence of the change in film curvature is a change in the relative amount of oil or water solubilized in the microemulsion. For these systems, the variation of volume fraction of oil with temperature has been measured, and the experimental results in Fig. 5 plotted against the corresponding change in volume fraction of oil, $\phi_o$, in Fig. 6. These results show a maximum in size when $\phi_o = 0.5$, or when there are equal amounts of oil and water in the microemulsion. In this figure, the solid continuous lines are calculated from the equation $\xi = \dfrac{6\phi_o\phi_w}{C_s\Sigma}$. In this case also, there is reasonably good agreement between experimental and calculated results.

## Conclusions

The interfacial properties and bulk structures of bicontinuous nonionic microemulsions have been studied. The renormalized bending elastic moduli for $C_8E_3$ and $C_{10}E_4$ surfactant films are determined to be $0.35\,kT$ and $0.76\,kT$ respectively. The persistence lengths evaluated from these values, and the characteristic sizes calculated using the cubic space-filling

Table 1. Measured and renormalized bending elastic moduli, and comparison of measured and calculated characteristic sizes of bicontinuous microemulsions

| System | Measured $K/kT$ | Scale $q_e$/cm$^{-1}$ | Renormalized $K(q)/kT$ at $q = 1.3 \times 10^7$ cm$^{-1}$ | Measured $\xi$/Å SANS | Calculated $\xi_k$/Å $2.117\,u$ $\exp(3.254\,K(q)/kT)$ | $\xi$/Å $\dfrac{6\phi_o\phi_w}{\Sigma C_s}$ |
|---|---|---|---|---|---|---|
| $C_8E_3$ | 0.31 | $2.8 \times 10^6$ | 0.35 | 58 | 50 | 69 |
| $C_{10}E_4$ | 0.51 | $8.2 \times 10^5$ | 0.76 | 187 | 188 | 211 |

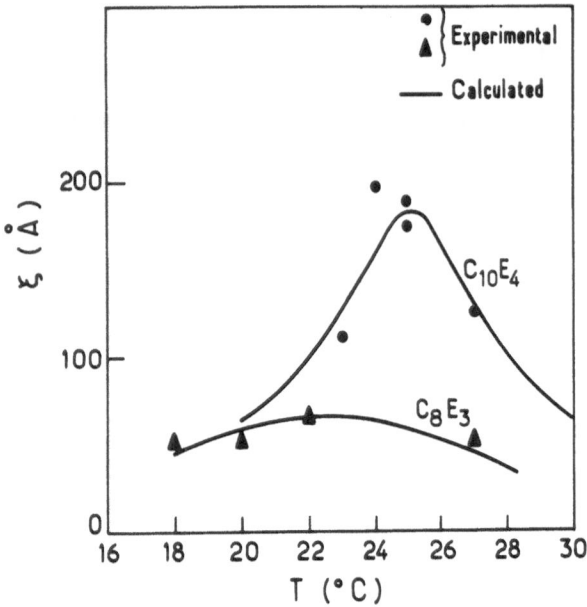

Fig. 5. Characteristic size versus temperature; solid lines are calculated from $\xi^2 \sim \dfrac{kT}{\gamma}$

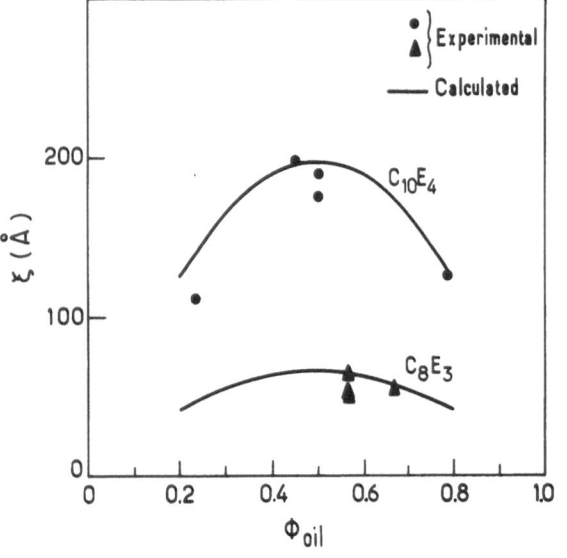

Fig. 6. Characteristic size versus oil volume fraction; solid lines are calculated from $\dfrac{6\phi_o\,\phi_w}{C_s\,\Sigma}$

model agree very well with the experimentally determined characteristic sizes. The interfacial tensions and the bulk structures of these microemulsions are found

to obey the simple relationship $\xi^2 \sim \dfrac{kT}{\gamma}$.

*Acknowledgements*

This is a good point to thank various people who have contributed towards the progress of this work: Dr. R. Strey with whom discussions have been very helpful, and from whom information on phase diagrams and sample preparation have been invaluable, and Dr. O. Abillon for discussions and assistance with the instruments.

## References

1. Kahlweit M, Strey R, Firman P (1986) J Phys Chem 90:671–677
2. Kahlweit M, Strey R, Firman P, Haase D, Jen J, Schomäcker R (1988) Langmuir 4:499–511
3. Kahlweit M, Strey R, Haase D, Firman P (1988) Langmuir 4:785–790
4. De Gennes PG, Taupin C (1982) J Phys Chem 86:2294–2304
5. Safran SA, Roux D, Cates ME, Andelman D (1986) Phys Rev Lett 57:491–4945
6. Langevin D (1988) Accounts of Chem Res 21:255–260
7. Jouffroy J, Levinson P, De Gennes PG (1982) J Physique 43:1241
8. Meunier J (1985) J Physique Lett 46:L1005–L1014
9. Langevin D, Meunier J (1977) In: Cummins HZ, Pike ER (eds) Photon correlation spectroscopy and velocimetry. Plenum Press, New York, pp 501–518
10. Meunier J (1987) J Physique 48:1819–1831
11. Beaglehole D (1980) Physica 100B:163–174
12. Drude P (1891) Ann Physique 43:126
13. Azzam RMA, Bashara NM (1977) Ellipsometry and polarized light. North Holland
14. Milner ST, Safran SA, Andelman A, Cates ME, Roux D (1988) J Physique 49:1065
15. Auvray L, Cotton JP, Ober R, Taupin C (1984) J Phys Chem 88:4586–4589
16. Helfrich W (1985) J Physique 46:1263
17. Peliti L, Leibler S (1985) Phys Rev Lett 54:1690–1693
18. Meunier J, Jerôme B (1986) In: Mittal KL (ed) Surfactants in solution: modern aspects. Plenum Press, New Delhi
19. Binks BP, Meunier J, Abillon O, Langevin D (1989) Langmuir 5:415–421
20. Guest D, Auvray L, Langevin D (1985) J Phys Lett 46:L1055–L1063
21. Paz L, Di Meglio JM, Dvolaitzky M, Ober R, Taupin C (1984) J Phys Chem 88:3415

Authors' address:

D. Langevin
Laboratoire de Physique Statistique de l'ENS
24, Rue Lhomond
75231 Paris Cedex 05
France

**Progress in Colloid & Polymer Science**                    Progr Colloid Polym Sci 81:215 (1990)

# Microemulsions — A qualitative thermodynamic approach

R. Strey and M. Kahlweit

Max-Planck-Institut für biophysikalische Chemie, Göttingen, FRG

*Key words:* Microemulsions; droplet size interfacial tensions; nonionic and ionic amphiphiles; phase behavior

Microemulsions are of growing interest in research and industry. In this review we have compiled some experimental facts about their properties that may serve as a basis for their application as well as for further theoretical work. It is shown that the phase behavior of multicomponent liquid mixtures of water, oils, amphiphiles and salts is essentially determined by the features of the corresponding binary mixtures. Studying the phase diagrams of the binary mixtures thus permits predicting qualitatively the mean temperature of the three-phase body within the pseudo-ternary phase prism at which one finds the highest efficiency of the amphiphiles with respect to homogenizing equal masses of water and oil. It is, furthermore, suggested to relate the efficiency of an amphiphile to the interfacial tension $\sigma$ between the water-rich and the oil-rich phase in the presence of a saturated interfacial monolayer of the amphiphile on the basis of the relation $(\sigma r^2/k_B T) \approx 1$, $r$ being the mean radius of the droplets in a stable dispersion. Because $\sigma$ decreases from $\approx 1$ mN m$^{-1}$ for short-chain amphiphiles to $<10^{-2}$ mN m$^{-1}$ for long-chain amphiphiles, one finds a gradual transition from weakly structured solutions ($r < 1$ nm) to stable colloidal dispersions of either oil or water droplets ($r > 10$ nm) in the corresponding solvent. It is, therefore, suggested defining microemulsions as stable dispersions sufficiently large for the solute to exhibit the properties of a bulk phase.

## References

Kahlweit M, Strey R, Busse G, J phys Chem, submitted

Authors' address:

R. Strey
Max-Planck-Institut für biophysikalische Chemie
Postfach 2841
D-3400 Göttingen
FRG

**Progress in Colloid & Polymer Science**                     Progr Colloid Polym Sci 81:216−221 (1990)

# High-pressure phase transitions in model biomembranes

R. Winter[1]) and P. Thiyagarajan[2])

[1]) Philipps-University of Marburg, Institute of Physical Chemistry, Marburg, FRG
[2]) IPNS Division, Argonne National Laboratory, Argonne, IL, USA

*Abstract:* By use of neutron small angle scattering (SANS) the structural properties of various saturated and unsaturated phosphatidylcholine model biomembrane systems have been investigated in the temperature range from 0 °C to 70 °C and at pressures from ambient up to 3.5 kbar. The analysis of the diffraction data allowed to determine the effect of chemical constitution (like the length and degree of unsaturation of the hydrocarbon chains) of the lipids on the thermotropic and barotropic phase behaviour, and the detection of different phases which lead to the evaluation of the temperature-pressure phase diagrams of the different phospholipids. The structural data are compared with recent results obtained from high pressure Raman, NMR and ultrasonic experiments on these model membrane systems. High pressure is also a characteristic feature of certain natural membrane environments. The biological relevance of the high pressure experiments on model biomembranes is also discussed in detail.

*Key words:* Model biomembranes; phosphatidylcholines; high pressure; neutron scattering; phase transition

*Abbreviations*

| | |
|---|---|
| DMPC | 1,2-dimyristoyl-sn-glycero-3-phosphocholine (di-C14:0) |
| DPPC | 1,2-dipalmitoyl-sn-glycero-3-phosphocholine (di-C16:0) |
| DSPC | 1,2-distearoyl-sn-glycero-3-phosphocholine (di-C18:0) |
| POPC | 1-palmitoyl-2-oleoyl-sn-glycero-3-phosphocholine (C16:0; C18:1, cis) |
| DEPC | 1,2-dielaidoyl-sn-glycero-3-phosphocholine (di-C18:1, trans) |
| DOPC | 1,2-dioleoyl-sn-glycero-3-phosphocholine (di-C18:1, cis) |

## Introduction

Aqueous dispersions of lipid bilayers, in particular the phosphatidylcholines[a]), provide valuable models for the investigation of biochemical and biophysical properties of membrane lipids, thus yielding valuable information for the understanding of the structure and function of complex biomembranes [1, 2].

In addition to the variety of lyotropic phases, the phosphatidylcholines also exhibit a rich thermotropic phase behaviour. A combination of calorimetric and diffraction studies on fully hydrated saturated 1,2-diacyl-phosphatidylcholine lipid bilayers have identified and characterized two temperature dependent phase transitions, corresponding to a lamellar gel-gel ($L_{\beta'} - P_{\beta'}$) and a gel-liquid crystalline ($P_{\beta'} - L_\alpha$) bilayer conversion at higher temperatures [1−8]. It has been shown that the $L_{\beta'}$ gel phase of DPPC converts even to a further low temperature gel form ($L_C$-phase) when held at low temperatures for prolonged periods of time [8−12].

In the liquid crystalline phase ($L_\alpha$) the hydrocarbon chains of the lipid bilayers are conformationally disordered ("melted"), i.e. the acyl chains undergo extensive trans-gauche isomerizations, but the average chain orientation is perpendicular to the bilayer surface. This liquid-like state has been proposed as a requirement of optimal biological function of biomembranes [2].

In the gel phases the hydrocarbon chains are more extended and relatively ordered, however, the lipid molecules can differ in bilayer surface structure and lipid chain packing [3−12]. For e.g. DPPC it has been shown that the $P_{\beta'}$ gel phase has a two-dimensional lattice structure in which the lipid bilayers are distorted by a periodic ripple in the plane of the lamellae, whereas the $L_{\beta'}$ gel phase exhibits a planar bilayer surface and its hydrocarbon chains, which are tilted by about 30°, are packed in a distorted hexagonal lattice structure [1−8]. Recently, Laggner has shown by time-resolved synchrotron radiation diffraction studies [13], that the main transition from the LC to the $P_{\beta'}$ gel state of DPPC is rather fast (< 1 s) whereas the $P_{\beta'} - L_{\beta'}$ transition is much slower and shows a strong hysteresis.

The complexity of the thermotropic phase behaviour presumably derives from the problems in simultaneously packing both the rather bulky phosphocholine polar head group and the two hydrocarbon chains. An optimal intermolecular packing can e.g. be accomplished by vertical displacement of neighbouring lipid molecules, chain tilting and bilayer rippling [1].

In addition to these thermotropic phase transitions, further pressure-induced phase transitions have been observed (for a review see [14, 15]). Besides the theoretical interest in high pressue phase behaviour of amphiphilic molecules, high pressure is also of considerable physiological interest [16, 17]. For example, pressure studies on lipid systems are of interest in understanding the physiology of deep sea organisms, the sensitivity of excitable cell membranes to pressure, and the antagonistic action of pressure to anesthetic action [17]. Deep sea is the largest ecological unit in the biosphere. About 70% of the surface of the earth is covered by the oceans. The average pressure on the ocean floor is about 400 bar and the highest pressures found e.g. in the Philippine Trench are about 1200 bar [16]. For these reasons, the study of pressure effects on the structure and dynamics of membranes is also of considerable interest.

Small modifications in the lipid molecule, like a change in length or degree of unsaturation of the fatty acid chains can have a drastic effect on the phase behaviour of membranes. In order to understand the influence of these parameters on the thermotropic and barotropic phase behaviour in detail, we performed high pressure SANS experiments for different phospholipids up to temperatures of 70°C and at pressures up to 3.5 kbar.

## Experimental

High purity phospholipids were obtained from Sigma Chemical Co. Fully hydrated lipid dispersions were prepared by vortexing the lipid/$D_2O$ mixture in a closed vial at a temperature well above the gel to liquid crystalline main transition temperature $T_m$ of the respective lipid. After immediate freezing of the sample in dry ice, the vortex-freeze cycle was repeated twice, leading to a homogeneous lipid dispersion. The liposomes resulting from the lipid dispersions have diameters ranging up to 5000 nm and contain multiple bilayers forming concentric shells.

Neutron diffraction has been applied for the structural investigations because neutrons can pass high pressure vessel materials [18] and neutrons also provide a simple way of changing contrast, i.e. the difference in scattering power of solute and solvent. In using $D_2O$ as solvent a factor of about 40 can be gained in scattering intensity relative to that in using $H_2O$ as solvent. We therefore used a solution of the lipid liposomes in pure $D_2O$.

The neutron scattering experiments were carried out on the D16 neutron diffractometer at the high flux reactor of the Institut Laue-Langevin, Grenoble, and on the SAD instrument at the Intense Pulsed Neutron Source at Argonne National Laboratory. The high pressure cell used for the diffraction experiments was made from an aluminium alloy of high tensile strength. $D_2O$ was used as pressure transmitting fluid. The pressure was applied by means of a hand pump and recorded by a Budenberg gauge. Temperature control was achieved by circulating a water/glycerol mixture from a thermostat through two outside jackets, which are located above and below the neutron beam window. For more experimental details see [15].

## Results and discussion

As an example for a saturated phospholipid, Fig. 1 shows the neutron scattering intensity $I(Q)$ of multilamellar DPPC (di-C16:0) vesicles in excess $D_2O$ at $T = 63$°C as a function of pressure. With increasing pressure, the first order Bragg peak around 0.1 Å$^{-1}$ is slightly shifted towards smaller $Q$-values. The corresponding bilayer repeat unit $d$, which can be calculated from the Bragg-equation, consists of the bilayer thickness including the water layer around the lipid headgroup. This $d$-spacing increases from 66 Å at 1 bar to 67.5 Å at 861 bar as a result of the inplane compression of the hydrocarbon chains. Therefore, the order parameter of the acyl chains should increase upon pressurization. Recently, we started to apply one and two dimensional high pressure NMR techniques to study the structure and dynamics of model biomembranes. We were able to show that the order parameter of the chains increases considerably with increasing pressure in the LC phase (e.g. from 0.36 at 1 bar to 0.6 at 2 kbar for DMPC at 64°C [19]).

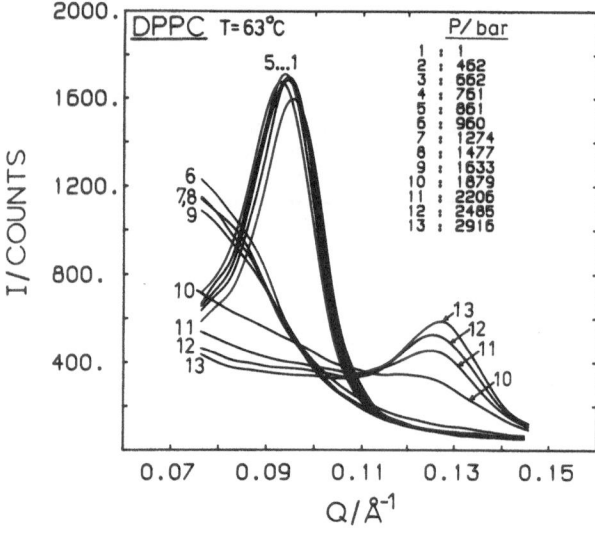

Fig. 1. SANS diffraction pattern of DPPC at $T = 63\,°C$ as a function of pressure

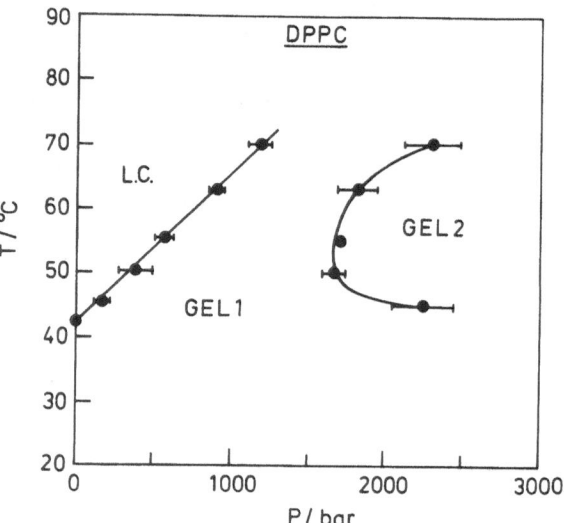

Fig. 2. The temperature-pressure phase diagram of DPPC multilamellar vesicles

Above 861 bar, the Bragg peak of DPPC at 63 °C disappears and the transition from the liquid crystalline (LC) phase to the gel-phase (Gel1) occurs, thus showing that increasing pressure shifts the LC-gel transition to higher temperatures ($T_m = 42\,°C$ at 1 bar). At pressures above about 1800 bar, the diffraction pattern changes again, a new Bragg-reflection develops, which corresponds to a $d$-spacing of 50 Å. This high pressure gel phase might be attributed to an interdigitation of the hydrocarbon chains opposing each other in the lipid bilayer. By monitoring the different changes in the diffraction pattern for different temperatures, a temperature-pressure phase diagram has been obtained (see Fig. 2). Within the experimental error, the LC to gel transition temperature increases linearly with increasing pressure with a slope $dT_m/dP = 22\,°C/kbar$. A further gel-phase (Gel2) appears at pressures higher than about 1.6 kbar and the pressure at which this transition takes place, varies nonlinearly with temperature. Such a high pressure gel phase has also been observed in oriented layers of DPPC and DSPC (di-C18:0) [20]. Recent high pressure ultrasonic experiments [21] have revealed a distinct increase in ultrasonic absorption at 10 MHz not only at the LC-gel transition but also a drastic increase in ultrasonic absorption in going from the Gel1 to the Gel2 state.

A similar interdigitated structure has also been observed by the addition of alcohols, salts, drugs, anesthetics, other surface active and amphiphilic molecules, and in other lipids (see e.g. [14], [22—25]). The biological significance of the interdigitated structure remains still to be clarified. However, the existence of this phase has already stimulated research to characterize their physical properties, and to determine the physical principles which govern their formation and stability.

Rather little is known about the high pressure properties of lipids containing cis unsaturated acyl chains, although they constitute a major part of the plasma membranes. Especially high pressure structural investigations have been very much in need.

A biologically very relevant case of unsaturated membrane lipids are the mixed chain phospholipids, each lipid containing one saturated and one unsaturated chain, like POPC (C16:0; C18:1, cis). Fig. 3 exhibits the SANS diffraction curves $I(Q)$ for POPC at $T = 10\,°C$ as a function of pressure up to about 2 kbar. It is evident from the diffraction data that a high degree of lamellar order is present. Contrary to DPPC, the transition from the LC to the gel state is indicated by a large shift of the Bragg peak to a lower $Q$-value, i.e. larger $d$-spacing, which occurs around 700 bar. In both phases, the LC and the gel phase, the $d$-spacing increases with increasing pressure, about $1-2\,°C/kbar$. This implies, that a considerable amount of disorder exists not only in the LC phase, but also in the gel state, which contrasts the behaviour observed for saturated phospholipids [20, 27, 28]. A similar result has been obtained for DEPC multilamellar vesicles, a lipid containing two trans monounsaturated acyl chains [26]. The temperature-pressure

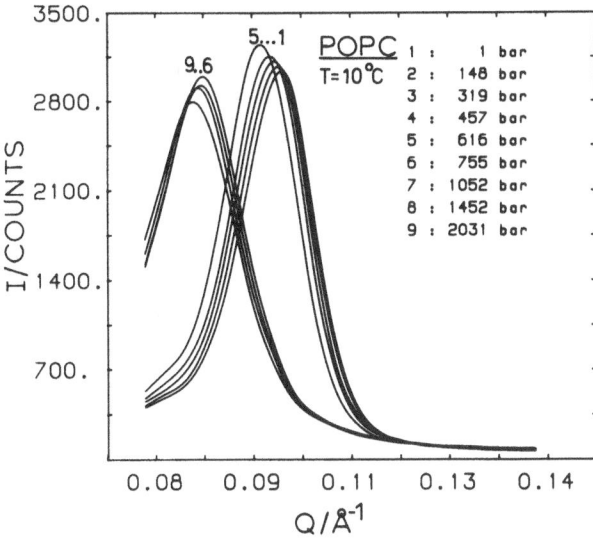

Fig. 3. SANS diffraction pattern of POPC at $T = 10\,°C$ as a function of pressure

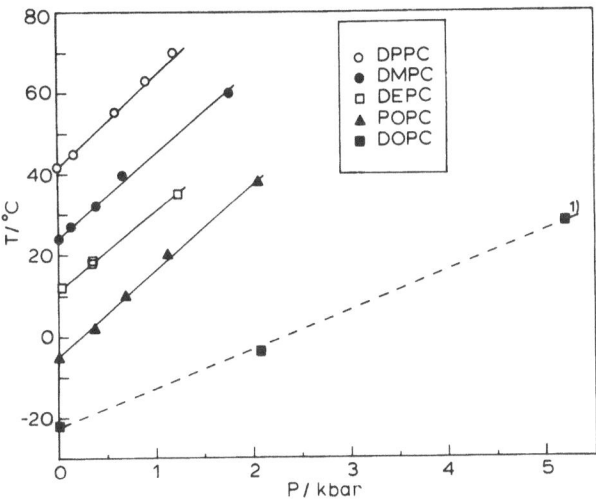

Fig. 4. The temperature-pressure phase diagram for the liquid crystalline-gel phase transition of different saturated and unsaturated phospholipid bilayers ([1] [30])

phase diagram for POPC is shown in Fig. 4. The slope of the LC to gel phase transition is about $dT_m/dP = 21\,°C/kbar$. Interestingly, a common value for the LC-gel transition slope $dT_m/dP$ of about $21\,°C/kbar$ has been obtained for the saturated phosphocholines DMPC [15], DPPC and the unsaturated lipids DEPC [26] and POPC.

As a representative of a biologically very important cis-dimonounsaturated phospholipid, Fig. 5 presents the SANS diffraction pattern of DOPC at $T = -4\,°C$

as a function of pressure. At a pressure of about 2070 bar, the first order Bragg peak disappears. Similar to the saturated phospholipid DPPC (see Fig. 1), the disappearance of the Bragg reflection probably reveals the transition from the LC to the gel state and indicates that the phase transition is accompanied by considerable bilayer stacking disorder. As has been observed from measurements of the pressure dependence of the $d$-spacing at several temperatures, $(\Delta d/\Delta P)_T$ is about 1 Å/kbar in the LC phase of DOPC, similar to what has been observed for the other unsaturated phospholipids [15, 26].

The temperature-pressure phase diagram for DOPC is also shown in Fig. 4. The data for the LC-gel transition temperature of DOPC at ambient pressure have been obtained from differential scanning calorimetry (DSC) measurements [29] and the transition pressure at $T = 28\,°C$ has been obtained from Wong et al. [14, 30] by employing high pressure Raman experiments. The high pressure Raman spectra of DOPC below about 5.2 kbar at $T = 28\,°C$ are typical of conformationally and orientationally disordered chains. Above about 5.2 kbar the reorientational fluctuations of the lipid molecules are largely damped and the orientation of the lipid chains is highly ordered, thus indicating the transition to the gel state. The slope of the LC-gel transition line of DOPC, as has been obtained from these different experiments discussed above, is about $10\,°C/kbar$, which drastically contrasts the behaviour observed for the other phospholipids so far.

An interdigitated gel phase has not been observed in DMPC or DEPC, POPC and DOPC up to pressures of about 3 kbar or 2 kbar, respectively.

Interestingly, the SANS experiments on DOPC revealed, that by cooling down DOPC at 1 bar a drastic change in $d$-spacing, from 65 Å to 58 Å, occured around $-8\,°C$, i.e. above the LC-gel transition temperature (see Fig. 4), which might be attributed to ice formation. To provide confirmation that the LC-gel phase transition in DOPC is independent of that ice formation, we performed DSC measurements also on DOPC/glycerol mixtures in order to suppress ice formation. Within the experimental error of about 3 degrees the LC-gel transition temperature remained at about $-22\,°C$ [29] and is therefore independent of the ice formation.

The about 7 Å decrease in layer periodicity during the ice formation can be explained in terms of a partially elimination of the water layer which initially separated the lipid bilayers [31–33]. These conclu-

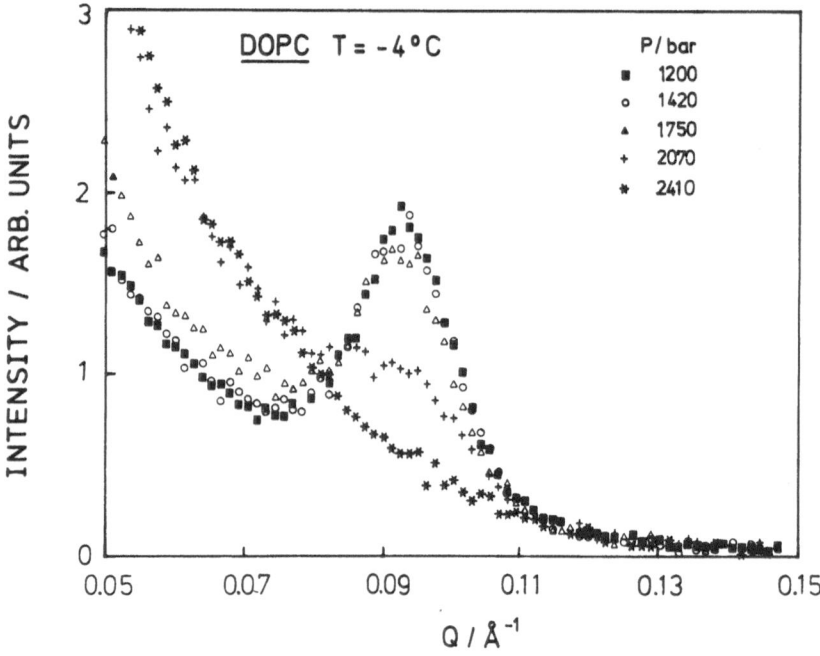

Fig. 5. SANS diffraction pattern for DOPC at $T = -4°C$ as a function of pressure

Table 1. The LC-gel phase transition slope $dT_m/dP$ of different unsaturated phospholipids and the LC-gel phase transition temperature $T_m$ at 1 bar relative to that of the corresponding saturated phospholipid, $T_m$ (sat.) [1, 2, 35]

| Phospholipid | $dT_m/dP$ (°C/kbar) | $T_m - T_m$ (sat.) (°C) |
|---|---|---|
| POPC | 21 | −55 |
| DEPC | 20 | −44 |
| DOPC | 10 | −77 |

sions are in good agreement with results obtained from X-ray diffraction experiments [31].

Besides the increased theoretical interst in phase transition phenomena of amphiphilic molecules, the effect of pressure, temperature and chemical constitution of the lipid molecules on the phase behaviour of membranes is also of considerable physiological interest. Living organisms have to compensate for the direct effect of temperature or pressure variation by compensatory adjustment of membrane structural order. As we have seen, deep sea pressures are sufficient to exert a significant ordering effect on membrane bilayers. Therefore, deep sea membrane bilayers require compensating changes in fluidity, i.e. "homeoviscous adaptation" to high pressure, in addition to low temperature, in order to maintain their fluid-like state. Table 1 demonstrates that the introduction of cis double bonds in the fatty acid acyl chains is probably one

of the most efficient ways in increasing the fluidity of a bilayer, as it causes the greatest decrease of the liquid crystalline-to-gel transition temperature $T_m$ and it causes the smallest fluidity depression under pressure, i.e. exhibits the smallest LC-gel transition slope $dT_m/dP$. The cis double bond imposes a kink in the linearity of the acyl chains, creating significant free volume in the bilayer, thus reducing the ordering effects of low temperature and high pressure and thus increases the fluidity of the membranes which is necessary for optimal physiological function. Actually, MacDonald et al. found an increased acyl chain cis unsaturation ratio in membranes of deep sea organisms, recently [17, 34]. Thus, by regulating the lipid composition of the cell membrane through changes in fatty acyl chain length and cis unsaturation, biological organisms are provided with a mechanism for efficiently modulating the fluidity of their membranes in response to changes in the external environment.

*Acknowledgements*

This work benefited from the use of the neutron beam facilities at the Institut Laue-Langevin in Grenoble and at the Intense Pulsed Neutron Source at Argonne National Laboratory (this facility is funded by the Department of Energy, BES-Materials Science, under contract W-31-109-Eng-38).

## References

1. Cevc G, Marsh D (1987) Phospholipid Bilayers. John Wiley & Sons, New York
2. Gennis RB (1989) Biomembranes. Springer-Verlag, Berlin
3. Tardieu A, Luzzati V, Reman FC (1973) J Mol Biol 75:711−733
4. Janiak MJ, Small DM, Shipley GG (1976) Biochemistry 15:4575−4580
5. Cullis P, deKruiff B (1979) Biochim Biophys Acta 559:393
6. Büldt G, Gally HU, Seelig A, Seelig J, Zaccai G (1978) Nature 271:182−184
7. Wiener MC, Suter RM, Nagle JF (1989) Biophys J 55:315−325
8. Ruocco MJ, Shipley GG (1982) Biochim Biophys Acta 684:59−66; ibid 691:309−320
9. Chen SC, Sturtevant JM, Gaffney BJ (1980) Proc Natl Acad Sci 77:5060−5063
10. Füldner HH (1981) Biochemistry 20:5707−5710
11. Nagle JF, Wilkinson DA (1982) Biochemistry 21:3817−3821
12. Stümpel J, Eibl H, Nicksch A (1983) Biochim Biophys Acta 727:246−254
13. Laggner P (1988) In: Synchrotron Radiation in Chemistry and Biology. Springer-Verlag, Berlin
14. Wong PTT, Siminovitch DJ, Mantsch HH (1988) Biochim Biophys Acta 947:139−171
15. Winter R, Pilgrim W-C (1989) Ber Bunsenges Phys Chem 93:708−717
16. Jaenicke R (1983) Naturwissenschaften 70:332−341
17. Jannasch HW, Marquis RE, Zimmerman AM (eds) (1987) Current Perspectives in High Pressure Biology. Academic Press, London
18. Winter R, Bodensteiner T (1988) High Pressure Research 1:23−37
19. Jonas J, Winter R, Grandinetti PJ, Driscoll D, submitted
20. Braganza LF, Worcester DL (1986) Biochemistry 25:2591−2596, ibid 25:7484
21. Böttner M, Winter R, to be published
22. Ruocco MJ, Siminovitch DJ, Griffin RG (1985) Biochemistry 24:2406−2411
23. McDaniel RV, McIntosh TJ, Simon SA (1983) Biochim Biophys Acta 731:97−114
24. Wilkinson DA, Tirrell DA, Turek AB, McIntosh TJ (1987) Biochim Biophys Acta 905:447−453
25. Boggs JM, Rangaraj G, Watts A (1989) Biochim Biophys Acta 981:243−253
26. Winter R, Xie CL, Jonas J, Wong PTT (1989) Biochim Biophys Acta 982:85−88
27. Stamatoff J, Guillon D, Powers L, Cladis P, Aadsen D (1978) Biochem Biophys Res Comm 85:724
28. Winter R, Thiyagarajan P, private communication
29. Brauns T, Winter R, private communication
30. Wong PTT, Mantsch HH (1988) Biophys J 54:781−790
31. Finean JB, Hutchinson AL (1988) Chem Phys Lip 46:63−71
32. Caffrey M (1987) Biochim Biophys Acta 896:123
33. Elkes J, Finean JB (1983) Exp Cell Res 4:69
34. MacDonald AG (1986) In: Klein R, Schmitz B (eds) Topics in Lipid Research. Royal Society of Chemistry, London
35. Kotyk A, Janacek K, Koryta J (eds) (1988) Biophysical Chemistry of Membrane Functions. John Wiley, New York

Authors' address:

Dr. R. Winter
Philipps-Universität FB14
Institut für Phys. Chemie
D-3550 Marburg
FRG

**Progress in Colloid & Polymer Science**          Progr Colloid Polym Sci 81:222−224 (1990)

# Photovoltages in bilayer lipid membranes incorporating cadmium sulfide particles

R. Rolandi and D. Ricci

Department of Physics, University of Genova, Italy

*Abstract:* Cadmium sulfide particulate films are formed at the water-membrane interfaces of monoolein bimolecular planar membrane (BLM) by spontaneous electrostenolysis. Hydrogen sulfide, added to one side, permeates the membrane and reacts with cadmium chloride on the opposite membrane interface, forming cadmium sulfide particles, whose formation and eventual condensation into a continuous film were observed by optical microscopy. − Changes in the electrical properties of the membrane were measured and related to the particle formation mechanism. Trans-membrane voltage were elicited by illumination of BLM supported CdS particles showing light driven charge separation. No external electric field was necessary for obtaining the photovoltage breakout. − The photovoltage as a function of time and the photovoltage evolution during particle formation are reported.

*Key words:* Bilayer lipid membranes (BLM); cadmium sulfide particles

## Introduction

Bimolecular lipid leaflets are the basic structure of photosynthetic centers occurring in animals and vegetables. The bimolecular lipid layers are able to separate electrical charges and chemical species, therefore, they are a suitable barrier for charge separation, the basic process of light energy harvesting in the living world.

Mimicking photosynthesis systems can be created taking advantage of the coupling of these organic structures with inorganic semiconductors.

Black lipid membranes (BLM) are a good structural model of the bimolecular lipid leaflets occurring in living cells and are particularly suitable for electrical measurements since they are accessible by macroscopic electrodes.

The spontaneous deposition of metallic crystals on the surface of a membrane separating different solution is termed electrostenolysis and was observed in bimolecular lipid membrane by H. Ti Tien [1]. It was later utilized to produce photosensitive and electrically active membranes [2−6].

## Experimental methods

### Materials

Glyceryl monooleate (GMO), $CdCl_2$ (Sigma Chemical Co.), KCl (Carlo Erba), $H_2S$ (purity >99%) (Alphagaz SIO) were used as received. Decane (BDH) was deoxygenated by bubbling nitrogen for 20 minutes and then used to make the BLM-forming solution (50 mg/ml of GMO in decane). Water was purified with a Millipore Milli-$Q$ system.

### BLM formation

Membranes were prepared according to the technique of Mueller et al. [7] by depositing a droplet of the lipid solution on a 0.7 mm circular hole connecting two chambers in a Teflon solid block, each filled with 3 ml or 5 mM KCl solution at ambient temperature (22−23 °C). At each end of the cell there was a glass window facing the hole in the septum for observation and illumination of the membrane. The BLM diameter was measured with a calibrated graticule mounted on the microscope.

### Semiconductor formation

The first step towards CdS formation on the BLM was obtained by adding 75 μl of 0.4 M $CdCl_2$ solution to one side (cis-side) and simultaneously adding the same amount of 5 mM KCl solution to the other side (trans-side), in order to equalize the hydrostatic pressure across the BLM. After about 10 minutes, $150-200$ µl of $H_2S$ gas were slowly injected over 2 minutes on the trans-side. Within $5-10$ minutes, CdS particles become visible on the BLM surface and soon the entire membrane surface became covered by an apparently continuous film of yellow crystalline particles.

### Electric measurements

All electrical measurements were taken using two Ag/AgCl electrodes immersed in the solutions on opposite sides of the BLM cell. Membrane capacitance was measured with a precision AC impedance bridge. The test signal used was a square wave of 20 mV amplitude at 500 Hz. Conductance measurements were made, in voltage clamp conditions, reading the current flowing through the membrane on both an oscilloscope and a HP7402A oscillographic paper recorder. For voltage and photoelectric effect measurements the Ag/AgCl electrodes were connected to the differential input of a Keithley Model 604 electrometer (input impedance = $10^{14}$ ohm).

Photoeffects were initiated by directing the light from a 150 W filament lamp (SHOTT KL 1500) via a glass optical fiber guide into the chamber on the semiconductor-containing BLM. The opening and closing time of the light shutter were $\leqslant 2$ ms, limiting to this value the shortest observable time constant of the photovoltage. The irradiation energy between 339 and 550 nanometers falling on the membrane area was 43 mW/cm$^2$. For wavelenghts greater than 550 nanometers the absorbance of CdS becomes negligible. Typical membrane area was $2 \cdot 10^{-3}$ cm$^2$. Both Ag/AgCl electrodes were covered with sleeves to avoid the generation of spurious photovoltages.

## Results and discussion

The formation of CdS microcrystalline particles is accompanied by transmembrane potential changes. The addition of CdCl$_2$ into the cis-side brings about a 30 mV voltage across the membrane with the cis-side at lower potential. This voltage measured in total darkness is the potential difference arising from the concentration difference of chloride ions (concentration E.M.F.), with which Ag/AgCl electrodes react reversibly. Cis-side potential increases after the H$_2$S addition in trans-side. This can be ascribed to an H$^+$/HS$^-$ diffusion potential adding linearly to the concentration E.M.F.

No significant membrane conductance changes were observed for CdCl$_2$ and H$_2$S additions. After the hydrogen sulfide addition, the membrane capacitance began decreasing. Within a few minutes the capacitance value reaches a steady state about one half of the initial value. The capacitance was already stabilized when CdS particles become optically observable in reflected light by a stereomicroscope (mag. 40 ×).

As already described by Zhao et al. [4] bright dots growing in size and merging in islands appear on the membrane surface. If enough H$_2$S is added, a uniform film is formed that usually does not extend to the Plateau-Gibbs border where, instead, bright dots are visible. Membranes supporting dense particles or uniform films last several hours with almost constant capacitance and resistance.

This picture of the particle formation does not change in its outline if different lipids are used to prepare membranes.

Figure 1 shows the membrane voltage response to a step-like luminous stimulus. Upward signals indicate that the trans-side becomes positive in respect to the cis-side. As described in the particle formation section, there is a dark voltage that is the superposition of the concentration E.M.F. and H$^+$/HS$^-$ diffusion potential. The photovoltage rises to a peak value in fractions of a second, then it relaxes to a steady state in a few seconds. When the light is switched off the voltage returns monotonically to the dark value.

The presence of a steady state different from zero in the photopotential indicates that a net charge transport through the membrane occurs during illumination.

The dependence of the photopotential on the particle size and distribution is shown by the peak and steady state increase in time after the addition of precursors in the solution (Fig. 2).

Baral and co-workers [5], on the basis of experiments showing that oxygen or another electron acceptor in cis-side and H$_2$S in trans-side are necessary for the photopotential outbreak, have proposed the following mechanism. Oxygen molecules, adsorbed to CdS particle surface, can capture electrons promoted by light to the conduction band, enriching the cis-side solution with electrons, while H$_2$S (or HS$^-$), coming from the trans side, hands over electrons to the semiconductor. A dark potential difference through the membrane should be the driving force of the process. We have observed that substantially unaltered photovoltages appear also when dark voltage is zero: therefore the system has an intrinsic asymmetry promoting charge separation. We propose that such an intrinsic asymmetry is due to the fact that H$_2$S, the electron donor, has a partition coefficient favourable to the lipidic core of the BLM with respect to the water solution [8]. Positive charges are then preferentially located in the membrane side of the CdS-membrane interface while negative charges are on the CdS side.

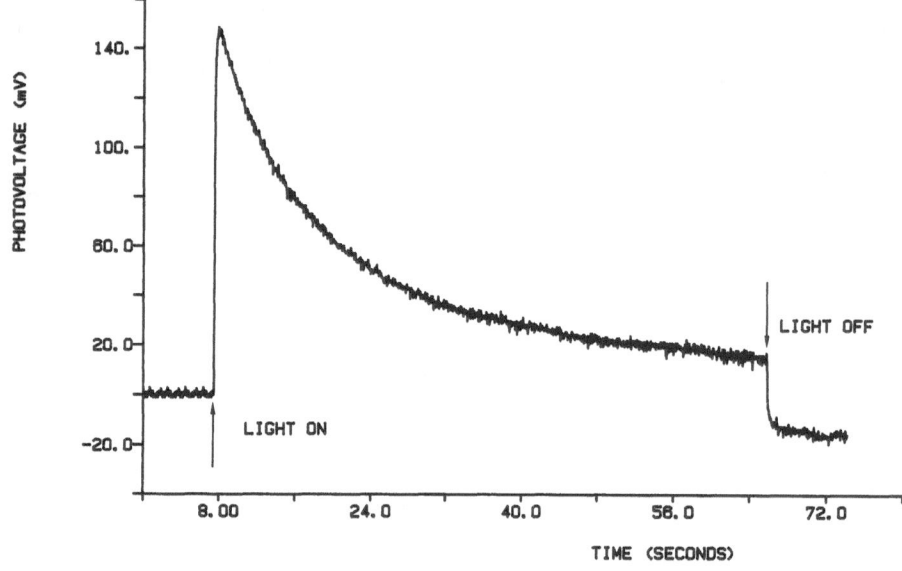

Fig. 1. Photovoltage signal recorded under open circuit condition. The electrode on the cis-side (CdS-containing side) became negative

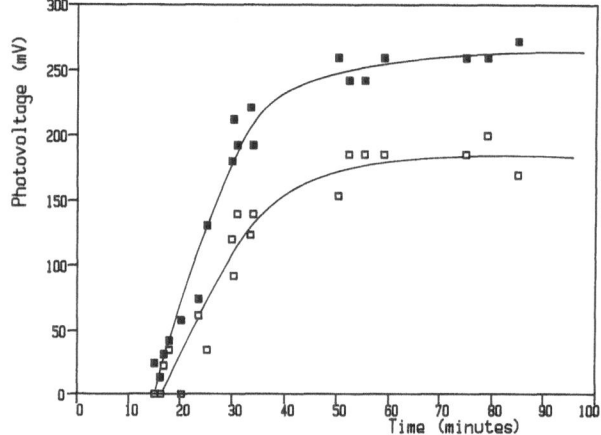

Fig. 2. Dependence of the magnitude of the photovoltage on the time lapsed from the injection of the ionic precursors in the solution bathing the BLM. The values are means from three normalized experiments. Peak value (■), steady state value (□). The continuous lines have been drawn to emphasize the trend of the data

Their capability to localize chemical reaction sites seems crucial in the photopotential formation mechanism.

*Acknowledgements*

The authors thank Prof. Ogden Brandt for reading the manuscript and for the benefit or his productive criticism. This work was partially supported by the CNR grant "Progetto Finalizzato Chimica Fine II".

## References

1. Tien HT (1976) Photochem Photobiol 24:97−116
2. Kutnik J, Ti Tien H (1987) Photochem Photobiol 46:413−419
3. Baral S, Zhao XK, Rolandi R, Fendler JH (1987) J Phys Chem 91:2701−2704
4. Zhao XK, Baral S, Rolandi R, Fendler JH (1988) J Am Chem Soc 110:1012−1024
5. Baral S, Fendler JH (1989) J Am Chem Soc 111:1604−1614
6. Zhao XK, Hervé PJ, Fendler JH (1989) J Phys Chem 93:908−916
7. Mueller P, Rudin DO, Tien HT, Wescott WC (1962) Nature 194:979−980
8. Solubilities of Inorganic and Organic Compounds, Stephen H, Stephen T (eds) (1963) Pergamon Press, Oxford

Authors' address:

Dr. Ranieri Rolandi
Dipartimento di Fisica, Università di Genova
Via Dodecaneso 33
16146 Genova, Italy

The experimental results reported confirm the finding of other groups [2, 5] that semiconductor particles and films can be formed on the surface of bimolecular membrane.

Photopotential observation in such a system confirms that bimolecular membrane are a suitable barrier in charge separation.

**Progress in Colloid & Polymer Science**                    Progr Colloid Polym Sci 81:225−231 (1990)

# Small-angle neutron scattering from aqueous mixed colloids of lecithin and bile salt

R. P. Hjelm, Jr.[1]), P. Thiyagaragan[2]), D. S. Sivia[3]), P. Lindner[4]), H. Alkan[5]) and D. Schwahn[6])

[1]) Los Alamos Neutron Scattering Center, Physics Division, Los Alamos National Laboratory, Los Alamos, New Mexico, USA
[2]) Intense Neutron Source Division and Chemistry Division, Argonne National Laboratory, Argonne, Illinois, USA
[3]) Los Alamos Neutron Scattering Center and Theoretical Division, Los Alamos National Laboratory, Los Alamos, New Mexico, USA
[4]) Institut Laue-Langevin, Grenoble, France
[5]) Department of Pharmaceutics, University of Illinois at Chicago, Chicago, Illinois, USA
[6]) Institut für Festkörperforschung, Kernforschungsanlage Jülich/KFA, FRG

*Abstract:* The morphology of particles in the isotropic phases of mixed aqueous colloids of lecithin and glycocholate are studied using small-angle neutron scattering. At the highest concentrations globular mixed micelles are present and have a radius of gyration of 24 Å. At lower concentrations rod-like structures are present. In each and every case the rod radii are about 27 Å. It is proposed that the rods form by aggregation of disk-like micelles about 27 Å radius and height 50 Å. The rods grow larger as the solutions are diluted and appear to associate into loose networks just before the occurrence of a concentration-induced transition to vesicles. The vesicles have walls that are a single lipid bilayer of lecithin and glycocholate.

*Key words:* Neutron scattering; glycocholate; lecithin; mixed micelles; vesicles

## Introduction

Aqueous mixed colloids of bile salts and lecithin, have for a number of years served as models for the structure and action of bile [1−6]. The rational is that bile salts and lecithin, as the major components of bile, should largely determine the physical chemical characteristics of the system, even though bile is a complex mixture of different bile salts and lecithin with bilirubin, cholesterol and other components (see the review by Cary [7] and references there in). The effects of the other components then can be studied as variations on the main theme determined by the bile salts and lecithin. This is our approach.

The physiological function of bile is in the transport of lipophilic products from the liver into the upper intestine, and in solubilization of fats in the digestive tract. From the liver bile passes down the bile duct, where it is concentrated and stored in the gall bladder. It is released from the gall bladder into the duodenum on hormonal stimulus from the stomach. Bile is essential for the action of lipases and in the adsorption of the hydrolysis products across the intestinal wall. Thus the information obtained from these studies of aqueous bile salt-lecithin mixtures will improve our understanding of important physiological processes. Furthermore, there is interest in the application of this system in the production of liposomes in drug delivery systems. Finally, this problem is important to a more general understanding of mixed aqueous colloid systems.

Here, as in our earlier studies [8, 9], we use the bile salt, glycocholate, and lecithin from egg yolk (Fig. 1). The characteristics of the aqueous mixtures are determined by the amphiphilic nature of the components. For egg yolk lecithin (Fig. 1), the long chain palmitic and oleic fatty acids are hydrophobic, whereas the charged head group, phosphatidylcholine, is hydrophilic. Glycocholate (Fig. 1) is one of a family of cholic acid conjugates which comprise the bile salts. The cholesterol chore of the cholic acid moiety is hydrophobic. The hydroxyl groups at C-3, C-7 and C-12 confer some degree of hydrophilicity to one side of the core, however. The glycine head group is, of course, hydrophilic.

As a consequence of its geometry, lecithin forms lamellar phases in dilute solution. The detergent action of the bile salt causes this to be broken up into other phases, including isotropic phases, consisting of discrete particles [1, 2]. Further, the bile salt is by far the more soluble of the two components; thus dilution will tend to repartition the bile salt from the particles into the aqueous environment. It is already well established that an isotropic phase exists which contains mixed micelles of bile salt and lecithin, and that the micelle size is a strong function of the total lipid concentration: the particles are larger as the concentration of lipids becomes smaller [3–6, 8, 9]. At sufficiently low total lipid concentration a transition occurs to another isotropic phase containing mixed vesicles [3–6, 8–10]. Our interest here is in the structure of different particles present in the isotropic phases.

The conclusions of our previous work [8, 9] using small-angle neutron scattering (SANS) have not been in accord with the predominantly held view that the increase in particle size with decreasing total lipid concentration in the mixed micelle phase is due to growth of the particles into ever larger disks [3–6]. We find, rather, that the morphology is rod-like. Figure 2 shows the compositions of samples studied in our previous reports [8, 9], indicated as points numbered 1–8 on a map of glycocholate versus lecithin concentration, and summarizes the salient findings for samples having lecithin to bile salt molar ratios (L/BS) of 0.9 and 0.56. At the highest concentrations (*, Fig. 2) interparticle effects preclude detailed analysis at this time. At lower concentrations (points 1 and 2, Fig. 2) the particles are elongated and eventually form long rods as the total lipid concentration is reduced (points 3 and 4, Fig. 2). The rod radii are about 27 Å. At sufficiently small concentrations a transition to single lipid bilayer vesicles occurs; though at point 5 (Fig. 2) there was ambiguity as to whether this was an extended bilayer sheet or a large vesicle. The size of the vesicles becomes smaller as one moves away from the transition region to smaller total lipid concentrations (points 5–8, Fig. 2).

In the further SANS studies presented in this paper, we present measurements on samples shown as points A-G of Fig. 2. The data are used to characterize the morphology of the rod-like particles further for lecithin to bile salt molar ratios of 0.5 and 0.8. We present evidence that rod-like micelles are formed by association of small, globular mixed micelles into linear arrays. Data are shown which suggests that when the

lipid concentration is reduced sufficiently there results an association of rods segments to form a loose network just before a concentration-induced transition to large vesicles consisting of a single lipid bilayer.

## Methods

### *Sample preparation:*

Lecithin-bile salt mixtures were produced at lecithin to bile salt molar ratios (L/BS) of 0.5 and 0.8 using the coprecipitation method described elsewhere [8, 9]. The dried samples were taken up in buffer containing 0.15 M NaCl and 10 mM tris (pH 7.0) to make stock solutions at 50 g/l. The stock was diluted to the final concentration, then incubated for 48 hors.

### *Neutron scattering:*

SANS was conducted on the time-of-flight (TOF) instruments SAD at IPNS and LQD at LANSCE, and the SANS cameras, D11 at the Institut Laue-Langevin (ILL), and KWS-I at the Kernforschungsanlage (KFA), Jülich. Data reduction for the TOF instruments was carried out as previously described [11, 12]. Shear-alignment was done using the shear cell designed by Lindner and Oberthuer [13].

### *Data analysis:*

Data is expressed as $S(Q)$, the differential cross section per steradian per unit mass, $cm^2\ mg^{-1}$. Very low-$Q$ data are analyzed using the Guinier approximation as applied to globular particles, rods and sheets. According to this approximation one writes,

$$S(Q) = [B_0^2] \exp(-Q^2 R_g^2/3) \tag{1}$$

for the scattering for a globular particle, where $[B_0^2]$ is the particle contrast factor per unit mass of material ($cm^2\ mg^{-1}$) and $R_g^2$ is the particle radius of gyration squared taken about the scattering mass centroid. For a rod-like particle of infinite length and uniform cross section the form is

$$S(Q) = [b_0^2] \exp(-Q^2 R_c^2/2) Q^{-1} \tag{2}$$

where $[b_0^2]$ is the contrast factor per unit length per unit mass of material ($cm^2\ Å^{-1}\ mg^{-1}$), and $R_c^2$ is the squared cross sectional radius of gyration taken about the rod axis. The analogous form for a sheet-like object is given as

$$S(Q) = [\beta_0^2] \exp(-Q^2 R_d^2) Q^{-2} \tag{3}$$

where $[\beta_0^2]$ is the contrast factor per unit area per unit mass of material ($cm^2\ Å^{-1}\ mg^{-1}$), and $R_d^2$ is the squared radius of gyration taken along a line normal to the sheet surface. Equation (2) has been shown to be good approximation for rod of finite length and variable cross section [14]. Likewise Eq. (3) has been shown to work even when the sheets are closed into vesicles, provided the vesicle is large [15].

Evaluation of data by maximum entropy (Maxent) was done using the algorithm developed by Skilling and Bryan [16] as modified by Skilling and Gull [17–18]. Calculations

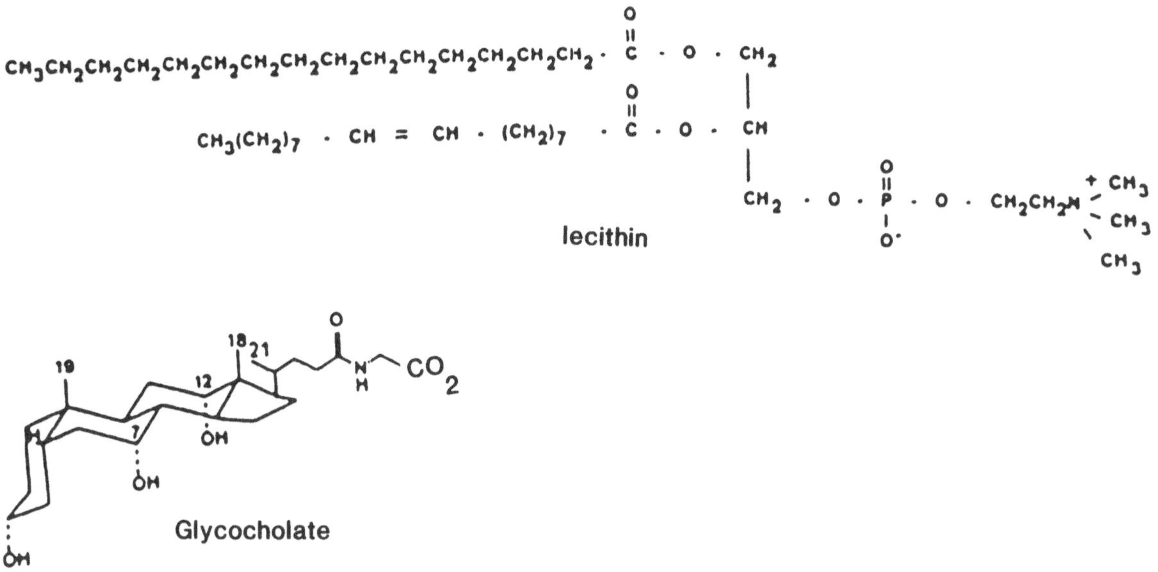

Fig. 1. Chemical Structures of Lecithin and Glycocholate

use the form factor for cylinders of radius, $R$, and height, $H$, given as

$$S(Q) = 16 \int_0^{2\pi} \frac{\sin^2(QH/2\cos\theta)}{Q^2 H^2 \cos^2\theta}$$

$$\times \frac{J_1^2(QR\sin\theta)}{Q^2 R^2 \sin^2\theta} \sin\theta \, d\theta. \qquad (4)$$

This is convoluted with the instrument resolution function [11] to give the numeric form used. The same form is used for the non-linear least squares procedures [8, 9].

Analysis of shear-orientated samples is done using the program ANISK [19] which uses Fourier analysis to reconstruct the scattering along the direction perpendicular and parallel to the shear orientation. Comparisons are made with scattering calculated from shear oriented rods using the formalism of Hayter and Penfold [20].

## Results

We derive further structural aspects of the system with some detailed analysis of the data, starting with the scattering from a sample at point $A$ in Fig. 2, which is part of the dilution series at L/BS = 0.5, at a total lipid concentration of 16.7 g/l. The scattering curve is fit to a cylindrical model using nonlinear least squares methods, allowing the height, $H$ and the radius $R$ to be the fitted parameters using Eq. (4), convoluted with the instrument resolution function [11]. In this case we express our belief that the particles are solid but that we are otherwise ignorant of the shape. The results show that there are two solutions which fit the data (Fig. 3A). One is an elongated cylinder,

$H \simeq 62$ Å and $R \simeq 24$ Å; the other is a slightly squat object, $H \simeq 42$ Å, $R \simeq 30$ Å. The data cannot be used to distinguish between these two possibilities using this method. Guinier analysis for the same data (Fig. 3B) using Eq. (1) gives an $R_g$ of $24.1 \pm 0.3$ Å.

Data taken for samples at lower total lipid concentrations, points $B$ and $C$ in Fig. 2, show scattering patterns characteristic of elongated particles. This can be shown by non-linear least squares analysis, but the fits to monodispersed systems are not particularly satisfying [8, 9]. This may be a consequence of not including polydispersity in the model. Maxent (see Methods) can be used to treat the putative polydispersity of the system. In this approach the particles are modelled as cylinders according to the form derived from Eq. (4) [11], considering all reasonable possibilities from squat to elongated cylinders, allowing, now, that a continuous distribution of particle sizes can be present. The results for points $A$, $B$ and $C$ of Fig. 2 (L/BS = 0.5) are shown in Fig. 4. Figures 4A–C show data points, the solid lines being the fit of the results of the Maxent calculation to the data. The Maxent maps in Figs. 4a–c illustrate the most probable distribution of particle sizes in the sample, given the data.

Starting with the most concentrated sample, which is similar to that shown in Fig. 3 (point A, Fig. 2), the Maxent distribution shows a single population, with particles having dimensions $R \simeq 25$ Å and $H \simeq 50$ Å being the most frequent. The distribution probably says nothing about the real distribution of particles in

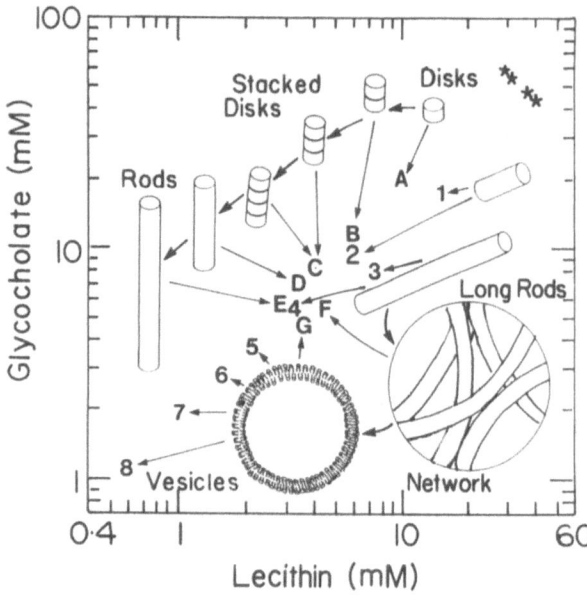

Fig. 2. Summary of Samples Measured and Interpretation of Structures in the Isotropic Phases of Mixed Aqueous Colloids of Lecithin and Glycocholate: The * represent the compositions of stock solutions (50 g/l) used in each of the dilution series. Numbers 1−8 indicate the compositions of some of the samples previously reported [8,9]. Molar ratios and total lipid concentrations are: 1, L/BS = 0.9, 20 g/l; 2, L/BS = 0.56, 10 g/l; 3, L/BS = 0.9, 10 g/l; 4, L/BS = 0.56, 5 g/l; 5−8, L/BS = 0.56, 3.3 g/l, 2.5 g/l, 1.7 g/l, 1.0 g/l, respectively. Letters $A−G$ indicate compositions of samples reported in this work: $A−E$, L/BS = 0.50, 16.7 g/l, 10 g/l, 7.1 g/l, 6.1 g/l and 5.0 g/l, respectively; $F$ and $G$, L/BS = 0.8, 6.3 and 5.0 g/l. Icons represent the models of the particle structures. Small arrows indicate appropriate samples. Large arrows indicate general nature of changes which occur with dilution

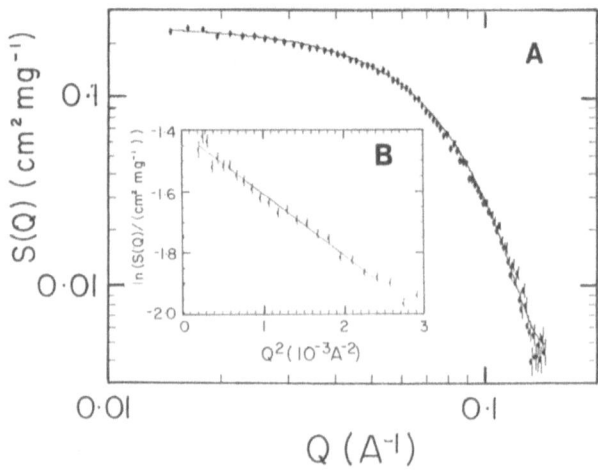

Fig. 3. Analysis of a Globular Structure: Data for a sample represented by point $A$ in Fig. 1 (L/BS = 0.5, 16.7 g/l total lipid concentration). *A:* Points are the data. The solid line are fits for cylinders described in the text. *B:* Guinier analysis of the data in $A$ according to Eq. (1)

solution as illustrated by the fact that simulated data for scattering from a homogeneous population of cylinders with $R = 25$ Å and $H = 50$ Å gives the same distribution. Thus, given the data, and the corresponding precision in $Q$ and $S(Q)$, the maps cannot be constrained to a sharper distribution. Comparison of this distribution with the models fitted by least squares (Fig. 4a) show the models to be at the edge of the distribution. The differences are a consequence of assuming that there is no polydispersity in the least squares model. The loci of points for $R_g$ of a cylinder $(R_g = \sqrt{(H^2/12 + R^2/2)})$ on the map (Fig. 4a) passes through the center of the distribution at $R \simeq 27$ Å, $H \simeq 50$ Å.

Consider next, figures 4B and b. The Maxent solution reveals two distinct classes of particles: One is virtually identical with that seen in the Fig. 4a; the other has the same radius but with $H \simeq 100$ Å —

twice that of the smaller particles. We attempt to follow the changes further in Figs. 4C and c. However, the data concerning the lengths of the particles becomes sparse due to limitations imposed by the minimum accessible $Q$ (about 0.006 Å$^{-1}$ for these data), and although a third class of particle is observed at $R \simeq 25$ Å and $H \simeq 170$ Å, the position of the peak is not highly reliable. Simulated data having the same statistical errors as the experimental results in Fig. 4C for a population of particles with $R = 25$ Å and $H$ values of 50, 100, 150 and 200 Å give the same result shown here. Again, the spread of the distribution of each size class reflects an inability to constrain the probability distribution further, given the data. These results suggest that growth of the rods occurs discontinuously by the aggregation of $\approx 50$ Å units.

The radial dimensions of rod-like particles can be extracted using Eq. (2) when they are sufficiently long [14]. The analyses for points $C$, $D$ and $E$ of Fig. 2 are given in Fig. 5, and show that the scattering approaches the form described by Eq. (2) as the concentration of lipids in the solution is decreased from 7.1 g/l to 5.0 g/l. The value for $R_c$ calculated from this plot is 19.1 $\pm$ 0.3 Å, and is the same as the values reported previously [8,9] for L/BS = 0.56 and 0.9, points 3 and 4 of Fig. 2, respectively. This value corresponds to a radius of 27 Å if the particle is modelled as a uniform rod, which agrees with the radial values determined from Maxent (Fig. 4).

Shear-orientation of the rod-like particles was used in an attempt to further characterize the axial and

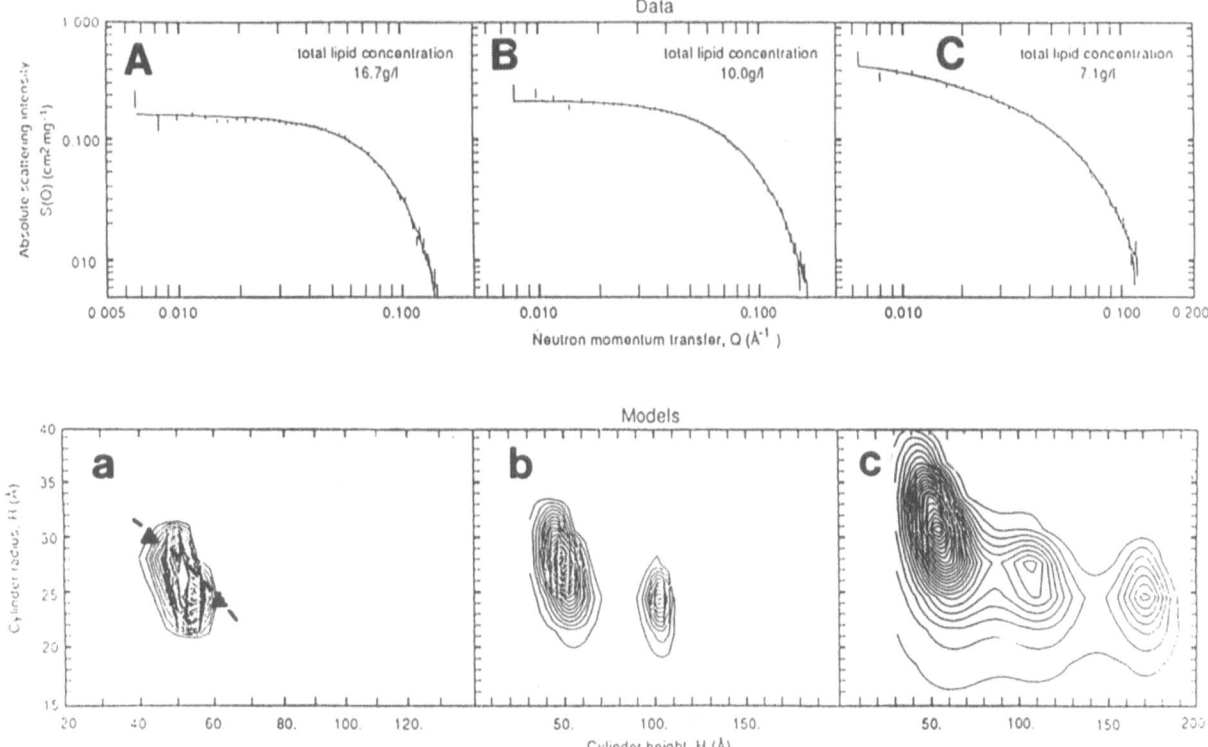

Fig. 4. Data and Maximum Entropy Analysis: Points in *A, B* and *C* are data for samples represented by points *A, B* and *C* in Fig. 2. *A*: L/BS = 0.5, 16.7 g/l. *B*: L/BS = 0.5, 10.0 g/l. *C*: L/BS = 0.5 7.1 g/l. Solid lines are fits to the data for the distributions of cylinders of height, *H*, and radius *R* given in the corresponding figures a,b and c, respectively. Lowest contour in figures a,b and c are 0.04 of the maximum frequency. In a: ▲ corresponds to the values from the least squares analysis shown in Fig. 3A; in 4a – – –, is the loci of values satisfying the radius of gyration derived from the Guinier analysis in Fig. 3B

radial structure. A sample similar to that analyzed in Fig. 5 (L/BS = 0.5, 5 g/l: point *E* in Fig. 2) was subjected to a shear gradient $G = 5000\ s^{-1}$. Only slight orientation was evident. Comparison of the data with simulated data shows that this result implies an upper limit on the rod length of roughly 1000 Å. The natural polydispersity of the samples can also account for a lack of observable orientation effects [21].

We conclude from this that at sufficiently high total lipid concentrations and lecithin to bile salt molar ratios, the globular mixed micelles are present as a homogeneous population. On dilution these aggregate into linear mixed micelles of radius $\approx 25-27$ Å. This appears to be a universal aspect of the rods throughout the phase map. Under conditions where the rigid rod-like morphology is most obvious, the rods are sufficiently long and are likely to be too polydisperse to orient in the shear gradient used.

The rigid rod-like morphology does not persist as the lipid concentration is diluted further. We illustrate

this with the series at L/BS = 0.8. The scattering for point *F* in Fig. 2 is shown in Fig. 6. The scattering appears "fractal", as it is characterized by a having a large domain of *Q* (0.001 to 0.05 Å⁻¹ over which it obeys a simple power law in *Q* spanning very nearly 2 decades in intensity. The slope of the double log plot suggests the fractal dimension is 1.2 [22]. This could be the result of random contact between segments of different rods or of the same (long, flexible) rod. We describe this as a loose network in the vignetted picture given by the data.

As the L/BS = 0.8 dilution series is continued further to 5 g/l (see point *G*, Fig. 2) the scattering becomes characteristic of a large vesicle approximately 670 Å radius (Fig. 7A). Previously [8, 9], we evaluated the structures as sheet-like (point 5, Fig. 2), but pointed out that they could be closed into vesicles as the minimum *Q* (0.005 Å⁻¹) available in the earlier measurements vignetted the particles. This result removes the ambiguity. We use Eq. (3) to determine the

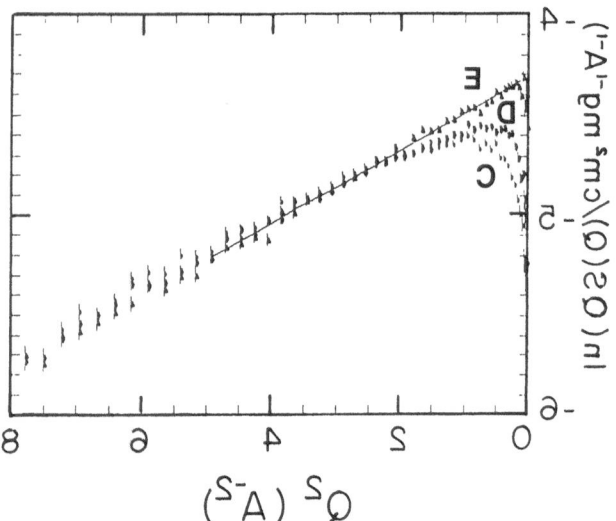

Fig. 5. Guinier Analysis for Rod-Like Forms: $C$, $D$ and $E$ are data for points $C$ (L/BS = 0.5, 7.1 g/l), $D$ (L/BS = 0.5, 6.1 g/l) and $E$ (L/BS = 0.5, 5.0 g/l) in Fig. 2 plotted according to Eq. (2). Solid line is the fit implied by Eq. (2) for $E$

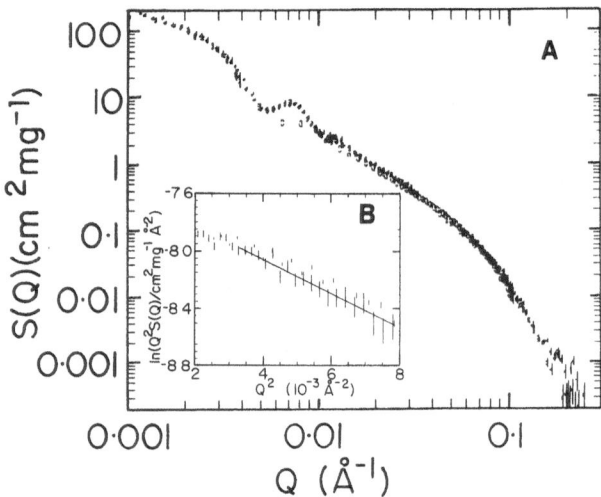

Fig. 7. Scattering Characteristic of a Large Vesicle Made up of a Single Lipid Bilayer: A: Scattering from a sample represented by point $G$ (L/BS = 0.8, 5.0 g/l) in Fig. 2. B: Guinier analysis for a sheet-like form according to Eq. (3). Solid line is the fit implied by Eq. (3)

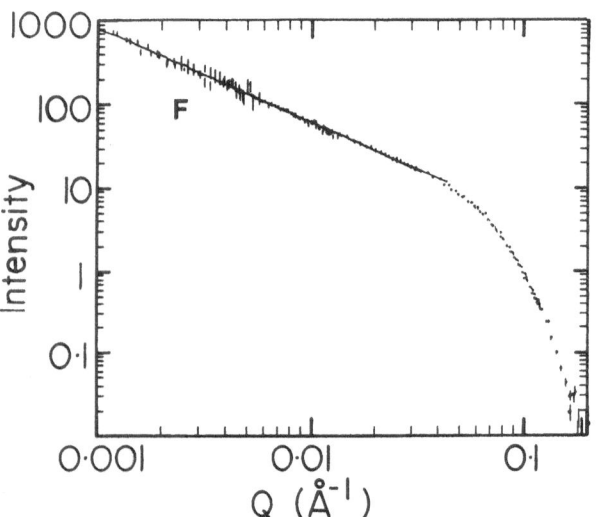

Fig. 6. Evidence for Extended Networks as a Transition Intermediate: Scattering data for sample represented by point $F$ (L/BS = 0.8, 6.3 g/l) in Fig. 2. Points are data, Line is a fit to $I \sim Q^{-12}$. $I$, the intensity labelled on the ordinate is in arbitrary units

value of $R_d = 10.3$ Å (Fig. 7B), corresponding to an apparent thickness of 36 Å for the vesicle wall, which is consistent with our earlier finding [8, 9]. This value strongly suggests that the vesicles consist of a single lecithin bilayer. This is consistent with our earlier result that the mass of the vesicles scales as the vesicle radius squared [8].

## Discussion

The results presented above support our earlier conclusion that when particles are diluted at constant lecithin to glycocholate ratios the initially globular particles become elongated along a single axis to form rod-like particles [8, 9]. The radius of the rods is 25 to 27 Å, and is the same regardless of the lecithin to bile salt molar ratio or total lipid concentration. This model is different from that derived from quasi-elastic light scattering [3–6] in which it was proposed that growth occurs by aggregation of material into ever larger disks. There is precedent for the formation of mixed micelle rods in bile salt-lecithin aqueous colloids, at least in the ordered phases, as it was found some time ago that hexagonal phases of rods are present in aqueous mixtures of sodium cholate with lecithin at high L/BS ratios [1]. Figure 2 summarizes our conclusions on the structures of the particles in the isotropic phases and the general nature of the changes which occur on dilution.

From our analysis we propose a mechanism of rod formation by association of small, globular, disk-like mixed micelles 25–27 Å radius and 50 Å height. Guinier (Fig. 3B) and Maxent (Fig. 4) analyses are consistent with the basic unit being the particles observed at point $A$ in Fig. 2. We do not have direct evidence on a detailed molecular picture of the basic mixed micelle units. We propose, however, that the basic unit is a mixed micelle disk like that proposed by Small [2] several years ago. In this model the lecithin is arranged into a simple bilayer approximately 50 Å thick. The glycocholate surrounds the disk as a ribbon with the more hydrophilic hydroxy residues projecting into the aqueous environment. Our data constrain the radius of the disk to around 25−27 Å. We cannot distinguish between Small's model and the one proposed by Mazer and colleagues [3], in which bile salt dimers are inserted into the lecithin bilayer as well. Certainly, the basis for the Mazer model as an explanation for particle growth is no longer valid. If we take the basic structure to be disk-like, it seems most plausible that assembly into rods occurs as a result of association between the flat disk surfaces. The result would be a cylinder consisting of stacked bilayers surrounded by a sheath of bile salt. The driving force for the aggregation has to be the repartitioning of bile salt from the particles to the bulk aqueous environment, in which case we must hypothesis that the maintenance of a hydrophilic surface needed for particle stability occurs only if the basic mixed micelles assemble into a rod. There are two mechanisms: either hydrophobic areas exposed by removal of bile salt from the particle associate, or the interaction of the bile salt with lecithin is stabilized in the rods. Within the context of the model assumed here, the former mechanism is difficult to understand, barring some internal rearrangement in the rods, as it is not clear where bile salt can be removed to promote association between the disk faces: the flat surfaces of the disk are already hydrophilic. This leaves us with stabilization of the bile salt-lecithin interactions, hence shifting of the equilibrium toward lecithin-bile salt association in the particles.

As the phase transition between rod-like and vesicle forms is approached, the rods appear to form a ramified, loose, gossamer network involving intra- and/or inter-rod association. Extending the idea of stabilization of lecithin-bile salt interactions by the formation of rods, it is expected that stabilization will become less and less important as the rods lengthen in response to dilution. Thus, at some point bile salt will be removed from the periphery leading to the formation of a network by lateral interactions between, or within, rods. Presumably, as this process continues the associations would become tighter leading to the direct formation of vesicles containing a single bilayer of lecithin and incorporating bile salt.

*Acknowledgements*

This work benefited from the use of the Low-$Q$ Diffractometer at LANSCE (USA), the SAD at IPNS (USA), KWA-1 at KFA, Jülich (FRG), and D11 at the ILL. Both United States instruments are supported by the United States Department of Energy, Office of Basic Energy Sciences and other Department of Energy Programs under contracts W-7405-ENG-32 to the University of California (LANSCE) and W-31-109-ENG-38 to the University of Chicago (IPNS).

## References

1. Small DM, Bourgès MC, Dervichian DG (1966) Biochim Biophys Acta 137:157−167
2. Small DM (1967) Gastroenterology 52:608
3. Mazer NA, Benedek GB, Cary MC (1980) Biochemistry 19:601−615
4. Mazer NA, Cary MC (1983) Biochemistry 22:426−442
5. Mazer NA, Schurtenberger P, Cary MC, Preisig R, Weigand K, Kanzig W (1984) Biochemistry 23:1994−2005
6. Mazer NA, Schurtenberger P (1985) Proc Int Sch Phys, Enrico Fermi 90:587−606
7. Cary MC (1988) Lipid Solubilisation in Bile, in Bile Acids in Health and Disease, Northfield T, Jazrawi R, Zentler-Munro P (eds) pp 61−82, Kluwer Academic Press, Boston
8. Hjelm RP, Thiyagarajan P, Alkan H (1988) J Appl Cryst 21:858−863
9. Hjelm RP, Thiyagarajan P, Alkan H (1990) Mol Cryst Liq Cryst 80A:155−164
10. Schurtenberger P, Mazer NA, Kanzig W (1985) J Phys Chem 89:1042−1049
11. Hjelm RP (1988) J Appl Cryst 21:618−628
12. Hjelm RP, Seeger PA (1989) Physics Conference Series 97:367−387, IOP Publications, Bristol
13. Lindner P, Oberthuer (1984) Revu Phys Appl 19:759−763
14. Hjelm RP (1985) J Appl Cryst 18:452−460
15. Knoll W, Haas J, Stuhrmann HB, Füldner H-H, Vogel H, Sackmann E (1981) J Appl Cryst 14:191−202
16. Skilling J, Bryan RK (1984) Mon Not R Ast Soc 211:111−124
17. Skilling J (1988) Classic Maximum Entropy in Maximum Entropy and Bayesian Methods, ed J Skilling, Cambridge 1988 Kluwer, pp 45−52, Academic Press, Boston
18. Gull S (1988) Developments in Maximum Entropy Data Analysis in Maximum Entropy and Bayesian Methods, Cambridge 1988 ed J Skilling, pp 53−57, Kluwer Academic Press, Boston
19. Lindner P, Hess S (1989) Physica B156:512−514
20. Hayter JB, Penfold J (1984) J Phys Chem 85:4589−4593
21. Cummins PG, Staples E, Hayter J, Penfold J (1987) J Chem Soc Faraday Trans 83:2773−2786
22. Martin JE (1986) J Appl Cryst 19:25−27

Authors' address:

Dr. R. P. Hjelm, Jr.
H805
Los Alamos National Laboratory
Los Alamos, New Mexico 87545, USA

**Progress in Colloid & Polymer Science**

Progr Colloid Polym Sci 81:232 – 237 (1990)

# A study of the swelling of n-butylammonium vermiculite in water by neutron diffraction

L. F. Braganza [1]), R. J. Crawford [2]), M. V. Smalley [2]) and R. K. Thomas [2])

[1]) Institut Laue-Langevin, Grenoble, France
[2]) Physical Chemistry Laboratory, Oxford, UK

*Abstract:* The swelling of an n-butylammonium vermiculite in solutions of n-butylammonium chloride has been studied as a function of temperature, hydrostatic pressure, uniaxial pressure, and the concentration of the soaking solution, by neutron diffraction. On heating the swollen clays a transition to the crystalline phase takes place at a well defined temperature, which depends on the external electrolyte concentration. The application of hydrostatic pressure induces macroscopic swelling of the vermiculite in the region of 1 kbar. The transition is completely reversible and also depends on temperature. — The dependence of interparticle distance on applied uniaxial pressure varies semiquantitatively with ionic strength in a similar way to the forces between mica plates in electrolyte solution. — Approximate dimensions of the Stern layer have been obtained from isotopic substitution of the counterion. The Stern layer is 5.5 ± 1 Å thick.

*Key words:* Clays; swelling; colloids; interparticle forces

## Introduction

The swelling of vermiculites in water is an unusual phenomenon. Some single crystals swell anisotropically, perpendicular to the plane of the silicate layers, leading to increases in volumes of as much as 50 times. The resulting silicate layer spacings are sufficiently large that the system behaves as a one dimensional colloid. The one dimensional nature of the swollen vermiculite makes it possible to do experiments not otherwise practicable on a three dimensional colloid. In particular, it is possible to apply a uniaxial stress to the crystal to derive force-distance information about the interparticle potential, and interference effects from a one dimensional structure give a much stronger signal than their three dimensional counterparts, allowing measurements out to a higher value of the momentum transfer.

The macroscopic swelling of vermiculite was first reported by Walker [1] and investigated for a variety of vermiculites by Garrett and Walker [2]. They found the expansion to be most uniform for a Kenya vermiculite which had a layer charge of 1.3 univalent cations per $O_{20}(OH)_4$ unit of structure. They also commented that for some alkylammonium vermiculite samples the swelling was dependent on temperature. Measurements of the expansion of vermiculites as a function of swelling pressure have also made by Rausel-Colom [3], Norrish & Rausel-Colom [4], and more recently by Viani et al. [5, 6].

In the present report we have used neutron diffraction to investigate a variety of phenomena associated with the swelling of a Brazilian (Eucatex) vermiculite. We have extended the range of previous measurements of the interparticle forces as a function of distance. In addition we have made some preliminary measurements of the scattering from the counterion distribution in the aqueous solution. In two recent publications [7, 8] we have made a thorough study of the effects of temperature and hydrostatic pressure on the swelling process and we start by reviewing the phase diagram with respect to these two variables.

## Experimental

All the experiments were done on crystals about 50 mm² by 1 mm thick and the details of their preparation and mounting in the cell have been given elsewhere, as have the details of the diffractometers used [7, 8]. The vermiculite used was the n-butylammonium form of Eucatex vermiculite.

For the force-distance measurements the pressure was applied via a rigid quartz block attached to a metal spring, on which a set of strain detectors were mounted. The applied force could be varied by using a screw to move the spring and quartz plate. The interparticle distance was recorded by neutron diffraction. The range of pressures covered using this apparatus was approximately $2-900 \times 10^3$ dynes $cm^{-2}$.

## Results

### The effect of temperature on swelling

The swelling of clays is conveniently divided into two regions, the first of which corresponds to the uptake of a relatively small proportion of water, possibly up to about 4 molecular layers, and the second of which involves the uptake of much larger volumes of water. The two types of swelling are quite distinct and we distinguish them as crystalline and macroscopic swelling respectively [9]. For a sample soaking in 1 M n-butylammonium chloride the equilibrium structure is a crystalline phase of spacing 19.4 Å. This phase is readily distinguished by its diffraction pattern (Fig. 1a), which consists of a series of sharp (00$l$) peaks. The range of the instrument used to obtain the diffraction pattern was only sufficient to show the first seven orders of the (00$l$) reflection but many more can be observed if a shorter incident wavelength is used. Only this series of peaks was observed because the experiment was done in reflection with the momentum transfer $\varkappa$ perpendicular to the plane of the crystal, and because the mosaic spread of the crystal was small ($\simeq 6^0$ F.W.H.M.).

Macroscopic swelling completely changes the appearance of the pattern and a typical result is shown in Fig. 1b for an external solution of 0.1 M n-butylammonium chloride.The three orders of diffraction correspond to an interparticle spacing of 122 Å and there is no trace of the (001) diffraction peak of the crystalline phase. The swelling is perfectly homogeneous and perhaps more remarkable is that the orientation of the platelets, as determined by the mosaic spread, is largely retained in the expansion. As found by other authors [2−6] the interparticle separation depends on the concentration of the external electrolyte. Thus, it increases to about 700 Å at an n-butylammonium concentration of 0.001 M and we were even able to obtain a value of approximately 900 Å at 0.0001 M. However, at these large spacings there is a considerable spread in the values obtained, as shown in Fig. 2. There are two possible reasons for this; that there is some variability in the samples, and

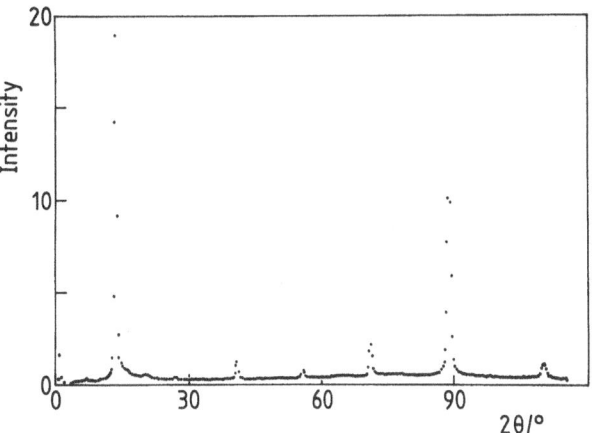

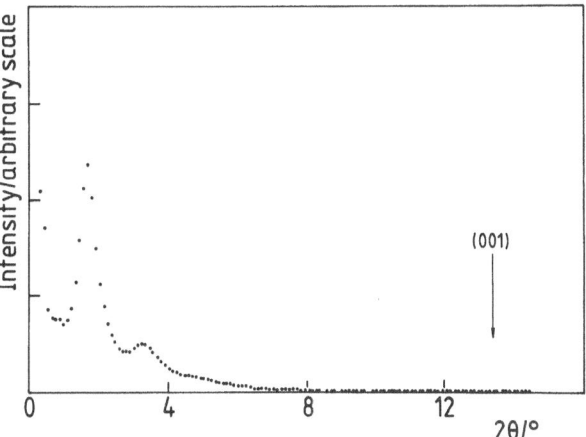

Fig. 1. (a) Diffraction pattern of the fully hydrated crystalline phase of n-butylammonium vermiculite measured in reflection geometry. $\lambda = 4.52$ Å. (b) Low angle diffraction pattern of n-butylammonium vermiculite at $T = 6.2\,°C$ and an external concentration of 0.1 M

that it is extremely difficult to avoid applying very small pressures when a sample is loaded into the quartz cell. This would not have a significant effect on the smaller spacings but may become important at large spacings. It is not easy to establish conclusively which of these two effects is significant.

Heating the macroscopically swollen samples to a well defined temperature brings about a transition to the crystalline phase. For example, at a concentration of external solution of 0.1 M, no change was detected in the diffraction pattern on heating to 13.5°C, but at 14.5°C the low-angle pattern of Fig. 1b collapsed into a broad envelope of scattering, and the pattern of the crystalline phase reappeared. Further scans at higher temperatures were identical to that at 14.5°C, showing that the phase change was complete in the inverval 13.5°−14.5°C. The transition was shown to be re-

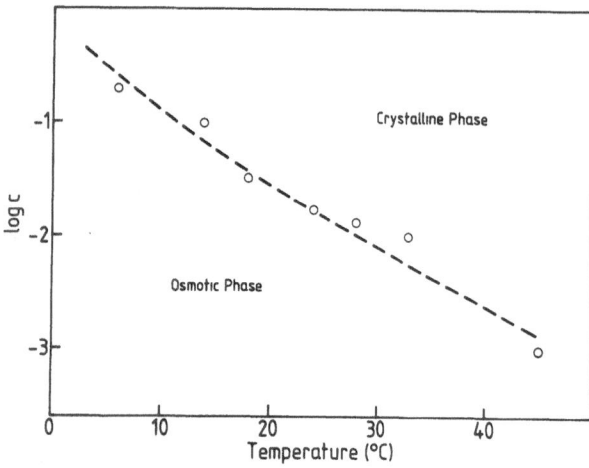

Fig. 2. The temperature-concentration phase diagram of n-butylammonium vermiculite

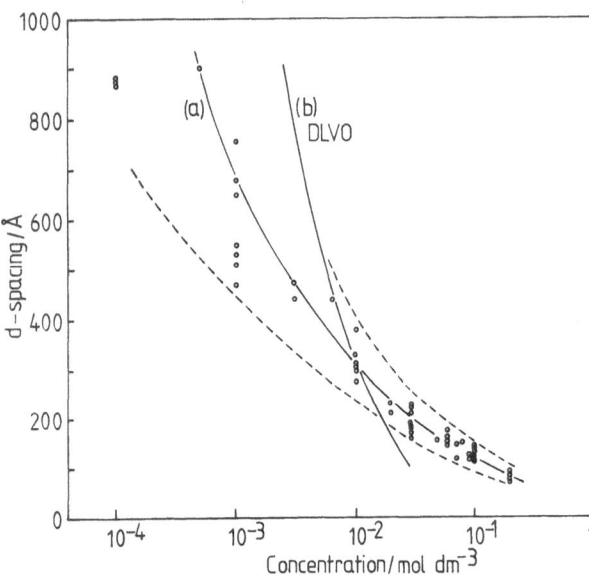

Fig. 3. Comparison between experimentally observed *d*-spacings and those calculated using DLVO theory with the single adjustable parameter (proportional to the Hamaker constant) chosen to fit the data at $10^{-2}$ M

versible by cycling the temperature between 10 and 20°C. The *c*-axis spacing for this particular sample was always found to be in the range 120–125 Å below 13.5°C, and the full diffraction pattern of the crystal was always recovered above 14.5°C. The transition temperature increased as the concentration of the external electrolyte *decreased*. The phase diagram for the transition is shown in Fig. 2.

The reversibility, sharpness, and reproducibility of this unusual phase change from crystalline to osmotically swollen gel show that it must be a true thermodynamic transition. In DLVO theory all hydrophobic sols are thermodynamically unstable because the minimum in the potential energy of interaction at short distances is deeper than the weak secondary minimum at large distances, which arises from the interplay of electrostatic and van der Waals forces [10]. It is therefore not possible for there to be a reversible change between crystalline and swollen phases within the framework of DLVO theory and our results disagree qualitatively with the theory. An additional short range repulsion, which modifies the relative depths of the two potential energy minima, would have to be introduced to account for this discrepancy.

The DLVO model postulates that the energy of interaction between two flat plates can be described in terms of a repulsive term $V_r$, resulting from the overlap of the electrical double layers, and an attractive van der Waals interaction, $V_a$. At equilibrium the two forces must balance and at large separations we have

$$B \exp(-\varkappa d_0) = A/d_0^4$$

where $d_0$ is the equilibrium separation, and $A$ and $B$ are constants. By choosing to fit this condition to a point in the middle of the experimental curve shown in Fig. 3 we can then obtain the DLVO prediction of the whole curve. The diagram then shows, in agreement with other authors, that the DLVO theory does not account for the variation of experimentally determined $d_0$-values with concentration in this system.

*The effect of hydrostatic pressure*

The unusual direction of the effect of temperature must be associated with an enhancement of the adsorption of the n-butylammonium ion with increasing temperature. This, in turn, may be a consequence of the hydrophobic effect, which also increases with temperature. The hydrophobic effect is known to be reduced by the application of hydrostatic pressure [11]. If this explanation is correct it would be expected that the application of hydrostatic pressure would favour the transition to the macroscopically swollen phase. This is somewhat contrary to the usual association of volume changes and pressure effects.

The initial measurement of the effect of hydrostatic pressure was made at an external salt concentration of 0.06 M at 20°C, 9°C above the transition temperature to the crystalline phase at atmospheric pressure. At an applied pressure of 1050 bars there was a sharp

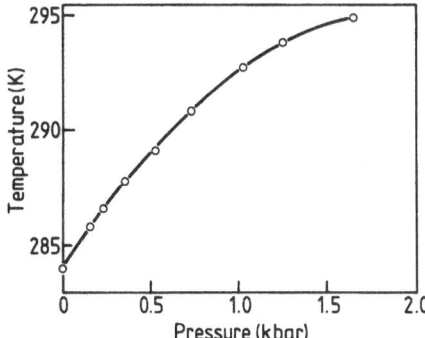

Fig. 4. Phase diagram for the swelling of n-butylammonium vermiculite in water as a function of temperature and hydrostatic pressure. The osmotic phase lies in the *lower* half of the diagram

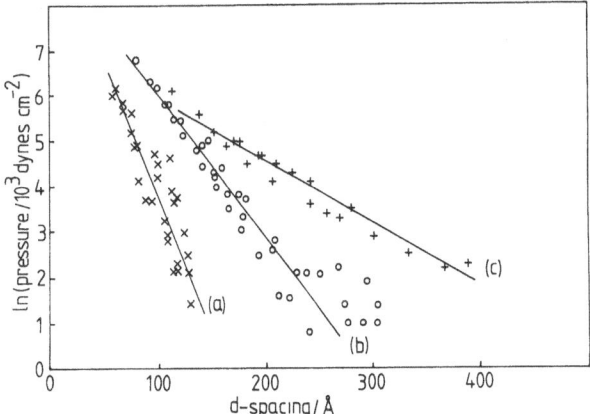

Fig. 5. Pressure/distance curves for n-butylammonium vermiculite in (a) 0.1, (b) 0.01, and (c) 0.001 M n-butylammonium chloride solutions

transition to the swollen phase, the transition being monitored by simultaneous observation of the intensities of both the crystalline (001) peak and the peaks at low angles corresponding to the 120 Å spacing. The transition was completed over a range of 100 bars and was found to be completely reversible with a time constant of approximately 10 minutes. As the pressure was increased further to 2000 bar, the c-axis spacing decreased by only about 1 Å per 300 bar. There was insufficient beam time to study the effect of concentration but we did measure the variation of transition pressure with temperature. The resulting *P-T* phase diagram of the 0.1 M n-butylammonium system is shown in Fig. 4.

These results give interesting information about the nature of the solution between colloidal particles. The slope of the line in Fig. 4, $dP/dT$, is related to the entropy and volume changes of the phase transition by the standard equation:

$$dP/dT = \Delta S/\Delta V.$$

$dP/dT$ was approximately constant at low applied pressures at 85 atm K$^{-1}$ ($\equiv$ 8.5 $\times$ 10$^6$ J m$^{-3}$ K$^{-1}$). From measurements of the heat capacity across the phase transition we were able to determine the entropy of the transition, which was found to be 0.0183 J K$^{-1}$ g$^{-1}$ of dry clay at 10.1 °C. The combination of these two values gave a volume change for the transition of $-2.15$ cm$^3$ kg$^{-1}$ of dry clay. Because the density of the vermiculite is approximately 2.3 g cm$^{-3}$, this volume change corresponds to a contraction of 4.9 cm$^3$ dm$^{-3}$ of dry clay for a five to six-fold expansion. If this contraction is converted to a change in the water volume from the bulk solution to between the clay

plates, it is equivalent to a fractional decrease of slightly less than 0.1%. As far as we know, such a contraction has not previously been observed.

*The effect of uniaxial pressure*

As is well known the application of pressure along the direction of the c-axis of a vermiculite decreases the interparticle spacing. The effect has been measured several times, using either X-ray diffraction to determine the interparticle spacing or measurements of the macroscopic thickness [3–6, 12, 13]. The use of neutrons to determine the interparticle separation has two advantages; larger samples may be used, allowing better control of the applied pressure, and the longer incident wavelengths allow much larger spacings to be observed.

Figure 5 shows the pressure/distance curves obtained for three different electrolyte concentrations. For each concentration we have made measurements on more than one sample. There is generally a greater variation from sample to sample than within a single set of measurements on one sample and this variation is shown by the scatter of points in the figure. It is clear, however, that the pressure/distance plot is approximately linear in all three cases and that the slopes are quite distinct. The range of pressures and distances covered is much larger than has previously been achieved.

There are a number of qualitative points of interest in these pressure distance curves. In comparison with the measurements of Barclay & Ottewill [13] on montmorillonites these are true equilibrium measurements; there was no need for pressure cycling, al-

*Progress in Colloid & Polymer Science, Vol. 81 (1990)*

though it was often necessary to wait for times of the order of an hour for the spacing to reach a steady value. Viani et al. [5, 6] found that their samples crystallised when subjected to a high enough pressure. On no occasion did we produce the crystalline phase by applying uniaxial pressure. This may have been because we were not able to reach a sufficiently high pressure. The lowest spacing we achieved was 39 Å at an external concentration of 0.2 M for which the zero pressure spacing was 83 Å. The overall appearance of the set of lines in Fig. 5 is similar to those obtained in mica force experiments [14] where, if the potential is dominated by the repulsive electrostatic interactions, the slopes of the lines should be the reciprocals of the Debye length. The figure shows the mean of the straight lines from each separate set of data and their slopes give reciprocal Debye lengths of 0.07, 0.03, and 0.013 $\text{Å}^{-1}$, to be compared with the values of 0.1, 0.032, and 0.010 $\text{Å}^{-1}$ calculated from the concentration of the external solution. The discrepancies between the two sets of values are comparable with those obtained in the mica force balance experiments. However, the difficulty in interpreting the results in this way is that our earlier observations of equilibrium swelling distances are such as to require the slopes of the graphs to tend towards $-\infty$ from the lowest pressures displayed on the figure. This is because, for example, the equilibrium spacing at an external concentration of 0.1 is, from Fig. 3, 130 $\pm$ 15 Å, which is immediately below the lowest pressure measurement on the graph. We believe this to be an important point and we are presently doing further experiments to establish the shape of the pressure/distance curve in this region.

*Ion distribution*

In most attempts to give a quantitative explanation of experimental observations of interparticle forces it is usually necessary to introduce additional parameters describing the nature of ion adsorption at the surface. A typical example is the introduction of a Stern layer characterised by an adsorption parameter and a thickness. In this way any defects in the model of the interparticle potential may be concealed by arbitrary adjustment of other parameters. The ideal would be to measure the parameters of the Stern layer independently and, preferably the ion distribution in the diffuse layer. This is, in principle, possible in neutron diffraction experiments on the vermiculites and we here present some preliminary results.

The method depends on changing the scattering length density of either the counterion or the solvent

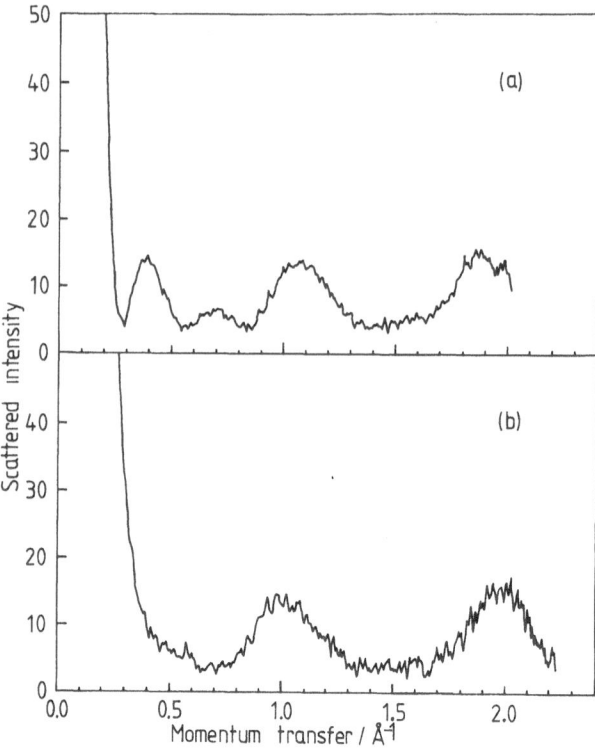

Fig. 6. Neutron diffraction patterns of n-butylammonium vermiculite in n-butylammonium chloride solution in $D_2O$, (a) 25% and (b) 75% of the n-butylammonium ion deuterated. The solvent and cell backgrounds have been subtracted

so that two different diffraction patterns are obtained for a single chemical structure. The most direct method is to change the isotopic composition of the counterion. Figure 6 shows the diffraction patterns for (a) 25% of the fully deuterated and 75% of the fully protonated counterion at a concentration of 0.09 M and (b) 75% of the fully deuterated and 25% of the fully protonated counterion at a concentration of 0.03 M, both in $D_2O$. The patterns depend very little on concentration in this concentration range and therefore the substantial differences in the two patterns depend only on the ion distribution, since nothing else has been altered by the substitution.

The steep rise in intensity at values of momentum transfer $\varkappa$ below about 0.25 $\text{Å}^{-1}$ is the diffraction pattern from the whole array of plates and which is shown on a reduced scale in Fig. 1. Above $\varkappa = 0.25$ $\text{Å}^{-1}$ the interparticle scattering is completely damped out and the remaining pattern depends only on the Fourier transform of the unit cell contents, i.e. the clay plate and the aqueous layer. Careful comparison of this region of the two patterns shows that the feature at approximately $2 \text{ Å}^{-1}$ is unchanged by the isotopic substitution and therefore results from the internal structure of the clay plate. This is confirmed by calculation of the structure factor of the plate itself. The plate has a thickness of $9.6 \text{ Å}$ including the van der Waals radius of the outer layer of oxygen atoms. The remaining features of the two patterns are approximately accounted for by the addition of a Stern layer of thickness $5.5 \pm 1 \text{ Å}$, most of the plate charge being neutralised by the adsorbed ions. The two strong features at 0.4 and $0.7 \text{ Å}^{-1}$ in Fig. 6a are diffraction effects from the clay plate plus Stern layer but in Fig. 6b these features disappear because the scattering length densities of Stern layer and clay are more closely matched.

The resolution of the experiment is sufficient for us to obtain more detailed information on the structure within the Stern layer and on the possible incorporation of water either into the Stern layer or in between the plate and the Stern layer. However, we have not yet been able to fit a single model to all the data. Unfortunately it seems unlikely that at this stage the experiment has sufficient sensitivity to observe the diffuse part of the double layer. This is because the signal is dominated by the large fraction of ions in the Stern layer.

# References

1. Walker GF (1960) Nature 187:312
2. Garrett WG, Walkger GF (1960) Clays & Clay Minerals 9:557
3. Rausel-Colom JA (1964) Trans Far Soc 60:190
4. Norrish K, Rausel-Colom JA (1963) In: Swineford, Franks PC (eds) Clays and Clay Minerals, Proc. 10th Natl. Conf., Austin, Texas, 1961, Ada. Pergamon Press, New York, 123
5. Viani BE, Low PF, Roth CB (1983) J Colloid Interface Sci 96:229
6. Viani BE, Roth CB, Low PF (1985) Clays & Clay Minerals 33:244
7. Crawford RJ, Smalley MV, Thomas RK (in press) Clays & Clay Minerals
8. Smalley MV, Thomas RK, Braganza L, Matsuo T (1989) Clays & Clay Minerals 37:474
9. van Olphen H (1977) An Introduction to Clay Colloid Chemistry, 2nd Edition, Wiley, New York, 150
10. Derjaguin BV, Churaev NV, Muller VM (1987) Surface Forces: Plenum, New York, 293
11. Braganza LF, Worcester DL (1986) Biochemistry 25:2591
12. Rausel-Colum JA, Saez-Aunon J, Pons CH (1989) Clays & Clay Minerals 24:459
13. Barclay LM, Ottewill RH (1970) Spec Disc Far Soc 1:138
14. Israelichvili JN, Adams GE (1978) J Chem Soc Far Trans I, 74:975

Authors' address:

Dr. R. K. Thomas
Physical Chemistry Laboratory
South Parks Road
Oxford, OX1 3QZ, United Kingdom

**Progress in Colloid & Polymer Science**          Progr Colloid Polym Sci 81:238–241 (1990)

# Global fluorescence analysis in micellar systems

A. Malliaris

N.R.C. "Demokritos", Athens, Greece

*Abstract:* The method of simultaneous analysis of multiple fluorescence decay curves, also referred to as global fluorescence analysis, is presented. By taking into account existing relationships among the various decay curves, the method allows better parameter recovery and model testing, compared to the usual single curve analysis. The application of the method, using simulated fluorescence decays, in micellar systems is discussed.

*Key words:* Micelle; fluorescence; simultaneous; global; multiple

## Introduction

The time-resolved fluorescence technique is a powerful tool for the study of electronically excited molecular states, with wide applications to fields such as photophysics, photochemistry and photobiology [1 – 2]. In its present state the technique involves high repetition rate lasers [4, 5], fast detecting devices [6], powerful computing machines and sophisticated methods of data analysis [7]. Under these conditions fluorescence lifetime measurements of only few picoseconds are not uncommon, while the intensity of the emission which is possible to be detected by the method spans up to about five orders of magnitude.

Note that the fluorescence decay data obtained in a real experiment, do not give the true fluorescence decay profile, since the response of the particular experimental set up affects the shape of the decay curve. Therefore, several deconvolution methods which separate the true fluorescence decay from the experimental artifacts have been developed [8, 9]. In the present account the problem of deconvolution will not be dealt with, and all decay curves will be treated as delta response functions.

## Analysis of fluorescence decay curves

### Single curve analysis

The extraction of fluorescence decay parameters, or the deduction of photophysical mechanisms, usually involves the recording of several fluorescence decay curves under changing experimental conditions, e.g. temperature, solvent, wavelength etc. Each curve is then individually analysed, by means of a certain decay law, to obtain the decay parameters. Finally, using either graphical or numerical methods the best value of the decay parameters can be determined. Since from the statistical criteria of the goodness of the fit, it is possible to draw conclusions concerning the validity of the decay law, several photophysical models can be examined, until the best one is obtained.

Let us consider a trivial example of single fluorescence decay curve analysis, e.g. a very dilute solution of pyrene $P$, in a solvent $S$, and in the absence of quenchers. The question is to determine the fluorescence lifetime $\tau_1$. Such fluorescence obeys the monoexponential decay law of Eq. (1), where

$$F(t) = a_1 \exp(-t/\tau_1) \qquad (1)$$

stands for fluorescence intensity at time $t$ after the exciting pulse. The preexponential factor $a_1$ depends on the amount of $P$ in $S$.

Using $n$ different dilute solutions of $P$ in $S$, and recording the fluorescence decay from each one of them, $2n$ parameters $a_{1n}$ and $\tau_{1n}$ are obtained, by separately fitting Eq. (1) to each individual decay curve

$$
\begin{array}{lll}
\text{decay curve 1} & a_{11}, & \tau_{11} \\
\text{decay curve 2} & a_{12}, & \tau_{12} \\
\text{decay curve 3} & a_{13}, & \tau_{13} \\
\cdots & \cdot & \cdot \\
\text{decay curve } n & a_{1n}, & \tau_{1n}.
\end{array}
\qquad (2)
$$

Finally, as the best experimental value for the fluorescence lifetime of $P$, one can take for instance, the numerical average of the $\tau_{1n}$ values from the $n$ different decays, i.e.

$$\tau_1 = (\tau_{11} + \tau_{12} + \tau_{13} + \ldots + \tau_{1n})/n. \tag{3}$$

Moreover, if there are other possible decay laws, except that of Eq. (1), it is feasible to chose the one with the best fit to the experimental data, from the goodness of the fit, by means of statistical criteria such as $x_2$, Durbin-Watson test, etc.

This popular and usually effective method, has a rather serious handicap. Namely, it does not take into account relationships which may exist among the individual decay curves. In the previous example the relationship among the different curves is that the magnitude of $\tau_{1n}$ must be the same in all solutions, i.e. $\tau_{11} = \tau_{12} = \ldots = \tau_{1n} = \tau_1$. Therefore, the set of Eq. (2) becomes

$$
\begin{array}{lll}
\text{decay curve 1} & a_{11}, & \tau_1 \\
\text{decay curve 2} & a_{12}, & \tau_1 \\
\text{decay curve 3} & a_{13}, & \tau_1 \\
\ldots & . & . \\
\text{decay curve } n & a_{1n}, & \tau_1
\end{array} \tag{4}
$$

where the $2n$ independent parameters are reduced to $n + 1$.

## Global fluorescence analysis

The global fluorescence analysis is a method which takes into account the fact that, as in the previous example, there may be certain parameters ($\tau_1$) which do not change independently among the various decays. These parameters, although allowed to take any value, they are nevertheless restricted by the condition that this particular value must be the same among all decays. Such parameters are called "linked". Therefore in global fluorescence analysis one can distinguish between free-running ($a_{1n}$) and linked ($\tau_1$) parameters. Note that it is also possible to have constant parameters, the magnitudes of which are known from independent measurements.

Fitting a theoretical curve to experimental decay data consists of consecutive iterative cheques until an acceptable minimization of deviations between theory and experiment is achieved. In global fluorescence analysis the same procedure is followed, except that in this case, instead of a single curve a set of decay curves is involved in the minimization of deviations. This is obtained by means of global mapping vectors which link a global parameters to the same parameter in each single curve (local parameter).

The simplest way to relate global to local parameters is to assign a global parameter $g_m$ to every local parameter in the order in which the local parameters appear in the set of the decay curves. Thus, the index $m$ of the global parameters will assume the values $1, 2, 3 \ldots$ correspondingly, for the 1st local parameter of the 1st decay curve ($a_{11}$), the 2nd local parameter of the 1st decay curve ($\tau_{11}$), the 1st local parameter of the 2nd decay curve ($a^{12}$), the 2nd local parameter of the 2nd decay curve ($\tau^{12}$) etc, as shown below:

$$
\begin{array}{ll}
\text{decay curve} & \text{local parameters} \\
1 & a_{11}, \tau_{11} \\
2 & a_{12}, \tau_{12} \\
3 & a_{13}, \tau_{13} \\
. & \quad . \quad . \\
n & a_{1n}, \tau_{1n}.
\end{array} \tag{4}
$$

$$
\begin{array}{l}
\text{global parameters} \\
g_1, \quad g_2 \\
g_3, \quad g_4 \\
g_5, \quad g_6 \\
. \quad . \\
g_{n+2}, g_{n+3}.
\end{array}
$$

However, as pointed out earlier, the global parameters $g_2$, $g_4$, $g_6 \ldots g_{n+3}$ are all equal to the fluorescence life time of the fluorophore, therefore they all should be set equal to $g_2$. Consequently, the assignment of global to local parameters, i.e. the determination of the indices $m$, changes as shown below:

$$
\begin{array}{ll}
\text{decay curve} & \text{local parameters} \\
1 & a_{11}, \tau_{11} \\
2 & a_{12}, \tau_{12} \\
3 & a_{13}, \tau_{13} \\
. & \quad . \quad . \\
n & a_{1n}, \tau_{1n}
\end{array} \tag{5}
$$

$$
\begin{array}{l}
\text{global parameters} \\
g_1, \quad g_2 \\
g_3, \quad g_2 \\
g_4, \quad g_2 \\
. \quad . \\
g_{n+1}, g_2.
\end{array}
$$

To simplify the process of transforming local to global parameters $a$ mapping matrix $A$ can be contructed [10], by simply copying the indices of the global parameters as they appear in Eqs. (5), i.e.

decay curve:

$$A = \begin{vmatrix} 1\ 2\ 3\ 4 & n \\ 1\ 3\ 4\ 5\ ...\ n+1 \\ 2\ 2\ 2\ 2\ ...\ 2 \end{vmatrix} \begin{matrix} \text{parameter} \\ a_{1n}(\text{free-running}) \\ \tau_1(\text{linked}) \end{matrix} \quad (6)$$

In the above matrix the two rows correspond to the two local parameters $a_{1n}$ and $\tau_1$, while the $n$ columns correspond to the $n$ decay curves. Therefore, having constructed the matrix $A$ is immediately evident which global parameter replaces any local one. For example, the first local parameter of the fourth decay curve, i.e. $a_{14}$, must be replaced by the global parameter with index given by the intersection of the first row and the fourth column, which is $g_5$. Likewise the global parameter corresponding to the second local parameter of the same fourth decay curve will $g_2$. In this way one can replace local by global parameters and then fit the decay Eq. (7) to the data of the fourth decay curve

$$F(t) = g_5 \exp(-t/g_2). \quad (7)$$

Evidently, after the replacement of local by global parameters any appropriate algorithm can be applied for the fitting.

An important point concerning global fluorescence analysis is the choice of the appropriate parameter which, although changes among the various experiments, does not affect the equation of the decay law. In other words, it does not affect the photophysical system. For instance, in the previous example the parameter which was changed from one decay to the other was the concentration of the fluorophore $[P]$. Such variation of $[P]$ influences only the parameter $a_{1n}$ and has no affect on Eq. (1), provided $[P]$ does not become very large. Consequently, $[P]$ at low concentration, is an appropriate parameter for the determination of $\tau_1$ by means of global analysis. If, on the other hand, $[P]$ were to assume large values, then the decay law should be changed all together because of excimer formation which alters the fluorescence mechanism. Therefore $[P]$ at high concentration is not a good parameter for the determination of $\tau_1$ by global analysis, since the lifetime itself changes.

## Global analysis in micellar media

The application of global fluorescence analysis to a surfactant solution of micellar concentration $[M]$, in which the immobile [11] fluorophore $[F]$ and quencher $[Q]$ have been solubilized, will be discussed. The concentration $[F]$ of the fluorophore is negligible.

Under these conditions the fluorescence decay rates have a Poisson distribution, and the fluorescence decay follows Eq. (11) [12, 13]. $A_1$, $A_2$, $A_3$ and $A_4$ are time independent parameters

$$F(t) = A_1 \exp(-A_2 t)$$
$$\times \exp\{-A_3[1 - \exp(-A_4 t)]\} \quad (11)$$

to be determined from the fitting of Eq. (11) to the experimental decay data. Furthermore, $A_1$ reflects the fluorescence intensity at the highest point of the decay, $A_2$ is the first order fluorescence decay rate in the absence of $Q$, $A_3$ is equal to the ratio $[Q]/[M]$ and $A_4$ is the intramicellar pseudo-first order fluorescence quenching rate for only one quencher in the micelle. If Eq. (11) is rewritten in the form of Eq. (12), it becomes obvious that in the equation

$$F(t) = A_1 \exp(-A_2 t)$$
$$\times \exp\{-[Q]/[M][1 - \exp(-A_4 t)]\} \quad (12)$$

the parameter which can be changed without changing the physical system, is $[Q]$. Indeed, when the quencher concentration changes, within certain limits (e.g. $A_3 < 2$), the physical system remains practically unperturbed. Therefore the important parameters $A_2$, $[M]$ and $A_4$, are the same in all decays, and consequently global analysis can be applied.

In this case there will be one free-running parameter, namely $A_1$, and three linked parameters $A_2$, $[M]$ and $A_4$. The corresponding mapping matrix will be the following:

decay curve:

$$A = \begin{vmatrix} 1\ 2\ 3\ 4\ ...\ n \\ 1\ 5\ 6\ 7\ ...\ n+3 \\ 2\ 2\ 2\ 2\ ...\ 2 \\ 3\ 3\ 3\ 3\ ...\ 3 \\ 4\ 4\ 4\ 4\ ...\ 4 \end{vmatrix} \begin{matrix} \text{parameter} \\ A_1(\text{free-running}) \\ A_2(\text{linked}) \\ [M](\text{linked}) \\ A_4(\text{linked}). \end{matrix}$$

The superiority of the global over the single curve analysis is remarkable and it has been clearly demonstrated by single and multiple curve fitting of Eq. (12) to simulated [14], as well as real [15, 16] fluorescence decay data in micellar systems. Thus, micellar parameter were recovered, by means of single curve analysis, with accuracies 2, 17 and 30% for $A_2$, $[M]$ and $A_4$ respectively. Note that the above accuracies correspond to long accumulation times ($10^4$ counts in the peak channel), long decay times (512 channels) and rather small $[Q]$ values ($A_3 < 0.2$). Under the same conditions global analysis produced values with corresponding errors ca. 0.7, 1.6 and 2%. The acceptability of all fits was based on statistical fitting criteria [14].

More interesting than parameter recovery, is the application of global analysis to model testing. The case of the micellar decay of Eq. (12), at very low $A_3$ values, is characteristic. Indeed, when $A_3 \ll 1$ the approximation $e^{-A} \simeq 1 - A$, simplifies Eq. (12) to the biexponential expression of Eq. (13),

$$F(t) = a_1 \exp(-t/\tau_1) + a_2 \exp(-t/\tau_2) \qquad (13)$$

where the new parameters $a_1$, $a_2$, $\tau_1$ and $\tau_2$ are function of the old ones $A_2$, $A_3$ and $A_4$ [14]. In practice it has been shown that even for $A_3$ values close to 1, i.e. values ordinarily used in experiments, single curve analysis gives statistically acceptable fits for both Eqs. (12) and (13). Therefore, a choice between the mechanisms of Eq. (12) and (13) can not be made by means of single curve fitting. In fact simulated decay data gave equally good fits with both equations. Similarly, when simulated decay curves corresponding to different non-zero $[Q]$ values, were treated globally, again both Eqs. (12) and (13) gave acceptable fits. However, when in the global analysis a simulated decay with $[Q] = 0$ was added, the two equations gave very different results. Thus, the fit of Eq. (12) gave an acceptable $Z$ chi square value equal to $-0.13$, whereas the fit of the biexponential decay gave $Z$ chi square $= 7.6$.

In conclusion, the ability of global fluorescence analysis to produce very accurate decay parameters on the one hand, and to test successfully different decay mechanisms on the other, must be emphasized. Moreover, taking into account the simplicity of the method, it becomes evident that it constitutes a powerful tool for photochemical and photophysical research.

*Acknowledgements*

This article is based on the work of the author did during his visit to the Chemistry Dept. of the Catholic Univ. (Leuven, Belgium), for which he is deeply indebted to Professor F. C. DeSchryver and to Drs. N. Boens and M. van der Auweraer.

## References

1. Thomas JK (1987) J Phys Chem 91:267
2. Malliaris A (1988) Intern Rev Phys Chem 7:95
3. Boens N, Van den Zegel M, DeSchryver FC, Desie G (1987) Photochem Photophys Suppl 93
4. Spears KG, Cramer LE, Hoffland LD (1978) Rev Sci Instr 49:255
5. Koester VJ, Dowben RM (1978) Rev Sci Instr 49:1186
6. Yamazaki IN, Tamai K, Kume H, Tsuchiya H, Oba K (1985) Rev Sci Instr 56:1187
7. Van den Zegel M, Boens N, Daems D, DeSchryver FC (1986) Chem Phys 101:311
8. Zuker M, Szabo AG, Bramal L, Krajcarski DT, Selinger B (1985) Rev Sci Instr 56:14
9. Boens N, Ameloot M, Yamazaki I, DeSchryver FC (1988) Chem Phys 121:73
10. Knutson JR, Beecham JM, Brand L (1983) Chem Phys Lett 102:501
11. Infelta P (1979) Chem Phys Lett 61:88
12. Infelta P, Gratzel M, Thomas JK (1979) J Phys Chem 78:190
13. Tachiya M (1975) Chem Phys Lett 33:289
14. Boens N, Malliaris A, Van der Auweraer M, Luo H, DeSchryver FC (1988) Chem Phys 121:199
15. Luo H, Boens N, Van der Auweraer M, DeSchryver FC, Malliaris A (1989) J Chem Phys 93:3244
16. Boens N, Luo H, Van der Auweraer M, Reekmans S, DeSchryver FC, Malliaris A (1988) Chem Phys Lett 146:337

Author's address:

Dr. A. Malliaris
N.R.C. "Demokritos"
Agia Paraskevi, Attiki
Athens 153 10
Greece

**Progress in Colloid & Polymer Science**

Progr Colloid Polym Sci 81:242−247 (1990)

# Characterization of silver colloid stabilization by surface enhanced Raman scattering (SERS)

F. Zimmermann and A. Wokaun

Physical Chemistry II, University of Bayreuth, Bayreuth, FRG

*Abstract:* The binding and exchange of stabilizing ligands on the surface of colloidal silver particles is studied by surface enhanced Raman spectroscopy. Primary stabilizers present during colloid synthesis, e.g. the dianion of ethylene-diamine-tetraacetic acid, can be replaced by the cationic surfactant cetyl-tri-methyl-ammonium bromide. To further elucidate this process, model systems are investigated. Choline ions are found to bind to colloidal silver particles through the trimethyl-ammonium head group; the molecule exclusively exhibits the gauche conformation with respect to the $C_\alpha-C_\beta$-bond. Coadsorbed anions are important for the binding to the silver surface, which is covered by a layer of $Ag^+$ ions: SERS intensities are increasing in the sequence $Br^- < Cl^- < OH^-$. − Polyvinyl alcohol is an excellent stabilizer for small ($\sim 3$ nm radius) colloidal silver particles prepared by $\gamma$-irradiation in the group of A. Henglein. Surface enhanced Raman spectroscopy reveals that the polymer is bound to the silver surface via the hydroxyl groups. Strong signals observed for $CH_2$ deformational vibrations are consistent with an adsorption geometry in which segments of the polymer chain are running parallel to the colloid surface in a disordered conformation. Acetate groups present on the polymer due to incomplete hydrolysis of the polyvinyl acetate precursor give rise to intensive SERS signals, indicating that the carbonyl groups participate in the bonding to the colloid surface.

*Key words:* Silver colloids, stabilization by polymers; Raman spectroscopy; surface enhanced Raman scattering

## Introduction

Colloidal dispersions of nanometer-sized metallic particles have recently received renewed attention in view of potential applications in submicron technologies [1]. The unusual optical properties of noble metal colloids [2], which are due to resonant surface plasmon excitation [3], are investigated due to their importance in the development of "nanometer-scale electrodynamics" [4]. Single colloidal particles are being considered as probing tips in near field optics [5] and in the emerging technique of scanning near field optical microscopy [6, 7]. Technical applications of metallic colloids range from the traditional coloring of glass panels [8] to the metallization of electronic circuits [9]. In biology, gold colloids are used for the labelling of biopolymers [10], in biochemical and in medical applications [11].

In all of these applications, colloids are being prepared as a stabilized dispersion. Ligand binding to the colloidal surface is an important factor for the stability of the colloid with respect to oxidation and aggregation. This is particularly relevant for nanometer-sized colloids prepared by $\gamma$-irradiation of aqueous metal salt solutions [12]; in this process reduction is initiated by solvated electrons.

Recently Wiesner et al. have reported [13] on the use of surface enhanced Raman scattering (SERS) [14−19] to study the geometry of stabilizer adsorption on the colloid surface, and to monitor the exchange of the primary ligand, employed during colloid preparation, by secondary stabilizers. For silver colloids prepared by reduction of $AgNO_3$ solutions with ethylene-diamine-tetraacetic acid (EDTA, dinatrium salt) [20], the negatively charged EDTA ligand is clearly detected in the SERS spectrum of the as-prepared colloid [13]. It was found [13] that EDTA can be replaced by the cationic surfactant cetyl-trimethyl-ammonium bromide (CTAB) when a concentrated ($10^{-2}$ M) solution of CTAB is added to the EDTA stabilized silver sol (total silver concentration $1.2 \times 10^{-4}$ M). Strong signals were detected for vibrations involving the trimethyl-ammonium head group of the surfactant molecule, indicating that the latter was directed towards the colloid surface [13]. Information derived from the red-shift of the absorption maximum [13, 21, 22] revealed that a submonolayer of $Ag^+$ ions is present on the surface of the Ag particles. Upon the addition of CTAB a thin AgBr layer is formed, with the bromide ions in a bridging position between the positively charged silver surface and the cationic surfactant.

Detailed analysis of the SERS spectrum showed [13] that both *gauche* and *trans* conformations are detected for the $C_\alpha - C_\beta$ bond of the $C_{14}$ alkyl chain bound to the ammonium head group. This result is in contrast to Raman studies of the crystals and ordered layers where predominantly *trans* conformations are found. The presence of *gauche* conformations with respect to $C_\alpha - C_\beta$ is an important indicator of the adsorption geometry. To further elucidate this point, the adsorption of small model compounds is investigated in this study. For this purpose, the choline cation ($HO - CH_2 - CH_2 - N^+(CH_3)_3$) has been adsorbed to the colloid surface. As the conformation-dependent Raman spectrum of choline has been assigned in detail [23], this test molecule is well suited to derive detailed information on the adsorption geometry. Furthermore, we shall report on SERS investigations of colloids that have been prepared by γ-irradiation, and are stabilized by polyvinyl alcohol (PVA) and by polyphosphates [24].

## Experimental

Two types of silver colloids have been employed in this study. For the choline adsorption experiments, silver sols were prepared according to the procedure of Creighton [14]. 10 ml of a $10^{-3}$ M $AgClO_4$ solution were added to 40 ml of a solution containing $NaBH_4$ at a concentration of $1.5 \times 10^{-3}$ M. After the preparation, the silver sols were aged for two days. Then 2 ml of choline chloride solution ($10^{-2}$ μ) were added to 2 ml of the colloidal dispersion. Subsequently, 0.1 M solutions of KBr, KCl or KOH were added in quantities of $0.6-1$ ml until a visible color change was observed (see below). SERS spectra were recorded within $2-6$ hours after the addition of the salt solution.

Polymer-stabilized silver colloids were prepared by the procedure of Henglein and Lilie [12, 24]. A solution containing $AgClO_4$ ($10^{-4}$ M), isopropanol ($10^{-2}$ M) and the polymer {polyvinylalcohol, $10^{-2}$ M, or polyphosphate, $(1-5) \times 10^{-4}$ M with respect to $(NaPO_3)_6$} was irradiated for 20 minutes by a $^{60}Co$ source emitting a γ-dose rate of $1 \times 10^5$ rads/h. Silver particles produced in this manner are characterized by an average radius of $\sim 3$ nm [12].

Raman spectra were excited with the 514.5 nm line of an $Ar^+$ laser. 10 mW of power were focussed onto an $8 \times 0.1$ mm line on the front surface of a quartz cuvette. The Raman emission was dispersed in an 0.75 m double monochromator set at 5 $cm^{-1}$ resolution, and was detected using photon counting.

## Results and discussion

### Choline

Surface enhanced Raman spectra of choline ions were recorded on silver sols prepared by reduction with $NaBH_4$, as described above. The absorption maximum of the silver colloid does not change upon choline addition. For the recording of SERS spectra, we found it essential to add KBr, KCl, or KOH solutions up to a final concentration of $(1-2) \times 10^{-2}$ M. While the salt solution is being added the visible color of the colloid changes from yellow through red to green; a corresponding absorption spectrum is shown in Fig. 1. The shoulder at longer wavelengths points to beginning aggregation, i.e. to the formation of doublets and strings of colloidal particles. These aggregates are known [14, 15] to provide much stronger enhancement of the surface local field than small spheres, due to the so-called lightning rod effect [25].

The SERS intensities of the choline signals show a significant dependence on the type of anion which is added in the salt solution, and increase in the sequence $Br^- < Cl^- \ll OH^-$. When bromide is used, residual signals arising from species adsorbed on the silver surfaces from the preparation ($NaBH_4$ reduction) of the sol are comparable in intensity with the choline bands.

The bands not related to choline are easily identified by recording a SERS spectrum of the silver sol in the absence of choline. By comparison with the IR- and Raman-spectra of $B(OH)_4^-$ solutions [26, 27], and of crystalline $B(OH)_3$ [28], the bands observed at 1030 $cm^{-1}$ and 1165 $cm^{-1}$ are assigned to $B - O - H$ deformational motions. The 1361 $cm^{-1}$ vibration, observed as a shoulder on the 1392 $cm^{-1}$ band of choline, is either due to $\nu_{as}(BO_3)$ (reported at 1365 $cm^{-1}$ in Ref. [28]) or due to adsorbed carbonate species [29]. The presence of $CO_3^{2-}$ is also indicated by the observation of a $C=O$ stretching vibration at 1590 $cm^{-1}$.

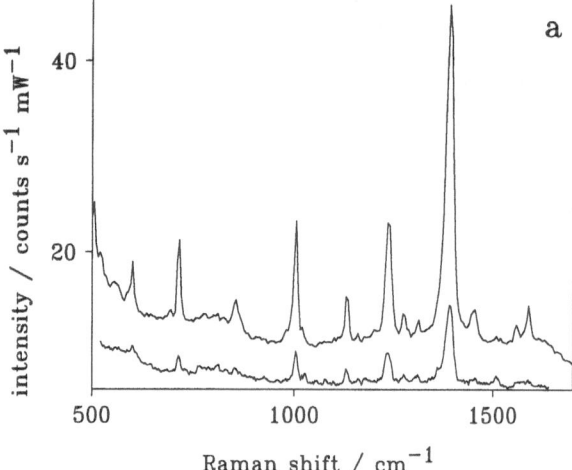

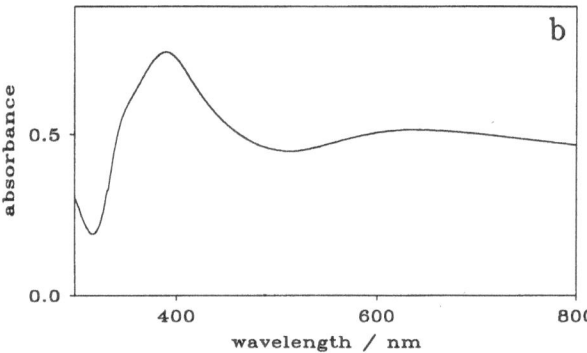

Fig. 1. (a) SERS spectra of choline adsorbed on a silver colloid prepared by NaBH₄ reduction [14]. Spectra were excited by 50 mW of power at 514.5 nm. Prior to recording the **top** spectrum, NaOH solution was added until a pH of 11.5 was established. For the spectrum shown in the **bottom** trace, KCl was added to a final concentration of $1.7 \times 10^{-2}$ M. (b) Absorption spectrum of the silver sol subsequent to NaOH addition

The remaining signals in the SERS spectra with KBr and KCl addition (Fig. 1, bottom) can all be assigned to adsorbed choline ions. The identification of the choline-related bands is facilitated by the fact that in the presence of OH⁻ ($3 \times 10^{-3}$ M), the latter vibrations are so intensive in the SERS spectrum that the borate impurity bands are hardly observable at all, as illustrated by the top spectrum in Fig. 1. Assignments of the bands in this spectrum will be based on a detailed analysis of the experimental Raman spectra of both *trans* and *gauche* conformations of choline [23]; in this reference normal coordinate analysis was used to identify the atomic motions that are dominantly contributing to the respective normal vibrations.

Bands are discussed in the order of increasing frequency. The three low-frequency vibrations at 715 cm⁻¹ (strong), 855 cm⁻¹ (medium) and 1005 cm⁻¹ (strong) are all sensitive to the conformation of the molecule with respect to the $C_\alpha - C_\beta$-bond, and are observed at the frequencies that are characteristic for the *gauche* conformation [23]. The band at 715 cm⁻¹ is assigned to the tetrahedrally symmetric N−C₄ stretching vibration. The band at 855 cm⁻¹ corresponds to the totally symmetric N−(CH₃)₃ stretching vibration of the head group. The strong band at 1005 cm⁻¹ is assigned to the rocking mode of the α-CH₂-group, again in *gauche* conformation.

Bands at 1132 cm⁻¹ (medium), 1237 cm⁻¹ (strong, broad) and 1275 cm⁻¹ (weak) are assigned to γ(CH₃) rocking vibrations of the methyl groups in the positively charged head group, of symmetries *E*, *E*, and $A_1$, respectively. The remaining two bands at 1392 cm⁻¹ (very strong) and 1450 cm⁻¹ (weak) are attributed to symmetric and asymmetric deformational motions within the methyl groups, respectively [23].

As mentioned above, the first conclusion derived from this spectrum is that the choline ions are adsorbed in a *gauche* conformation. Second, the observed SERS intensities can be used to derive information on the adsorption geometry. SERS selection rules [30, 31] imply that vibrations corresponding to the $\alpha_{zz}$ element of the polarizability tensor experience the strongest enhancement, with $z$ denoting the direction normal to the surface. Prominent bands in the SERS spectrum are $\nu_s(N-C_4)$, $\nu_s(N-(CH_3)_3)$, $\gamma_r$(CH₂), γ(CH₃), and $\delta_s$(CH₃). This suggests that choline is bound to the surface via the −N⁺(CH₃)₃ head group, with the C₂ alkyl chain pointing away from the surface. Note that the free choline molecule in solution exhibits an intensive band at 955 cm⁻¹ due to an N−(CH₃)₂ vibration which is antisymmetric with respect to the plane defined by $C_\alpha$, $C_\beta$, and the nitrogen atom. The latter vibration is not observed in the SERS spectrum. This result is consistent with the adsorption geometry proposed above, as the induced dipole moment of the antisymmetric vibration would be close to parallel with the silver surface, and therefore only weakly enhanced.

*Polyvinyl alcohol*

Polyvinyl alcohol (PVA) has been successfully used by Henglein [24] to stabilize small silver sols prepared by the γ-irradiation technique [12]. From the preparation, the solutions contain isopropanol which acts

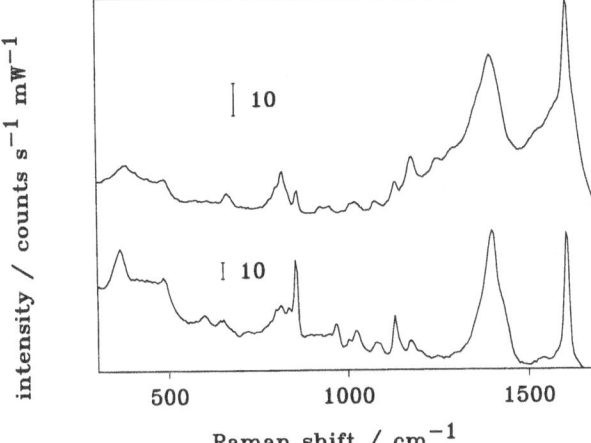

Fig. 2. SERS spectra of polyvinyl alcohol (PVA) adsorbed to silver colloids prepared by γ-irradiation [24]. Spectra were excited by 50 mW of power at 514.5 nm. The solution concentration was $10^{-2}$ M for the bottom spectrum, and $10^{-3}$ M for the spectrum shown in the top trace. Note that the SERS intensities are determined by the product of surface coverage and Raman enhancement, and do not directly reflect solution concentrations. A likely explanation for the more intensive SERS signals observed with the lower PVA concentration is a higher tendency to form aggregates of colloidal particles, which are capable of sustaining higher surface enhancements

as the reduction agent, at a concentration of $2 \times 10^{-2}$ M.

In this work, several sets of preparations have been investigated in which the concentration of the stabilizer, PVA, was varied between $10^{-3}$ M and $10^{-2}$ M; SERS spectra are shown in Fig. 2. Isopropanol from the solution, although present at a higher concentration, is apparently not adsorbed; therefore, only weaker bands of isopropanol are observed in the SERS spectrum.

Lee and Meisel [32] have previously recorded SERS spectra of PVA on Creighton sols. The spectrum of Fig. 2 will be discussed with reference to the IR spectrum of PVA [33], and the Raman spectra of polyvinyl acetate [34] and isopropanol [35].

Prominent PVA-related bands in the SERS spectrum are: the $\gamma_r(CH_2)$ rocking vibration at 855 cm$^{-1}$; the $\nu(C=O)$ stretching vibration at 1130 cm$^{-1}$; and the intensive asymmetric band at 1400 cm$^{-1}$ which apparently consists of two contributions, i.e. the deformational motions of backbone CH$_2$-groups and of the OH-groups. Weaker bands at 665 cm$^{-1}$ and 1023 cm$^{-1}$ also have their counterparts in the published IR spectrum of PVA. Bands of weak and medium intensity observed at 485, 815, 967, and

1175 cm$^{-1}$ are assigned to isopropanol in agreement with the literature [35]. In addition to the signals discussed so far, a strong band at 365 cm$^{-1}$, a weak vibration at 1034 cm$^{-1}$, and a very intensive band at 1605 cm$^{-1}$ are discernible in the spectra of Fig. 2. The latter frequency must be due to C=C or C=O double bonds. In agreement with Lee and Meisel [32], who observed a similar band at 1640 cm$^{-1}$, we attribute these bands to acetate groups present on the PVA backbone due to incomplete hydrolysis of the precursor, polyvinyl acetate. The 365 and 1023 cm$^{-1}$ bands match the literature spectrum of polyvinyl acetate [34]; other vibrations expected from this spectrum [34] (at 630, 1127, 1355, 1373, and 1439 cm$^{-1}$) would coincide with the most intensive bands of PVA. The carbonyl stretching vibration, which occurs at 1730 cm$^{-1}$ in polyvinyl acetate, is down-shifted to 1605 cm$^{-1}$ in our system. This points to a strong interaction of the carbonyl oxygen with the colloid surface, in agreement with FTIR studies of adsorbed aldehyde and carbonyl species [36].

From the most intensive bands observed in the SERS spectrum, some conclusions on the geometry of PVA adsorption can be drawn. Intensive vibrations observed for $\nu(C-O)$ and $\delta(O-H)$ indicate that the polymer adsorbs to the positively charged [37] silver surface through the OH groups, as expected on chemical grounds. Observation of SERS signals for both $\delta(CH_2)$ (vibrational dipole perpendicular to the H$-$H interconnecting line) and $\gamma_r(CH_2)$ (vibrational dipole approximately parallel to H$-$H) suggests that segments of the PVA chain are running parallel to the surface in a disordered conformation. Other authors have supported this view from flocculation experiments [38], neutron scattering measurements and evanescent wave spectroscopy [39]: it has been shown that for polymers adsorbed to latex particles, segments adsorbed to the surface of a latex sphere alternate with segments that are looping out into the solution. Most remarkable is the intensity of the 1605 cm$^{-1}$ vibration which we assigned to unhydrolyzed acetate groups on the polymer. Carbonyl groups have a large vibrational dipole moment familiar in IR spectroscopy; an adsorption geometry with the C=O bond perpendicular to the surface ($\alpha_{zz}$ element of the polarizability tensor) would give rise to strong SERS enhancements. These statements may explain the high intensity observed for $\nu(C=O)$. Taken by itself, the fact that acetate vibrations are observed clearly demonstrates that these side groups are strongly involved in the binding of the polymer to the silver colloid surface.

*Other systems studied*

In an earlier communication [13] Wiesner et al. had reported on SERS observations of polyvinyl sulfate stabilizers adsorbed onto silver colloids prepared by γ-irradiation. We have attempted to more fully assign this spectrum by studying the adsorption of model compounds: SERS spectra of silver sols containing sodium propyl sulfonate have been recorded. However, it was found that propyl sulfonate is too weak a ligand to *replace* another stabilizer already present on the colloid surface. A weak SERS spectrum of propyl sulfonate was observed [13] only when $Na^+C_3H_7SO_3^-$ was added already *during* the preparation of the silver sol by $NaBH_4$ reduction. However, these solutions exhibited a greenish-gray tint and a strong tendency to aggregation. In agreement with results from numerous other techniques, we conclude that the monomeric ligand propyl sulfonate is a much inferior stabilizer, as compared to the polymeric ligand polyvinyl sulfate which cooperatively binds through many ionic groups.

Recently, oligophosphates have been successfully used to stabilize small colloidal silver particles prepared by γ-irradiation, as well as silver clusters consisting of 3–4 silver atoms [24]. We have therefore attempted to record SERS spectra of a silver sol synthesized in the presence of "hexametaphosphate". This commercial product is a mixture of oligophosphates $Na_{15}P_{13}O_{40} - Na_{20}P_{18}O_{55}$; the concentration used in the synthesis was $1.5 \times 10^{-4}$ M with respect to $Na_{17}P_{15}O_{46}$.

The absorption spectrum of the as-prepared silver sols exhibits a narrow band centered between 380 and 390 nm, which indicates small spherical silver particles with a narrow size distribution [22]. No surface enhanced Raman signals from the colloid have been detected. Attempts to initiate coalescence by the addition of various salt solutions, and to record SERS spectra of adsorbates on the growing aggregates which would provide a higher lightning rod effect, have not been successful.

In an attempt to study the competitive adsorption of polyphosphate and PVA, a silver sol was prepared by γ-irradiation in the presence of $2 \times 10^{-5}$ M $Na_{17}P_{15}O_{46}$ and $1 \times 10^{-2}$ M PVA. No SERS signals have been detected in this system as well. The only conclusion that can be drawn from this failure to observe SERS spectra is that small colloidal silver spheres are excellently stabilized by polyphosphates. Even in a solution containing a thirtyfold excess of $(-CH_2-CHOH-)$ repeat units with respect to $(-PO_3^- -)$, no SERS signals of PVA are detected.

Apparently, the use of polyphosphate during colloid synthesis initiated by γ-irradiation results in spherical particles that are too small to sustain sizeable surface enhanced local fields.

## Conclusions

The results of this study confirm that the surface of silver particles prepared by the specified techniques is covered by a (sub)monolayer of $Ag^+$ ions, in agreement with earlier investigations [13, 22]. Cations like choline or the cetyl-trimethyl-ammonium ion (CTA) can only be adsorbed through the intermediacy of polarizable negative ions. The increase of choline SERS intensities for the sequence of anions, $Br^- < Cl^- < OH^-$, parallels the electron donor properties of these ligands. Choline is adsorbed with the trimethyl-ammonium head group directed toward the surface, in agreement with the analogous findings for CTA [13]. The hydroxyl group of choline is in a gauche conformation with respect to the adsorbed head group.

Polyvinyl alcohol is found to bind to the surface via the OH functional groups; here probably the electron-rich oxygen atom represents the binding site. Segments of the polymer chain are running parallel to the surface in a disordered conformation. Acetate groups which are present in the polyvinyl alcohol chain due to incomplete hydrolysis of the polyvinyl acetate precursor are definitely involved in the binding.

Propyl sulfonate is a poor stabilizer which gives only weak SERS signals, while polyvinyl sulfate excels by both good stabilizing properties and high SERS intensities [13]. This illustrates the advantage of using polymeric adsorbates where the surface binding energies of many repeat units are adding, a point which has been amply emphasized in other contexts.

A further example to this point are polyphosphate ligands, which are the most promising stabilizers with respect to long-term stability for silver sols prepared by γ-irradiation [24]. Apparently the polyphosphates stabilize very small spherical silver particles which are not capable of providing surface enhanced fields. These polyelectrolytes are bound to the surface so strongly that they can not be displaced by other ligands.

*Acknowledgements*

All silver colloids prepared by γ-irradiation have been kindly provided by A. Henglein (Hahn-Meitner-Institut, Berlin). Generous help with the colloid preparation by F. Henglein is gratefully acknowledged. The authors are indebted to J. Wiesner for his help and advice during the early parts of the SERS experiments. Financial support by the

Deutsche Forschungsgemeinschaft (SFB 213) is gratefully acknowledged.

## References

1. Deckman HW, "Microfabrication of Molecular Scale Microstructures", IBM Europe Institute Seminar, Garmisch (BRD), Aug. 14–18, 1989
2. Kerker M (1985) J Colloid Interface Sci 105:297–314
3. Raether H (1980) Excitation of Plasmons and Interband Transitions by Electrons. Springer, Berlin
4. Leitner A (1990) Mol Phys (in press); Aussenegg FR, Leitner A (1989) Optics Lett 14 (in press); Aussenegg FR, Leitner A, Pedarnig JD (1989) Appl Phys B 49:279–282
5. Ash EK, Nichols G (1972) Nature 237:510–512; Pohl DW, Denk W, Lanz M (1984) Appl Phys Lett 44:651–653
6. Courjon D, Sarayeddine K, Spajer M (1989) Optics Commun 71:23–27
7. Fischer UCh, Pohl DW (1989) Phys Rev Lett 62:458–461
8. Doremus RH (1964) J Chem Phys 40:2389; (1965) J Chem Phys 42:414, and references therein
9. Henglein A, Hahn-Meitner Institute, D-1000 Berlin, private communication
10. Verkleij AJ, Leunissen JLM (eds) (1989) Immuno-Gold Labelling in Cell Biology. CRC Press, Boca Raton, Florida
11. Schwab ME, Thoenen H (1978) J Cell Biology 77:1–13; for a review, see also: Roth J (1983) In: Immunocytochemistry. Vol 2, Academic Press, London, pp 218–284
12. Henglein A, Lilie J (1981) J Am Chem Soc 103:1059–1066
13. Wiesner J, Wokaun A, Hoffmann H (1988) Progr Colloid Polym Sci 76:271–277
14. Creighton JA, Blatchford CG, Albrecht MG (1979) J Chem Soc Faraday II 75:790–798
15. Chang RK, Furtak TE (eds) (1982) Surface Enhanced Raman Scattering. Plenum, New York
16. Otto A (1983) In: Cardona M (ed) Light Scattering in Solids. Vol 4, Springer, Berlin, pp 289–418
17. Wokaun A (1985) Mol Phys 56:1–33
18. Moskovits M (1985) Rev Mod Phys 57:783–826
19. Sandroff CJ, Garoff S, Leung KP (1983) Chem Phys Lett 547–551
20. Fabrikanos A, Athanassiou S, Lieser KH (1963) Z Naturforsch 18b:612–617
21. Barnickel P, Wokaun A (1989) Mol Phys 67:1355–1372
22. Barnickel P, Wokaun A (1990) Mol Phys 69:1–9
23. Rihak P (1979) Ph. D. Thesis # 6393, ETH Zürich; Fringeli UP, Günthard HsH (1981) In: Grell E (ed) Membrane Spectroscopy. Springer, Berlin, 270–332
24. Henglein A (1989) Chem Phys Lett 154:473–476
25. Wokaun A (1984) Solid State Phys 38:223–294
26. Goulden JDS (1959) Spectrochim Acta 9:657–671
27. Edwards JO, Morrison GC, Ross VF, Schultz JW (1955) J Am Chem Soc 77:266–268
28. Weidlein J, Müller U, Dehnicke K (1982) Schwingungsspektroskopie. Thieme, p 102
29. Otto A (1978) Surf Sci 75:L392–396
30. Moskovits M (1982) J Chem Phys 77:4408–4416
31. Creighton JA (1988) In: Clark RJH, Hester RE (eds) Spectroscopy of Surfaces. Wiley, Chichester, pp 37–89
32. Lee PC, Meisel D (1983) Chem Phys Lett 99:262–265
33. Hummel DO, Scholl F (1984) Atlas of Polymer and Plastics Analysis. Vol 2, part a/II, Hanser, Munich, p 465
34. Schrader B, Meier W (1974) Raman/IR Atlas. Vol 2, VCH Weinheim, Spectrum N-07
35. Schrader B, Meier W (1974) Raman/IR Atlas. Vol 1, VCH, Weinheim, Spectrum A3–04
36. Jobson E, Baiker A, Wokaun A (1989) Ber Bunsenges Phys Chem 93:64–70
37. Koglin E, Sequaris JM, Valente P (1983) In: Aussenegg FR, Leitner A, Lippitsch ME (eds) Surface Studies with Lasers, Springer, Berlin, pp 64–71
38. Cabane B, Wong K, Wang TK, Lafuma F, Duplessix R (1988) Colloid & Polymer Sci 266:101–104
39. Auvray L, Cotton JP, Daoud M, Farnoux B, Ausserre D, Cauchete I, Hervet H, Rondelez F (1988) Abs Pap ACS 195:223–223; Edwards J, Ausserre D, Hervet H, Rondelez F (1989) Appl Opt 28:1881–1884

Authors' address:

Prof. A. Wokaun
University of Bayreuth
P.O. Box 101251
D-8580 Bayreuth, FRG

**Progress in Colloid & Polymer Science**                    Progr Colloid Polym Sci 81:248 (1990)

# Abstracts

## Structure of perfluorinated ionomer solutions

P. Aldebert*, B. Dreyfus, G. Gebel, N. Nakamura**, M. Pineri and F. Volino*

Groupe Physico Chimie Moléculaire, Service de Physique, Département de Recherche Fondamentale, Centre d'Etudes Nucléaires de Grenoble, Grenoble, France
which seems to remain unchanged, follows Virk's ultimate

*Abstract:* A hexagonal packing of rod like structures is proposed in solutions and gels of a perfluorinated ionomer. The diameter of the rods has been obtained by two different approaches either geometrical from shifts of the interference peak versus concentrations in small angle scattering experiments or direct from analysis of the structure factor in diluted solutions. Consistent results give values between 18 and 31 Å for the radius of the rods, depending on the solvent. The rods have a perfluorinated core with the charges on the surface and the diameter depends on the surface tension rather than on the dielectric constant of the solvent. The structure may result from a balance between elastic and interfacial energies as it is shown by calculations.

Authors' address:

P. Aldebert
Groupe Physico Chimie Moléculaire,
Service de Physique
Département de Recherche Fondamentale
Centre d'Etudes Nucléaires de Grenoble
85 X-38041 Grenoble Cedex, France

---

\* CNRS
\*\* On sabbatical leave from Ritsumeiukan University, Chemistry Department, Japan.

## Drag reduction in surfactant solutions

Hans-Werner Bewersdorff

Department of Chemical Engineering, University of Dortmund, FRG

*Key words:* Drag reduction; surfactant solutions; friction behaviour; velocity profiles; shear-induced state

*Abstract:* Drag reduction is the reduction of the friction factor of a turbulent flow. Since the friction factor is proportional to the pressure drop, less energy is required for pumping drag reducing fluids. Therefore drag reducing surfactants solutions can be used for saving energy in practical applications, like central heating systems, cooling circuits, and air-conditioning systems. Drag reducing surfactant systems are characterized by the presence of rod-like micelles which are formed by single surfactant molecules when the surfactant concentration is above a characteristic value which strongly depends on temperature. In the concentration range for drag reduction, typically 200–2000 ppm by weight, the length of the rod-like micelles is less than 250 nm. In this concentration range stiff rods of this length would rotate freely without hindering each other in shear flows. Shear viscosity measurements show that surfactant solutions in this concentration range behave like normal Newtonian fluids at low shear rates. At shear rates above a critical value, however, the viscosity suddenly increases due to the formation of a shear induced state [1, 3]. The critical shear rate depends on the temperature, the surfactant and additional electrolyte concentration, properties which affect the length of the micelles.

The turbulent friction behaviour of drag reducing surfactant solutions is characterized by a critical wall-shear stress. Beyond this critical wall shear stress a reversible loss of drag reduction is observed, or i.e. drag reduction recovers when the wall-shear stress is lowered again below the critical value. Even in pipes of larger diameters Virk's empirical maximum drag reduction asymptote [2] can be approached. Turbulent velocity profile measurements [1, 3] show that the dimensionless velocity profile beyond the viscous sublayer, which seems to remain unchanged, follows Virk's ultimate profile [2] in the buffer zone. In the core region of the flow the velocity profiles can be parallel to the Newtonian core region as it was found for drag reducing dilute polymer solutions. Sometimes, however, an increased slope was observed in the core region indicating a changed turbulence structure. In the case of Virk's maximum drag reduction the velocity profiles in the buffer zone are "S-shaped" meaning that they deviate from Virk's logarithmic ultimate profile to lower dimensionless velocity values at smaller wall distances, to higher ones in the medium range, and again towards lower ones at the largest wall distances. Therefore the shape of the dimensionless velocity profiles at maximum drag reduction conditions is more similar to a laminar than to a logarithmic turbulent profile. A possible explanation for the observed changes in the turbulent structure is based on an increased effective viscosity probably caused by the shear-induced state [3].

## References

1. Bewersdorff HW, Ohlendorf D (1988) Colloid Polym Sci 266:941
2. Virk PS (1975) AIChEJ 21:625
3. Bewersdorff HW (1990) In: Gyr A (ed) Structure of Turbulence and Drag Reduction, Springer Verlag, Berlin (in print)

Authors' address:

Dr. Hans-Werner Bewersdorff
Universität Dortmund, Fachbereich Chemietechnik
Postfach 500 500
D-4600 Dortmund 50, FRG

# Waterless ternary microemulsions: Viscosity, dielectric relaxation and conductivity

C. Mathew, Z. Saidi, J. Peyrelasse and C. Boned

Université de Pau et des Pays de l'Adour — Centre Universitaire de Recherche Scientifique
Laboratoire de Physique des Matériaux Industriels, Pau, France

*Key words:* Waterless microemulsion; percolation

*Abstract:* A waterless microemulsion glycerol/AOT/isooctane has been studied as a function of temperature $T$ ($10\,°C \leqslant T \leqslant 25\,°C$), of the (glycerol + AOT) $\phi$ volume fraction, the molar ratio $n = $ [glycerol]/[AOT] ($1 \leqslant n \leqslant 4$) and of the amount $p_s$ of salt in glycerol (up to 1.1% in weight of NaCl). The measured properties are electric conductivity $\sigma$, dynamic viscosity $\eta$, and dielectric relaxation.

At given $T$, $n$ and $p_s$ an increase of conductivity and dynamic viscosity is observed as the volume fraction $\phi$ increases. The dielectric relaxation may be fitted by a generalized Davidson-Cole distribution of relaxation times [1]. The quantities $\frac{1}{\sigma}\frac{d\sigma}{d\phi}$ and $\frac{1}{\eta}\frac{d\eta}{d\phi}$ go through a maximum as well as static permittivity $\varepsilon_s$. At the same time the quantity $1/v_R$ ($v_R$: characteristic frequency of dielectric relaxation) goes through a maximum.

The results are discussed within the framework of the theory of percolation. For the conductivity [2] we have $\sigma \propto (\phi_c - \phi)^{-s}$ if $\phi < \phi_c$, and $\sigma \propto (\phi - \phi_c)^\mu$ if $\phi > \phi_c$ where $\phi_c$ is the percolation threshold. These laws are valid if $\sigma_2/\sigma_1 \ll 1$ where $\phi$ is the volume fraction of constituent 1. In the case of the glycerol/AOT/oil systems the ratio $\sigma_2/\sigma_1 = \sigma_{\mathrm{oil}}/\sigma_{\mathrm{glycerol}}$ does not satisfy $\sigma_2/\sigma_1 \ll 1$. However we have checked that $\frac{1}{\sigma}\frac{d\sigma}{d\phi}$ presents a maximum which corresponds to the percolation threshold. But the equations above cannot be used to provide a quantitative description of the conductivity. It is the same for the dielectric aspect, however the theory predicts [3] a maximum of $\varepsilon_s$ and $1/v_R$ as it was observed experimentally.

For the viscosity, in a previous paper [4] we proposed that $\eta \propto (\phi_c - \phi)^{-s'}$ for $\phi < \phi_c$ and $\eta \propto (\phi - \phi_c)^{\mu'}$ if $\phi$
$> \phi_c$, valid when $\eta_2/\eta_1 \ll 1$. Here we have $\eta_{\mathrm{oil}}/\eta_{\mathrm{glycerol}} \ll 1$ and the analysis of the results gives a very good quantitative comparison with $s' = 1.20 \pm 0.20$ and $\mu' = 200 \pm 0.25$ which are the values expected by the dynamic percolation theory [4, 5].

Investigating the $\eta(\phi)$ curves allows us to know the variations of $\phi_c(T)$ at $n$ and $p_s$ constants, $\phi_c(n)$ at $T$ and $p_s$ constants, $\phi_c(p_s)$ at $T$ and $n$ constants. We observed that $\phi_c$ decreases when either $T$ increases, or $n$ increases, or $p_s$ decreases. This corresponds to a growth of the interactions [3] between the droplets in the system.

## References

1. Peyrelasse J, Boned C, Saidi Z (1989) Progr Colloid Polym Sci 79:263−269
2. Moha-Ouchane M, Peyrelasse J, Boned C (1987) Phys Rev A 35, 7:3027−3032
3. Peyrelasse J, Moha-Ouchane M, Boned C (1988) Phys Rev A 38, 2:904−917
4. Peyrelasse J, Moha-Ouchane M, Boned C (1988) Phys Rev A 38, 8:4155−4161
5. Saioi Z, Mathew C, Peyrelasse J, Boned C (1990) Phys Rev A, (in press), Phys Rev A, (in press)

Authors' address:

J. Peyrelasse and C. Boned
Laboratoire de Physique des Matériaux Industriels
Centre Universitaire de Recherche Scientifique
Avenue de l'Université — 64000 PAU, France

**Progress in Colloid & Polymer Science**                    Progr Colloid Polym Sci 81:250 (1990)

# Role of stretching and bending energies in the transition from micelles to swollen microemulsion droplets: Interfacial tensions, size distributions and shape fluctuations

M. Borkovec

Institut für Lebensmittelwissenschaft, ETH-Zentrum, Zürich, Switzerland

*Abstract:* Adding a surfactant into a two phase oil-water system results in a rapid decrease of the interfacial tension of the macroscopic interface. This decrease stops abruptly at the critical micelle concentration (CMC) and upon further increase of the surfactant concentration the interfacial tension remains essentially constant [1]. The resulting tension can be related to the extend of swelling of the micelles formed. A surfactant aggregating in dry (nonswollen) micelles results in a moderately low tension while surfactants aggregating in strongly swollen micelles lead to ultra-low tensions characteristic for microemulsions. The extend of swelling of the micelles at the CMC is determined by the bending energy [1−3] of the saturated surfactant monolayer. Increasing the splay moduli or decreasing the spontaneous curvature favours swollen micelles (microemulsion droplets) and ultra-low tensions.

Such conclusions follow from a theoretical description of the aggregation equilibrium of dry and swollen micelles [4]. While we specialize on spherical, non-interacting aggregates, we fully treat the competition between phase separation, dissolution in monomers and formation of micelles with variable extend of swelling. The present model is based on a interfacial free energy of the surfactant monolayer which includes stretching and bending contributions. This free energy describes the macroscopic interface in a two phase system as well as the internal interface surrounding the aggregates. This allows to evaluate differences between the macroscopic and internal interfaces and the deviations of these monolayers from the saturated state. Furthermore, we address size distributions of the aggregates, the concept of a CMC in a microemulsion. Studying shape fluctuations of the aggregates [3] we conclude that spherical micelles may be stabilized by small interfacial tension alone. On the other hand, stability of strongly swollen microemulsion droplets requires a finite splay modulus of the monolayer. Within the present framework we recover several known results for globular microemulsions [1−3].

## References

1. de Gennes PG, Taupin C (1982) J Phys Chem 86:2284
2. Huse DA, Leibler S (1988) J Physique 49:605
3. Safran SA (1983) J Chem Phys 78:2073
4. Borkovec M (1989) J Chem Phys 91:6268

Author's address:

M. Borkovec
Institut für Lebensmittelwissenschaft
ETH-Zentrum
CH-8092 Zürich, Switzerland

# Magnetoviscosity of colloidal suspensions

G. Bossis, E. Lemaire, C. Mathis, Z. Mimouni and C. Paparoditis

Laboratoire de Physique de la Matière Condensée, Université de Nice-Sophia Antipolis, Nice, France

*Abstract:* We show that the viscosity of a colloidal suspension of magnetic particles can be increased considerably by the application of a moderate magnetic field ($H \approx 1000\,\text{Œ}$). Unlike a ferrofluid which is composed of very small magnetic grains ($\langle d \rangle = 100\,\text{Å}$) this colloidal suspension is formed of polystyrene spheres of micronic dimension $\langle d \rangle = 1.8\,\mu\text{m}$ containing inclusions of magnetic (23% or 62% in volume). The ratio of the magnetic energy of attraction between two particles to the thermal energy is proportional to the cube of the radius of the particles. So the larger the particles (if they do not sediment) the more easily they interact and form linear structures which are responsible for the increase in viscosity. This explains why this effect is much lower in ferrofluids than in this suspension of micronic latexes.

The viscosity was measured in a capillary flow experiment with a plastic tube wound into a coil and placed in a uniform magnetic field.

The behaviour of the suspension has been analysed on the basis of a Bingham fluid. It appears that this model is not satisfied primarily because we have two different yield stresses associated with two different regimes. In the first regime the fibrinated structure sticks to the wall and the suspending medium (water) flows through this structure as in a porous medium. In the second regime when the applied wall stress is high enough, there is no more difference between the velocities of the solid and the liquid phases, but we have a plug flow whose size will vary depending on a second yield stress necessary to overcome the magnetic force between the particles. The modelization is complicated by the breakdown of the radial symetry in thepresence of the field. Further study in a plane Poiseuille flow will help us to understand the rheological law of these magnetic suspensions.

Authors' address:

Dr. G. Bossis
Laboratoire de Physique de la Matière Condensée
Université de Nice-Sophia Antipolis
Parc Valrose
06034 Nice Cedex, France

# Percolation in water-in-oil microemulsions. Low-frequency electrical conductivity measurements

C. Cametti[1], P. Codastefano[1], P. Tartaglia[1], J. Rouch[2] and S. H. Chen[3]

[1]) Dipartimento di Fisica, Universita' di Roma *La Sapienza,* Rome, Italy
[2]) Laboratoire d'Optique Moleculaire, Universite' de Bordeaux I, Talence, France
[3]) Nuclear Engineering Department, Massachusetts Institute of Technology, Cambridge, MA, USA

*Abstract:* Water in oil (W/O) microemulsions, a collection of surfactant-coated water particles of high conductivity dispersed in a continuous low-conductivity medium and interacting via an attractive potential, are ideal systems for studying electrical conductivity percolation phenomena with dielectric spectroscopy techniques. The electrical conductivity of Sodium di-2-ethylhexyl-sulfosuccinate (AOT)/water/decane microemulsions at the molar ratio [water/AOT] = 40.8 at a frequency of 10 kHz has been measured for the whole range of temperatures from 10°C to 50°C and the volume fraction $\phi$ from 0.098 to 0.65. The conductivity follows a sigmoidal curve with the lower portion converging to a common straight line and reaching, as $\phi$ is increased, a value corresponding to the limit of the static percolation. In the regime of very dilute microemulsions, the conductivity $\sigma$ is interpreted on the basis of the charge fluctuation model of Eicke et al. [1]. At higher values of $\phi$, where conductivity percolation phenomena become important, the percolation treshold $T_p$ has been located from the inflection point of the curve $\sigma$ vs $T$ and by means of a non-linear least squares analysis. The scaled conductivities have been interpreted in terms of power-law behavior with indices related to the dynamic percolation picture ($s' = 1.2$) below the treshold and to the static one ($t = 1.94$) above the treshold. The locus of the lower cloud points and the percolation line in the $\phi - T$ plane (Fig. 1), from the vicinity of the lower consolute

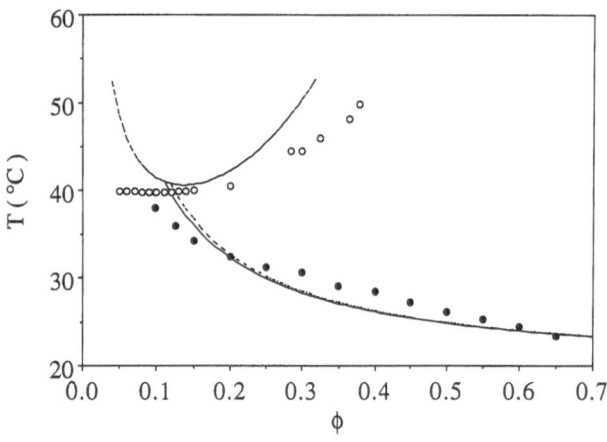

Fig. 1

point up to $\phi = 0.65$ have been calculated using the models II and III of percolation in probability of the theory of Xu and Stell [2], with parameters obeying the empirical relation

given by Lin and Chen [3]. The agreement with our data is very good, over the whole interval of the $\phi$ values.

### References

1. Eicke HF, Borkovec M, Das-Gupta B (1989) J Phys Chem 93:314–317
2. Xu J, Stell G (1988) J Chem Phys 89:1101–1111
3. Chen SH, Lin TL, Huang JS (1988) In: Safran SA, Clark NA (eds) Physics of complex and Supramolecular Fluids, Wiley, New York

Authors' address:

Dr. C. Cametti
Dipartimento di Fisica
Universita' di Roma La Sapienza
Rome, Italy

# Rheological behavior of copolymer latexes prepared in inverse microemulsions

D. Collin[1]), F. Kern[1]) and F. Candau[2])

[1]) Laboratoire de Spectrométrie et d'Imagerie Ultrasonores, Université Louis Pasteur, Strasbourg, France
[2]) Institut Charles Sadron, Strasbourg, France

*Abstract:* The rheological properties of poly (acrylamide-co-sodium-acrylates) microlatexes prepared in inverse nonionic microemulsions have been investigated. These inverse microlatexes provide interesting models for rheological studies. They differ from the more conventional aqueous latexes by a lower particle size and by a large swelling of the polymer core (e.g. 50% water, 50% polymer).

From the study of the viscosity versus the shear stress for different volume fractions $\Phi$, it appears that these materials follow a behavior similar to that observed for colloidal dispersions, that is a newtonian behavior at low $\Phi$ and a shear thinning effect for the largest $\Phi$. However this modification appears for volume fractions much higher (0.45) than those observed for aqueous latexes (0.25 to 0.30); this notable difference may be partially due to the small size of the particles (diameter <100 nm).

The divergence of the zero shear stress viscosity with $\Phi$ is discussed in the framework of the Krieger-Dougherty model. The value of 2.5 found for the intrinsic viscosity shows that the microlatex particles behave like hard and undeformable spheres suspensions. However the close-packing volume fraction obtained is significantly lower than that of 0.64 expected for a random close-packing of hard spheres. A study performed on different compounds shows that the close-packing volume fraction decreases as the content in sodium acrylate increases, which seems to indicate the presence of repulsive interactions between particles.

Authors' address:

Dr. D. Collin
Laboratoire de Spectrométrie et d'Imagerie Ultrasonores
Université Louis Pasteur
4, rue Blaise Pascal
67070 Strasbourg Cedex, France

**Progress in Colloid & Polymer Science**

Progr Colloid Polym Sci 81:253 (1990)

# Cosurfactant structure and solubilization properties of middle-phase microemulsions. Application of the pseudophase model

J. P. Canselier, J. L. Pellegatta and J. P. Monfort

Ecole Nationale Supérieure d'Ingénieurs de Génie Chimique, Toulouse, France

*Abstract:* Most of the multicomponent systems of interest in the so-called chemical process for enhanced oil recovery contain not only brine, oil and surfactant(s), but also at least one alcohol as a cosurfactant. Phase behavior studies of some of such systems have already shown that the solubilization power of middle-phase microemulsions (in Winsor III systems) along the optimum salinity curve is higher when secondary and/or branched middle-chain length alcohols are used rather than straight-chain, primary ones, all other things being equal [4]. On the other hand, the pseudophase model [1, 3] allows prediction of microemulsion composition provided that the overall mixture composition and the number of phases are known. In principle, this approach requires the determination of four equilibrium constants. In fact, the partition constant of water between organic and aqueous phases or pseudophases, $\varepsilon$, always very low, may be neglected. The other three constants are: the self-association constant of alcohol in oil ($K$) and the distribution constants of monomeric alcohol between aqueous and organic (pseudo)phases ($K_w$), and between organic and membrane pseudophases ($K_M$). They require the study of the binary oil-alcohol, the pseudoternary brine-oil-alcohol and the complete multicomponent systems respectively.

To mixtures containing sodium chloride brine, decane and a laboratory-made $C_{18}$ α-olefinsulfonate [2] various cosurfactants were added: with 1-butanol as a starting structure, the goal was to investigate the effect of lengthening the chain, shifting the OH-group and/or methyl branching. Our results, presented in Table 1, let appear:

i) the stronger self-association of 1-butanol vs. others, especially at 25 °C and the levelling effect of a temperature rise;

ii) the following order of decreasing hydrophilicity of the alcohols [5]:

$C_4OH$-2 > $C_4OH$-1 > Me-3 $C_4OH$-2 > $C_5OH$-1 ~ Me-3 $C_4OH$-1 ;

iii) a less easy incorporation of the alcohols into the pseudophase membrane [5] along the series: $C_4OH$-1 > $C_4OH$-2 > $C_5OH$-1 ~ Me-3 $C_4OH$-1 ~ Me-3 $C_4OH$-2.

Therefore, the lower solubilization properties of 1-butanol, though less hydrophilic than its secondary isomer, seem to be related to higher $K$ and $K_M$ values.

Table 1. Constants of the pseudophase model

| Alcohol | $t$ (°C) | $K$ | $K_w$ | | $K_M$ |
|---|---|---|---|---|---|
| $C_4OH$-1 | 25 | 123 | 8.5 | | |
| | 50 | 45 | 3.4 | | 39 |
| $C_4OH$-2 | 25 | 81 | 12 | w/o NaCl | |
| | | | 11.5 | NaCl 10 g/l | |
| | | | 11 | NaCl 20 g/l | |
| | 50 | 32 | 3.6 | | 33 |
| $C_5OH$-1 | 25 | 85* | 2.1 | | |
| | 50 | 37 | | | 32 |
| Me-3 $C_4OH$-1 | 25 | 82 | 2.1 | w/o NaCl | |
| | | | 1.8 | NaCl 15 g/l | |
| | | | 1.65 | NaCl 30 g/l | |
| | 50 | 27 | 1.0 | NaCl 20 g/l | 27 |
| Me-3 $C_4OH$-2 | 25 | 78* | 4.8 | | |
| | 50 | 33 | | | 26 |

(* at $t = 22$ °C)

## References

1. Biais J, Bothorel P, Clin B, Lalanne P (1981) J Disp Sci Technol 2(1):67–95
2. Castro V (1986) Thesis (Doctorat d'Etat), Inst Nat Polytech Toulouse, Jan 23, no 105
3. Lalanne P, Clin B, Biais J (1984) Chem Eng Comm 27:193–208
4. Voca BR, Canselier JP, Noïk C, Bavière M (1988) Progr Colloid Polym Sci 76:144–152
5. with respect to monomeric alcohol species in oil

Authors' address:

J. P. Canselier
Laboratoire d'Etude et d'Analyse des Procédés
Ecole Nationale Supérieure d'Ingénieurs de Génie Chimique
Chemin de la Loge
F 31078 Toulouse, France

# Effect of D-propanolol on neutral phospholipidic liposomes

A. Cao, E. Hantz-Brachet and E. Taillandier

Laboratoire de Spectroscopie Biomoléculaire, UFR de Médecine, Université Paris XIII, Bobigny, France

*Abstract:* We have investigated the effect of D-propanolol, a cationic amphiphilic drug, on dimyristoyl phosphatidyl-choline liposomes. Two techniques have been used, Quasielastic Light Scattering (QLS) and Fourier Transform Infrared Spectroscopy (FT-IR).

With QLS, the variation of the scattered intensity and of the vesicles size has been followed as function of temperature. In the presence of drug, a shift of the transition point of DMPC large unilamellar vesicles was observed. Moreover, compared with the case in the absence of drug, there is an alteration of the vesicles area in the low-temperature phase.

These results are consistent with FT-IR data showing a frequency variation of the acyl $CH_2$ stretching modes which reflects a trans/gauche conformers ratio change.

On the other hand, FT-IR showed a strong modification of the $PO_2$ stretching modes frequencies at 1088 and 1231 cm$^{-1}$. It is very probable that drug molecules are bound to the polar head groups of the lipid molecules and alter the inner region.

Authors' address:

Dr. A. Cao
Laboratoire de Spectroscopie Biomoléculaire
UFR de Médecine, Université Paris XIII
74 rue Marcel Cachin
93012 Bobigny Cedex, France

# Dynamical computer simulation of concentrated hard sphere suspensions

B. Cichocki[1]) and K. Hinsen

Institut für Theoretische Physik, RWTH Aachen, FRG

*Abstract:* A computer simulation study of hard sphere suspensions in the diffusive limit and without hydrodynamic interactions was performed to find the coherent and incoherent dynamic structure factors of the system and the mean-square displacement of a sphere. A strong dependence of the results on the average displacement of a particle in an elementary step $\Delta t$ was observed. To solve this problem an extrapolation procedure for $\Delta t \to 0$ was proposed. The simulation was performed for various values of the volume fraction of the suspension between 0.1 and 0.5.

The results were compared with predictions of the low-density, Enskog and mode-mode coupling theories. The theory combining in a proper way the Enskog term with the

mode-mode coupling term leads to almost perfect agreement with the computer simulation data.

It was also found that the temporal behaviour of the time derivative of the meansquare displacement of a particle is very well described by the stretched exponential $\exp(-$const. $\sqrt{t})$.

Authors' address:

Dr. B. Cichocki
Institut für Theoretische Physik
RWTH Aachen, FRG

*) On leave of absence from Institute of Theoretical Physics, Warsaw University, ul. Hoza 69, Warsaw, Poland

# Alkanals versus alkanols as microemulsion cosurfactants

A. Meziani, D. Touraud and M. Clausse

URA CNRS N° 858, Département de Génie Biologique, Université de Technologie de Compiègne, Compiègne, France

*Abstract:* For the aim of evaluating the potential of aldehydes to act as microemulsion cosurfactants, phase diagram studies have been carried out to determine the ability of various n-alkanals (cthanal to heptanal) to form with water and the anionic surfactant sodium dodecylsulfate stable single-phase ternary solutions. It comes out from the data obtained that the alkanals considered are globally less efficient than the corresponding n-alkanols, in particular as concerns the formation of ternary solutions of the inverse kind. However, as far as ternary solutions of the direct kind are concerned, the ability of the alkanals is comparable to or even better than that of the corresponding n-alkanols, which is consistent with the fact that the values of the sodium dodecylsulfate CMC in water/alkanol and water/alkanal mixtures do not differ noticeably. These results are of interest for the formulation of new chemical specialities and the composition design of reaction media usable for enzymatic catalysis.

Authors' address:

Prof. M. Clausse
URA CNRS N° 858, Département de Génie Biologique
Université de Technologie de Compiègne
BP 649, F-60206 Compiègne Cedex, France

# Realm-of-existence and electroconductive behavior of water/ionic surfactant/alkanol/hydrocarbon microemulsions. Influence of the hydrocarbon nature

M. Clausse and D. Touraud

URA CNRS N° 858, Département de Génie Biologique, Université de Technologie de Compiègne, Compiègne, France

*Abstract:* The influence of the hydrocarbon nature upon some of the characteristics of water/sodium dodecylsulfate/1-pentanol/hydrocarbon microemulsion-type single phases has been investigated, at $T = 25\,°C$, by varying the n-alkane number of carbon atoms or by substituting aromatic hydrocarbons for aliphatic ones. From the results thus obtained, it comes out that microemulsion tridimensional domain configuration type and electroconductive behavior type are unaffected by changes in the hydrocarbon molecular structure that consequently does not seem like a key composition parameter. This confirms a general conclusion derived from previous studies, namely that the basic prop-erties of water/ionic surfactant/cosurfactant/hydrocarbon single-phase microemulsions are mainly predetermined by those of the corresponding hydrocarbonless ternary mono-phasic solutions.

Authors' address:

Prof. M. Clausse
URA CRNS N° 858, Département de Génie Biologique
Université de Technologie de Compiègne
BP 649, F 60206 Compiègne Cedex, France

**Progress in Colloid & Polymer Science**                    Progr Colloid Polym Sci 81:256 (1990)

# Structural effects of polydispersity in charged colloidal suspensions: A comparison of experimental results with integral equations theories

B. D'Aguanno and R. Klein

Fakultät für Physik, Universität Konstanz, FRG

*Abstract:* Experimental information on the microscopic structure of charged colloidal suspensions is usually extracted from the static light scattered intensity $I(k)$. With the use of the scattering relations [1] $I(k)$ is linked to the static structure factor $S^M(k)$ that is interpreted in terms of macroion-macroion correlations and resembles that of simple liquids. Therefore, $S^M(k)$ can be computed from integral equation theories [2] and contrasted with experimental data. Our main goals are the reinterpretation of given experimental results for $S^M(k)$ in terms of an explicitly polydisperse model and the systematic analysis of the polydispersity effects on $S^M(k)$ itself.

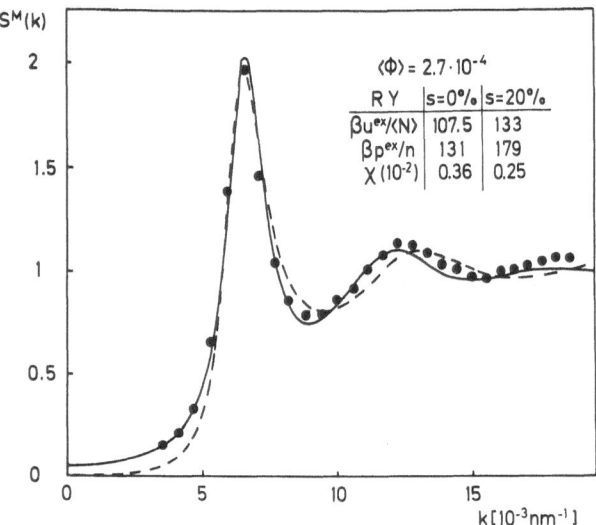

Fig. 1. Static structure factor $S^M(k)$. ●: exp. data from sample 2 of Ref. [4]; – – –: RY results for a monodisperse suspension; ———: RY results for a polydisperse suspension with $s = 20\%$. The mean particle diameter is $\langle\sigma\rangle = 80$ nm and the mean particle charge is $\langle Z\rangle = 499$ e

In our model, the interaction between macroions is assumed to be of a Yukawa type and the polydispersity is characterized by a Schulz distribution, with standard deviation, $s$, from 10% to 40%. The continuous Schulz distribution is replaced by a discretized one, having $N$ components ($N$ from 3 to 9) and then the corresponding partial structure factors, $S_{\alpha\beta}(k)$, are evaluated solving the multicomponent Ornstein-Zernike (OZ) equations in connection with the thermodynamically selfconsistent closure of Rogers-Young [3] (RY). The measured structure factor is given by the relation [1]: $S^M(k) = (\overline{f^2}\,\overline{P(k)})^{-1} \sum_{\alpha=1}^{N} \sum_{\beta=1}^{N} f_\alpha f_\beta B_\alpha(k) B_\beta(k) S_{\alpha\beta}(k)$ and the results are shown in Fig. 1. The agreement with experimental data is quantitative for all $k$ values and, in particular, in the range of small $k$ in which all one-component models are inaccurate. We stress the fact that $S^M(0)$ is *not* the normalized compressibility, $n k_B K_T$, of the system, since it depends on the single particle scattering quantities $f$ and $B(k)$.

## References

1. Pusey PN, Tough RJA (1985) In: Pecora R (ed) Dynamic Light Scattering, Plenum, New York, p 85
2. Hansen JP, McDonald IR (1986) Theory of Simple Liquids, Academic Press, London
3. Rogers FJ, Young DA (1984) Phys Rev A 30:999
4. Krause R, Nägele G, Karrer D, Schneider J, Klein R, Weber R (1988) Physica A 153:400

Authors' address:

Dr. B. D'Aguanno
Fakultät für Physik
Universität Konstanz
Postfach 5560
D-7750 Konstanz 1, FRG

# The gas-liquid phase separation in colloidal dispersions

C. G. de Kruif

Van't Hoff Laboratorium, Universiteit van Utrecht, Utrecht, The Netherlands

*Abstract:* When colloidal silica particles (covered with alkane chains) are suspended in marginal solvents (like toluene or long chain alkanes), a phase transition of the gas-liquid type takes place upon lowering the temperature. The effective pair potential is modelled as an "adhesive" hard sphere.

The phase transition is highly pressure dependent if toluene is used as a solvent. This novel phenomenon cannot be explained in terms of (small) volume dissimilarities between solvent and stabilizing chains. We have evidence that a phase transition of the stabilizing alkyl chains causes the effective change in interaction between the particles.

Author's address:

Dr. C. G. de Kruif
Van't Hoff Laboratorium
Universiteit van Utrecht
Padualaan 8
3584 CH Utrecht, The Netherlands

# Spherical and rodlike micelles from non-ionic surfactants with oligosaccharide head-groups

P. Denkinger, M. Kunz and W. Burchard

Institute of Macromolecular Chemistry, University of Freiburg, Freiburg, FRG

*Abstract:* Non-ionic surfactants of the sugar-lipid hybrid type of the general formula $C_n G_m$ have been prepared by coupling an alkane chain $C_n$ with a maltooligosaccharid over an amide linkage.

Aldonolactones **2** were obtained by electrolytic oxidation of the oligosaccharides **1**. Coupling have been performed with maltonolactone ($m = 2$) and n-alkylamine chains $C_n$, with $n = 8$, 10, 12, 14 and 16, and with different oligosaccharides ($m = 2$, 3, 4 and 6) onto a $C_{16}$-chain. The case of $m = 4$ and 6 are of particular interest since these chains can be elongated by enzymatic synthesis.

The solution properties of the various products have been studied by means of static and dynamic light scattering and by electron-microscopy. The results are as follows: For $n < 14$, small spherical micelles of about 3 nm in diameter are observed. These micelles aggregate further in time and form increasingly larger spherical clusters which eventually precipitate.

If $n \geq 14$, long filamentous fibres are formed for $m = 2$, 3. Contour length and chain stiffness have been determined applying theories of semiflexible chains. A transition from rod-like to spherical micelles was observed when the hydrophillic group was increased beyond $m = 3$. Electron microscopy confirmed these results.

Authors' address:

Dr. P. Denkinger
Institute of Macromolecular Chemistry
University of Freiburg
D-7800 Freiburg, FRG

**Progress in Colloid & Polymer Science**                    Progr Colloid Polym Sci 81:258 (1990)

# Flow equations of real colloidal systems derived by the chemical kinetics methods

Lj. Djaković, I. Šefer and T. Djaković

Faculty of Technology, University at Novi Sad. Yugoslavia

*Abstract:* Structural theories for the non-Newtonian flow, usually assume some model for the system chemical structure and for the kinetics of its change during the flow. As incide the flowing system, in fact, only physico chemical structure and particle interactions are changed, chemical kinetics could be applied considering the rate of a rheological change directly, avoiding thus arbitrary assumptions on the possible chemistry of it. If dependence of the shearing stress $S$ (Pa), or viscosity $\eta$ (Pa s), on the shear rate $D$ (s$^{-1}$) is a consequence of different internal particle interaction changes, whose overall kinetic reaction order can turne out to be a fractional, then the following rheological rate expressions could be of more general importance:

$$\frac{dS}{dD} = k + k_d n D^{n-1} \qquad (1)$$

where for the pseudoplastic flow $0 < n < 1$, and

$$-\frac{d\eta}{dD} = Q m (\eta - \eta_\infty)^2 D^{m-1} \qquad (2)$$

where $m > 0$. $k, k_d$ and $Q$ are rheological change rate constants.

From expressions (1) and (2), various flow equations can be derived. Parameters of these equations can be easily evaluated by simple viscometric measurements, making use of well known experimental methods of the chemical kinetics. Different non-Newtonian systems (concentrated emulsions and macromolecular gels) were investigated by a rotational viscometer (with coaxial cylinders). A method of a computer program formation for systematic investigations of experimental data has been developed.

Some theoretical conclusions on the main rheological effects acting in a pseudoplastic system during the flow, have been derived. Besides the known rheological parameters ($\eta_0$ and $\eta_\infty$), some other important physical constants can be calculated, as are the rheological change rate constant, specific internal resistance energy and relaxation time.

Authors' address:

Prof. Dr. Lj. Djaković
Colloid chemistry dept.
Faculty of Technology, POB 340
21000 Novi Sad, Yugoslavia

# Influence of 17a-ethinylestradiol on the fluidity, polarity and size of DDAB vesicles

A. S. Domazou and A. E. Mantaka-Marketou

National Research Center "Demokritos", 15310 Aghia Paraskevi Attikis, Greece

*Key words:* Vesicles; Fluidity; Polarity; Estrogen; Chemiluminescence

*Abstract:* The structural changes in Didodecyldimethylam-monium Bromide (DDAB) vesicular aggregates caused by 17a-Ethinylestradiol [1] were investigated. Size modifications of the vesicles were followed by electron microscopy; the estrogen itself was used as a fluorescent probe to detect the polarity alterations of the vesicular polar region [2]. In both cases no pronounced effect was detected. In addition, the effect of 17a-Ethinylestradiol on the quantum yields of the Lucigenin chemiluminescent (CL) reaction in DDAB vesicular systems was examined. The presence of 17a-Ethinyl-lestradiol in concentrations higher than $0.5 \times 10^{-4}$ M resulted in reduced quantum yields (the maximum decline being 15%) while the system was insensitive to lower concentrations. Quenching of the fluorescence of N-Methylacridone, the primary emitter of the light reaction, by 17a-Ethinylestradiol was not observed. The alterations in the CL

quantum yields were rationalised in terms of the effect of 17a-Ethinylestradiol on the fluidity of these membrane mimetic agents [3].

## References

1. Mantaka-Marketou AE, Vassilopoulos G, Nikokavouras J (1985) Monatsh Chem 116:973−978
2. Domazou AS (1989) Doctorate thesis
3. Varveri FS, Mantaka-Marketou AE, Vassiopoulos G, Nikokavouras J (1988) Monatsh Chem 119:703−710

Authors' address:

Dr. A. E. Mantaka-Marketou
National Research Center "Demokritos"
15310 Aghia Paraskevi Attikis, Greece

# Interaction of basic peptides with negatively charged biological model membranes. A $^{31}$P- and $^{2}$H-NMR study

E. J. Dufourc[1]), A. Campos[3]), C. Abad[3]), G. Laroche[2]), M. Pézolet[3]) and J. Dufourcq[1])

[1]) CRPP, CNRS, Château Brivazac, Pessac, France
[2]) Dept. Chimie, Univ. Laval, Québec, Canada
[3]) Dept. de Química Física y de Bioquímica y Biol. Mol., Fac. Ciencias, Univ. Valencia, Spain

*Abstract:* The action of Melittin (a bee venom toxin) and of Poly-L-lysines (PLL) of different molecular weight on negatively charged model membranes (dimyristoyl phosphatidic acid (DMPA)-water dispersions) has been monitored by Phosphorus-31 ($^{31}$P) and Deuterium ($^{2}$H) solid state Nuclear Magnetic Resonance (NMR).

Melittin induces a macroscopic restructuration of the membrane leading to new supramolecular entities (lipid-toxin complexes), a decrease of the transition temperature $T_c$, and promotes a phase separation. NMR spectra show a superimposition of a sharp isotropic line on a powder spectrum characteristic of large bilayer structures. Comparison with recent work on zwitterionic lipids [1, 2] allows the attribution of the sharp NMR line to the presence of small complexes (few hundred Å diameter) which tumble rapidity to average the angular dependent NMR interactions. Unlike zwitterionic lipids [1, 2], these sharp lines appear both in the gel and fluid phases, at low toxin concentrations.

On the other hand, both short (m.w. = 3300−4000) and long (m.w. = 180000−200000) PLL do not modify the macroscopic lamelar structure of the membrane but rather lead to an increase in the degree of molecular ordering of the lipid acyl chains, both in the gel and fluid phases. In

addition, a 20°C upshift in $T_c$ is promoted by long PLL whereas no changes is observed with short PLL. These effects are related to temperature-induced conformational changes of the polypeptides.

A rationale for the action of Melittin and of short and long PLL on membranes can be provided by considerations of all driving energies (hydrophobic interactions, electrostatic effects, changes in peptide secondary structure, lipophylic effect) involved in lipid-protein interactions.

### References

1. Dufourc EJ et al. (1986) FEBS Letters 201:205−209
2. Dufourc EJ et al. (1988) Biochimie 71:117−123

Authors' address:

Dr. E. J. Dufourc
CRPP, CNRS
Château Brivazac
33600 Pessac, France

# Mercury NMR. A tool to investigate the chemical structure of mercury compounds in solution and to follow binding phenomena to membrane systems

M. Delnomdedieu[1,2]), A. Boudou[2]), D. Georgescauld[1]) and E. J. Dufourc[1])

[1]) CRPP, CNRS, Domaine Universitaire, France
[2]) Laboratoire d'Ecologie Fondamentale et Ecotoxicologie, Université de Bordeaux I, Talence, France

*Abstract:* Mercury compounds such as HgCl$_2$ are known to be highly toxic for living organisms and their interaction with biological membranes appears to be involved in the toxicity mechanism.

Mercury chloride exhibits a complex chemical speciation, i.e. hydroxydes, oxydes, di-, tri- and tetrachlorides are observed depending on pH and pCl.

High resolution $^{199}$Hg-NMR has been performed on mercury chloride solutions as a function of pH, pCl and various chemical ligands. This technique appears to be very sensitive to the chemical structure and to the chemical environment of mercury. Chemical shift variations up to thousands of ppm allow to identify mercury species in

solution and are being used to monitor binding to a variety of chemical compounds (EGTA, MOPS, HEPES, TRIS, BORAX) or phospholipid membranes.

Mercury NMR thus appears to be a powerful mean for studying interactions between membranes and mercury compounds, from the point of view of the perturbating agent.

Authors' address:

Dr. M. Delnomdedieu
CRPP, CNRS
Chateau Brivazac
33600 Pessac, France

**Progress in Colloid & Polymer Science**　　　　　　　　　　　　　Progr Colloid Polym Sci 81:260 (1990)

# Rigidities and dynamics of shape fluctuations in dilute AOT-microemulsions by Kerr effect measurements

D. Bedeaux[3]), M. Borkovec[1,2]), H.-F. Eicke[1]), R. Hilfiker[1]) and E. van der Linden[3])

[1]) Institut für Physikalische Chemie, Universität Basel, Basel, Switzerland
[2]) Institut für Lebensmittelwissenschaft, ETH-Zentrum, Zürich, Switzerland
[3]) Gorlaeus Laboratoria, Department of Physical and Macromolecular Chemistry, University of Leiden, Leiden, The Netherlands

*Abstract:* We report on static and dynamic Kerr effect measurements in dilute water/AOT/alcane microemulsions for various water-to-surfactant ratios ($w_0 = [H_2O]/[AOT]$) and surfactant concentrations. The specific Kerr constant ($K$) extrapolated to infinite dilution yields information on the elastic properties and intrinsic birefringence of the AOT-monolayer surrounding the core of the nanodroplet. The observed negative birefringence at small $w_0$ [1] can be explained on the basis of the different optical polarizabilities parallel and perpendicular to the AOT monolayer [2]. At large $w_0$, the form birefringence (always positive) outweighs the negative contribution. A quantitative theory [2, 3] yields the following relation between the specific Kerr constant and the mean square $l = 2$ fluctuation amplitudes ($\langle | u_{2m} |^2 \rangle$),

$$K = C \langle | u_{2m} |^2 \rangle \tag{1}$$

where $C$ is an explicitly known constant depending on the dielectric properties and the size of the droplet and the dielectric properties of the solvent. $\langle | u_{2m} |^2 \rangle$ is related to the

monolayer rigidity ($\varkappa$). A fit of Eq. (1) to the experimental data gives in hexane as well as in i-octane a value of $0.5\ k_B T$ for $\varkappa$.

The time dependence of the birefringence relaxation was investigated with decaline as a solvent because of its high viscosity. This leads to slower relaxation times and increases thus the experimental accuracy. We obtain excellent fits of the birefringence response assuming a Gaussian distribution (using a polydispersity of 25% observed by both QELS and SAXS [4] in these systems) of the droplet radii and a $r^3$ dependence of the relaxation times. We were able to separate a slow and a fast contribution to the signal (i.e. a bimodal Gaussian distribution). The slow contribution originates from the (dis)orientation of dimers. The fast contribution, being four times faster than the Debye relaxation time of the monomers, is due to the relaxation of shape fluctuations of the droplets. This interpretation is strongly substantiated by the concentration dependences of the amplitudes of the slow and fast processes (Fig. 1). The findings are consistent with observed rigidities of $0.5\ k_B T$.

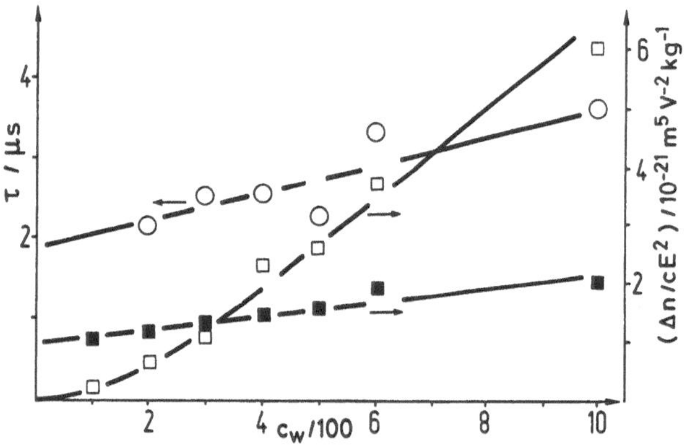

## References

1. Hilfiker R, Eicke H-F, Hammerich H (1987) Helv Chim Acta 70:1531
2. van der Linden E, Geiger S, Bedeaux D (1989) Physica A 156:130
3. Borkovec M, Eicke H-F (1988) Chem Phys Lett 147:195
4. Hilfiker R, Eicke H-F, Sager W, Steeb CH, Hofmeier U, Gehrke R (1990) Ber Bunsenges Phys Chem in press

Fig. 1. Slow relaxation time (circles) and contribution of the slow (open squares) and fast (solid squares) processes to the birefringence signal versus concentration (mass/mass) of the droplets. System: water/AOT/decaline (cis/trans mixture), $w_0 = 50$, $T = 318$ K

Authors' address:

Prof. H.-F. Eicke
Institut für Physikalische Chemie, Universität Basel
Klingelbergstrasse 80
CH-4056 Basel, Switzerland

**Progress in Colloid & Polymer Science**                                      Progr Colloid Polym Sci 81:261 (1990)

# Small-angle neutron scattering of ethoxylated sodium alkylcarboxylates surfactants

C. Fagotti[1]) and R. K. Heenan[2])

[1]) Eniricerche, San Donato Milanese, Italy
[2]) Rutherford Appleton Laboratory, Chilton, Great Britain

*Abstract:* Ethoxylated ionic surfactants are a class of amphyphylic compounds that are industrially very promising. Their wide applicability is mainly due to the special properties arising from the presence of the ionic group adjacent to the oxyethylene groups, that make these surfactants usable in more demanding conditions (e.g. high salinity).

In this poster we present SANS data from LOQ at ISIS obtained for compounds of this class: sodium dodecylpoly(oxyethylene) carboxylates ($C_{12}H_{25}(OCH_2CH_2)_n OCH_2CO_2Na$).

Results have been obtained for a different numbers of ethoxy groups ($n = 7, 8, 9, 12$) at different concentrations in pure $D_2O$ and in the presence of a bivalent added salt ($CaCl_2$).

Comparison with SANS measurements performed on other ethoxylated ionic surfactants (Sodium dodecylpoly(oxyethylene) Sulfates [1−3]) without added salt are in general agreement, confirming some trends of this class of surfactants. They are spherical micelles with low polydispersity.

Differences present in the fractional charge (around 0.55), may be explained as due to a larger number of ethoxy groups compared to [1−3].

Unusual results are present in some of the measurements with added salt. These can probably be ascribed to the limits of the analytical models now available to calculate $S(Q)$ in colloidal systems [4, 5].

## References

1. Triolo R, Caponetti E, Graziano V (1985) J Phys Chem 89:5743
2. Triolo R, Caponetti E (1986) J Solution Chemistry 15:377
3. Minero C et al. (1986) J Phys Chem 90:1620
4. Belloni L (1986) J Chem Phys 85:519
5. Hayter JB, Penfold J (1981) Molecular Phys 42:109

Authors' address:

Dr. C. Fagotti
Eniricerche
Maritano Via Maritano 26
I-20097 S. Donato Milanese, Italy

# Langmuir Blodgett films containing metal ions

S. Bettarini, F. Bonosi, G. Gabrielli and M. Puggelli

Department of Chemistry, University of Florence, Firenze, Italy

*Abstract:* Monolayers and multilayers of Behenic Acid ($C_{22}$) at the water/air interface on substrated containing $Cu^{2+}$ and $Mn^{2+}$ ions were studied.

The spreading isotherms at different temperatures and the corresponding two-dimensional equations of state and compressional moduli were determined for monolayers on subphases containing or not bivalent ions.

The transfer conditions of the ions from the aqueous subphase to the multilayers obtained using the Langmuir Blodgett method were determined.

The presence of ions in collapsed films and in built-up LB films were determined by means of Electron Spectroscopy for Chemical Analysis (ESCA) and Electron Spin Resonance (ESR).

Attainment and properties of Langmuir Blodgett multilayers were very dependent on thermodynamic and rheologic properties and on molecular orientation in monolayers.

Authors' address:

Dr. G. Gabrielli
Department of Chemistry, University of Florence
Via Gino Capponi 9
50121 Firenze, Italy

**Progress in Colloid & Polymer Science**    Progr Colloid Polym Sci 81:262 (1990)

# Monolayers and bilayers of lipids as models of biological membranes

E. Margheri, A. Niccolai, P. Lo Nostro and G. Gabrielli

Dep. of Chemistry, University of Florenco, Firenze, Italy

*Abstract:* Different models of natural membranes were studied, using Ceramides (CER), Monoolein (MON) and Dioleoylphosphatidylcholine (DOPC).

Spreading isotherms of CER and MON alone were recorded at the water/air interphase, using NaCl 0.1 M solutions as substrate, thus determining the equations of state which characterize the film phases, the compressional moduli values and, with respect to mixtures, the miscibility at the interphase.

Langmuir-Blodgett monolayers and bilayers of CER were prepared, and their thickness and refractive index determined by ellipsometry.

BLMs constituted of MON and MON/CER mixtures were prepared at room temperature, in presence of $Na^+$ and $Ca^{2-}$ ions and their electrical properties were studied.

Finally, vesicles constituted of DOPC and DOPC/CER mixtures were prepared and their dimensions were determined by Scanning Electron Microscopy (SEM).

The experimental results allow us to deduce that the studied substances have their hydrophobic chains in the same orientation in monolayers and in bilayers and, for this reason, the bilayer can be considered to be formed by two overlaid monolayers.

Authors' address:

Dr. G. Gabrielli
Dep. of Chemistry, University of Florence
Via Gino Cappioni 9
50121 Firenze, Italy

# Small angle neutron scattering (SANS) investigations of sodium alkyl sulfate micelles

S. Borbely, L. Cser, I. A. Gladkih, Yu. M. Ostanevich and Sz. Vass

Hahn-Meitner Institut, Berlin

*Abstract:* The average aggregation number ($n$) of sodium alkyl sulfate micelles in heavy water solution depends on the amphifilic concentration ($c$), on the temperature ($T$) and on the length of the hydrocarbon chain ($n_c$). As it is well known, three competing effects determine the equilibrium value of the average aggregation number of a given micellar solution. Namely, the hydrophobic effect, the repulsion electrostatic force inside the micelles acting between the charged head groups, and electrostatic interaction between the micelles. The aim of the present work is to find the simplest model which still gives physically meaningful description of the concentration, temperature and the hydrocarbon chain length dependence of the aggregation number obtained from SANS data. The essential features of this model are:

- the hydrophobic interaction is characterized by a constant multiplied by the number of carbon atoms;
- the polar head group interaction is described by an average value which depends only on the aggregation number;
- the intermicellar repulsive interactions was reduced to Coulomb force and in the case of sufficiently diluted micellar solution this term was simply neglected.

The use of these above conditions leads to differential equations having analytical solutions. These solutions prescribe the $n^4 \sim c$ and $n \sim n_c^2$ relations which are in good agreement with the observations [1, 2]. The temperature dependence of the aggregation number values can be expressed only numerically from the equation

$$T = An - Bn^{-\frac{7}{3}} + C.$$

The evaluation of $n$ values describes the experimental data well again [1]. The estimation of the relative variation of the aggregation number also gives reasonable values.

## References

1. Bezzobotnov VYu, Borbely S, Cser L, Farago B, Gladkih IA, Ostanevich YuM, Vass Sz (1988) J Phys Chem 92:5738–5743
2. Borbely S, Cser L, Ostanevich YuM, Vass Sz, submitted to J Phys Chem

Authors' address:

I. Gladkih
Hahn-Meitner Institut
Pf. 390128
Glienicker 100
1000 Berlin 39

# A comparative study on different techniques and evaluation methods for sizing of colloidal systems using light-scattering techniques

O. Glatter, H. Schnablegger and J. Sieberer

Institute of Physical Chemistry, University Graz, Austria

*Abstract:* Quasi-elastic light scattering is a well established method for particle sizing in the sub-micron range. A recently developed method for particle sizing using elastic light scattering can be applied to systems in the size range from about 100 nanometers up to several microns. Fraunhofer diffraction starts at a level of a few microns and reaches up to several hundred microns.

The aim of our work was to compare different scattering methods in terms of resolution and range of applicability and to compare the different inversion techniques available for data evaluation for the different experimental techniques.

Our results show that elastic scattering techniques have a rather high resolution (minimum peak separation about 25%) but cover a small size range (about one and a half decade). Quasi-elastic experiments can be performed in a size range of about three decades but the resolution is low (minimum peak separation 200–300%). These results are nearly independent of the data evaluation method used.

There exists a large variety of different methods for the solution of the inverse scattering problem. The most important techniques are: regularization techniques (RT) with many different constraints, singular value decomposition (SVD) and maximum entropy methods (MEM).

Our results show that SVD is of the same quality like RT without positivity constraint. RT results are highly improved by this and other new constraints. MEM results depend on the type of the transform and on the numerical routine.

Authors' address:

Otto Glatter
Institute of Physical Chemistry
Heinrichstrasse 28
A-8010 Graz, Austria

# Coupling of radial and orientational orders between colloidal particles in asphaltenes solutions

P.-G. Gottis[1]) and J.-R. Lalanne[2])

[1]) Ciba-Geigy AG, Basel, Switzerland
[2]) CNRS – Paul Pascal, Domaine universitaire, Talence, France

*Key words:* Asphaltene; light scattering; infrared; colloidal solutions; petroleum

*Abstract:* As well known, Asphalt and Asphaltene solutions are black, even at low concentration, and strongly absorbing the visible light. However, both refractive index measurements and a complete Rayleigh scattering study of

Petroleum asphaltene (ex SAFANYA) colloidal solutions were achieved thanks to the development of infra-red Laser techniques. As asphaltene solutions display a window of low absorption near 1000 nm, it was possible to obtain fluorescence free light scattering measurements of both Vv an Hv components up to concentrations of 4 g · l$^{-1}$. For diluted solutions (lg · l$^{-1}$), the molecular mass at infinite dilution was for the first time measured. Correlatively, the mean value and the anisotropy of the first order polarizability at 1064 nm are determined.

For higher values of the concentration, a strong coupling between asphaltene particles can be seen both in the isotropic (radial correlations) and anisotropic (orientational correlations) components. The observed non-linear increase of the anisotropic component with the concentration implies a paratropic (parallel configuration) order in the solution. As expected, interactions between particles occur, leading at higher concentrations to the well-known colloidal macrostructure of asphaltenes. But an important – and rather original – observation must be emphasized: As shown in Fig. 1, where a log-log plot was prefered for a better clarity, a striking similarity affects the increase of both

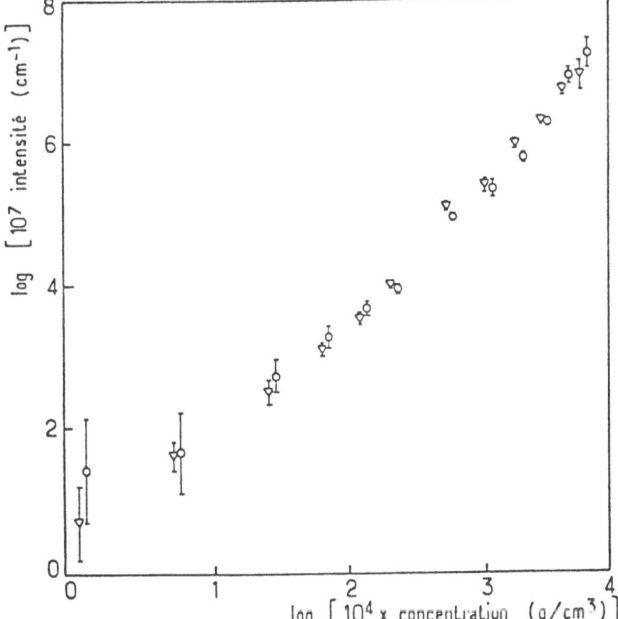

Fig. 1. Safanya asphaltene/orthoxylene solutions isomorphism of isotropic and anisotropic Rayleigh components: ▽, isotropic component ($\Delta R_{iso}$): ○, anisotropic component ($\Delta R H_v / 0.17$)

isotropic and anisotropic component intensities with the concentration, suggesting the following conclusions:

(i) by using a microscopic picture, it seems that the increasing position correlations between asphaltene entities do not affect the markedly paratropic short range orientational order.

(ii) from a more macroscopic description, our experimental results suggest a strong coupling of the orientational and radial orders within these solutions, rarely observed in the field of chemical physics.

## References

1. Gourlaouen C (1984) Ph D Thesis No 32035 IFP-ENSPM, Paris VI University
2. Gottis P-G (1988) Ph D Thesis IFP-ENSPM, Paris VI University
3. Gottis P-G, Lalanne J-R (1989) Infrared Phys 29:511−516
4. Gottis P-G, Lalanne J-R (1989) Fuel 68:804−805

Authors' address:

P.-G. Gottis
Ciba-Geigy AG
K-401.1.34
CH-4002 Basel, Switzerland

# Interaction of gas-blocking foam with oil in model porous media

J. E. Hanssen, T. Meling and K. R. Jakobsen

Rogaland University Centre, Stavanger, Norway

*Abstract:* Gas-blocking foams may potentially improve production from certain oil reservoirs [1]. Our present work is aimed at contributing to a better understanding of the flow behaviour of fluids in foam-filled porous media, particularly in the presence of oil. Direct visual observations of pore-level phenomena are made in an etched-glass micromodel, gas permeability through foam-filled glass-bead packs is measured, and the results are modelled using a mathematical network model. A model fluid system consisting of $C_{13}$ to $C_{17}$ fractions of a *sec*-alkyl sulfonate (SAS) in distilled water, n-alkanes, and nitrogen, is studied at 20 °C and pressures up to 20 bar.

In the micromodel, SAS foam was observed to exist as a network of liquid films which did not move through the pore channels. This, and oscillations seen upon each pressure increase, indicates that gas flow occurs by a film breaking and reforming process.

In bead packs, a wide range of gas-blocking efficiency was obtained with combinations of SAS-fraction foam and oil. Clear carbon-number compatibility trends were observed and are reported separately [2].

The porous media were modelled as a two-dimensional square network, each element characterized by a threshold pressure gradient $(\Delta p/\Delta x)_t$, taken to be uniformly distributed. This yields for gas velocity $u_g$:

$$u_g = 0 \qquad \text{for } (\Delta p/\Delta x) < (\Delta p/\Delta x)_t$$

$$u_g \sim [(\Delta p/\Delta x) - (\Delta p/\Delta x)_t]^a \quad \text{for } (\Delta p/\Delta x) > (\Delta p/\Delta x)_t$$

$$u_g \sim (\Delta p/\Delta x) \qquad \text{for } (\Delta p/\Delta x) \gg (\Delta p/\Delta x)_t$$

The network model was found to describe well the experimental relationship between gas velocity and applied pressure drop in glass-bead packs for all SAS-fraction foams

studied in the absence of oil, with an experimental value for the exponent $a$ of 2.5. This compares well with the theoretical

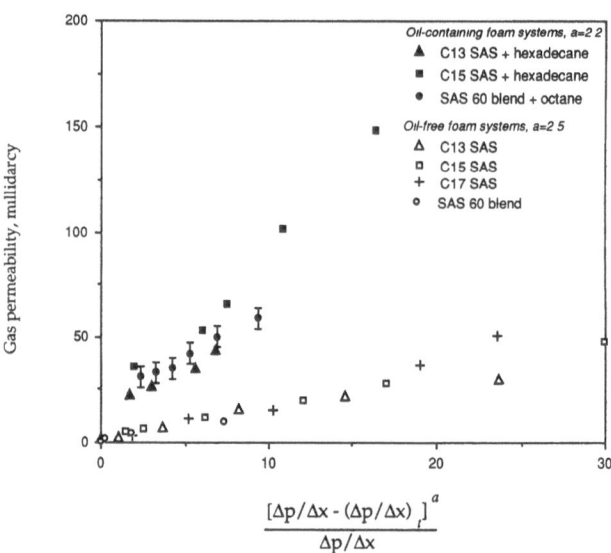

Fig. 1. Permeability to nitrogen of 1-metre long, foam-filled glass-bead pack of 8500 millidarcy absolute permeability, measured at apparent steady state flow, generalized by plotting as functions of the normalized pressure gradients obtained from network modelling. Temperature 21 °C, back pressures ≈ 6 bar, differential pressures up to 12 bar. Distilled-water foams made with secondary alkane sulfonate (SAS) foams of indicated carbon-number fraction. Typical error bars shown. For more complete experimental data see Ref. [2]

value of $a = 2.0$, considering that the bead packs are not strictly two-dimensional.For oil-containing systems, a value for $a$ of 2.2 was obtained.

## References

1. Hanssen JE et al (1989) Foam barriers against gas coning in thin oil zones, Proc 5th European Symposium on Enhanced Oil Recovery, Budapest

2. Meling T, Hanssen JE, Gas-blocking foams in porous media: Effects of oil and surfactant hydrophobe carbon number, Progr Colloid Polymer Sci vol 81, in press

Authors' address:

Jan Erik Hanssen, M. Sc.
Institute of Petroleum Engineering
Rogaland University Centre
P.O. Box 2557, Ullandhaug
N-4004 Stavanger, Norway

# Iridescent colours of a surfactant in water at very high dilution

K. Hiltrop

Universität Paderborn, FRG

*Abstract:* More detailed information about the Schiller Phase was achieved by wide angle x-ray scattering. At large Bragg anlges observed by means of a Guinier camera always *four sharp reflexes* were detected at different $C_{18}PySal$ concentrations in the range 0.80 wt%. The position of the reflexes did not change with concentration. The corresponding spacings are 0.480 nm, 0.452 nm, 0.387 m, and 0.376 nm. This pattern evidences the *two dimensional crystallinity* of the structure. — The Schiller Phase is *birefringent.* Even without the action of external fields the optical axis orientation usually is uiform within domains of 10 mm width or more; the domains can be *oriented by a magnetic field.* A 0.40 wt% sample in a 1 cm cuvette was exposed to 2.5 T. After 30 min the sample was taken out of the field and showed a very uniform optical axis orientation parallel to the magnetic field *all over the cuvette.* The colorful light scattering platelets were uniformly oriented normal to the field direction.

*Experimental results:* At an aqueous phase of octadecylpyridiniumsalicylate ($C_{18}PySal$) an intensive colourful light scattering can be observed in the very low concentration region of about 0.5 wt% surfactant.

The influence of some parameters is described briefly:

*Temperature:* There is only a small range of about 2 K on heating and 5 K or more on cooling within which colours can be observed. This range lies between 18 °C and 26 °C, depending on the surfactant concentration. Exceeding the upper temperature of that intervall results in an isotropic transparent solution ("high temperature phase"), beneath the lower border temperature white light scattering occurs ("low temperature phase"). The wavelength of scattering is *blueshifted with increasing temperature;* the brilliance passes through a maximum.

*Concentration:* Wavelength selective light scattering in the visible region is restricted to a concentration range from below 2 wt% to beyond 0.05 wt% $C_{18}PySal$. At the lower concentrations it can easily be recognized that the scattering effect originates from many *discrete platelets* in the sample. With increasing concentration the number of platelets per volume increases whilst the scattering is only slightly redshifted. All of the platelets are very regular in shape, some of them exhibit a *sixfold symmetry.*

*Scattering angle:* The wavelength of the scattered light is *blueshifted with decreasing grazing angle $\vartheta$* (i.e. increasing angle between incident light and observation direction).

*Further properties:* A $^1H$-nmr experiment suggested the *crystalline* structure of the low temperature phase.

A 1.13 wt% $C_{18}PySal$ solution in $D_2O$ shows at 28 °C only one sharp peak which originates from some $H_2O$ in the sample: the surfactant protons are "frozen". Raising the temperature to 30 °C results in a drastically changed nmr spectrum showing additional peaks of the aliphatic and aromatic protons.

Small angle x-ray data taken at different concentrations between 100 wt% and 13 wt% $C_{18}PySal$ gave up to *five reflex orders* which leads to the assumption of a *crystalline lamellar* structure. A linear dependence of the layer period $D$ vs. $(1 - c)/c$ as to be expected for a *swelling lamellar phase* was found ($c$ = weight fraction of $C_{18}PySal$).

*IDEA:* "There are thin plane parallel crystals (may be swollen), which give rise to *interference colours.*"

*Comment:* This idea is supported by the angular dependence of the light scattering. The evaluation of the data yields platelet thicknesses of 100 to 140 nm and refractive indices of 1.35 to 1.40. Particularly the refractive indices are somewhat too small to be explained by $C_{18}PySal$ crystals; the values measured for the $L_\alpha$ phase of pure $C_{18}PySal$ are 1.489 and 1.512, for the crystalline phase 1.61 was observed. Therefore the light scattering platelets *can not be pure crystalline* $C_{18}PySal$ but must consist of very swollen layers. The fairly monochromatic scattering (at fixed angle, temperature, concentration) requires monodisperse platelet thicknesses.

*CONCLUSION:* The assumption of a *crystalline, swollen lamellar structure* is most evident.

Author's address:

Dr. K. Hiltrop
Universität Paderborn
Paderborn, FRG

**Progress in Colloid & Polymer Science**                              Progr Colloid Polym Sci 81:266 (1990)

# The aggregation behaviour of triblock copolymers of ethylene oxide and propylene oxide in aqueous solutions

G. Wanka, H. Hoffmann and W. Ulbricht

Lehrstuhl für Physikalische Chemie I, Universität Bayreuth, Bayreuth, FRG

*Abstract:* Light scattering, SANS, rheological and interfacial tension measurements were carried out on aqueous solutions of triblock copolymers of Ethyleneoxide and Propyleneoxide. The compounds were commercial samples and had the following approximate composition $EO_{20}PO_{70}EO_{10}$, $EO_{18}PO_{58}EO_{18}$ and $EO_{106}PO_{69}EO_{106}$. All three compounds formed micelles above a critical concentration. The size of the micellar core is determined by the length of the Propyleneoxide block. The transfer energy of a Propyleneoxide unit from the aqueous to the micellar phase is about $0.3\ kT$ at room temperature.

The aggregation number of the micelles increases strongly with increasing temperature while the hydrodynamic radius remains about constant. The SANS-data show a strong correlation peak. Both the SANS and the light scattering data can be interpreted on the basis of the theory of hard sphere particles. Solutions with a volume fractions $> 0.2$ gellifie when the temperature is raised above room temperature. The position of the correlation peak of the SANS data is not effected by the gel formation. Some samples showed however clear evidence of long range order in the system. These gels seem to consist of liquid crystalline cubic phases. The shear moduli of the system can be understood on the basis of hard sphere models. For the investigated samples the moduli were in the range of $10^4$ Pa. Some of our conclusions were similar as the ones reached by B. Chu et al. [1].

## References

1. Zhou Z, Chu B (1988) J Coll and Interf Sci 126:171

Authors' address:

Prof. H. Hoffmann
Lehrstuhl für Physikalische Chemie I, Universität Bayreuth
Universitätsstr. 30
D-8580 Bayreuth, FRG

# Structure of microemulsion-based organo-gels

P. J. Atkinson[1]), R. K. Heenan[3]), M. J. Grimson[2]), A. M. Howe[4]) and B. H. Robinson[1])

[1]) School of Chemical Sciences, University of East Anglia, Norwich, UK
[2]) AFRC Institute of Food Research, Colney Lane, Norwich, UK
[3]) Rutherford-Appleton Laboratory, Chilton, Didcot, Oxon, UK
[4]) Surface Science Laboratory, Research Division, Kodak Limited, Harrow, Middlesex, UK

*Abstract:* Water and oil can form thermodynamically-stable, optically-transparent dispersions of aqueous microdroplets in an oil-continuous medium. These water-ion-oil microemulsions are only formed in the presence of suitable surfactants. It may come as a surprise that such fluids can be gelled on addition of the water-soluble (bio)polymer gelatin [1−3]. The thermo-reversible "organo-gels" thus formed have useful properties, such as high electrical conductivity, which suggest a range of novel applications.

A small-angle neutron scattering (SANS) study was carried out on the LOQ instrument at the SERC Rutherford-Appleton Laboratory. Broadly, the scattering from an organo-gel of composition 3.5% w/v gelatin, 10% $D_2O$, 0.1 mol $dm^{-3}$ AOT, may be described by two power-law regions: at low $Q$ $(0.006 − 0.02\ Å^{-1})$ the exponent is $−1$ (typical of long rigid rods) which tends to $−4$ at higher $Q$ (consistent with scattering from a smooth interface), whence it follows the behaviour of the parent microemulsion. A detailed fit to the data provides evidence that the aqueous content of the organo-gel consists of a rigid network of surfactant-coated gelatin/water rods coexisting with microemulsion droplets.

## References

1. Haering G, Luisi PL (1986) J Phys Chem 90:5892
2. Quellet C, Eicke H-F (1986) Chimia 40:233
3. Howe AM, Katsikides A, Robinson BH, Chadwick AV, Al-Mudaris A (1988) Prog Colloid Polym Sci 76:211

Authors' address:

Dr. A. M. Howe
Surface Science Laboratory
Research Division
Kodak Limited
Harrow, Middlesex, HA1 4TY

# Interfacial behavior of binary surfactant mixtures in aqueous solution

K. Huber and M. Hall

Ciba-Geigy AG, Dyestuffs and Chemicals Division, Basel, Switzerland

*Abstract:* Solutions of mixed surfactants have gained increasing interest over the past two decades. Critical micelle concentration and interfacial tension of such solutions are often significantly lower than for the corresponding pure components. This nonideal behavior is both of scientific interest and of considerable industrial importance.

The present work deals with aqueous solutions of a variety of binary surfactant mixtures. Micellation and interfacial tension between the solution and air or an apolar solid (poly-tetra-fluoro-ethylene) were investigated as a function of the surfactant mixing ratio, using four different combinations of surfactant components. The nonionic surfactants $CH_3(CH_2)_n-(-OCH_2-CH_2)_m-OH$ with $n = 12/m = 4$, 6, 8 and $n = 8/m = 4$ are mixed with sodium dodecylsulfate.

For each surfactant mixture, concentration versus composition curves are determined for three distinct properties: critical micellation, fixed value for surface tension and fixed value for interfacial tension at the solid/liquid interface. The concentration versus composition curves are discussed in terms of a regular solution model [1−3], resulting in a single adjustable parameter, which is interpreted as a free energy of mixing parameter.

The present selection of surfactant components yields information on structure − property relations between the structure of the nonionic components and the interaction parameter in its mixture with sodium dodecylsulfate. It is shown that the interaction parameter behaves differently in the three surfactant domains i.e. micelles, solid/liquid interface and liquid/air interface.

## References

1. Hildebrand JH (1929) J Am Chem Soc 51:66
2. Rubingh DN (1979) In: Mittal KL (ed) Solution Chemistry of Surfactants, Plenum Press, NY 1:337
3. Hua XY, Rosen MJ (1982) J Coll Interf Sci 90:212

Authors' address:

Dr. K. Huber
Ciba-Geigy AG
Dyestuffs and Chemicals Division
K-410.306
CH-4002 Basel, Switzerland

# SANS studies on the distribution of surfactants in latex films

E. Hädicke[1]), K. Hahn[1]), G. Ley[1]), J. Streib[1]) and P. Lindner[2])

[1]) Kunststofflaboratorium, BASF AG, D-6700 Ludwigshafen, FRG
[2]) Institut Laue-Langevin (ILL), Grenoble, France

*Key words:* Poly-n-butylmethacrylate; lauric acid; sodium dodecyl sulphate; SANS; latex film; surfactant; latex film structure; dynamics of latex film formation

*Abstract:* Latices often are stabilized by surfactants. The dispute on the fate of surfactants in latex films is still open [1]. Because the properties of these films depend on their morphology it is essential to determine the spatial distribution of the surfactants.

As has been shown by SANS studies on poly-n-butylmethacrylate (P[nBMA]) latex films [2,3] the film formation process is connected with an interdiffusion of the polymers originally belonging to adjacent latex particles. In order to see how the compatibility between surfactants and the polymer matrix influences the final morphology, two systems have been investigated with SANS:

a) P[nBMA] latices plus deuterated lauric acid (or its ammonium salt respectively) (D-LA) (compatible system)
b) P[nBMA] latices plus deuterated sodium dodecyl sulphate (D-SDS) (incompatible system).

From our experimental results (LIMI, ELMI and SANS) we found that after the film formation process D-LA is dispersed homogeneously in the matrix. This distribution is also stable on tempering at temperatures of $50-150\,°C$.

In the films with D-SDS as surfactant there are scattering entities with a mean Rg of 12 nm which are most probably randomly distributed. Tempering at $80\,°C$ yields a distinct increase in Rg (18 nm). At higher tempering temperatures there is an indication of clustering of the D-SDS into very large particles ($> 100$ nm).

All results will be published soon in detail [4].

## References

1. Bindschaedler C, Gurny R, Doelker E (1987) J Appl Polymer Sci 34:2631−2647
2. Hahn K, Ley G, Schuller H, Oberthür R (1986) Colloid Polym Sci 264:1092−1096
3. Hahn K, Ley G, Oberthür R (1988) Colloid Polymer Sci 266:631−639
4. Hädicke E, Hahn K, Ley G, Streib J, Lindner P, to be published in Colloid Polymer Sci

Authors' address:

Dr. E. Hädicke
Kunststofflaboratorium BASF Aktiengesellschaft, ZKM-Bi
D-6700 Ludwigshafen, FRG

**Progress in Colloid & Polymer Science**                    Progr Colloid Polym Sci 81:268 (1990)

# Rotational diffusion of a tracer particle in a magnetic field

F. Alavi and R. B. Jones

Department of Physics, Queen Mary College, London, England

*Abstract:* We study the rotational diffusion of a magnetic tracer particle in a low-density suspension of non-magnetic particles in the presence of an external magnetic field. For the case of spherical particles with pairwise hard sphere direct interactions and including hydrodynamic interactions we explore the magnetic field dependence of the orientational time correlation functions at short and intermediate times. At short times we present results to first order in volume fraction but for arbitrarily strong magnetic field. At intermediate times we use an approximate perturbative technique to evaluate the lowest order effect of the magnetic field on the decay of the orientational time correlation functions.

Authors' address:

Dr. R. B. Jones
Department of Physics
Queen Mary College
Mile End Road
London E1 4NS, England

# Electric double layer forces in the presence of polyelectrolytes

T. Åkesson, C. Woodward and B. Jönsson

Physical Chemistry 2, Chemical Centre, Lund, Sweden

*Abstract:* An electric double layer is studied by means of Monte Carlo simulations and mean field theory. The counterions of the uniformly charged surfaces are modelled as flexible polyelectrolytes. For this particular model system it turns out that the traditional double layer repulsion becomes attractive for a wide range of systems. The main reason for this attraction is an entropically driven bridging mechanism and its magnitude is significant compared to ordinary double layer or van der Waals forces. The polyelectrolyte Poisson-Boltzmann theory developed here behaves in a qualitatively correct manner, also predicting an attractive interaction extending over several nanometers. These results may have some relevance to technical and biological systems, where sometimes puzzling force behaviour is seen in the presence of polyelectrolytes.

Authors' address:

Dr. B. Jönsson
Physical Chemistry 2, Chemical Centre
POBox 124
S-22100 Lund, Sweden

# Structural pathways and short-lived intermediates in phospholipid phase-transitions. Millisecond synchrotron x-ray diffraction studies

P. Laggner[1]), M. Kriechbaum[1]), G. Rapp[2]) and J. Hendrix[3])

[1]) Institut für Röntgenfeinstrukturforschung, Austrian Academy of Sciences, Graz, Austria
[2]) MPI für Med. Forschung, Heidelberg, FRG
[3]) EMBL, Hamburg Outstation at DESY, FRG

*Abstract:* Small-angle synchrotron X-ray powder diffraction was used to investigate the structural changes during phospholipid phase transitions triggered by a single pulse from an Erbium IR-laser (energy: 2 joules, width: 2 ms) leading to a 10 degrees $T$-jump. The time-evolution of the ensuing powder patterns was resolved into 1 ms-periods. Several novel facts concerning the mechanisms and kinetics of these processes were established. A hitherto unknown intermediate lattice was detected in the "pretransition" $(L'_\beta - P'_\beta)$ of dipalmitoyl-lecithin (DPPC). This lamellar 58 Å-lattice coexists for about 100 ms after the $T$-jump with the original $L'_\beta$-lattice (66 Å), and subsequently the two structures merge to build up the 74 Å − average repeat of the known monoclinic sawtooth − or "ripple"-structure.

The $L_\beta - L_\alpha$ transition of phosphatidylethanolamines was found to be as fast as the experimental time constant of the $T$-jump ($<2$ ms) suggesting a strongly correlated transition mechanism of the diffusionless martensitic type along equivalent lattice planes.

In contrast, the lamellar — hexagonal phase transition is considerably slower (half life: several seconds). There, the results show a fast initial approach of adjacent bilayers followed by the slow fusion into tubular structures.

## References

1. Laggner P, Kriechbaum M, Hermetter A, Paltauf F, Hendrix J, Rapp G (1989) In: Bothorel P (ed) Trends in Coll Interface Sci III, Proceedings of the II. European Colloid and Interface Society Conference, Bordeaux (1988), Prog Colloid Polymer Sci 79:33–37 (1989)
2. Kriechbaum M, Rapp G, Hendrix J, Laggner P (1989) Rev Sci Instrum 60:2541–2544

Authors' address:

Prof. P. Laggner
Inst. für Röntgenfeinstrukturforschung
Oestr. Akademie der Wissenschaften
Steyrergasse 17
A-8010 Graz, Austria

# A new maximum entropy algorithm for the analysis of dynamic light-scattering data: Application to particle distributions and DNA internal motions

J. Langowski[1]) and R. Bryan[2])

[1]) EMBL, Grenoble Outstation, Grenoble, France
[2]) EMBL, Heidelberg, FRG

*Abstract:* We have developed a new maximum-entropy fitting procedure (MEXDLS) which is particularly suitable for oversampled data such as obtained in dynamic light scattering (DLS). A full account of this algorithm will be published (R. Bryan, Eur. Biophys. J., in press). It exploits the fact that regardless of the precision of the measured data, the experiment carries only a limited amount of information; thus the transformation matrix that connects data space and model space has only a small number of nonzero singular values, typically 20 for a DLS measurement of 140 data points that is analyzed by a multiexponential distribution of 100 model points, both model and data space ranging over 2–3 decades in time. The fitting problem is then expressed in the space of the singular vectors of the transformation matrix. The advantage of our procedure over previous applications of maximum entropy to DLS data analysis (e.g. A. K. Livesey, P. Licinio, M. Delaye (1986) J. Chem. Phys. **84**, 5102) is the greatly increased computational speed, the automatic determination of the most probable value of the experimental baseline A, and the determination of the scaling of the noise statistics.

We have applied MEXDLS to simulated DLS spectra of three-component mixtures, measured DLS data from two-component mixtures of polystyrene latex spheres, and DLS data from the plasmid DNA pACL29 (5400 base pairs) in its superhelical and linear form. The data sets were analyzed using MEXDLS and Provencher's CONTIN program. The latter program was used for comparison since it is one of the most frequently applied methods of analysis for DLS data.

We find that our program offers higher resolution than CONTIN for the simulated as well as the measured bi- and trimodal distribution data. For the simulated three-exponential data, MEXDLS resolves components of equal amplitude separated by a factor of 3, while CONTIN needs a factor of 4 for resolution; similarly, a mixture of 64 and 196 nm polystyrene spheres is separated into two peaks by MEXDLS, while CONTIN can resolve only a 64/250 nm mixture reliably. For the DNA, the existence of a component of rotational diffusion at small scattering vectors could be established unambiguously for the supercoiled form of pACL29.

Authors' address:

Dr. J. Langowski
EMBL, Grenoble Outstation
c/o ILL, 156X
F-38042 Grenoble Cedex, France

**Progress in Colloid & Polymer Science**

Progr Colloid Polym Sci 81:270 (1990)

# Theory and small-angle neutron scattering studies of the rod-to-sphere transition of diheptanoyl-PC micelles solubilizing tributyrin

T.-L. Lin[1]), S.-H. Chen[2]) and M. F. Roberts[3])

[1]) Department of Nuclear Engineering, National Tsing-Hua University, Hsin-Chu, Taiwan
[2]) Department of Nuclear Engineering, Massachusetts Institute of Technology, Cambridge, MA, USA
[3]) Department of Chemistry, Boston College, Chestnut Hill, USA

*Abstract:* The solubilization of tributyrin by rodlike diheptanoyl-PC micelles has been studied by using small-angle neutron scattering techniques. A rod to sphere transition is observed from the measured neutron scattering spectra. This transition can be explained by the formation of globular mixed micelles which have tributyrin molecules at the center surrounded by the diheptanoyl-PC molecules. When the molar ratio of tributyrin to diheptanoyl-PC is increased beyond 0.18, all the rodlike diheptanoyl-PC micelles are transformed into globular mixed micelles. The small-angle neutron scattering data were analyzed to obtain the structural parameters of these mixed micelles. They are found to grow with increasing tributyrin concentration. At the tributyrin to diheptanoyl-PC molar ratio less than 0.18, rodlike micelles coexist with globular mixed micelles. By fitting the neutron scattering data in the middle $Q$-range, it is possible to determine the size and number density of the mixed micelles in the coexistence region. The results of analyzing the small-angle neutron scattering data in the coexistence region show that some tributyrin molecules are solubilized by the rodlike diheptanoyl-PC micelles and only part of the added tributyrin form globular micelles. In the coexistence region, the size of these mixed micelles do not grow with increasing tributyrin concentration, which is in contrast to the results found in the simple mixed micelle region. Theoretical explanations are given to explain these findings.

Authors' address:

Dr. T. L. Lin
Dept. of Nuclear Engineering
National Tsing Hua University
Hsin-Chu
Taiwan 30043
ROC

# Hydrodynamic instabilities in microemulsions

M. A. López Quintela, A. Fernández Nóvoa and L. Liz

Grupo Biodinámica Física, Dpt. de Química Física, Universidad de Santiago, Santiago de Compostela, Spain

*Key words:* Microemulsions; synergetics; convection; chaos; nucleation

*Abstract:* Microemulsions constitute a very broad field to study the Rayleigh-Bènard problem because of their great variety of behaviours depending on the type of microemulsions.

To perform the study of hydrodynamic instabilities in microemulsions, we have deviced a new experimental setup [1] which avoids Marangoni effect because of its both rigid surfaces; it also allows both transversal and lateral observations, and it is possible the pattern observation both in the direct space and in the reciprocal one as well as performing spot light intensity measurements.

The theoretical treatment of the problem has been carried out by means of the algorithm developed by Swift and Hohenberg [2], based on Ginzburg-Landau equations:

$$\dot{\Psi} = \left[\varepsilon - (k_0^2 + \Delta)^2\right]\Psi(x,t) + \delta\Psi^2(x,t) - \Psi^3(x,t). \qquad (1)$$

By means of this algorithm the observation of a kind of nucleation [3] in the development of patterns in subcritical conditions detected in some experiments performed with microemulsions composed by Aerosol-OT, n-decane, water and KCl has been interpreted. The results have been compared with numerical calculations performed according to Eq. (1).

The appearance of chaos when the control parameter of the system, i.e., the distance from the critical Rayleigh number, achieves values large enough [3], observed in experiments with microemulsions has been interpreted by means of the delay coordinate method [4].

## References

1. López Quintela MA, Fernández Nóvoa A (1989) Rev Sci Instr (submitted)
2. Swift J, Hohenberg PC (1977) Phys Rev A 15:319
3. López Quintela MA, Liz L (1989) Phys Rev A (submitted)
4. Packard NH, Crutchfield JP, Farmer JD, Shaw RS (1980) Phys Rev Lett 45:712

Authors' address:

Prof. M. Arturo López Quintela
Departamento de Química Física
Universidad de Santiago
E-15706 Santiago de Compostela, Spain

**Progress in Colloid & Polymer Science**

Progr Colloid Polym Sci 81:271 (1990)

# The resolution of water-in-crude-oil emulsions by the addition of low molar mass demulsifiers

R. Aveyard, B. P. Binks, P. D. I. Fletcher and J. R. Lu

School of Chemistry, University of Hull, Hull, England

*Abstract:* We have measured the rates of demulsification of water-in-crude oil emulsions using a series of octylphenyl-polyethoxylates and sodium bis-2-ethylhexylsulphosuccinate (AOT) as demulsifiers [1]. The hydrophilic-lipophilic balance (HLB) of the surfactant systems were varied systematically by changing the number of ethoxy groups in the case of the nonionic surfactants and by changing the concentration of added NaCl in the case of AOT. For all surfactants and conditions, the demulsification rate increased with surfactant concentration up to the onset of surfactant aggregation in the oil, the water or a third surfactant-rich phase. The highest rate reached is estimated to be close to the diffusion controlled value. Higher concentrations of the surfactants produced either a decrease in demulsification rate ("overdosing") or a slight increase to a plateau value of the rate. Overdosing was observed for solutions where the surfactants formed aggregates in the oil phase ("hydrophobic surfactants"). "Hydrophilic surfactants" (for which aggre- gation occurred in the water phase) showed overdosing be- haviour only for cases where a high viscosity aqueous phase was produced. Higher concentrations of hydrophilic surfac- tants which gave low viscosity aqueous phases on demulsi- fication gave a plateau value in the demulsification rate.

## References

1. Aveyard R, Binks BP, Fletcher PDI, Lu JR, to appear in J Coll Interf Sci

Authors' address:

Dr. J. R. Lu
School of Chemistry, University of Hull
Hull, HU6 7RX, England

# Viscosity measurements in dense microemulsions, evidence of aggregation processes

D. Majolino[1]), F. Mallamace[1]), P. Migliardo[1]), N. Micali[2]) and C. Vasi[2])

[1]) Dipartimento di Fisica dell' Universita', Vill. S. Agata, Messina
[2]) Istituto di Tecniche Spettroscopiche del C.N.R., Messina

*Abstract:* The phase in which microemulsion is of the water in oil type may be viewed, as shown by the analysis of Small angle neutron scattering data, as a colloidal suspension. The pair potential presents an hard core repulsion plus an attractive (London-van der Waals) contribution. According to different theories and Molecular Dynamics experiments on dense packing of hard spheres and in the Lennard-Jones glass, light scattering results in AOT microemulsions at high volume fractions clearly shown that in this system large fractal structures are built up by a well defined aggregation process. Such an aggregation process for colloidal system is explained by the DLVO theory in terms of the interaction potential among two monomers. The pair potential results by a long range repulsive $V_R(r)$ plus an attractive short ranged van der Waals-London interaction $V_A(r)$. The explained by the DLVO theory in terms of the interaction with a barrier and its form depends on the particle size. When the particles are able to overcome the barrier, the primary minimum is reached giving rise to the aggregation.

Following such suggestions, that at high concentrations large clusters can be originates from the droplets aggregation, we have studied the viscosity as a function of the concentration at different temperatures in a AOT-water-oil microemulsion, particularly in the region of high concentration where the system shows a glass-like behaviour and a fractal structure. The obtained results are analyzed in terms of a recent two fluid model developed for colloidal systems in which the measured viscosity $\eta$ of the dispersion is taken as the sum of the solvent viscosity $\eta_0$ and the viscosity of interacting system of colloidal particles $\eta_1$, together with a coupling $\eta_2$ which contains all hydrodynamic interactions, namely $\eta = \eta_0 + \eta_1 + \eta_2$. In such a way we can give a detailed analysis of the behaviour of the system putting in evidence the role of the interparticle interactions. Therefore we consider two different contributions to the viscosity: one is directly connected to the repulsive interaction potential "hard core", and easily calculated, while the second one due to the attractive interaction, contributes indirectly through the formation of aggregates. This last short-ranged attractive contribution gives as a result a well defined maximum in the viscosity that is directly connected with the formation of fractal clusters among the spherical droplets, constituent the microemulsion. The results agree with those obtained previously by light scattering measurements, where we have observed for $T = 25\,°C$, in the concentration region in which the system shows the glass transition ($\phi_G \simeq 0.56$), the presence of large clusters with fractal structure. The measured fractal dimension $D \simeq 2.1$ gives the indication that the kinetic growth mechanism is reaction limited.

The experimental evidence that the aggregation phenomenon is thermally activated is given by calculating the probability $k$, that particles initially in the second minimum escape over the potential barrier (we have used the well know Chandrasekhar expression for colloids). The obtained value, for the potential barrier height, agrees with the one estimated for the interaction of two spherical particles in a colloidal solution.

Authors' address:

Prof. F. Mallamace
Dipartimento di Fisca dell' Universita'
Vill. S. Agata C.P. 55, 98166 Messina

# Use of reverse micelles for the extraction of proteins

G. Marcozzi[1]), M. Caselli[2]) and P. L. Luisi[3])

[1]) Eniricerche, S.p.A., Rome, Italy
[2]) Dipartimento di Chimica, Università degli Studi, Bari, Italy
[3]) Institut für Polymere, ETH-Zentrum, Zürich, Switzerland

*Abstract:* In recent years, the interest in the development of efficient methods for the extraction and purification of proteins and other bioproducts is growing up. Several research groups are applying the microemulsion w/o system: reverse micelles.

A reverse micelle is a surfactant-stabilized small water drop dispersed in organic solvent. It is able to solubilize, within its polar core, many different hydrophilic species, including proteins without exposing them to the organic environment. The protein extraction by reverse micelles can be made specific by controlling the water content, the type and concentration of surfactant, the type and concentration of salt and pH.

The complete purification procedure requires the optimization of two processes:

I° STEP — Liquid-liquid forward transfer for the solubilization of α-chymotrypsin (CT) and cytochrome $c$ (Cyt. $c$) by reverse micelles.

II° STEP = Liquid-liquid backward transfer for the recovery of a previously solubilized protein.

We have optimized protein selectivity extraction by reverse micelles using AOT (Bis 2-ethylhexyl sodium sulfosuccinate) as surfactant in function of: 1) shaking speed; 2) pH; 3) type and concentration of salt.

The protein solubilization occurs in a definite range of salt concentration, with a yield up to 98%. The protein yield by the liquid-liquid backward transfer is about 90%, with a residual activity of 70%.

We applied a new methodology for the backward transfer too: we carried out the transfer of protein from micellar solution by adding silica powder to the micellar phase (1:5 v/v). The absorption of water and protein on the solid phase is complete. The preliminary results using silica powder for the recovery of protein from reverse micelles are encouraging to solve the troublesome backward step. It is worth to note the complete retention of biological activity of the processed protein.

Authors' address:

Dr. G. Marcozzi
IfP, ETH-Zentrum
Universitätsstr. 6
CH-8092 Zürich

# Dynamic light scattering in middle phase microemulsions: Detection of the oil-in-water/bicontinuous transition

G. Marion, S. El Ahmadi, A. Graciaa and J. Lachaise

LTEMPM Université de Pau et des pays de l'Adour, Pau, France

*Abstract:* Dynamic light scattering is used to probe structural changes in microemulsions prepared with water, isooctane and polyethoxylated octylphenol as surfactant. The surfactant concentration is increased to obtain successively Winsor I and Winsor III systems.

In the microemulsions of the Winsor I systems the droplet size increases as the surfactant concentration does. A droplet concentration gradient is observed in the lower phase near the WI — WIII transition. This gradient affects the dynamical structure factor but has no influence on the effective diffusion coefficient $D_{eff}$.

At the beginning of the Winsor III zone, the lower phase is a diluted O/W microemulsion, the middle phase a concentrated O/W microemulsion and the upper phase a

**Progress in Colloid & Polymer Science**

Progr Colloid Polym Sci 81:274 (1990)

# Emulsion formation: A stochastic model to forecast the drop size distribution

B. Mendiboure, A. Graciaa, J. Lachaise, G. Marion and J. L. Salager*

LTEMPM, Université de Pau et des Pays de l'Adour, 64000 Pau, France
* also Lab FIRP, Universidad de Los Andes, Mérida 5101, Venezuela

*Abstract:* The drop size distribution of an emulsion is a characteristic of utmost importance, specially whenever surfactant adsorption and emulsion stability are concerned.

The purpose of the present model is to forecast the drop size distribution from the information that describes the emulsification process and the composition of the system.

On one hand the theory of isotropic turbulence is used to bracket the distribution range between feasible extreme diameters.

On the other hand a stochastic description of the emulsion is based on a number of states, each of them corresponding to a diameter interval, i.e., a size class. At each time step, a size transition is allowed for each drop, and the new distribution is computed. Physicochemical variables may be made to influence the result through the transition probability matrix elements. The time scaling is rendered by the number of steps required to reach a stationary distribution within the limits of accuracy.

Several alternate selections are investigated. A markovian model that features a simple up/down alternative, a time-independent matrix and linear variations of the probability through the range, is studied in details.

The distribution results are compared with experimental data obtained from light scattering sizing of heavy hydrocarbon O/W emulsions.

Authors' address:

Dr. B. Mendiboure
LTEMPM, Université de Pau et des Pays de l'Adour
64000 Pau, France

# Small-angle neutron scattering from polysaccharide gels

H. D. Middendorf[1]), F. Cavatorta[2]) and A. Deriu[2])

[1]) Department of Biochemistry, University of Edinburgh Medical School, and Clarendon Laboratory, University of Oxford, UK
[2]) Dipartimento di Fisica, Università di Parma, Italy

*Abstract:* Aqueous biopolymer gels consist of networks of hydrophilic fibres which provide a "molecular scaffolding" capable of retaining and structuring pure water or weak electrolytes (biochemical buffers) over a very wide concentration range. Such gels are increasingly being used to investigate fundamental aspects of biopolymer-water interactions in disordered media, to study biologically important polymerisation and transport processes, and to separate proteins or nucleic acids by electrophoresis.

We are using neutron diffractometers at the Institut Laue-Langevin (D11 and D17, steady state reactor) and at the Rutherford Appleton Laboratory (LOQ, pulsed source) to study the large-scale reticulate structure of polysaccharide gels (agarose) as a function of concentration, H/D contrast, and temperature. Our $S(Q)$ data on agarose gels extend over $2-3$ decades in both $Q$ and $C$, but a uniform coverage of this large parameter domain has not been achieved. We focus here on SANS data for fully $D_2O$-exchanged gels at $T = 293$ K, and on concentrations $C \leqslant 0.1$ ($C = g$ agarose/g water) which are of great interest in many applications (Fig. 1). At very low $Q$ ($< 0.005$ Å$^{-1}$, region I), we observe Guinier behaviour characteristic if a distribution of long rods only for $C \leqslant 0.01$; above this value the $\log(Q \cdot S(Q))$ vs. $Q^2$ curves slope upwards as $Q \to 0$ and there is not simple interpretation. In regions II ($0.005 \leqslant Q \leqslant 0.009$ Å$^{-1}$) and III ($0.009 \leqslant Q \leqslant 0.075$ Å$^{-1}$) we find extended linear segments in $\log S$ vs. $\log Q$ plots. It is then possible to derive certain scaling parameters which relate to the exponent $D$ in $S(Q) \propto Q^{-D}$ and reflect both the fractal properties of such gels and their inherent polydispersity. The $D$ values we have derived are close to 1.0 and almost concentration independent in region I while they markedly depend on concentration in region III, and decrease from about 2.0 to 1.1 as $C$ increases from 0.01 to 0.4. In region IV ($Q > 0.075$ Å$^{-1}$) there is a change in slope again, accompanied by intensity changes that reflect structural features with scale lengths below 100 Å.

These results relate closely to current work on the description of partially disordered, hierarchically organised systems in terms of fractal concepts and parameters characterising the polydispersity. It seems clear already from the limited $S(Q)$ data obtained so far that we are dealing here with three or four levels in the $Q$-dependence, each described by a scale length but "washed out" differently at different concentrations by the polydispersity inherent in a random network made up of bundles of helical rods connected through junction zones. A question of central interest addressed in this work is whether one can describe

**Progress in Colloid & Polymer Science**                    Progr Colloid Polym Sci 81:273 (1990)

pure oil phase, $D_{eff}$ takes the same value in the lower and middle phases; it remains constant until the complete solubilization by the middle phase of the lower phase droplets. The light scattered correlation function of the lower phase is characteristic of a brownian diffusor while that of the middle phase is representative of an adapted linked diffusor.

At higher concentrations, a genuine Winsor III system appears with a middle phase in equilibrium with two pure excess phases. $D_{eff}$ brokenly changes, reflecting the strong structural modification linked with the establishment of the bicontinuity in the medium. In accordance with the turbidity

decay and the geometrical feature of a Voronoi model for a bicontinous phase, the size of the typical domain is smaller than the droplet one. The main change in $D_{eff}$ may be due to the evolution of the static structure factor corresponding to the ordering of the interfacial network together with a slowing down of the velocity fluctuations.

Authors' address:

Dr. G. Marion
LTEMPM Université de Pau et des payes de l'Adour
64000 Pau, France

# The role of hydrocolloids in the stabilization of oil-in-water emulsions for enteral nutrition

G. Masson and F. Iseli

Nestlé Research Centre, Lausanne, Switzerland

*Abstract:* The influence of hydrocolloids addition on the stability of peptides and electrolytes containing oil-in-water emulsions for enteral feeding has been explored.

As shown in Fig. 1 B, steric stabilization, by formation of a complex structure including peptides and emulsifier molecules which coat the oil droplet surface, was proposed

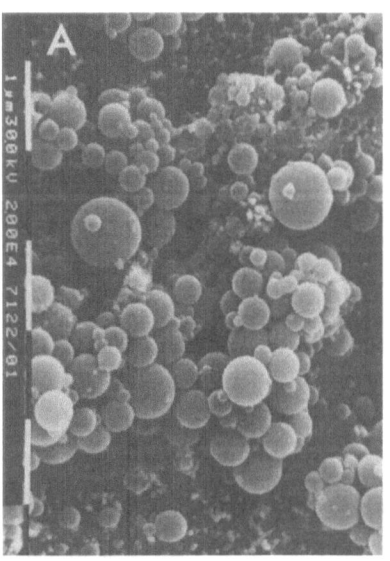 

Fig. 1. Scanning electron microphotographs of the oil-in-water emulsion (A), just before mixing with the colloidal solution containing peptides and electrolytes, and also before addition of polysaccharides (B) shows the oil droplets coated with hydrocolloids.

At low concentrations, a balanced mixture of polysaccharides was found to be an effective stabilizer of the complex solution.

It was observed that the presence of hydrocolloid molecules in the aqueous phase of the emulsions resulted in an improvement of their stability with respect to sedimentation, creaming and serum separation. Moreover, the addition of hydrocolloids did not increase greatly the continuous phase viscosity of the emulsions.

as the most probable mechanism by which hydrocolloids stabilize the emulsions.

Authors' address:

Dr. G. Masson
Nestlé Research Centre
Vers-chez-les-Blanc
CH-1000 Lausanne 26, Switzerland

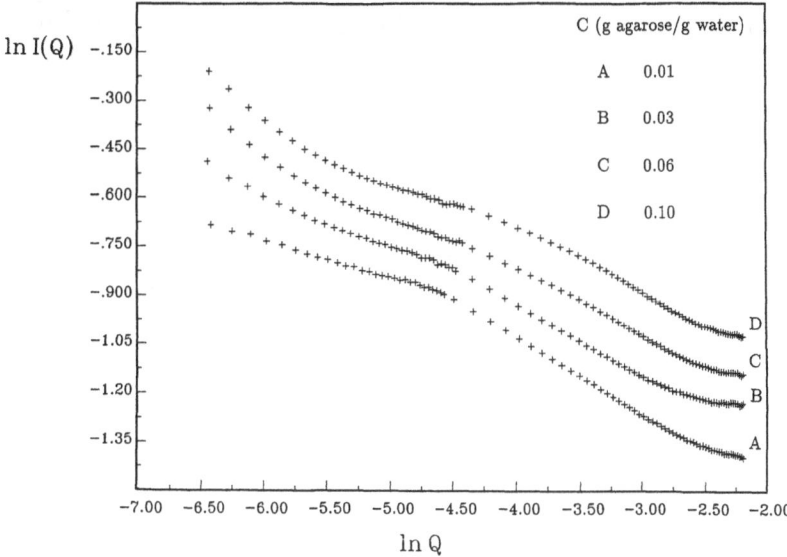

an agarose gel as a fractal network of interpenetrating structures within limits given by a smallest and a largest mesh size, or whether a (say) Gaussian distribution of mesh sizes in conjunction with a similar distribution of bundle cross-sections could equally well explain the $Q$ and $C$ dependent patterns observed.

Authors' address:

Prof. A. Deriu
Dip. di Fisica
Università di Parma
I-43100 Parma, Italy

# Liposomes as bioreactors: Transport/kinetic phenomena

G. Mossa[1]), M. C. Annesini[2]), A. Di Giulio[3]) and A. Finazzi-Agrò[4])

[1]) (GM) Istituto Medicina Sperimentale del CNR, Roma, Italy
[2]) (MCA) Dip. Ing. Chimica, Università La Sapienza, Roma, Italy
[3]) (ADG) Dip. Scienze Biomediche, Università L'Aquila, Italy
[4]) (AFA) Dip. Medicina Sper., Università Tor Vergata, Roma, Italy

*Abstract:* This study concerns the biochemical and physico-chemical behaviour of liposomes as enzymatic bioreactors. The aim is to study a quantitative treatment of transport/kinetic phenomena in these bioreactors. The experimental system studied is the enzyme ascorbate oxidase (AAO) entrapped into dipalmitoyl-phosphatidylcholine unilamellar vesicles. AAO is a plant enzyme catalyzing very efficiently *in vitro* the reaction of oxidation of ascorbate to dehydroascorbate.

The vesicles were prepared by two different methods, namely by controlled detergent dialysis technique or by reverse phase technique. Liposomes obtained by controlled dialysis were found to be more homogeneous when observed by negative staining in electron microscopy. The size of the vesicles was lower and the distribution profile was narrower for the enzyme loaded than for the empty liposomes.

AAO seems to be distributed both inside an on the outside of vesicles as demonstrated by immunochemical and kinetic methods. A kinetic model which considers the presence of AAO both on the surface and in the aqueous core of the liposomes has been developed. The model takes into account the effect of the limited permeability of the lipid bilayer on the overall kinetics of the process. Both the percentage of the enzyme contained within the vesicles and kinetic and diffusional parameters of the process were obtained by analyzing the kinetic experiments based on ascorbate oxidizing activity in solution and in liposomes treated with proteinase K and detergent.

Authors' address:

Dr. G. Mossa
Istituto Med. Sperimentale
2a Università di Roma
I-00173 Roma

**Progress in Colloid & Polymer Science**                    Progr Colloid Polym Sci 81:276 (1990)

# Chemiluminescence of 9-(p-methyl)-benzylidene-9,10-dihydro-10-methylacridine in CTAC micelles

J. Hadzianestis and J. Nikokavouras

National Research Center Democritos, Aghia Paraskevi Attikis, Greece

*Key words:* Chemiluminescence; lucigenin; micelles

*Abstract:* Working with the Lucigenin (L) light reaction we have shown [1] that chemiluminescence (CL) in micellar media results in increased quantum yields. This effect is further intensified in membrane mimetic agents such as the didodecyldimethylammonium bromide bilayer lamelar aggregates [2]. Furthermore, factors affecting the rigidity and fluidity of biological membranes such as cholesterol, steroid sex hormones, vitamins etc also affect the CL quantum yields in the above media [3]. In order to employ chemiluminescence as a tool in the study of organized systems we have sought to replace L by other chemiluminescent compounds lacking the dissadvantages of L which is cationic and very soluble in water. The title compound was synthesized and its CL was studied in cetyltrimethylammonium chloride (CTAC) micelles on reaction with singlet oxygen ($NaOCl + H_2O_2$) as a function of CTAC concentration. Below the CMC, the quantum yields were very low; then started increasing at the CMC to reach a plateau at *ca* $5 \times 10^{-3}$ M. The maximum quantum yield was $2 \times 10^{-4}$ Einstein $mol^{-1}$. The quantum yield was $7 \times 10^{-6}$ in methanol and $8 \times 10^{-6}$ in methanol-water. As with L, the excited product was N-methylacridone formed by the decomposition of an intermediate dioxetan. Autoxidation at higher surfactant concentrations gives rise to light emission also characteristic of the N-methylacridone fluorescence. The title compound is suitable for CL studies in organized systems.

## References

1. Paleos CM, Vassilopoulos G, Nikokavouras J (1982) J Photochem 13:327 – 334
2. Nikokavouras J, Vassilopoulos G, Paleos CM (1981) JCS Chem Commun 1082 – 1083
3. Varveri FS, Mantaka-Marketou AE, Vasilopoulos G, Nikokavouras J (1988) Mh Chem 119:703 – 710

Authors' address:

Prof. J. Nikokavouras
National Research Center "Democritos"
Aghia Paraskevi Attikis
15310, Greece

# Phase diagram of the water-$C_{12}E_5$ system: Extension and properties of the $L_3$ and dilute lamellar phases

R. Strey[1]), R. Schomaecker[1]), D. Roux[2]), F. Nallet[2]) and U. Olsson[2])*

[1]) Max Planck Institut für Biophysikalische Chemie, Göttingen, FRG
[2]) Centre de Recherche Paul Pascal, Domaine Universitaire, Talence, France

*Abstract:* We have studied the binary phase diagram water-$C_{12}E_5$. Of particular interest, are the $L_3$ and dilute lamellar phases, that were found to swell to approximately 99.5 and 98.8 wt% of water, respectively, much further than what has been previously reported. Focusing on these two phases, we present results from static light and small angle neutron scattering and electrical conductivity measurements. The experimental results will be discussed in the light of recent theoretical developments.

Authors' address:

Dr. U. Olsson
Centre de Recherche Paul Pascal
Domaine Universitaire
F-33405 Talence Cedex, France

*) On leave from Div. of Physical Chemistry 1, University of Lund, Sweden

**Progress in Colloid & Polymer Science**                    Progr Colloid Polym Sci 81:277 (1990)

# Investigation of the structure of reversed micelles in a AOT + $C_6(H,D)_6$ + $D_2O$ system by small-angle scattering (SANS)

N. I. Gorski and Yu. M. Ostanevich

Joint Institute for Nuclear Research, Moscow, USSR

*Abstract:* The small-angle neutron scattering (SANS) seems to be the most powerful method to get detailed information on the structure of micelles. We present the results of SANS investigations of reversed micelles in the system water + AOT + $C_6(H,D)_6$ where AOT is sodium di-2-ethylhexylsulfosuccinate. By isotopic contrast variation of the solvent we have determined the mean values of scattering-length density $\bar{\varrho}$ for dry micelles and also for those in the presence of water at the molar ratio of $X = [D_2O]/[AOT]$ 4.3 and 8.6. The aggregation numbers and the volume of the micelle — $V$ and of their ingredient parts (the volume of the water core $V_1$ and of the organic part $V_2$ ($V = V_1 + V_2$) are obtained. The specific volume per water molecule is estimated $V_{D_2O} = 87$ Å$^3$ for $X = 4.3$ and $V_{D_2O} = 48$ Å$^3$ for $X = 8.6$.

It was possible to estimate the radius of gyration $R_\infty$ and the $\alpha$-value of the micelle by Stuhrmann's method. The comparison of volumes, obtained from the absolute intensities of zero-angle scattering $(d\Sigma/d\Omega)_{q \to 0}$ and from the $R_\infty$-value, shows that some penetration of the solvent into the outer sphere of the micelle takes place.

Authors' address:

Dr. Yu. M. Ostanevich
Joint Institute for Nuclear Research
Head Post Office, P.O. Box 79
101000 Moscow, USSR

# Experimental survey of the dynamic of the electrical charges in a water-in-oil microemulsion

M. Paillette

Groupe de Physique des Solides Université Paris 7, Paris, France

*Key words:* Microemulsion; electrical conductivity; permittivity

*Abstract:* The low frequency (1 KHz) electrical conductivity $\kappa$ ($10^{-12} - 10^{-4}$ Sm$^{-1}$) and permittivity $\varepsilon$ of an B.H.D.C./water/benzene water-in-oil microemulsion have been measured, at room temperature, as a function of the molar water-to-surfactant ratio $W_0$ (ten series from 5 up to 39) and of the volume fraction $\phi$ ($10^{-4}$ up to percolation threshold $\phi \simeq 0.09$).

Preliminaries results of a detailed analysis are shown [1].

At very low values of $\phi$, the observed $\kappa \propto \phi^2$-type dependence seems indicate a dimer-type submicellar aggregation. In the linear domain ($\kappa \propto \phi$), we observe that the effective charge number $|Z|$ increases with increasing $W_0$. A new relation for $|Z|$ leads to a good fitness up to $W_0 = 18$. As $\phi$ tends towards $\phi_p$, a clustering threshold is revealed. In the early stage of clustering a $W_0$-dependent power law for volume fraction dependence of $\kappa(\kappa \propto \phi^n)$ is shown such as:

$n = 2$ for $W_0 \leqslant 11$; $3/2$ for $15 \leqslant W_0 \leqslant 21$
and $2/3$ for $W_0 \geqslant 25$.

As observed earlier in other systems, the conductivity, at constant surfactant concentration, as a function of $W_0$ exhibits a maximum for $W_0 = 10$ [2].

From the first crude results of the $\phi$-dependent permittivity values, we have tried to extend the model proposed by D. Bedeaux et al. in terms of the Clausius-Mossotti equation [3]. The nature of the system or the volume fraction measurement range raise some difficulties in the determination of the value of the polarizability $\alpha_p$ of a single microemulsion droplet. The function $I(\phi_p, T)$ giving the volume fraction- and temperature-dependent correction to Clausius-Mossotti exhibits a minimum around the values $18 - 20$ of $W_0$.

These detailed experimental approaches provide new insights allowing further improvements in the understanding of the behaviour of the $W_0$-dependent Kerr constant.

## References

1. Paillette M (in preparation)
2. Eicke HF, Denss A (1969) In: Mittal KL (ed) Solution Chem Surfactant 2, Plenum Press, New York, pp 699 – 706
3. Van Dijk MA, Joosten JGH, Levine YK, Bedeaux D (1989) J Phys Chem 93:2506 – 2512

Authors' address:

Dr. M. Paillette
Groupe de Physique des Solides
Université Paris 7
Tour 23, 4 ème étage, 2 place Jussieu
75251 Paris Cedex 05, France

**Progress in Colloid & Polymer Science**

Progr Colloid Polym Sci 81:278 (1990)

# Electrophoretic light scattering in ordered colloidal suspensions

M. Deggelmann[1]), T. Palberg[1]), P. Leiderer[1]), H. Versmold[2]) and R. Weber[1])

[1]) University of Konstanz, Konstanz, FRG
[2]) RWTH Aachen Inst. f. phys. Chem. II, Aachen, FRG

*Key words:* Latex spheres; electrophoresis; light scattering ordered suspensions

*Abstract:* Electrophoretic light scattering (ELS) in the presence of electroosmosis is performed on PS- and PMMA-latex spheres with several experimental setups and different cell geometries. Theoretical expressions for the electrophoretic spectra and correlation functions are presented, which can be extended to include spectral broadening mechanisms.

For non-interacting systems both theory and data from all ELS experiments are found to be consistent and to agree quantitatively with data from conventional microelectrophoresis [1]. For interacting systems changes in the characteristic flow profile are reported. Though in these systems the theory cannot account for the exact intensity distribution of the ELS spectra, it still allows an evaluation of electrophoretic mobilities via depth dependent measurements. The mobility is found to increase dramatically from $3-5 \times 10^{-8}$ $m^2 V^{-1} s^{-1}$ to a maximum value of over $12 \times 10^{-8} m^2 V^{-1} s^{-1}$ as the total ion concentration, monitored via conductivity, is decreased from $10^{-5}$ mol $l^{-1}$ to $6 \times 10^{-7}$ mol $l^{-1}$. The values seem to be independent of particle number density and structure of the suspensions within the experimental error of ca. 5% [2].

## References

1. Palberg T, Versmold H (1989) J Phys Chem 93:5296−5301
2. Deggelmann M, Palberg T, Weber R et al., J Coll Interface Sci, submitted

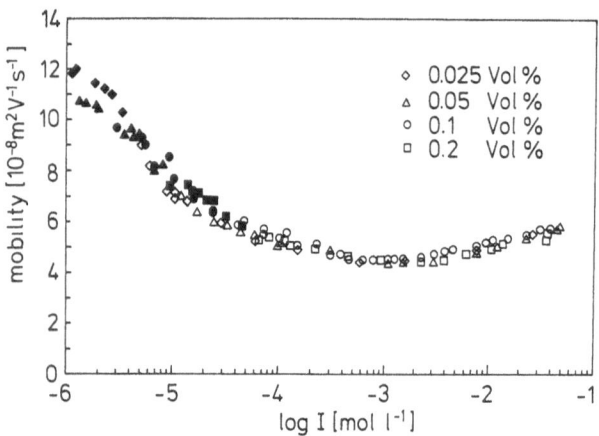

Fig. 1. Electrophoretic mobility versus total concentration of small ions for 100 nm PMMA-latex spheres at four different volume fractions. Filled symbols represent either fluid-like or crystalline ordered systems

Authors' address:

Dr. T. Palberg
Universität Konstanz
Postfach 5560
D-7750 Konstanz, FRG

**Progress in Colloid & Polymer Science**                    Progr Colloid Polym Sci 81:279 (1990)

# Small-angle neutron scattering from anisotropic micelles aligned by shear flow

J. Penfold[1]), E. Staples[2]) and P. G. Cummins[2])

[1]) Neutron Science Division, Rutherford Appleton Laboratory, Chilton, Didcot
[2]) Unilever Research, Port Sunlight Laboratory, Bebington, Wirral

*Abstract:* Shear flow is now an established technique for aligning anisotropically shaped micelles in small angle neutron scattering studies.

We report here some preliminary measurements on the interacting anisotropic micelles of the mixed cationic-nonionic surfactants cetyltrimethyl ammonium bromide $C_{16}TAB$, and hexaethylene glycol monohexdecyl ether, $C_{16}E_6$.

At low $C_{16}TAB$ concentrations ($< 0.01$ mol ratio), in the absence of shear the long micelles present in this system show no evidence of intermicellar interactions. On shearing the system an interaction peak is observed, the position of the peak is consistent with a rod length $\sim 4000$ Å, which compares favourably with the estimate (3700 Å) based on the coupling of an isolated rod with the shear field. At much higher $C_{16}TAB$ concentrations ($\sim 0.2$ mol ratio) the headgroup interaction prevents the formation of long rod species. The small anisotropy is reflected in the invariance of the interaction dominated curves with shear. The peak in the scattering curve can be identified with the mean interparticle spacing, with an associated micelle axial ratio of 2:1 at 30 °C and 3:1 at 48 °C.

At $C_{16}TAB$ concentrations intermediate between these two extremes the situation is more complex. A strong interaction peak is observed, and in the absence of shear, the peak position can still be identified with the mean interparticle spacing: that dimension is consistent with a mean rod length of 270 Å. This approach remains valid unless oriented domains of long rods exist. In this system we assume that there are domains containing relatively few small aligned micelles, such that a broad interaction peak results. In the presence of increasing shear, these domains grow reflecting an increase in the average micelle length and a subsequent increased definition of the interparticle spacing. The existence of such large domains negates the above simple approach for the determination of the micelle geometry, although the absence of scattering in the $q$ parallel direction confirms that micelle dimensions exceed several thousand angstroms.

Authors' address:

Dr. J. Penfold
Neutron Science Division
Rutherford Appleton Laboratory
Chilton, Didcot
Oxon OX11 0QX

**Progress in Colloid & Polymer Science**                                     Progr Colloid Polym Sci 81:280 (1990)

# The nature of adsorbed layers of nonionic surfactants on sol particles

J. Penfold[1]), E. Staples[2]) and P. G. Cummins[2])

[1]) Neutron Science Division, Rutherford Appleton Laboratory, Chilton, Didcot
[2]) Unilever Research, Port Sunlight Laboratory, Bebington, Wirral

*Abstract:* The nature of adsorbed layers of hexaethyleneglycol monododecyl ether ($C_{12}E_6$) on ludox silica sols has been determined by small angle neutron scattering.

The coated sols show critical opalescence, and the onset of attractive interactions have been observed for both concentrated and dilute sols.

The small angle neutron scattering studies have been carried out on the LOQ spectrometer at the ISIS pulsed neutron source (in the $Q$ range 0.002 to 0.2 $\text{Å}^{-1}$) and on the D11 spectrometer at the Institute Laue Langevin (in the $Q$ range 0.004 to 0.08 $\text{Å}^{-1}$). Both TM (diameter $\sim 300$ Å) and HS (diameter $\sim 150$ Å) ludox silica sols at a concentration $\sim 4\%$ have been studied in an aqueous solvent; variations in both pH and temperature have been studied. The scattering measurements have been carried out using a solvent scattering density matched to the silica (61% $D_2O$/ 39% $H_2O$): the scattering then arises only from the adsorbed layer, and maximises the sensitivity to the structure of the adsorbed layer.

The scattered intensity for a 4% TM sol at a pH of 8.5 has been analysed in detail for partial to saturation (in the adsorption isotherm) coverage of $C_{12}E_6$, and is reported in full elsewhere [1]. The interpretation of the data is consistent with the formation of small islands of surfactant bilayers on the surface, where the thickness of the adsorbed layer is $\sim 40$ Å. At saturation coverage only 70% of the adsorbed layer volume is occupied by surfactant.

On raising the temperature of the nonionic coated sols a marked increase in the solution turbidity is observed, reminiscent of the critical scattering observed in nonionic micellar solutions as the cloud point is approached [2].

Raising the solution temperature beyond the turbid region causes reversible phase separation, and two clearly defined regions of reduced turbidity are formed. Small angle scattering measurements on the two distinct phases, after

phase separation, show that there is no change in the shape of the scattering profile at high values of momentum transfer, $Q$: this indicates that there is no significant change in the nature of the adsorbed surfactant layer.

However, in the more dense phase the scattering at low $Q$ is suppressed due to excluded volume contributions: a marked increase in scattering is now observed as the clouding temperature of this dense phase is approached, indicating the onset of attractive interactions.

The onset of attractive interactions have been observed at lower sol concentrations when measurements have been made to lower $Q$ values. It has been shown that for the $C_{12}E_6$ coated HS sol (with a clouding temperature $\sim 47\,°C$ at saturation $C_{12}E_6$ coverage) at $25\,°C$ the interactions are repulsive, at $35\,°C$ they are hardsphere and at $45\,°C$ they can be described by an attractive soft potential [3].

Such dispersions provide a suitable model system which is "geometry invariant" for the investigation of critical phenomena in nonionic surfactants.

## References

1. Penfold J, Cummins PG, Staples E, J Phys Chem (in press)
2. Zulauf M, Weckstrom K, Hayter JB, Degiorgio V, Corti M (1985) J Phys Chem 89:3411
3. Hayter JB, Zulauf M (1982) Colloid Polymer Sci 260:1023

Authors' address:

Dr. J. Penfold
Neutron Science Division
Rutherford Appleton Laboratory
Chilton, Didcot
Oxon, OX11 0QX

**Progress in Colloid & Polymer Science**                                    Progr Colloid Polym Sci 81:282 (1990)

# Time-correlated photon counting and self-diffusion in concentrated microemulsions

J. Ricka[1]), M. Borkovec[2,3]), U. Hofmeier[2]) and H.-F. Eicke[2])

[1]) Institute of Appl. Physics, University of Berne, Bern, Switzerland
[2]) Inst. of Phys. Chemistry, Univ. of Basel, Basel, Switzerland
[3]) Present address: Institute of Food Sciences, ETH-Zentrum, Zürich, Switzerland

*Abstract:* In a typical water-in-oil microemulsion, such as the system water-AOT-hexane, water is dispersed in the form of small nearly spherical droplets coated with surfactant. This structure accounts for many interesting properties of the system and manifests itself also in the results of time correlated photon counting techniques such as quasielastic light scattering (QLS) and fluorescence correlation (FC). It offers, for example, the possibility to measure *pure* self diffusion in concentrated microemulsions by QLS, despite the relatively low accessible *q*-range. This fact may appear

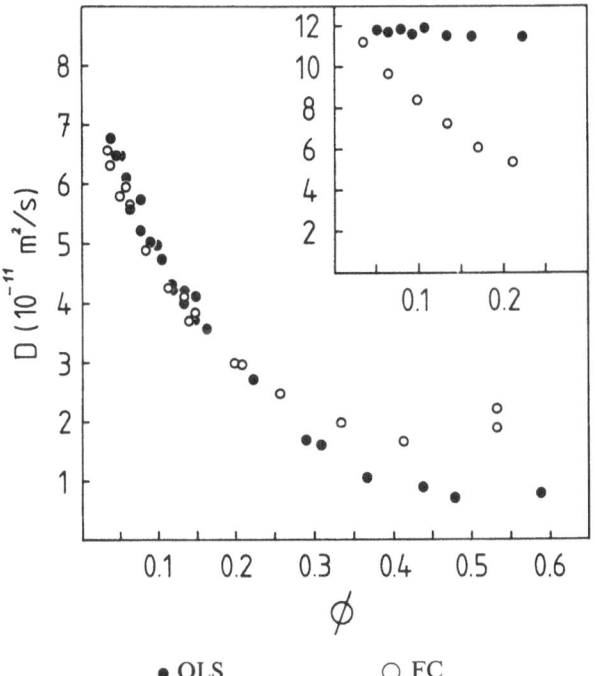

● QLS          ○ FC

somewhat surprising — usually QLS measures a collective diffusion or a mixture of the self and the collective modes which are difficult to separate. The measurability of pure self diffusion arises, however, quite straightforwardly from two features of microemulsion droplets:

- Firstly, the droplets are optically inhomogeneous, i.e. their polarizability does not obey the simple $r^3$ law. In

fact the scattering amplitude of an individual droplet may be positive or negative depending on its radius. Consequently, one finds a particular microemulsion composition where the average polarizability $\langle \alpha \rangle$ and also the refractive index increment of the suspension vanishes — the optical match condition.

- Secondly, the droplet radius is slightly polydisperse. Therefore, the average of the squared polarizability $\langle \alpha^2 \rangle$, i.e. the average scattering power of the droplets, always remains finite. A microemulsion scatters even at the optical matching.

The self-contribution to the dynamic structure factor is proportional to $\langle \alpha^2 \rangle$ whereas the distinct contribution is proportional to $\langle \alpha \rangle^2$ — at the optical matching one measures therefore a pure self term. We demonstrate this fact by comparing the QLS measurements with the self diffusion measurements by FC. With the latter technique self diffusivity is obtained from the correlation function of intensity fluctuations caused by motion of fluorescent particles through the profile of a strongly focused laser beam. In the case of microemulsions the necessary labelling is particularly simple; it consists of adding a small amount of a fluorescent dye which is soluble only in the droplet phase of the microemulsion. With both techniques we are able to measure the self diffusivity in a range of volume fractions $\Phi$ between 0.03 and 0.6. The data at optical matching shown in the main frame of the figure are in excellent agreement up to the volume fraction of 0.3. (For comparison data obtained far from optical matching are shown in the inset.) The linear coefficient $k_1$ in the expansion $D(\Phi) = D_0(1 + k_1\Phi + ...)$ amounts to $-5 \pm 1$; the $\Phi$ dependence is steeper than predicted for a hard sphere model.

## References

Ricka J, Borkovec M, Hofmeier U, Eicke H-F (1990) Europhys Lett 11:379–385

Authors' address:

Dr. J. Ricka
Institut für angew. Physik
Sidlerstr. 5
CH-3012 Bern, Switzerland

# Structure of apolar gels at the sol/gel transition point

C. Petit[1]), T. Zemb[2]) and M. P. Pileni[1,2])

[1]) Universite P. et M. Curie Laboratoire S.R I., Paris, France
[2]) D.Ph.G.-S.C.M., Gif sur Yvette, France

*Abstract:* The formation of apolar optically clear gels has been shown by the addition of gelatin to reverse micelles [1]. Different model has been proposed to explain the structure [1, 3, 4]. For a gel close to the saturation point (situated at the common limit of the four types of macroscopic behaviour observed for this microemulsion: sol, gel, precipitate or demixion: $[AOT] = 0.15$ mol $\times$ $1^{-1}$, $W = 30$, $[H_2O] = 4.5$ mol $\times$ $1^{-1}$ and 10% (W/V) gelatin), a complete structural study has been carried out.

densities are known. A good agreement between this model and the experiment is obtained. This model is also consistent with one of the structure proposed previously [4]. Each AOT micelle which contains a part (30%) of the gelatin molecules and water is connected via one gelatin strand to a neighbouring micelle, ensuring that the extent of the connections is large enough to induce a macroscopic gelation of the solution, since the percolation threshold is 1.1 cylinder per sphere [5]. Then it is also possible to

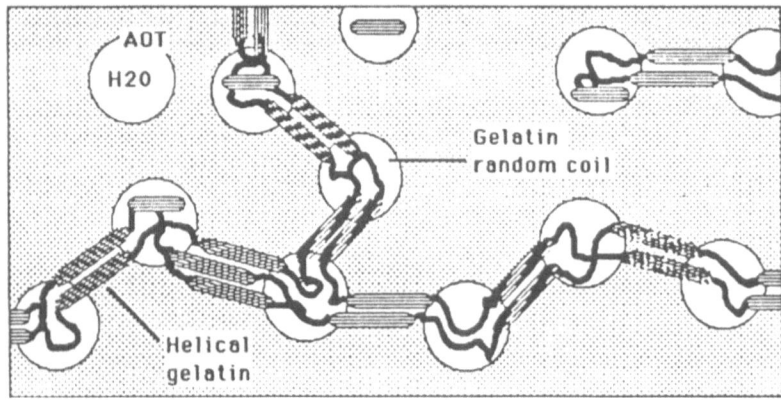

Fig. 1. Our model of micelles connected by strands of gelatin

The total specific interface (water/AOT) was found to be $5.6 \times 10^{-3}$ $Å^2/Å^3$ from S.A.N.S. This corresponds to a surface per polar head of AOT of $55 \pm 5$ $Å^2$. This is in good agreement with the data obtained in AOT reverse micelles [2]. If the gelatin were totally covered by AOT and water the total surface according to the gelatine concentration would be equal to $3.4 \times 10^{-2}$ $Å^2/Å^3$ which is five time greater than that obtained. SANS experiments indicate the presence of droplets with a radius close to 60 Å which is greater than that obtained in empty droplets corresponding to 30% of gelatin in the water pool.

S.A.X.S. shows the presence of independent 25 Å radius cylinders. This is not observed in AOT reverse micelles in the absence of gelatin.

To understand the difference between the data obtained by SANS and SAXS, a model is proposed in which spheres are connected by cylinders: assuming that 70% of the gelatin is implicated in the 25 Å radius cylinders, we can deduce the total length cylinders which is found to be the sum of the length needed to connect once each micelles to one of the neighbours.

This model is simulated on an absolute scale without any parameter, since molecular volumes and scattering length

simulate, with our model, the experimental phase diagram, a good agreement is obtained with the experimental data.

## References

1. Haering G, Luisi PL (1986) J Phys Chem 90:5982
2. Pileni MP, Zemb T, Petit C (1985) Chem Phys Lett 118:414
3. Capitani D, Segre AL, Haering G, Luisi PL (1988) J Phys Chem 92:3500
4. Quellet C, Eicke HF (1986) Chimia 40:22
5. Barnes IS, Hyde ST, Ninham BW, Derian PJ, Drifford M, Zemb T (1988) J Phys Chem 92:2286

Authors' address:

Dr. C. Petit
Laboratoire S.R.I.
11 Rue P. et M. Curie
75005 Paris, France

**Progress in Colloid & Polymer Science**                    Progr Colloid Polym Sci 81:283 (1990)

# Solution structures of catanionic surfactant systems

A. S. Sadaghiani and A. Khan

Physical Chemistry 1, Chemical Center, University of Lund, Sweden

*Abstract:* Catanionic surfactants are uncharged and composed of equimolar quantities of both anionic and cationic surfactants. Like lecithins, this type of surfactants are insoluble in water but swell in water yielding lamellar liquid crystalline phase [1]. In the ternary system containing an oil there appears an isotropic solution phase (microemulsion) which extends from the oil axis to almost near the surfactant axis, but the phase has limited capability to incorporate water. Addition of few percent of an ionic surfactant, the isotropic solution phase extends substantialy to the oil-water axis [2]. The structures of the microemulsion systems are studied by the multicomponent pulsed gradient spin echo FT NMR technique and the experimental data are analysed with the available theoretical models. The preliminary results indicate the formation of water droplets with very high oil content, but a clear cut droplet structure appears to be absent when the water content in the system is increased. Some results obtained in isotropic solution phases for the system catanionic surfactant-water-alcohol will also be presented.

## References

1. Jokela P, Jönsson B, Khan A (1987) J Phys Chem 91:3291
2. Jokela P, Jönsson B, to be published

Authors' address:

Dr. A. S. Sadaghiani
Physical Chemistry 1, Chemical Center
University of Lund, Sweden

# Phase behaviour and viscoelastic properties of lecithin reverse micelles

R. Scartazzini, P. Schurtenberger and P. L. Luisi

Institut für Polymere, ETH Zentrum, Zürich, Switzerland

*Abstract:* Reverse micellar solutions of lecithin in a variety of organic solvents can be transformed into a gel-like, viscoelastic solution by the addition of a small amount of water [1].

Upon the addition of water, the viscosity increases dramatically to values as high as $10^4$ poise and the viscosity passes through a distinct maximum within a narrow range of added water. Upon further addition of water, the viscosity decreases sharply and phase separation into two macroscopically separated, optically transparent phases can be observed.

Polarization microscopy, NMR ($^{31}$P, $^2$H), small-angle X-ray (SAXS) and neutron (SANS) scattering indicate the absence of liquid crystalline order and support the presence of a transient network formed by entangled cylindrical reverse micelles.

We have characterized the rheological properties of the viscoelastic solutions. Measurements of zero shear viscosity ($\eta_s$), elastic modulus ($G$) and stress relaxation time ($\tau$) were carried out as a function of lecithin concentration and temperature at different water to lecithin molar ratios ($w_0$).

The results were compared with recent theoretical and experimental work on rheological properties of semi-dilute polymer solutions and viscoelastic ionic micellar solution [2,3]. The high-frequency elastic modulus, measured at a time scale much shorter than the lifetime of the transient network, shows a power law dependence upon the concentration analogous to polymer solutions, and it is weakly dependent on $w_0$ and temperature. However, $\eta_s$ and $\tau$ depend strongly on temperature and $w_0$. The concentration dependence of $\eta_s$ and $\tau$ differs remarkably from the behaviour found in polymer solutions and viscoelastic ionic micellar solutions.

Our results are interpreted using a recent theoretical model for the dynamics of living polymers by taking into account the finite lifetime of micellar aggregates [4, 5].

Drugs, enzymes and other guest molecules can be solubilized in this gel-like system. Therefore, they are potentially useful for applications in pharmaceutical, medical and chemical formulations.

## References

1. Scartazzini R, Luisi PL (1988) J Phys Chem 92:829
2. Candau SJ, Hirsch E, Zana R, Adam M (1988) J Colloid Interface Sci 122:430
3. Rehage H, Hoffmann H (1988) J Phys Chem 92:4714
4. Cates ME (1987) Macromolecules 20:2289
5. Cates ME (1988) J Phys France 49:1593

Authors' address:

Dr. R. Scartazzini
Institut für Polymere, ETH Zentrum
8092 Zürich, Switzerland

**Progress in Colloid & Polymer Science**                                            Progr Colloid Polym Sci 81:284 (1990)

# $^1$H magnetic resonance relaxation study — low-resolution technique — of water-in-hexadecane microemulsions

L. Lendinara[1]), D. Senatra[2]) and M. G. Giri[1])

[1]) Department of Physics, University of Bologna, Bologna, Italy
[2]) Department of Physics, University of Florence, Florence, Italy

*Key words:* Microemulsions; $^1$H NMR; low resolution NMR; free water; DSC: Differential scanning calorimetry

*Abstract:* A W/Hexadecane microemulsion, with K-Oleate and n-Hexanol as surface active components, was studied by $^1$H NMR technique. Investigations with Differential Scanning Calorimetry (DSC) carried out against increasing water concentration ($C$, mass fraction) had shown that in the above microemulsion several forms of water coexist, namely: bound, interphasal and free, distinguished by the melting temperatures of 233 K, 263 K and 273 K [1, 2]. Purpose of the research was to establish whether the water configurations detected by DSC-endothermal analysis, could give rise to different or characteristic $^1$H NMR relaxation processes.

NMR measurements were performed with a Minispec P 20 instrument (Bruker), equipped with a system for control and data acquisition (Stelar) and a thermostat. The spin-lattice relaxation curves were detected by the inversion recovery sequence and the spin-spin ones by the Carr-Purcell-Meiboom-Gill sequence [3]. Both $H_2O$/oil and $D_2O$/oil microemulsions were tested at 310 K and 281 K against $C$ in the interval $0.03 < C < 0.35$. Single exponential functions did not adequately describe the relaxation curves in all the investigated situations. The bi- and triple exponentiality of the spin-lattice and spin-spin relaxation

curves was analyzed by means of iterative procedures supplied by Minuit, CERN Library. A differentiation between the degrees of hydration of the microemulsion was observed; the main change occurring as $C \geqslant 0.2$, that is in samples exhibiting a free water DSC endotherm at $T = 273$ K.

## References

1. Senatra D, Cabrielli G, Guarini GGT (1987) Europhys Lett 2:455–463
2. Senatra D, Guarini GGT, Gabrielli G, Zoppi M (1984) J Physique 45:1159–1174
3. Farrar TC, Becker E (1971) Pulse and Fourier transform NMR, Academic Press, New York

Authors' address:

Prof. D. Senatra
Department of Physics
University of Florence
Largo Enrico Fermi, 2 (Arcetri)
50125 Florence, Italy

# Influence of the presence of non-adsorbing polymer on the crystallization phenomena of colloidal dispersions

C. Smits, J. K. G. Dhont and H. N. W. Lekkerkerker

Van't Hoff Laboratory, Utrecht, The Netherlands

*Abstract:* In colloidal systems the transition from a liquid-like structure to a crystalline arrangement of the particles can be influenced by the addition of non-adsorbing polymer [1–3]. In this study [4] experiments have been performed on silica spheres, coated with stearylalcohol, and non-adsorbing polystyrene molecules in cyclohexane at the theta-temperature (34.5 °C). The silica particles have a radius of $160 \pm 10$ nm and their volume fraction ranges from 25 to 55 per cent. The weight averaged molecular weight of polystyrene is 100000 and its concentration is varied from 0 to about 4 g/dm$^3$.

Without added polymer crystallization takes a very long time after the required concentration has been reached by sedimentation under gravity. Large crystallites appear on top of the sediment after a period ranging from one to three months.

The addition of non-adsorbing polymer has a pronounced effect on the crystallization phenomena. At relatively low polymer concentration (e.g. $0 - 2.5$ g/dm$^3$) the addition of polymer does not have much effect. The particles sediment due to gravity and after about one month large crystallites appear on top of the sediment. At somewhat higher polymer concentrations (e.g. $2.5 - 3$ g/dm$^3$) a slow phase separation takes place and small crystallites can be observed in the entire lower phase after about two days. The crystals and particles sediment afterwards due to gravity. At relatively high polymer concentration (e.g. $3 - 4$ g/dm$^3$) a fast phase separation occurs and small crystallites can be observed on top of the lower phase after about six hours. The experiments clearly show that at definite combinations of silica and polystyrene concentration crystallization throughout the entire lower phase can be observed and that the rate of crystallization increases dramatically.

At present we can only speculate on the mechanism for these effects. The addition of polymer causes an effective attractive depletion interaction between the Brownian particles, which may lead to a spinodal-like decomposition, driving small clusters of particles into the crystalline state. Clearly experiments (e.g. light scattering) are needed to analyse the mechanism further.

## References

1. McPherson A (1976) J Biol Chem 251:6300
2. Kose A, Hachisu S (1976) J Colloid Interface Sci 55:487
3. Sperry PR (1984) J Colloid Interface Sci 99:97
4. Smits C, Briels WJ, Van Duijneveldt JS, Dhont JKG, Lekkerkerker HNW, Accepted for publication in Phase Transitions.

Authors' address:

Dr. C. Smits
Van't Hoff Laboratory
Padualaan 8
3584 CH Utrecht, The Netherlands

# Adsorption of model 4-alkylphenylamines at aqueous/organic interfaces and the rate of palladium (II) extraction

M. Wiśniewski and J. Szymanowski

Poznań Technical University, Institute of Chemical Technology and Engineering, Poznań, Poland

*Abstract:* Pure individual 4-alkylphenylamines containing from 6 to 16 carbon atoms in their straight alkyl were obtained and used to extract palladium (II) from 3 N HCl solutions. Interfacial tension isotherms were determined for various extraction systems, in which heptane, toluene and benzene were used as diluents. The adsorption parameters were determined using the Gibbs, Szyszkowski, Frumkin and Temkin isotherms, the polynomial and the spline function. The partition coefficients of 4-alkylphenylamines in extraction systems and their solubilities in 3 N HCl solutions were also determined.

The considered adsorption isotherms match quite well the experimental interfacial tension data. The surface excess isotherms computed according to the Gibbs isotherm, the polynomial and the spline function exhibit apparent maxima for amines containing $8 - 14$ carbon atoms in the alkyl. They are not observed for compounds soluble well in the aqueous phase and toluene, i.e. for 4-hexylphenylamine and 4-hexadecylphenylamine.

The adsorption/regression coefficients exhibit appropriate maxima or minima for 4-alkylphenylamines containing $10 - 14$ carbon atoms in the alkyl.

The maximum rate of palladium (II) extraction is observed for 4-decyl- and 4-dodecylphenylamine characterized by their best adsorption at the interface. Compounds well soluble in the aqueous phase (4-hexylphenylamine) or in toluene (4-hexadecylphenylamine) extract palladium (II) slower than 4-decyl- and 4-dodecylphenylamines.

Authors' address:

Prof. J. Szymanowski
Poznań Technical University
Pl. Sklodowskiej-Curie 2
Po-60965 Poznań, Poland

**Progress in Colloid & Polymer Science**

Progr Colloid Polym Sci 81:286 (1990)

# Oxidation of urushiol homologues by *Rhus* laccase in microemulsions

R. Oshima[2]), M. Takada, S. Tsuchiya[2]), T. Miyakoshi[3]), M. Seno[2]) and G. Ebert[1])

[1]) Institut für Polymere der Universität Marburg, Marburg/L., FRG
[2]) Institute of Industrial Science, University of Tokyo, Tokyo, Japan
[3]) Faculty of Engineering, Meiji University, Kawasaki, Japan

*Abstract:* A sap of the lacquer tree (*Rhus vernicifera*) consists of urushiol (60−70 wt%), water (20−25%), plant gum (7%), water-insoluble glycoprotein (2%) and copper glycoproteins (laccase and stellacyanin) (less than 1%), and as a whole the sap is in a fine w/o type emulsion, where laccase is present in an aqueous phase, and urushiol constitutes an oily phase. The initial oxidation may occur at the interface between the two phases, yielding urushiol-quinone which is the important intermediate of drying process of Japanese lacquer. However, no kinetic work has been explored on the laccase-mediated oxidation of urushiol, a natural substrate of laccase. Nonetheless, it seems very important to find the evidence of formation of the quinone and to clarify the mechanism of oxidation catalyzed by laccase.

In this work we describe the water-insoluble 3-alkylcatechols, which are analogous to urushiol, are oxidized by *Rhus* laccase in o/w microemulsion using hexadecyltrimethylammonium chloride (HTAC) as a surfactant at various pH to give corresponding quinones. Kinetic parameters are almost the same as those of the reaction of water-soluble substrates with laccase in the aqueous system, indicating the rationality of the use of the microemulsion. The second-order rate constants are dependent on the properties of the microemulsion and the size of substrates. For 3-methylcatechol the rate constants obtained in microemulsions resemble to that in an aqueous medium. It is assumed that 3-alkylcatechol is incorporated into microemulsions and interactions between the hydrophobic part of the substrate and HTAC and between the aromatic part of it and micellar head groups of HTAC contribute to the solubilization sites of the substrate in microemulsions. The ES-complex is formed between the substrate incorporated into microemulsions and the enzyme. The interactions between the substrate and surfactants and between the substrate and laccase are responsible for the stabilization of the ES-complex.

Authors' address:

Dr. M. Takada
Institut für Polymere der Universität Marburg
D-3550 Marburg/L., FRG

# Surfactant aggregation in hydrocarbons: solvent-induced structural variations in physical steroid gels

P. Terech

Institut Laue-Langevin, Grenoble, France

*Abstract:* We study the influence of the solvent type upon the surfactant aggregation in apolar media. Dipolar interactions between polar heads of the surfactant molecules are the driving force of the aggregation process in reversed configurations obtained in non-aqueous media. In various cases, the influence of the solvent is not restricted to the solvent polarisibility or solubility parameter but should include some specific sterical term.

We are interested here by the aggregation of a small steroid derivative in hydrocarbons. When put in hydrocarbons, the steroid molecules aggregate to give very long rod-like fibres of colloidal dimensions (100 Å diameter and up to several microns length). At a critical steroid concentration, the fibres are entangled and give thermoreversible gel phases, while for higher concentrations lyotropic phases are obtained.

We study the changes in both the microscopic structures and the macroscopic behaviors for the related steroid gels in various solvents from cyclohexane to decalin. On the one hand, small angle scattering experiments (neutron and X-rays) are used to detect a significative increase of the monodispersity of the rod-like aggregates and their related cross-section diameter (from 100 to 150 Å) from cyclohexane to ethyl-cyclohexane or decalin. Absolute intensity measurements, radii of gyration determination, model fittings of the full scattering curves and calculation of the distance distribution function by the so-called Hankel transformation yield a coherent picture of the structural modifications. On the other hand, the gel phase behaviors are evaluated by optical polarizing micrographs and bidimensional intensity contours for neutron scattering.

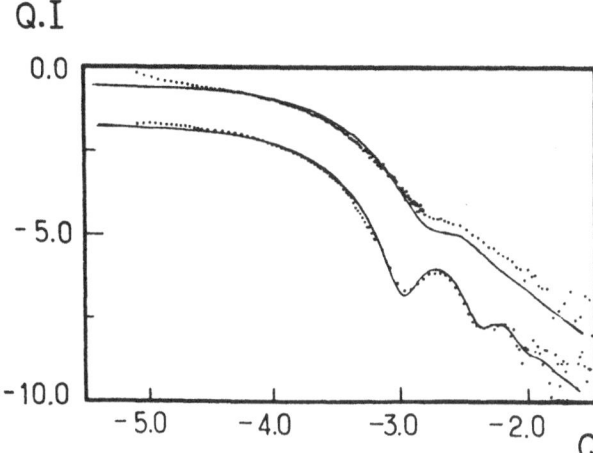

Fig. 1. X-ray small angle scattered intensity for cross-section of the steroid fibres. Upper curve: cyclohexane; Lower curve: ethyl-cyclohexane. Experimental data and simulations.

The results argue for an increase of the rigidity of the fibres from cyclohexane to solvents with only minor differences of their sterical volume. The microscopic observations are correlated to the macroscopic behaviors observed. A conformational correlation between solvent and solute fibres could be found out as for some polymeric physical gels in organic media.

### References

Terech P (1989) J Phys France 50:1967

Author's address:

Dr. P. Terech
Institut Laue-Langevin
156X
38042 Grenoble Cedex France

# Temperature dependence of solubility and the liquid crystal formation in aqueous solutions of metal-dodecylbenzenesulfonates

D. Tezak, F. Strajnar and I. Fischer-Palkovic

Department of Chemistry, Faculty of Science, University of Zagreb, Zagreb, Yugoslavia

*Abstract:* Thermochemistry of magnesium dodecylbenzenesulfonate precipitation, solubility conditions and formation of liquid crystalline phases were investigated in aqueous solutions between 5 and 100°C.

The calculated solubility products as well as thermodynamic parameters will be discussed in terms of predomination of particular ionic species in the solution depending on the precipitation conditions.

Authors' address:

Prof. D. Tezak
Department of Chemistry, Faculty of Science
University of Zagreb
P.O. Box 163
41001 Zagreb, Yugoslavia

**Progress in Colloid & Polymer Science**                    Progr Colloid Polym Sci 81:288 (1990)

# Mixed micelle formation and precipitation in mixture of anionic and cationic surfactants

N. Filipović-Vinceković and V. Tomašić

"Ruder Bošković" Institute, Department of Physical Chemistry, Laboratory of Radiochemistry, Zagreb, Yugoslavia

*Abstract:* Studies on anionic/cationic surfactant systems revealed formation of mixed micelles whose composition, size and shape depend on the total concentration and molar ratio of surfactants. At least three different concentration regions can be differentiated in the range of mixed micelles formation, concentration regions with anionic or cationic surfactant in the excess and equimolar concentration region. Mixed micelles formed in systems with anionic surfactant in the excess bound cations while those formed in systems with cationic surfactant in the excess bound anions as counterions in mixed micelle/solution interface. The equimolar concentration range is terminated by precipitation of almost uncharged particles.

The association equilibria in sodium dodecyl sulphate/ dodecyl ammonium chloride solutions were examined by measuring conductances and solubilities at different molar ratios of surfactants. The conductivity curves showed several changes in the slopes indicating structural changes in mixed micelles. These changes point to the existence of several "critical" concentrations. The precipitation diagram obtained provides evidence of the equilibrium in solution.

Authors' address:

Dr. V. Tomašić
"Ruder Bušković" Institute,
Department of Physical Chemistry,
Laboratory of Radiochemistry
Bijenička 54
41000 Zagreb, Yugoslavia

# Effect of linear and micelle-like polyelectrolytes on the kinetics of reduction of dialkyl-viologens by dithionite

C. Tondre, H. S. Kim and B. Claude

Laboratoire d'Etude des Solutions Organiques et Colloïdales (L.E.S.O.C.), Unité Associée au C.N.R.S. n° 406, Université de Nancy I, Vandoeuvre-lès-Nancy, France

*Abstract:* Polyelectrolytes and micellar systems have proved to be useful to retard the back electron transfer reaction in different photoredox processes. The efficiency of the charge separation appears to depend on a subtle balance between electrostatic and hydrophobic interactions. Whereas most experiments in this field rest on laser flash photolysis techniques, a novel approach has been used in this work which was aimed at clarifying the respective contributions of electrostatic and hydrophobic forces.

Using the stopped-flow technique, the electron transfer between sodium dithionite (a reducing agent widely used by biochemists) and a series of dialkyl-viologens (4,4'-dialkyl bipyridium) has been used here as a model reaction to investigate the effect of electrically charged polymers. Two polyelectrolytes widely differing by their global structure have been compared, one of them being characterized by its linear shape (polystyrene sulfonic acid, PSSA) and the other

one by its micelle-like structure (alternated copolymer of maleic acid and cetylvinylether, MA-CVE).

The electron transfer reactions leading to the cation radicals were in agreement with pseudo-first order kinetics from which apparent formation rate constants $k_{app}$ have been deduced. Both polyelectrolytes have been found to have retarding effects but their effectiveness is quite different, depending on the alkyl-chains carbon number of the viologens (see Fig.). Whereas the dimethyl compound was found to be the more efficient in the case of PSSA, it has appeared to be the less efficient with MA-CVE. For PSSA the retarding effect is only weakly affected by the change of hydrophobicity of the viologen and the rate increases with the alkyl-chain carbon number as in pure water. For MA-CVE, the situation is completely different since the rate of the electron transfer reaction is strongly decreasing when the hydrophobic character of the viologen molecules increases.

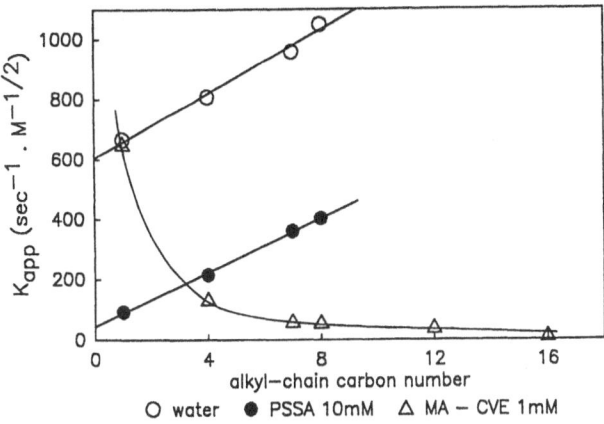

The reaction could be slown down as much as 45 times comparatively to that obtained with the dimethyl-viologen in pure water, when using the di-$C_{16}$ derivative.

It can be concluded from this preliminary report that the interactions between dialkyl-viologens and polyelectrolytes are mainly of electrostatic nature in the case of PSSA and of hydrophobic nature in the case of MA-CVE. The latter leading to a much more efficient charge separation for the model electron transfer reaction considered here, when a highly hydrophobic di-cation was involved.

Authors' address:

C. Tondre
Laboratoire d'Etude des Solutions Organiques et
Colloïdales (L.E.S.O.C.)
Faculté des Sciences
Université de Nancy I
Boîte Postale 239
F-54506 Vandoeuvre-lès-Nancy Cedex, France

# Structural investigation of liquid crystalline phases of ternary surfactant systems

S. Radiman[1], C. Toprakcioglu[1,2], J. O. Raedler[1], A. de Vallera and R. P. Hjelm Jr.[4]

[1]) Cavendish Laboratory, Cambridge, UK
[2]) Institute of Food Research (AFRC), Colney Lane, Norwich, UK
[3]) Department of Physics, University of Lisbon, Portugal
[4]) LANSCE, Los Alamos National Laboratory, Los Alamos, New Mexico, USA

*Key words:* Cubic phases; constant mean curvature surfaces; DDAB

*Abstract:* We have studied the isotropic cubic liquid crystalline phases formed by the surfactant didodecyl dimethyl ammonium bromide (DDAB) when mixed with water and octane, using the techniques of small-angle neutron scattering (SANS), freeze-fracture electron microscopy and rheology. Such liquid crystals are formed by the ordered arrangement of the aqueous and paraffinic microphases which are separated by a surfactant monolayer (or bilayer). The lattice constants of these liquid crystals are typically ca. 100 Å, so we may regard these structures as "macrocrystals" which produce Bragg reflections in the small-angle regime either with X-rays or neutrons. Rheological measurements indicate a drastic reduction in viscosity on heating the cubic phases. This is due to the "melting" of the cubic macrocrystals into a random microemulsion, leading to loss of long-range order, and with it the property of ringing. The ringing samples are characterised by a marked increase of the storage modulus $G'(\omega)$, coupled with a reduction in the loss modulus $G''(\omega)$ at audible frequencies $\omega$ which leads to a large quality factor $Q \gg 1$, where $Q = G'(\omega)/G''(\omega)$.

Our results are consistent with a periodic constant mean curvature topology for these phases, which has the property of minimising the oil/water interfacial area for a given volume fraction.

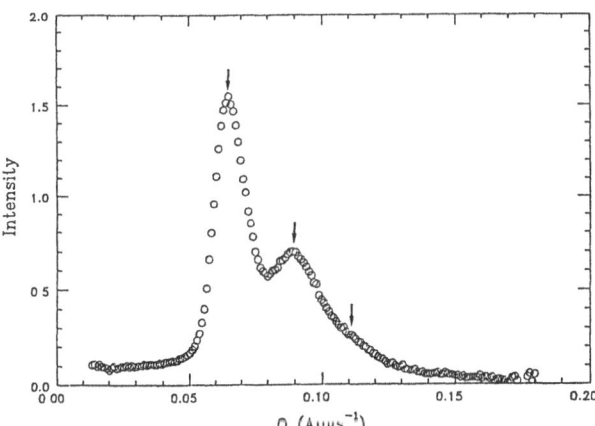

Fig. 1. Sample of DDAB/$D_2O$/Octane = 35.23/54.84/9.83 wt.% with water 50% deuterated and oil fully hydrogenated. The peaks correspond to a spacing of $1:\sqrt{2}:\sqrt{3}$ with $2\pi/Q_{max} = 96.7$ Å, for the principal peak

Authors' address:

Dr. C. Toprakcioglu
Cavendish Laboratory
Madingley Road
Cambridge CB3 0HB, UK

**Progress in Colloid & Polymer Science**                    Progr Colloid Polym Sci 81:290 (1990)

# Neutron reflection investigation of polymer layers adsorbed at the solid-lqiuid interface

J. B. Field[1]), C. Toprakcioglu[1,2]), R. C. Ball[1]), H. Stanley[3,5]), A. Rennie[4]), J. Penfold[5]) and W. Barford[5])

[1]) Cavendish Laboratory, Madingley Rd, Cambridge, UK
[2]) Institute of Food Research, AFRC, Colney Lane, Norwich, UK
[3]) I.C.I. plc, The Heath, Runcorn, Cheshire, UK
[4]) Institute Laue-Langevin, Grenoble, France
[5]) Rutherford-Appleton Laboratory, Chilton, Didcot

*Key words:* Neutron reflectivity; polymers; solid-liquid interace

*Abstract:* Neutron reflectometry techniques were employed to investigate the profile of polymers adsorbed at the solid-liquid interface. An optically flat, single crystal quartz surface was used as a substrate, in contact with a polymer solution. The geometry employed, allowed the neutron beam to travel through the quartz and be reflected from the quartz-liquid

from toluene (a good solvent at 21 °C). The results indicate a compact adsorbed layer of thickness ca. 5 nm in system (a). In system (b), however, the chains are terminally attached to the quartz surface (via the PEO block) as PS does not adsorb to the substrate from toluene, the latter being a very good solvent. The result for the $150 \times 10^3$ molecular weight block copolymer ($M_w/M_n = 1.16$, PEO content = 1.5%) is a highly extended adsorbed layer (ca. 40 nm) as indicated by the markedly different reflectivity profiles (see Fig. 1).

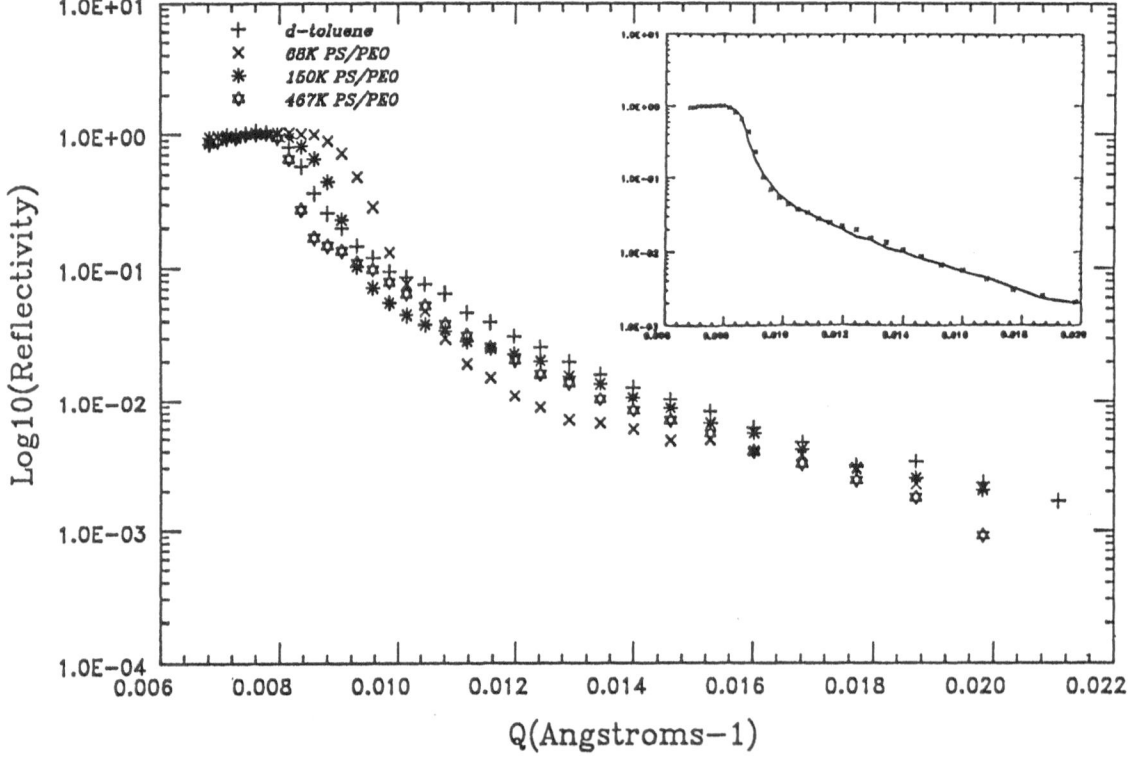

Fig. 1. Reflectivity profiles of the quartz/d-toluene interface (+) and the quartz/d-toluene interface with a PS/PEO block copolymer adsorbed from d-toluene at 21 °C of various molecular weights $68 \times 10^3$ (×), $150 \times 10^3$ (*), $467 \times 10^3$ (☆). The PEO content is ca. 1.5% in each case. Inset: Reflectivity profile of PS/PEO 150 K. The continuous curve is based on a polymer layer thickness of 40 nm

interface, the reflectivity of which was measured both before and after polymer adsorption. The systems studied were: a) polystyrene (PS) ($M_w = 156 \times 10^3$, $M_w/M_n = 1.03$) adsorbed on quartz from cyclohexane (a poor solvent at 21 °C) and b) polystyrene/polyethyleneoxide (PS/PEO) block copolymer of various molecular weights adsorbed on quartz

Authors' address:

Dr. C. Toprakcioglu
Cavendish Laboratory
Madingley Road
Cambridge CB3 0HE, UK

**Progress in Colloid & Polymer Science**                    Progr Colloid Polym Sci 81:291 (1990)

# Improvement of wicking flow of surfactant solutions in fabrics by the use of cosurfactants

R. H. Traber and M. Fehlbaum

CIBA-GEIGY AG, Dyestuffs and Chemicals Division, Basel, Switzerland

*Abstract:* A fast wicking flow of liquids in fibrous assemblies and in porous matrices is a prerequisite for high quality and fast processing rates in many industrial wet processes.

In the pretreatment of cotton, for example, which is the first step in wet processing, followed by dyeing, printing and finishing, the spontaneous uptake of the aqueous liquor by the hydrophobic raw fabric or yarn bobbin is made possible by the use of surfactants.

We have now observed that the addition of small amphiphiles (cosurfactants), like alcohols of medium chain length, to the surfactant solution increases considerably the rate and the extent of imbibition. (Mixtures of surfactants and cosurfactants are more efficient than the single components of the same mole concentration.)

The effect will be discussed by applying the "caterpillar vehicle motion" model of liquid spreading on a solid [1] to a capillary. From this model it becomes clear that the wicking flow must cause surfactant depletion at the water/air interface in the capillary, unless it is compensated by very fast surfactant diffusion. According to the Washburn equation surfactant depletion reduces the wicking rate by the decrease of the effective, dynamic adhesion tension.

A possible explanation can be deduced from this, namely that the cosurfactant, which is solubilized in the surfactant micelle, will enhance surfactant transport from the micellar reservoir to the advancing water/air interface by accelerating the exchange of the surfactant between the micelle and the surrounding solution, as was shown by chemical relaxation experiments by [2, 3].

### References

1. Dussan V EB, Davis SH (1974) J Fluid Mech 65:71
2. Yiv S, Zana R, Ulbricht W, Hoffmann H (1981) J Colloid Interf Sci 80:224
3. Lang J, Zana R (1986) J Phys Chem 90:5258

Authors' address:

Dr. R. H. Traber
CIBA-GEIGY AG
K-420.205
CH-4002 Basel, Switzerland

# Use of thermodynamic and nuclear/SANS, positron annihilation/ methods in the investigation of alkylsulphate micelles

Sz. Vass

Central Research Institute for Physics of the Hungarian Academy of Sciences, Budapest, Hungary

*Abstract:* SANS, positron annihilation/PA/and precise density/apparent molar volume/measurements have been carried out in normal- and heavy water solutions of different sodium alkylsulphates vs. temperature and surfactant concentration. Results obtained for the micellar structure and for orthopositronium/o-Ps/properties are summarized as follows:

1. Relative size of micelles can be deduced from positron lifetime spectra by utilizing a microscopic diffusion model; its trends vs. temperature and surfactant concentration correspond to those obtained from SANS experiments.
2. The lifetime of o-Ps in the micellar size can be separated from that in the aqueous phase. Micellar core densities calculated from apparent molar volumes of the aggregated alkyl chains and preliminary o-Ps lifetimes in the micellar phase indicate that the micellar core differs from bulk hydrocarbons.
3. Water content of the micellar core, estimated from apparent molar volumes of the aggregated alkyl chains, was found to be in a reasonable agreement with an upper limit set by SANS experiments.
4. A definite solvent isotopic effect on the apparent molar volumes and on o-Ps diffusivity was found.

Author's address:

Dr. Sz. Vass
Central Researche Institute for Physics of the Hungarian Academy of Science
P.O. Box 49
H-1525 Budapest, Hungary

**Progress in Colloid & Polymer Science**                    Progr Colloid Polym Sci 81:292 (1990)

# The rheology and microstructure of charged suspensions

N. J. Wagner and R. Klein

Fakultät für Physik, Universität Konstanz, Konstanz, BRD

*Abstract:* The effects of electric charge interaction and particle correlations on suspension rheology are examined. In particular, predictions of the low shear limiting rheology are made for a charged, monodisperse suspension of spheres using measured particle characteristics without adjustable parameters.

A one-component fluid analysis that includes many-body thermodynamic and pair hydrodynamic interactions is applied to charged suspensions of spherical colloids at high concentrations under shear [1]. The effective electrostatic interparticle potential is modelled as a screened coulombic interaction of the Yukawa form. Solution of the Ornstein-Zernike equation with the Rogers-Young closure yields the equilibrium microstructure as a function of particle and added salt concentrations. This structure is then perturbed by a weak shear as calculated through a Smoluchowski equation, which includes recent improvements in treating many-body thermodynamic interactions in dense suspensions.

Ensemble averaging the interparticle, Brownian, and hydrodynamic forces over these microstructures yields predictions for the rheology of the suspension as a function of particle concentration and added salt concentration. In the steady, weak shear limit the suspension is Newtonian while for time dependent deformations it is viscoelastic as the structure deformation lags the shear. The high frequency limiting modulus is found to scale with particle volume fraction $\phi$ and salt concentration $C_{salt}$ as $G_{\infty}^{'} \propto \phi^{2.4} C_{salt}^{-10}$.

Reasonable comparison of the low shear limiting, Newtonian viscosity is made with the measurements of Mitaku et al. [2], where the reported particle diameter of 125 nm, 1240 surface charges per particle, particle volume fraction $\phi = 0.117$, and salt concentration are the only inputs to the theory. The direct electrostatic interaction is found to drive the divergence in the shear viscosity near the phase transition. We note that uncertainties associated with the colloid charge, polydispersity effects, and inaccuracies in the assumed form of the interparticle potential could be resolved via direct comparison of equilibrium structure measurements with the theoretical predictions.

## References

1. Wagner NJ, Russel WB (1989) Physica 155 A:475
2. Mitaku S, Ohtsuki T, Kishimoto A, Okano K (1980) Biophys Chem 11:411

Authors' address:

Dr. Norman J. Wagner
B268, T12
Los Alamos National Laboratory
Los Alamos, NM 87545, USA

# Dependence of enzyme activity and conformation on water content in reverse micelles

D. Han, Q. Peng, P. Walde and P. L. Luisi

Institut für Polymere, ETH-Zentrum, Zürich, Switzerland

*Abstract:* For a better understanding of the behaviour of enzymes in living systems and for possible biotechnological applications, the activity and conformation of enzymes are investigated in reverse micelles. We have studied different hydrolases in two reverse micellar systems.

In isooctane containing 50 mM AOT [bis (2-ethylhexyl) sodium sulfosuccinate], the activity of *C. rugosa* lipase shows a bell-shaped dependence on $W_0$ (= [H$_2$O]/[Surfactant]) in reverse micelles. The conformation of this lipase in aqueous solution [as determined by circular dichroism (CD) spectroscopy] is altered rapidly after transferring the enzyme into reverse micelles. In the presence of 10 mM oleic acids however, the CD spectrum in reverse micelles is very similar with that in water, indicating that the substrate contributes to the stabilization of the lipase in the organic milieu.

In a system containing 50 mM di-heptanoyl phosphatdiylcholine in isooctane-hexanol (9:1, v/v), the activities of both $\alpha$-chymotrypsin and trypsin also show a bell-shaped dependence on $W_0$. The kinetic parameters obtained in that system were comparable with those obtained in bulk water. The CD spectra of both enzymes show that no significant conformational transition takes place on going from water to lecithin reverse micelles.

Fourier transform infrared spectroscopic and $^1$H-NMR studies indicate that several water molecules are consumed for the hydration of one surfactant head group. The lack of available water may be responsible for the low enzyme activities found at low $W_0$ values.

Authors' address:

Dr. P. Walde
Institut für Polymere, ETH-Zentrum
CH-8092 Zürich, Switzerland

**Progress in Colloid & Polymer Science**

Progr Colloid Polym Sci 81:293 (1990)

# Study of 2-butoxyethanol/water mixtures of critical composition in the vicinity of the lower critical point: measurements of light scattering and ultrasonic absorption

C. Baaken, L. Belkoura, S. Fusenig, T. Müller-Kirschbaum and D. Woermann

Institut für Physikalische Chemie, Universität Köln, FRG

*Abstract:* Considerable attention has been focused lately on the problem of critical phenomena in binary mixtures of water and nonionic surfactants of the type $C_i E_j$ (i.e. $CH_3[CH_2]_{i-1} \cdot O \cdot [CH_2CH_2O]_j H$; 2-butoxyethanol: $C_4E_1$). Several studies seemed to suggest that these systems do not belong to the 3D Ising universality class for which the divergence of osmotic susceptibility and correlation length of local concentration fluctuations are characterized by critical exponents $\gamma = 1.24$, $\nu = 0.63$ respectively. However, recently Dietler and Cannell (Phys. Rev. Lett. *60*, 1852 (1988)) found 3D Ising exponent values for and in critical mixtures of $C_{12}E_8/H_2O$ and $C_{12}E_8/D_2O$.

In this study results of static light scattering experiments which critical mixtures of $C_4E_1/H_2O$ and $C_4E_1/D_2O$ are reported which show that the experimentally determined values of $\gamma$ and $\nu$ are consistent with the 3D Ising exponent values. The critical amplitude $\xi_0$ of correlation length of local concentration fluctuations has values of $\xi_0(C_4E_1/H_2O) = (0.43 \pm 0.01)$ nm and $\xi_0(C_4E_1/D_2O) = (0.38 \pm 0.01)$ nm. They indicate that the system $C_4E_1$/water belongs to a group of binary liquid mixtures of molecules of low molar mass with comparatively large values of $\xi_0 (> 0.3$ nm) reflecting possibly the influence of association phenomena.

Measurements of the composition and temperature dependence of ultrasonic absorption (frequency range: $9.5$ MHz $\leq f \leq 45$ MHz) in the system $C_4E_1/H_2O$ show that the mixture with maximal ultra sound absorption contains less $C_4E_1$ than the critical mixture. This is unusual for binary liquid mixtures of components of low molar mass and miscibility gap. Approaching the critical temperature the maximum of ultra sound absorption is shifted to mixtures with even smaller concentrations of $C_4E_1$ and a second maximum of small amplitude appears at the critical composition. The total ultrasonic absorption of a critical mixture at a given frequency can be decomposed into three contributions: (a) a critical contribution described by the Ferrell-Bhattacharjee theory (1985), (b) a chemical contribution characterized by at least two relaxation times studied by Nishikawa et al. (1985) and (c) a temperature and frequency independent background contribution.

Authors' address:

Prof. D. Woermann
Inst. für phys. Chemie
Luxemburgerstr. 116
D-5000 Köln 41, FRG

**Progress in Colloid & Polymer Science**                    Progr Colloid Polym Sci 81:294 (1990)

# In situ viscosity changes and phase transitions in cationic and non-ionic aqueous surfactant systems triggered via photoreactions of solubilizates

T. Wolff, C.-S. Emming, B. Klaußer and G. von Bünau

Physikalische Chemie, Universität Siegen, Siegen, FRG

*Abstract:* When certain aromatic substances are solubilized in dilute micellar solutions of cetyltrimethylammonium bromide (CTAB) or of Triton X-100 the bulk viscosity of the solution may increase up to 20 times depending on the solubilizate. Other aromatic compounds do not show this effect. When the former class of compounds is photochemically transformed to the latter (and vice versa), the viscosity changes accordingly. Thereby in situ viscosity modulations are possible when photochromic solubilizates are used. Information on microscopic causes of the macroscopic viscosity changes was gained from low angle light scattering experiments and rheological studies revealing whether or not micellar sizes or shapes vary upon solubilizing the aromatic compounds and upon their photochemical transformations. Two examples are given: In CTAB solutions the viscosity increases when 9-n-butyl anthracene is added. Very large globular micelles are formed which do not change in aggregation number n upon photodimerization of the anthracenes whereas the viscosity decreases [1, 2]. In Triton X-100 solutions the viscosity increases when N-methyldiphenyl amine is solubilized while n does not change remarkably. Upon photoconversion of the amine to N-methyl carbazole the viscosity increases further [3] without affecting micellar sizes. In these cases we have to ascribe the photorheological effects to the varying extent of the electric double layer and/or the shell of hydration water of the micelles.

In more concentrated (20–30%) CTAB solutions phase transition temperatures pT between the nematic lyotropic liquid crystalline and the isotropic phase were changed by solubilizing substances which showed photorheological effects in dilute solutions. Therefore, phase transitions can be induced when the solubilized molecules are partly photoconverted. For instance, when 0.5% of N-methyldiphenyl amine is added, pT increases from 30 to 33.8°C in 23% CTAB-H$_2$O. The effect is reverted after photochemical formation of carbazole [4]. Another example is the addition of a crown ether bearing azobenzene, 0.3% of which decrease pT by 2.5°C. Trans-cis photoisomerization reversibly removes the observed pT decrease.

## References

1. Wolff T, Suck TA, Emming C-S, von Bünau G (1987) Progr Colloid Polym Sci 73:18–27
2. Wolff T, Emming C-S, Suck T, von Bünau G (1989) J Phys Chem 93:4894–4898
3. Wolff T, Schmidt F, von Bünau G (1989) J Photochem Photobiol A 48:435–456
4. Wolff T (1989) Colloid Polym Sci 267:345–348

Authors' address:

Thomas Wolff
Physikalische Chemie
Universität Siegen
Postfach 101240
D-5900 Siegen, FRG

**Progress in Colloid & Polymer Science**                    Progr Colloid Polym Sci 81:295 (1990)

# Nonionic microemulsions as model of biosystems studied by probing techniques

A. Xenakis, C. T. Cazianis and A. Malliaris

National Hellenic Research Foundation, Institute of Biological Research, Athens, Greece

*Abstract:* Certain molecular organizates such as micelles or microemulsions may be considered as model systems of biomembranes. One interesting model, under specific conditions, is the nonionic microemulsion system, consisting of water/decane/tetraethyleneglycoldodecylether ($C_{12}EO_4$). Therefore, studies aiming at characterizing some structural properties of this system would help to better understanding biomembranes.

In this study, spin and fluorescence probing techniques were used to investigate the state of the dispersed water phase, for different microemulsion compositions.

*ESR studies:* The ESR spectra of iodoacetamide (IA), 5-doxyl stearic acid (DSA) and FDNB derivative spin probes were recorded in microemulsions with various amounts of surfactant (from 5 to 25%) and water contents (from $R = 0$ to 30, where $R = [H_2O]/[C_{12}EO_4]$). Analysis of these spectra shows the existence of different spin states, depending on the exchange rate of the probe between the water core and the micromembrane separating the latter phase from the continuous oil phase [1−3]. The correlation times $\tau_c$ of the probes calculated from the line shapes [4], indicate that the $\tau_c$ values for FDNB and DSA probes increase from 2.1 $\times$ $10^{-10}$ s ($R = 0$) to 1.1 $\times$ $10^{-9}$ s ($R = 30$) and from 6.9 $\times$ $10^{-10}$ s ($R = 0$) to 1.3 $\times$ $10^{-9}$ s ($R = 30$), respectively, with an abrupt change between $R = 8$ and $12$, for all surfactant contents above 5%. In contrast the $\tau_c$ values for IA are lower and slightly vary from 2 $\times$ $10^{-10}$ s ($R = 0$) to 3.5 $\times$ $10^{-10}$ s ($R = 30$) for all surfactant contents.

The above results suggest the existence of two distinct states of the dispersed water phase. For $R = 0$ up to 8 all water molecules contribute only to the hydration of the ethylene-oxides of the surfactant molecules, while for higher $R$ values the appearance of a distinct water core indicates the formation of reverse micelles.

*Fluorescence quenching:* In another set of experiments the static fluorescence quenching method with $Ru(bPy)_3^{2+}$ as fluorophor and $Fe(CN)_6^{3-}$ as quencher was used to distinguish between the two microenvironments. Thus, for $R < 8−10$ the Stern-Volmer relationship gives a linear plot for $\ln(I_0/I_i)$ vs $[Fe(CN)_6]^{3+}$ (Fig. 1), indicating compartmentalization of spaces containing the reactants. On the contrary, for $R > 10$ the Stern-Volmer plot of $I_0/I_i$ vs. $[Q]$ is linear, therefore in this case the water pools are large enough for reactants to behave as in isotropic media.

In conclusion, it appears that, depending on the water content, these w/o microemulsions demonstrate two distinct

mean water pool sizes. The one, which is very small and comparable to the molecular dimensions of the reactants, and the other which is relatively large, allowing thus the fluorophor and quencher to behave as in isotropic solutions.

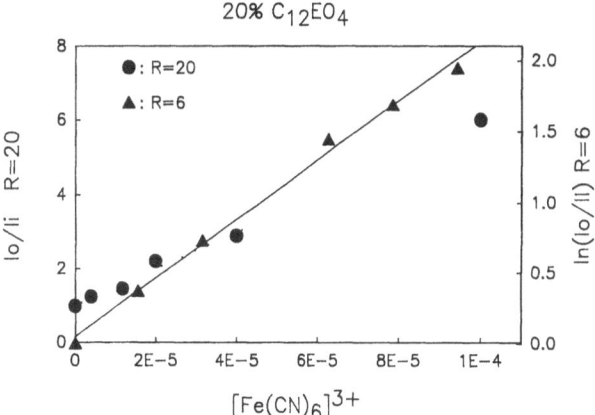

Fig. 1. Stern-Volmer plots of the fluorescence intensities ratios $I_0/I_i$ ($R = 20$) and $\ln(I_0/I_i)$ ($R = 6$) as a function of the quencher concentration

## References

1. Hubbell WL, McConnell HM (1968) Proc Natl Acad Sci 61:12
2. Xenakis A, Cazianis CT (1988) Progr Colloid Polym Sci 76:159
3. Cazianis CT, Xenakis A (1989) Progr Colloid Polym Sci 79:214
4. Martinie J, Michon J, Rassat A (1975) J Amer Chem Soc 97:1818

Authors' address:

Dr. A. Xenakis
National Hellenic Research Foundation
Institute of Biological Research
48, Vas. Constantinou Ave.
11635 Athens, Greece

**Progress in Colloid & Polymer Science**                                       Progr Colloid Polym Sci 81:296 (1990)

# Transition of a disordered random bilayer to a swollen lamellar liquid crystal

M. Dubois and Th. Zemb

Département des Lasers et de Physico-chimie, Gif sur Yvette, France

*Abstract:* We use binary solutions of brine and dihexadecyklammonium acetate (2C16N$^+$ Ac$^-$). The reverse micelles formed with this surfactant follow the predictions of the DOC model, showing that the interfacial layer is stiff: local bending radius fluctuation is forbidden. The phase diagram of this system shows an isotropic micellar solution, a Disordered connected Random Lamellar region ($L_3$) and a swollen lamellar liquid crystal phase ($L_\alpha$).

The $L_3$ phase shows strong flow birefringence behaviour whereas the $L_a$ phase is birefringent. This *binary* system can be swollen up to 0.5% weight without loosing liquid crystalline ($L_\alpha$) character, showing a strong Bragg reflection (period up to 300 nm) which can be followed by low angle neutron and light scattering.

This behaviour is not changed upon addition of up to $10^{-2}$ M of salt. This shows that the ordering of the lamellae into large crystallites is *not* due to the electrostatic repulsion, but to the intrinsic rigidity of the bilayer.

The structure of $L_3$ solution presents some anomaleous features: some birefringent solutions exhibit no Bragg peak in the scattering whereas some $L_3$ solutions show a broad correlation peak which has not yet been reported for other $L_3$ phases found in ternary or quaternary systems, where the bending constant $k_c$ is lower.

We observe that this strange behaviour of sterically interacting stiff bilayers is closely related to the stability of the *spontaneous emulsion* which is formed upon addition of oil to the $L_3$ phase.

Authors' address:

Dr. M. Dubois
Département de Physique Genérale, S.C.M.
CEN SACLAY Bat 125
91191 Gif sur Yvette, France

**Progress in Colloid & Polymer Science**                    Progr Colloid Polym Sci 81:297 (1990)

# Colloidal particles with dodecahedral symmetry formed spontaneously by lecithin, bile salts, cholesterol and a plant saponin

D. Schmidt, V. E. Colombo and M. Zulauf

F. Hoffmann-La Roche AG, Central Research Units, Basel, Switzerland

*Abstract:* Mixtures of lecithin, glycocholate and cholesterol taken up in $H_2O$ from a dry film in molar ratios of $10:x:2.5-7.5$ with $2 \leqslant x \leqslant 20$ yield mixed micelles or unilamellar vesicles with hydrodynamic radii between 30 and 500 Å, depending on the bile salt concentration. When the saponin Quil $A$ is added to these solutions in equimolar quantities to cholesterol, monodisperse particles with a radius of about 130 Å form spontaneously. These show morphological features under the electron microscope arranged according to pentagonal dodecahedra (Fig. 1).

Quil $A$ is a saponin extracted from the bark of the tree *Quillaja Saponaria Molina*. As judged by HPLC and other techniques, the extract contains a large number of components. Probably, the basic structure of these consists of two chains of branched sugars with varying composition and contains, as aglycones, quillaic acid and two terpenic acids [1]. Regular particles form only when the amphiphile molecular weight $\geqslant 2000$. They are not found with saponins from other plants or in the absence of cholesterol. Particles occasionally form in the absence of lecithin, when cholesterol is solubilized by detergents, notably MEGA-10, before addition of Quil $A$.

The particles have been discovered in connection with subunit vaccines [2] and are known under the name "iscom matrices"; Quil $A$ has been in use as a veterinary vaccine adjuvant for a long time [3, 4].

## References

1. Higuchi R et al. (1988) Phytochemistry 27:1165
2. Morein B (1988) Nature 332:287
3. Espinet RG (1951) Gaceta Veterinaria 13:268
4. Dalsgaard K (1978) Acta Vet Scand Suppl 69:1

Fig. 1. Particles stained with 2% phosphotungstic acid and visualized by TEM, $90\,000 \times$, diameter $\approx 250$ Å

Authors' address:

Dr. M. Zulauf
c/o F. Hoffmann-La Roche
ZFE 65-509
CH-4002 Basel, Switzerland

# Author Index

# Subject Index